T0245143

Book of Abstracts of the 55th Annual Meeting of the European Association for Animal Production

The EAAP Book of Abstracts is published under the direction of Ynze van der Honing

EAAP - European Association for Animal Production

The European Association for Animal Production wishes to express its appreciation to the
Ministero delle Politiche Agricole e Forestali (Italy) and the
Associazione Italiana Allevatori (Italy)
for their valuable support of its activities.

Book of Abstracts of the 55th Annual Meeting of the European Association for Animal Production

Bled, Slovenia, 5 -9 September 2004

Ynze van der Honing, Editor-in-chief

A. Hofer, E. Cenkvári, C. Fourichon, Y. Chilliard, C. Lazzaroni, L. Bodin, C. Wenk, W. Martin-Rosset and J.E. Hermansen

CIP-data Koninklijke Bibliotheek
Den Haag

ISSN 1382-6077
ISBN 9076998450

The individual contributions in this publication and any liabilities arising from them remain the responsibility of the authors.

Subject headings:
animal production,
book of abstracts

The designation employed and the presentation of material in this publication do not imply the expression of any option whatsoever on the part of the European Association for Animal Production concerning the legal status of any country, territory, city or area or of its authorities, or concerning the delimitation of its frontiers or boundaries.

Wageningen Academic Publishers
The Netherlands, 2004

Printed in The Netherlands

Preface

The 55th annual meeting of the European Association for Animal Production (EAAP) is held in Bled, Slovenia, 5 to 9 September 2004. The annual EAAP meeting gives the opportunity to present new scientific results and discuss their potential applicability in animal production practices. This year's meeting is of particular interest for participants from a wide range of animal production organisations and institutions. Discussions stimulate developments in animal production and encourage research on relevant topics. In Bled the main theme is: "**Sustainability of Livestock Systems**". Of the 38 sessions in total, 21 joint sessions of two or more Study Commissions are planned on:

- Society's needs in relation to animal production systems (Plenary session)
- Animal health and welfare in intensive and extensive systems
- Management of grasslands for production, environment and landscape benefit
- Management of genetic variation
- Ethical issues in animal agriculture
- Low-input systems
- Large-scale pig farming systems
- Indicators of environmental impact in livestock systems
- Genomics of lactation
- Genetics of disease resistance
- International genetic evaluation
- Longitudinal data in genetics
- Cattle and buffalo reproduction, production and breeding
- Herd modeling for improved management
- Feed evaluation and livestock nutrition
- Regulation of food intake
- Alternative protein sources
- Anti-nutritional factors and mycotoxins
- Pig metabolism, growth and production
- Sheep and Goat production
- Sheep and Goat behaviour - relevance to welfare and management

The book of abstracts is the main publication of the scientific contributions to this meeting; it covers a wide range of disciplines and livestock species. It contains the full programme and abstracts of the invited as well as the contributing speakers, including posters, of all 38 sessions. The number of abstracts submitted for presentation at this meeting (741) is a true challenge for the different study commissions and chairpersons to put together a scientific programme. In addition to the theatre presentations, there will be a large number of poster presentations during the conference.

Several persons have been involved in the development of the book of abstracts. All their input and support is much appreciated.

The programme is very interesting and I trust we will have a good meeting in Bled. During the annual meeting always a large number of people actively involved in livestock science and production enjoy to meet, exchange ideas and discuss developments in animal production. I hope that you will find this book a useful reference source as well as a reminder of a good meeting.

Ynze van der Honing
Editor-in Chief

EAAP Program Foundation

Aim

EAAP aims to bring to our annual meetings, speakers who can present the latest findings and views on developments in the various fields of science relevant to animal production and its allied industries. In order to sustain the quality of the scientific program that will continue to entice the broad interest in EAAP meetings we have created the "EAAP Program Foundation". This Foundation aims to support

- Invited speakers by funding part or all of registration and travel costs.
- Delegates from less favoured areas by offering scholarships to attend EAAP meetings
- Young scientists by providing prizes for best presentations

The "EAAP Program Foundation" is an initiative of the Scientific Advisory committee (SAC) of EAAP. The Foundation is aimed at stimulating the quality of the scientific program of the EAAP meetings and to ensure that the science meets societal needs . In its first year (2003), the "EAAP Program Foundation" concentrates on the program of the Genetics commission. For the coming years the activities will be broadened to entire meeting. The Foundation Board of Trustees oversees theses aims and seeks to recruit sponsors to support its activities.

Sponsorships

We distinguish three categories of sponsorship: Student award sponsor, Gold sponsor, and Sponsor. The sponsors will be acknowledged during the scientific sessions commission. of the Genetics commission. The names of Student Award sponsors will be linked to the awards given to young scientists with the best presentation. Gold Sponsors and Student Award sponsors will have opportunity to advertise their activities during the meeting and their support for EAAP.

Contact and further information

If you are interested in becoming a sponsor of the "EAAP Program Foundation" or want to have further information, please contact the secretary of the Foundation, Dr Andreas Rosati (e-mail: rosati@eaap.org).

Sponsors of 2003 meeting

The board of Trustees wants to thank the following organisations for their support:

Student award sponsor:
Labogena
www.labogena.fr
labogena@jouy.inra.fr

Sponsors:
Cavalor
www.cavalor.com

Intervet
www.intervet.com

European Association for Animal Production (EAAP)

President	Aimé L. Aumaitre
Executive Vice-President	Andrea Rosati
Address	Villa del Ragno, Via Nomentana 134, I-00161 Rome, Italy
Phone	+39 06 86329141
Telefax	+39 06 86329263
E-Mail	eaap@eaap.org

In cooperation with

University of Ljubljana, Biotechnical Faculty, Zootechnical Department
the Ministry of Agriculture, Forestry and Food
the Ministry of Education, Science and Sport

Organizing Committee

Drago Kompan	Biotechnical Faculty, Zootech. Depart., Domžale
Franc Habe	Biotechnical Faculty, Zootech. Depart., Domžale
Viktor Krek	Ministry of Agriculture, Forestry and Food, Ljubljana
Marija Klopčič	Biotechnical Faculty, Zootech. Depart., Domžale
Jože Antonič	The Bled Commune, Bled
Slavko Čepin	Biotechnical Faculty, Zootech. Depart., Domžale
Marko Čepon	Biotechnical Faculty, Zootech. Depart., Domžale
Milan Pogačnik	Ministry of Agriculture, Forestry and Food, Ljubljana
Branko Ravnik	Agricultural Chamber of Slovenia, Ljubljana
Jože Stopar	Biotechnical Faculty, Zootech. Depart., Domžale
Jože Čeh	Biotechnical Faculty, Zootech. Depart., Domžale

Scientific advisory Committee

Franc Habe	Biotechnical Faculty, Zootech. Depart., Domžale
Marija Klopčič	Biotechnical Faculty, Zootech. Depart., Domžale
Silvester Žgur	Biotechnical Faculty, Zootech. Depart., Domžale
Drago Kompan	Biotechnical Faculty, Zootech. Depart., Domžale
Milan Pogačnik	Ministry of Agriculture, Forestry and Food, Ljubljana
Peter Dovč	Biotechnical Faculty, Zootech. Depart., Domžale
Milena Kovač	Biotechnical Faculty, Zootech. Depart., Domžale
Andrej Lavrenčič	Biotechnical Faculty, Zootech. Depart., Domžale
Emil Erjavec	Biotechnical Faculty, Zootech. Depart., Domžale
Jože Osterc	Biotechnical Faculty, Zootech. Depart., Domžale
Andrej Orešnik	Biotechnical Faculty, Zootech. Depart., Domžale
Janez Salobir	Biotechnical Faculty, Zootech. Depart., Domžale
Jože Verbič	Agricultural Institute of Slovenia, Ljubljana
Anton Vidrih	Biotechnical Faculty, Zootech. Depart., Domžale

Organizing Committee

Biotechnical Faculty, Zootechnical Department
Groblje 3, 1230 Domžale, Slovenia
Tel. +386 1 7217 855 - Fax: +386 1 7211 701
e-mail: eaap2004@bfro.uni-lj.si

56th Annual Meeting of the European Association for Animal Production

Uppsala, Sweden, June 5 - 8, 2005

Organizing Committee

President: Bengt Everitt,
Swedish Dairy Association, Stockholm

Vice President: Hans Gustafsson,
Swedish Dairy Association, Uppsala

Secretary: Kerstin Johansson,
Swedish Dairy Association, Eskilstuna

Congress Secretariat

Academic Conferences
P.O. Box 7059
SE-750 07 Uppsala
Sweden
Phone: +46 18 67 20 84
Fax: +46 18 67 35 30
E-mail: EAAP2005@slu.se
Web: www-conference.slu.se

Abstract submission
Please find the abstract form on the internet:
www.WageningenAcademic.com/eaap
Submission of the completed abstract form before Januari 31, 2005 to:
EAAP2005@WageningenAcademic.com

Mediterranean Symposium

"Comparative avantages for typical animal products from the Mediterranean area"

Boa Fonte, Portugal, 25-27 September 2005

Contact person
J.M.C Ramalho Ribeiro
Estação Zootécnica Nacional
Fonte Boa
2005- 048 Vale de Santarém
Portugal

Tel: 243 767 300
Fax 243 767 307
Email: ezn.inia@mail.telepac.pt

35th ICAR Session and INTERBULL Meeting

June 3 - 10, 2005, Kuopio, Finland

Contact: ICAR Secretariat, Rome, Italy
Email: icar@eaap.org

Timetable

Thursday, September 2nd (Conference Venue)

09.00 - 18.00 **Satellite Workshop:**

- 12th International Symposium: Animal Science Days - ASD (Excursion)
- Annual meeting SAVE Foundation & Network

Friday, September 3rd (Conference Venue)

12.00 - 19.00 Registration of Participants

09.00 - 18.00 **Satellite Symposia, Workshops:**

- 12th International Symposium: Animal Science Days - ASD
- FAO - DAGENE - EAAP Workshop: Molecular Genetic Methods - "AnGR"
- DAGENE Meeting (together with FAO and EAAP)
- Annual meeting SAVE Foundation & Network

14.00 - 18.00 *FAO - DAGENE - ERFP - SAVE Excursion (afternoon)*

Saturday, September 4th (Conference Venue)

08.00 - 20.00 Registration of Participants

08.30 - 18.00 EAAP Scientific Advisory Committee Meeting
Joint Session of the Council with Presidents and Secretaries of Study Commissions, President of Working Groups and Representatives of International Organizations, Council Meeting

Satellite Symposia, Workshops:

- 12th International Symposium: Animal Science Days - ASD
- DAGENE Meeting (together with ASD)
- ERFP Workshop: Animal Genetic Resources
- International Symposium: "Sustainable Recultivation and Land Use on Karst and Mountainous Regions by Use of Animals"
- CEEC WG Workshop: "Dairy Farm Management and Extension Needs in CEE under the Restrictions of the EU Milk Quota"
- Academic Curricula Programs: "Animal Nutrition Teaching"
- Workshop: "Preparing and Presenting Scientific Papers"
- Annual meeting SAVE Foundation & Network

19.00 Exhibition and Tasting of Slovenian Local Products (Sport Hall)

Sunday, September 5th (Conference Venue)

07.00 - 20.00 Registration of Participants
08.30 - 09.00 Plenary Session
09.15 - 13.00 Session I of Study Commissions
14.00 - 15.00 The Show of Traditional Customs and Folk Culture
15.00 - 18.30 Session II of Study Commissions
18.30 - 20.30 Poster Presentation
20.30 - 22.00 Welcome Cocktail given by the Mayor of Bled

Monday, September 6th (Conference Venue)

08.00 - 20.00	Registration of Participants
08.30 - 12.30	Session III of Study Commissions
14.00 - 16.00	Round Table on "Europeanisation of the animal agriculture: opportunities and menaces"
16.00	Departure to Ljubljana (by bus)
18.00 - 19.30	Opening Ceremony // Awards Ceremony
19.30 - 22.00	Conference Party - Welcome by the Mayor of Ljubljana

Tuesday, September 7th (Conference Venue)

08.00 - 20.00	Registration of Participants
08.30 - 12.30	Council Meeting
08.30 - 12.30	Session IV of Study Commissions
14.00 - 18.00	Session V of Study Commissions
15.00 - 18.00	Meeting of the CEEC Working Group
18.00 - 19.30	General Assembly
20.00 - 23.00	Slovenian Social Evening

Wednesday, September 8th (Conference Venue)

08.00 - 14.00	Registration of Participants
08.30 - 10.30	Council Meeting
08.30 - 12.30	Session VI of Study Commissions
12.30 - 14.00	Scientific Advisory Committee Meeting
14.00 - 18.00	Conference Tour (Bled Castle)
20.00 - 22.00	Post Conference Dinner

Thursday, September 9th

08.00	Departure for Post-Conference Tours
09.00 - 18.00	EAAP -ASAS Workshop: "Biology of Lactation in Farm Animals" PhD-Course: Estimation of Covariance Components and Breeding Values with the VCE 5

Friday, September 10th

08.00	Departure for Post-Conference Tours
09.00 - 18.00	EAAP -ASAS Workshop: "Biology of Lactation in Farm Animals" PhD-Course: Estimation of Covariance Components and Breeding Values with the VCE 5

Saturday, September 11th

09.00 - 18.00	PhD-Course: Estimation of Covariance Components and Breeding Values with the VCE 5

Sunday, September 12th

09.00 - 18.00	PhD-Course: Estimation of Covariance Components and Breeding Values with the VCE 5

Monday, September 13th

09.00 - 18.00	PhD-Course: Estimation of Covariance Components and Breeding Values with the VCE 5

Scientific programme EAAP-2004

Plenary Sunday 5 Sept. 08.30 - 09.00	Session 1 Sunday 5 Sept. 09.15 - 13.00	Session 2 Sunday 5 Sept 15.00 - 18.30	Session 3 Monday 6 Sept. 08.30 - 12.30
'Society's needs in relation to animal production systems' **Prof M Gill (UK)** **Chair:Aumaitre (F)**	**Animal health and welfare in intensive and extensive systems** **(M*,C, L)** **Chair: Fourichon (F)**	**Management of grasslands for production, environment and landscape benefit.** **(L*, N, C, S,)** **Chair: Zervas (GR)**	**Management of genetic variation** **(G)** **Chair: Gandini (I)**
	Anti-nutritional factors and mycotoxins (P*, N, H) Chair:Torrallardona (ES)	**Large-scale pig farming systems** **(P*,M, L, N)** **Chair: Kovac (Slov), Edwards (UK)**	**Indicators of environmental impact in livestock systems** **(L*,P, M)** **Chair: Hermansen (DK)**
	Genomics of lactation (G*, Ph) Chair: Dovc (Slovenia)	Genetics of disease resistance (G*M) Chair: Bishop (UK)	Regulation of food intake (Ph*,N) Chair: Friggens (DK)
	Economics and profitability of sheep and goat production under new support regimes and market conditions (S) Chair: Gabina (ES)	Breeding schemes and functional genetics - Free Communications (G) Reinsch (D)	Endangered horse breeds and genetic distance (H* + Rare Breeds International) Chair: Bodo (Hungary)
	Effect of globalisation on livestock systems (L) Chair: Peters (D)	Mammary gland health (Ph) Chair: Fitzpatrick (UK)	Use of hormones for reproduction in sheep and goats - impact and alternatives. (S) Chair: Leboeuf (F)
		Growth and bone disorders in Horses (H) Chair: Bergero (I)	Implementation of new management practices + excursion. (C) Chair: Kuipers (NED), Osterc (Slov)

Key to Commissions (subject areas): (G – Genetics; N – Animal Nutrition; M- Animal Management and Health; Ph – Animal Physiology; C – Cattle Production; S – Sheep and Goat Production; P – Pig Production; H – Horse Production; L – Livestock Farming Systems)
* = Organising Commission

Scientific programme EAAP-2004

Session 4 Tuesday 7 Sept. 08.30 - 12.30	Session 5 Tuesday 7 Sept. 14.00 - 18.00	Session 6 Wednesday 8 Sept. 08.30 - 12.30
08.30 - 09.30 Business meetings *followed by* ——— 09.30 - 12.30 Simultaneous sessions of Free Communications: • Cattle and buffalo reproduction, production and breeding (C*, L and Ph) Chair: Keane (IRL) • Applied genetics (G); Chair: Strandberg (S) • Evaluation and performance of horses (H) Chair: Fleurance (FR) • Animal health (M) Chair: Sorensen (DK) • Feed evaluation and livestock nutrition (N*, Ph) Chair: Crovetto (I) • Pig metabolism, growth and production (P* and Ph) Chair: Chadd (UK) • Sheep and Goat production (S* and L) Chair: Gabina (ES)	**Ethical issues in animal agriculture** **(M*L)** **Chair: Marie (FR)** ——— Alternative protein sources (N*,C, S) Chair: Dourmad (F) ——— International genetic evaluation (H*,G) Chair: Koenen (NED) *followed by* • Genetics of disorders (G) Chair: Eggen (F) ——— Herd modeling for improved management (P*,L) Chair: Knap (D)	**Low-input systems** **(Ph*,L)** **Chair: Faverdin (FR)** ——— Sheep and Goat behaviour - relevance to welfare and management. (S*,M) Chair: Waterhouse (UK) ——— Longitudinal data in genetics (P*,G): Chair: Knap (D) Horse production in Slovenia (H) Chair: Habe (Slovenia) ——— Harmonisation of feeding standards (N) Chair: Newbold (B) *followed by* • Nutrition and metabolism - Free Communications Chair: Lindberg (S) ——— Developments in biometrical methods (G) Chair: Madsen (DK) ——— Indicators of milk and meat quality (C) Chair: Hocquette (F)

Key to Commissions (subject areas): (G – Genetics; N – Animal Nutrition; M- Animal Management and Health; Ph – Animal Physiology; C – Cattle Production; S – Sheep and Goat Production; P – Pig Production; H – Horse Production; L – Livestock Farming Systems)
* = Organising Commission

Agendas of Study Commissions

Commission on Animal Genetics (G)

President
Dr.V. Ducrocq
INRA
Centre National Recherche de Zootechnie
Domaine de Vilvert
78350 Jouy-en-Josas Cédex
FRANCE

Secretary 1
Dr A. Hofer
SUISAG
AG für Dienstleistungen in der Schweineproduktion
Allmend
6204 Sempach
SWITZERLAND
0041 41 462 65 56
0041 41 462 65 49
hofer@inw.agrl.ethz.ch

Secretary 2
Dr E. Martyniuk
Warsaw Agricultural University
ul. Przejazd 4
05840 Brwinów
POLAND
+48 22 729 62 14
+48 22 729 59 15
eshz@perytnet.pl

GENOMICS OF LACTATION

Session 1

Joint session with the commission on Animal Physiology

Date	05-sep-04
	9:15 - 13:00 hours
Chairperson	Dovc (SLO)

Theatre

Genetics

Poster

FREE COMMUNICATIONS IN GENETICS

Session 2

Date 05-sep-04
15:00 - 18:30 hours
Chairperson Reinsch (D)

Theatre

GENETICS OF DISEASE RESISTANCE

Session 2

Joint session with the commission on Animal Management and Health

Date	05-sep-04
	15:00 - 18:30 hours
Chairperson	Bishop (UK)

Theatre

Poster

MANAGEMENT OF GENETIC VARIATION

Session 3

Date 06-sep-04
8:30 - 12:30 hours
Chairperson Gandini (IT)

Theatre

Poster

Genetics

FREE COMMUNICATIONS

Session 4

Date	07-sep-04 8:30 - 12:30 hours
Chairperson	Strandberg (S)

Theatre

Poster

Genetics

Genetics

Genetics

GENETICS OF DISORDERS

Session 5

Date 07-sep-04
14:00 - 18:00 hours

Chairperson Eggen (F)

Theatre

Poster

INTERNATIONAL GENETIC EVALUATION

Session 5

Joint session with the commission on Horse Production

Date 07-sep-04
14:00 - 18:00 hours
Chairperson Koenen (NED)

DEVELOPMENTS IN BIOMETRICAL METHODS

Session 6

Date 08-sep-04
8:30 - 12:30 hours
Chairperson Madsen (DK)

Theatre

Poster

LONGITUDINAL DATA IN GENETICS

Session 6

Joint session with the commission on Pig Production

Date	08-sep-04
	8:30 - 12:30 hours
Chairperson	Knap (D)

Commission on Animal Nutrition (N)

President
Dr G.M. Crovetto
University of Milan
Faculty of Agriculture
Via Celoria 2
20133 Milano
ITALY
+39 02 7064 94 37
+39 02 7063 80 83
matteo.crovetto@unimi.it

Secretary 1
Dr. E. Cenkvári
Szent István University
Faculty of Veterinary Sciences
P.O. Box 2
100 Budapest
HUNGARY
+36 478 4124, 8608
+36 478 4128
ecenkvari@univet.hu

Secretary 2
Dr. G. Zervas
Agricultural University of Athens
Animal Nutrition
Iera Odos 75
11855 Athens
GREECE
gzervas@aua.gr

ANTI-NUTRITIONAL FACTORS AND MYCOTOXINS

Session 1

Joint session with the commissions on Pig Production and Horse Production

Date	05-sep-04 9:15 - 13:00 hours
Chairperson	Torrallardona (E)

MANAGEMENT OF GRASSLANDS FOR PRODUCTION, ENVIRONMENT AND LANDSCAPE BENEFIT

Session 2

Joint session with the commissions on Cattle Production, Sheep and Goat Production and Livestock Farming Systems

Date 05-sep-04
15:00 - 18:30 hours
Chairperson Zervas (GR)

	LNCS2	Page	Page
Papers	**1-9**	**340-344**	**XCVIII-XCIX**
Posters	**10-19**	**345-349**	**XCIX-C**

LARGE-SCALE PIG FARMING SYSTEMS

Session 2

Joint session with the commissions on Animal Management and Health, Pig Production and Livestock Farming Systems

Date 05-sep-04
15:00 - 18:30 hours
Chairperson Kovac (SLO) & Edwards (UK)

	PNML2	Page	Page
Papers	**1-9**	**267-271**	**LXXXI**

REGULATION OF FOOD INTAKE

Session 3

Joint session with the commission on Animal Physiology

Date 06-sep-04
8:30 - 12:30 hours
Chairperson Friggens (DK)

	PhN3	Page	Page
Papers	1-8	174-177	LVII
Posters	9-16	178-181	LVII-LVIII

FREE COMMUNICATIONS - FEED EVALUATION, NUTRITION AND METABOLISM

Session 4

Joint session with the commission on Animal Physiology

Date 07-sep-04
8:30 - 12:30 hours
Chairperson Crovetto (I)

Theatre

Nutrition

Poster

B. Ruminant nutrition and metabolism: Cattle

Nutrition

Nutrition

D. Nutrition of pigs

ALTERNATIVE PROTEIN SOURCES

Session 5

Joint session with the commissions on Cattle Production and Sheep and Goat Production

Date 07-sep-04
14:00 - 18:00 hours
Chairperson Dourmad (F)

Theatre

Nutrition

Poster

FREE COMMUNICATIONS - NUTRITION AND METABOLISM

Session 6

Date 08-sep-04
8:30 - 12:30 hours
Chairperson Lindberg (S)

Theatre

N6 no. Page

Poster

N6 no. Page

Commission on Animal Management and Health (M)

President
Prof.Dr E.H. von Borell
Martin Luther University of Halle
Adam-Kuckhoff-Str. 35
06108 Halle (Saale)
GERMANY
+49 345 55 22 332
+49 345 55 27 105
borell@Landw.uni-halle.de

Secretary 1
Dr. C. Fourichon
Veterinary School of Nantes - INRA
P.O. Box 40706
44307 Nantes Cedex 03
FRANCE
+33 240 68 77 86
+33 240 68 77 68
fourichon@vet-nantes.fr

Secretary 2
A. Formigoni
University of Bologna
Veterinary faculty
Via Tolara di Sopra 50
40064 Ozzana Emilia (BO)
ITALY

ANIMAL HEALTH AND WELFARE IN INTENSIVE AND EXTENSIVE SYSTEMS

Session 1

Joint session with the commissions on Cattle Production and Livestock Farming Systems

Date	05-sep-04 9:15 - 13:00 hours
Chairperson	Fourichon (F)

Theatre

MCL1 no. Page

Poster

LARGE-SCALE PIG FARMING SYSTEMS

Session 2

Joint session with the commissions on Animal Nutrition, Pig Production and Livestock Farming Systems

Date	05-sep-04 15:00 - 18:30 hours
Chairperson	Kovac (SLO) & Edwards (UK)

INDICATORS OF ENVIRONMENTAL IMPACT IN LIVE-STOCK SYSTEMS

Session 3

Joint session with the commissions on Pig Production and Livestock Farming Systems

Date	06-sep-04 8:30 - 12:30 hours
Chairperson	Hermansen (DK)

FREE COMMUNICATIONS - ANIMAL HEALTH

Session 4

Date	07-sep-04 8:30 - 12:30 hours
Chairperson	Sorensen (DK)

Theatre

Poster

Management and Health

ETHICAL ISSUES IN ANIMAL AGRICULTURE

Session 5

Joint session with the commission on Livestock Farming Systems

Date 07-sep-04
14:00 - 18:00 hours
Chairperson Marie (F)

Theatre

SHEEP AND GOAT BEHAVIOUR - RELAVANCE TO WELFARE AND MANAGEMENT

Session 6

Joint session with the commission on Sheep and Goat Production

Date	08-sep-04 8:30 - 12:30 hours
Chairperson	T.Waterhouse (UK)

Commission on Animal Physiology (Ph)

President
Dr. K. Sejrsen
Danish Institute of Agricultural Sciences (DIAS)
Research Centre Foulum
P.O. Box 50
8830 Tjele
DENMARK
0045 89 991 513
0045 89 991 564
kr.sejrsen@agrsci.dk

Secretary 1
Dr B. Kemp
Wageningen UR
Zodiac
Postbus 338
6700 AH Wageningen
NEDERLAND
0317 482 539
0317 485 006
Bas.Kemp@wur.nl

Secretary 2
Dr. Y. Chilliard
INRA
Centre of Research Clermont-Ferrand / Theix
63122 Saint Genès Champanelle
FRANCE
+33 473 62 41 14
+33 473 62 45 19
chilliar@clermont.inra.fr

GENOMICS OF LACTATION

Session 1

Joint session with the commission on Animal Genetics

Date	05-sep-04
	9:15 - 13:00 hours
Chairperson	Dovc (SLO)

Physiology

MAMMARY GLAND HEALTH

Session 2

Date 05-sep-04
15:00 - 18:30 hours
Chairperson Fitzpatrick (UK)

Theatre

Poster

REGULATION OF FOOD INTAKE

Session 3

Joint session with the commission on Animal Nutrition

Date 06-sep-04
8:30 - 12:30 hours
Chairperson Friggens (DK)

Theatre

Poster

FREE COMMUNICATIONS

Session 4

Joint session with the commission on Animal Nutrition

Date 07-sep-04
8:30 - 12:30 hours
Chairperson Crovetto (I)

FREE COMMUNICATIONS

Session 4

Joint session with the commission on Pig Production

Date 07-sep-04
8:30 - 12:30 hours
Chairperson Chadd (UK)

LOW-INPUT SYSTEMS

Session 6

Joint session with the commission on Livestock Farming Systems

Date 08-sep-04
8:30 - 12:30 hours
Chairperson Faverdin (F)

Theatre

PhL6 no. Page

Poster

PhL6 no. Page

Physiology

Commission on Cattle Production (C)

President
Dr S. Gigli
Animal Production Research Institute
Istituto Sperimentale per la Zootecnia
Via Salaria 31
00016 Monterotondo Scalo (Roma)
ITALY
0039 06 900 90 209
0039 06 906 15 41
isz@flashnet.it

Secretary 1
A. Kuipers
Wageningen UR
Expertise Centre for Farm Management and Knowledge
Agro Business Park 36
6708 PW Wageningen
NEDERLAND

Secretary 2
Dr C. Lazzaroni
University of Torino
Via Leonardo da Vinci 44
10095 Grugliasco (Torino)
ITALY
+39 011 670 85 64
+39 011 670 85 63
carla.lazzaroni@unito.it

ANIMAL HEALTH AND WELFARE IN INTENSIVE AND EXTENSIVE SYSTEMS

Session 1

Joint session with the commissions on Animal Management and Health and Livestock Farming Systems

Date	05-sep-04 9:15 - 13:00 hours
Chairperson	Fourichon (F)

MANAGEMENT OF GRASSLANDS FOR PRODUCTION, ENVIRONMENT AND LANDSCAPE BENEFIT

Session 2

Joint session with the commissions on Animal Nutrition, Sheep and Goat Production and Livestock Farming Systems

Date	05-sep-04
	15:00 - 18:30 hours
Chairperson	Zervas (GR)

	LNCS2	Page	Page
Papers	**1-9**	**340-344**	**XCVIII-XCIX**
Posters	**10-19**	**345-349**	**XCIX-C**

IMPLEMENTATION OF NEW MANAGEMENT PRACTICES + EXCURSION

Session 3

Date	06-sep-04
	8:30 - 12:30 hours
Chairperson	Kuipers (NL) & Osterc (SLO)

Theatre

Poster

FREE COMMUNICATIONS - CATTLE AND BUFFALO REPRODUCTION, PRODUCTION AND BREEDING

Session 4

Joint session with the commissions on Animal Physiology and Livestock Farming Systems

Date 07-sep-04
8:30 - 12:30 hours
Chairperson Keane (IRL)

Theatre

A. Reproductive efficiency - technology in cattle and buffalo

B. Growth and carcass quality of cattle

Cattle

C. Milk production by cattle

ALTERNATIVE PROTEIN SOURCES

Session 5

Joint session with the commissions on Animal Nutrition and Sheep and Goat Production

Date	07-sep-04 14:00 - 18:00 hours
Chairperson	Dourmad (F)

Cattle

	NCS5	Page	Page
Papers	1-8	132-136	XLV-XLVI
Posters	9-14	136-139	XLVI

INDICATORS OF MILK AND MEAT QUALITY

Session 6

Date 08-sep-04
8:30 - 12:30 hours
Chairperson Hocquette (F)

Theatre

C6 no. Page

Poster

Commission on Sheep and Goat Production (S)

President
Dr D. Gabiña
Mediterranean Agronomic Institute of Zaragoza
IAMZ-CIHEAM
Apdo. 202
50080 Zaragoza
SPAIN
0034 976 716 000
0034 976 716 001
gabina@iamz.ciheam.org

Secretary 1
Dr L. Bodin
INRA
Centre de recherches de Toulouse
P.O. Box 27
31326 Castanet-Tolosan Cedex
FRANCE
0033 5 61 28 52 73
0033 5 61 28 53 53
bodin@toulouse.inra.fr

Secretary 2
Dr M. Schneeberger
Swiss Sheep Breeders Association
P.O. Box 76
3360 Herzogenbuchsee
SWITZERLAND
0041 62 956 68 69
0041 62 956 68 79
schneeberger@pop.agri.ch

ECONOMICS AND PROFITABILITY OF SHEEP AND GOAT PRODUCTION UNDER NEW SUPPORT REGIMES AND MARKET

Session 1

Date	05-sep-04 9:15 - 13:00 hours
Chairperson	Gabina (ES)

Theatre

Poster

MANAGEMENT OF GRASSLANDS FOR PRODUCTION, ENVIRONMENT AND LANDSCAPE BENEFIT

Session 2

Joint session with the commissions on Animal Nutrition, Cattle Production and Livestock Farming Systems

Date	05-sep-04 15:00 - 18:30 hours
Chairperson	Zervas (GR)

USE OF HORMONES FOR REPRODUCTION IN SHEEP AND GOATS - IMPACT AND ALTERNATIVES

Session 3

Date	06-sep-04 8:30 - 12:30 hours
Chairperson	Lebeouf (F)

Theatre

S3 no. Page

FREE COMMUNICATIONS - SHEEP AND GOAT PRODUCTION

Session 4

Joint session with the commissions on Livestock Farming Systems

Date	07-sep-04 8:30 - 12:30 hours
Chairperson	Gabina (ES)

Theatre

Sheep and Goat

Poster

Sheep and Goat

ALTERNATIVE PROTEIN SOURCES

Session 5

Joint session with the commissions on Animal Nutrition and Cattle Production

Date 07-sep-04
14:00 - 18:00 hours
Chairperson Dourmad (F)

	NCS5	Page	Page
Papers	**1-8**	**132-136**	**XLV-XLVI**
Posters	**9-14**	**136-139**	**XLVI**

SHEEP AND GOAT BEHAVIOUR - RELAVANCE TO WELFARE AND MANAGEMENT

Session 6

Joint session with the commission on Animal Management and Health

Date 08-sep-04
8:30 - 12:30 hours
Chairperson T.Waterhouse (UK)

Theatre

Sheep and Goat

Poster

Commission on Pig Production (P)

President
Prof.dr. C.Wenk
ETH-Zentrum
Institute of Animal Science
Clausiusstrasse 50 , CLU
8092 Zürich
SWITZERLAND
+41 1 632 32 55
+41 1 632 11 28
caspar.wenk@inw.agrl.ethz.ch

Secretary 1
Dr M. Kovac
University of Ljubljana
Groblje 3
1230 Domzale
SLOVENIA
00386 1 721 78 70
00386 1 7241 005
milena@mrcina.brfo.luni-lj.si

Secretary 2
Dr. D.Torrallardona
IRTA
Center de Mas Bové
Apartat 415
43280 Reus
SPAIN

ANTI-NUTRITIONAL FACTORS AND MYCOTOXINS

Session 1

Joint session with the commissions on Animal Nutrition and Horse Production

Date	05-sep-04
	9:15 - 13:00 hours
Chairperson	Torrallardona (E)

Theatre

PNH1 no. Page

Pig

Poster

LARGE-SCALE PIG FARMING SYSTEMS

Session 2

Joint session with the commissions on Animal Nutrition, Animal Management and Health and Livestock Farming Systems

Date 05-sep-04
15:00 - 18:30 hours
Chairperson Kovac (SLO) & Edwards (UK)

Theatre

INDICATORS OF ENVIRONMENTAL IMPACT IN LIVESTOCK SYSTEMS

Session 3

Joint session with the commissions on Animal Management and Health and Livestock Farming Systems

Date	06-sep-04 8:30 - 12:30 hours
Chairperson	Hermansen (DK)

	LMP3	Page	Page
Papers	**1-7**	**350-353**	**C-CI**
Posters	**9-13**	**353-355**	**CI**

FREE COMMUNICATIONS - PIG METABOLISM, GROWTH AND PRODUCTION

Session 4

Joint session with the commission on Animal Physiology

Date	07-sep-04 8:30 - 12:30 hours
Chairperson	Chadd (UK)

Theatre

Poster

Pig

HERD MODELLING FOR IMPROVED MANAGEMENT

Session 5

Joint session with the commission on Livestock Farming Systems

Date	07-sep-04
	14:00 - 18:00 hours
Chairperson	Knap (D)

Theatre

Poster

LONGITUDINAL DATA IN GENETICS

Session 6

Joint session with the commission on Animal Genetics

Date 08-sep-04
8:30 - 12:30 hours
Chairperson Knap (D)

Theatre

PG6 no. Page

Pig

Poster

Commission on Horse Production (H)

President
Dr W. Martin-Rosset
INRA
Centre of Research Clermont-Ferrand / Theix
63122 Saint Genès Champanelle
FRANCE
+33 473 62 40 99
+33 473 62 42 73
wrosset@clermont.inra.fr

Secretary 1
Dr T. Jezierski
Polish Academy of Science
Inst.Animal Genetics & Breeding
Jastrzebiec
05552 Wolka Kosowska
POLAND
0048 22 756 17 11
0048 22 756 16 99
tjezierski@rocketmail.com

Secretary 2
Dr E. Gerber
Swedish University of Agricultural Sciences
P.O. Box 7023
75007 Uppsala
SWEDEN
0046 18 672 789
0046 18 672 648
elisabeth.olsson@hgen.slu.se

ANTI-NUTRITIONAL FACTORS AND MYCOTOXINS

Session 1

Joint session with the commissions on Animal Nutrition and Pig Production

Date	05-sep-04
	9:15 - 13:00 hours
Chairperson	Torrallardona (E)

	PNH1	Page	Page
Papers	**1-7**	**261-264**	**LXXIX-LXXX**
Posters	**8-13**	**264-267**	**LXXX**

GROWTH AND BONE DISORDERS IN HORSES

Session 2

Date 05-sep-04
15:00 - 18:30 hours
Chairperson Bergero (IT)

Theatre

RARE BREEDS: GLOBAL GENETIC DISTANCE

Session 3

Date 06-sep-04
8:30 - 12:30 hours
Chairperson Bodo (H)

Theatre

Poster

Horse

FREE COMMUNICATIONS - EVALUATION AND PERFORMANCE OF HORSES

Session 4

Date	07-sep-04
	8:30 - 12:30 hours
Chairperson	Fleurance (F)

Theatre

Poster

INTERNATIONAL GENETIC EVALUATION

Session 5

Joint session with the commission on Animal Genetics

Date 07-sep-04
14:00 - 18:00 hours
Chairperson Koenen (NED)

Theatre

Poster

HORSE PRODUCTION IN SLOVENIA

Session 6

Date 08-sep-04
8:30 - 12:30 hours
Chairperson Habe (SLO)

Theatre

Commission on Livestock Farming Systems (L)

President
Dr. A. Gibon
INRA
Centre de recherches de Toulouse
P.O. Box 27
31326 Castanet-Tolosan Cedex
FRANCE

Secretary 1
Dr. J. Hermansen
Danish Institute of Agricultural Sciences (DIAS)
Research Centre Foulum
P.O. Box 50
8830 Tjele
DENMARK
+45 8999 1233
+45 8999 1200
John.Hermansen@agrsci.dk

Secretary 2
A. Bernues Jal
Servicio de Investigacion Agroalimentaria
Gobierno de Aragon
Apartado 727
50080 Zaragoza
SPAIN

EFFECT OF GLOBALISATION ON LIVESTOCK SYSTEMS

Session 1

Date	05-sep-04 9:15 - 13:00 hours
Chairperson	Peters (D)

Theatre

Poster

ANIMAL HEALTH AND WELFARE IN INTENSIVE AND EXTENSIVE SYSTEMS

Session 1

Joint session with the commissions on Animal Management and Health and Cattle Production

Date	05-sep-04 9:15 - 13:00 hours
Chairperson	Fourichon (F)

	MCL1	Page	Page
Papers	1-8	143-146	XLVIII-XLIX
Posters	9-12	147-148	XLIX

MANAGEMENT OF GRASSLANDS FOR PRODUCTION, ENVIRONMENT AND LANDSCAPE BENEFIT

Session 2

Joint session with the commissions on Animal Nutrition, Cattle Production and Sheep and Goat

Date	05-sep-04 15:00 - 18:30 hours
Chairperson	Zervas (GR)

Poster

LARGE-SCALE PIG FARMING SYSTEMS

Session 2

Joint session with the commissions on Animal Nutrition, Animal Management and Health and Pig Production

Date 05-sep-04
15:00 - 18:30 hours
Chairperson Kovac (SLO) & Edwards (UK)

INDICATORS OF ENVIRONMENTAL IMPACT IN LIVE-STOCK SYSTEMS

Session 3

Joint session with the commissions on Animal Management and Health and Pig Production

Date 06-sep-04
8:30 - 12:30 hours
Chairperson Hermansen (DK)

Theatre

Poster

FREE COMMUNICATIONS

Session 4

Joint session with the commissions on Animal Nutrition and Animal Physiology

Date 07-sep-04
8:30 - 12:30 hours

Livestock farming systems

Poster

ETHICAL ISSUES IN ANIMAL AGRICULTURE

Session 5

Joint session with the commission on Animal Management and Health

Date 07-sep-04
14:00 - 18:00 hours
Chairperson Marie (F)

	ML5	Page	Page
Papers	**1-13**	**162-168**	**LIII-LIV**

HERD MODELLING FOR IMPROVED MANAGEMENT

Session 5

Joint session with the commission on Pig Production

Date 07-sep-04
14:00 - 18:00 hours
Chairperson Knap (D)

	PL5	Page	Page
Papers	1-6	293-296	LXXXVI
Posters	7	296	LXXXVI

LOW-INPUT SYSTEMS

Session 6

Joint session with the commission on Animal Physiology

Date 08-sep-04
8:30 - 12:30 hours
Chairperson Faverdin (F)

	PhL6	Page	Page
Papers	1-4	182-183	LIX
Posters	5-7	184-185	LIX-LX

ANIMAL GENETICS [G]

Theatre GPh1.1

Physiology of lactation: Old questions, new approaches

M. Ollivier-Bousquet, INRA, Unité Génomique et Physiologie de la Lactation, 78352 Jouy-en-Josas cedex, France

Questions regarding the mechanisms of growth, differentiation and function of the mammary gland have been raised a long time ago but are still not fully answered. Advances in the field of endocrinology have made it possible to describe the role of numerous lactogenic hormones. Cell and molecular biology have contributed to partly explain the mechanisms of action of these hormones. Presently, synergies between genetics and physiological approaches promote new scientific progress. For example, genetic maps and bacterial artificial chromosomes (BAC) have provided tools useful for understanding gene fonctionning in the nuclear environment. The creation of mice carrying gene mutations that act in multiple points of the prolactin signalling pathway provides new insight into prolactin action. Goat αS1-casein presents an extensive polymorphism which is associated with quantitative variations of αS1-casein in milk. This polymorphism is correlated with alterations in the mammary cell secretory pathway and sheds new light on the mechanism of casein secretion. New results on the effects of oxytocin on the milk secretory pathway suggest a putative role of this hormone on mammary epithelial cell functionning in addition to its classical role on milk ejection. In addition to these approaches, proteomics and genomics technologies are now available. Questions answerable in the immediate future by these technologies will be discussed.

Theatre GPh1.2

Genetic aspects of lactation

H. Geldermann and A.W. Kuss, Department of Animal Breeding and Biotechnology, University of Hohenheim, D-70593 Stuttgart, Germany

Development and function of the mammary gland are of crucial importance for survival and growth of the newborn mammals. Milk from domestic animals is one of the essential foods for humans, and genes encoding lactation traits, like differentiation, organogenesis and protein expression, are being intensively studied. In recent years the genomic sequences of important genes were analysed and their loci mapped. Effects of gene variants on milk production have been investigated by association studies, QTL mapping, DNA-protein binding assays, as well as transgenic cells or animals. Much of the work is oriented towards lactoprotein genes, showing that associations between polymorphic alleles and trait values depend on breed and many other influences, while the quantities of specific gene products seem to be largely influenced by the individual coding genes. New methods are now being used for a causal analysis of the effects of single nucleotide variants on the function of genes and specific lactation features. The knowledge thus gained can be applied to improve the composition and technological quality of milk, protection against pathogens as well as many other traits. Recent research regards the mammary gland as model for human diseases and oncogenesis. Mammary glands are used to produce recombinant proteins for pharmaceutical use and to develop Functional Foods for human nutrition. Basic and applied genetic research on lactation is therefore of central importance in animal genetics and breeding.

Theatre GPh1.3

Production of kappa-casein knock-out mouse chimera

M. Baranyi[1], E. Gócza[1], A.P.C. Lemos[1], L. Hiripi[1], B.V. Carstea[1], B. Whitelaw[2] and Zs. Bösze[1*], [1]Agricultural Biotechnology Center, Department of Animal Biology, Gödöllö, Hungary, [2]Roslin Institute, Department of Gene Expression and Development, Roslin, UK*

During secretion and storage, caseins are assembled into stable supra-macromolecular structures called calcium-phosphate micelles. Among the caseins, kappa-casein (κ-CN) plays an important role in micelle formation and stabilization in milk. It also has a key role in milk clotting. Phenotypic analysis of a κ-CN knock-out mouse might reveal other functions of this protein and the altered physico-chemical properties of milk without functional κ-CN.
The gene construct containing the mouse κ-CN gene disrupted by a positive selection marker sequence (PGKneo) was electroporated into the R1 embryonic stem cell line. From the G418 selected clones, 312 were frozen for long-term storage and their DNA analyzed for homologue recombination event. Among the 312 ES cell clones 3 were identified as the carrier of a non-functional null-mutant κ-CN gene. Those ES cell clones were aggregated to CD1 host-embryo to create aggregation chimeras and transferred to female recipients. Seven male chimeric mice were obtained so far with two out of the three ES cell clones. The founders were crossed with control CD1 females and the litters will be classified by the coat colour marker of the R1 ES to identify the germline chimeric founders. Homozygote and heterozygote kappa-casein knock-out mice will be analysed for lactational changes and milk quality.
Supported by OTKA grant: T34767

Theatre GPh1.4

In vitro study of beta and kappa casein promoters

P. Frajman, T. Lenasi, M. Debeljak, M. Narat and P. Dovc, Department of Animal Science, Biotechnical Faculty, University of Ljubljana, Groblje 3, SI-1230 Domzale, Slovenia*

Expression of casein genes during lactation underlies complex hormonal regulation. Sequence analysis of different casein promoters revealed a number of cis regulatory elements being responsible for timely coordinated and tissue specific expression. Potential binding sites for transcription factors as Stat5, YY1, C/EBP, NF1, GRE, Stat6, AP1 and AP2 were found in different casein gene promoters. We sequenced promoter regions from bovine and equine beta- and kappa casein genes. The functionality of some cis elements has been studied by electro mobility shift assay using nuclear extracts from lactating mammary gland and Stat5 overproducing COS7 cells. Different promoter length variants were cloned into the pGL3 enhancer vector containing luciferase reporter gene. These constructs were used for induction of transient luciferase expression in bovine mammary BME-UV $^1/_2$ cell line. Comparison of promoter activity was performed for 1.9 kb long equine, 2.1 kb and 0.9 kb long bovine kappa casein promoters, 1.9 kb and 0.8 kb long equine beta casein promoters, equine beta casein promoter with the exon 1 and 396 bp of the intron 1 and 0.9 kb long bovine beta casein promoter. Interestingly, the lowest expression was observed using equine beta casein promoter with exon 1 and part of the intron 1, suggesting the presence of negative regulatory element in the intron 1 of the equine beta casein gene.

Theatre GPh1.5

Sequence polymorphisms in the 5' region of the α_{S1}-casein gene (*CSN1S1*) in goat: A new perspective to explain quantitative variability in α_{S1}-casein content?

E.-M. Prinzenberg, K. Gutscher and G. Erhardt, Justus-Liebig-University Giessen, Institute for Animal Breeding and Genetics, Ludwigstr. 21 B, 35390 Giessen, Germany*

Quantitative variation in αS1-casein is well known in goat with high (3.6 g/L), intermediate (1.6 g/L) and low (0.6 g/L) αs1-casein content associated with *CSN1S1* alleles *A, B, C* and *E* or *D, F* respectively. Insertions and deletions within the gene have been identified as one, but probably not the only reason for this. The 5' flanking region of all casein genes contains promoter elements and different transcription factor binding sites and is highly conserved in mammals. Single strand conformation polymorphism (SSCP) and following DNA sequence analyses of the 5' flanking region of *CSN1S1* in the German breed "Bunte Deutsche Edelziege" revealed 12 polymorphic sites (SNP) within a 1kb fragment preceding the first exon. Comparison to three goat sequences available in GenBank (AJ504710-12) showed a TT deletion in all our sequences and two additional SNPs unique for two of the newly generated sequences. Another 66 polymorphic sites (SNPs, deletions and insertions) were identified in comparison to cattle.

Four sequence types result from the samples sequenced, and four different migration patterns were found in SSCP analysis. Computer based transcription factor binding motif search identified two SNPs in goat, that directly affect putative binding sites. A PCR-RFLP was used to verify one of these mutations and to determine allele frequencies goat.

Theatre GPh1.6

Goat β-casein C allele: characterization and detection at the DNA level

S. Chessa[1], E. Budelli[2], F. Chiatti[1], A.M. Cito[3], A. Caroli[3], [1]Department VSA, Via Trentacoste 2, 20134 Milan, Italy, [2]CERSA, Segrate, Milano, Italy, [3]Department SBA, Bari, Italy

Goat caseins show a complex qualitative and quantitative variability, characterised by several genetic polymorphisms and multiple post-translation modifications. Recently, a new β-casein protein variant was identified in goat milk by peptide mass fingerprinting and tandem mass spectrophotometry. The variant, named C, differed in a mono amino acid substitution, Ala_{177} to Val_{177}, from the A variant.

In order to get further information on goat β-casein (*CSN2*) variability at the DNA level, exon 7 was analysed by Polymerase Chain Reaction - Single Strand Conformation Polymorphism (PCR-SSCP) in different Italian goat breeds. An unknown SSCP pattern was found and sequenced, and the correspondence between the C protein variant and the SSCP polymorphism was demonstrated by identifying the nucleotide exchange responsible for the amino acid substitution Ala_{177} to Val_{177}. Since both amino acids are neutral, the mutation is not detectable by screening protein techniques such as milk isoelectric focusing.

The PCR-SSCP analysis allows an easy monitoring of the polymorphism, which was identified at different frequencies in most of the typed Italian goat breeds. Studies on the caprine casein locus variability could include the PCR-SSCP test here described to identify the *CSN2*C* allele at the DNA level.

Theatre GPh1.7

Associations of milk protein polymorphism and milk production traits and udder health traits in Finnish Ayrshire cows

M. Ojala, T. Seppänen, A.-M. Tyrisevä and T. Ikonen, University of Helsinki, Department of Animal Science, P.O.Box 28, FI-00014 Helsinki, Finland*

Milk protein genotypes could be used as a criterion in selection to improve cheese production properties of milk. A prerequisite for the previous is that the effects of the genotypes on important production traits have been established. The objectives of the study were to estimate the effects of β-κ-casein (β-κ-CN) and β-lactoglobulin (β-LG) genotypes on 305-day milk, fat and protein yield, fat and protein content, somatic cell score and treatment for mastitis. Complete information with respect to the model was available for 18686 cows with first, 16274 with second and 12636 with third lactation milk production traits. The linear univariate model included year-season of calving, calving age, days open, β-κ-CN and β-LG genotype as fixed effects, and herd, animal and residual as random effects. Based on β-κ-CN genotypes, there was a tendency that the A_2A haplotype was associated with high milk yield and low fat but moderate protein content, the A_1B haplotype with low milk yield and high fat and protein content, and the A_1E haplotype with low milk yield and high fat but low protein content. The haplotypes A_1B and A_2B tended to decrease whereas the haplotypes A_1E and A_1A tended to increase the somatic cell score. The β-LG BB was associated with high fat content and slightly decreased milk yield. No effect of the β-κ-CN and β-LG genotypes was observed on treatment for mastitis.

Theatre GPh1.8

Effects of genetic milk protein variants on the protein recovery in cheese

E. Hallén[1], A. Andrén[1], T. Allmere[1] and A. Lundén[2], [1]Department of Food Science; [2]Department of Animal Breeding and Genetics, Swedish University of Agricultural Sciences, Uppsala, Sweden*

The genetic milk protein variants are known to influence processing properties and product yield of milk. Previous results indicate that the β-casein E, a milk protein variant being common in the Nordic Ayrshire dairy cow populations, has unfavourable effects on milk coagulation properties and thereby cheese yield. Analysis on industrial scale of factors affecting cheese output per kilo of milk is hampered by the elaborate process of producing adequate numbers of full-size cheeses. Therefore, we instead focus on the loss of milk proteins to the whey during different steps of the cheese making process, measured on a micro scale.
Up till now, we have genotyped 127 cows of the Swedish Ayrshire and Swedish Holstein breeds in the university experimental herd for β-casein, λ-casein and β-lactoglobulin. Individual milk samples are clotted using chymosin and subjected to syneresis to form a "micro fresh cheese". The whey fractions and the curd are being analysed, both after clotting and after syneresis, for concentrations of α_{s1}-, α_{s2}-, β-, λ-casein, β-lactoglobulin and α-lactalbumin and the rheological properties of milk gels are monitored repeatedly during the coagulation process. Associations between milk protein genotype and the concentrations of individual proteins and the technological properties of milk are studied.

Theatre GPh1.9

Use of graphical gaussian models to infer relationships between genes

P. Sørensen and J.H. Badsberg, Danish Institute of Agricultural Sciences, Department of Animal Breeding and Genetics, Blichers Allé, P.O. Box 50, 8830 Tjele, Denmark*

One of the central problems of molecular biology is to uncover relationships between genes and/or their products. The relative level of these molecules in tissue samples can be measured using a variety of high-throughput molecular techniques. These techniques are generating massive data sets and provide both a great wealth of information and challenges in making inferences. Unsupervised clustering methods are often considered as the necessary first step in visualization and analysis of such data. Most of these methods cluster genes based on their Pearsons's correlation coefficients. Here we applied Graphical Gaussian Models (GGM) to infer relationships between the genes. GGM utilize both the Pearsons's and the partial correlation coefficients among genes. The partial correlation coefficient is measuring the correlation between two genes after the common effects of all other genes in the genome are removed. The partial correlation should therefore provide additional information about regulatory links between genes. Results from both simulated and real data will be presented and its implications discussed.

Poster GPh1.10

Frequency of the bovine *DGAT1* (K232A) polymorphism in selection lines with high and low milk fat content

J. Näslund, G. Pielberg and A. Lundén, Department of Animal Breeding and Genetics, Swedish University of Agricultural Sciences, Uppsala, Sweden

A quantitative trait locus (QTL) for milk production traits has been found on the centromeric region of the bovine chromosome 14. A dinucleotide substitution in the gene for acyl-CoA:diacylglycerol acyltransferase1 (DGAT1) was proposed to be the causative mutation giving rise to the QTL-effect. The substitution replaces lysine with alanine (K232A) which results in an increase in protein and milk yield and a decrease in fat yield, fat and protein percentage. The high yielding dairy herd at the Swedish University of Agricultural Sciences includes cows of the Swedish Ayrshire and the Swedish Holstein breeds. Since 1985 the Ayrshire cows have been selected for either high (HF) or low (LF) fat content, the two lines having an equal total milk energy production. In total, 280 cows were genotyped for the *DGAT1* polymorphism and gene frequencies were estimated for the selection lines and the two breeds. The frequency of the lysine variant was only 12% but was significantly higher in the HF line. Also the two breeds differed, with the lysine variant being more frequent in the Swedish Holstein breed than in the Swedish Ayrshire breed. These initial results constitute part of a study which aims at investigating the effect of *DGAT1* genotype on the detailed composition of milk fat.

 Poster GPh1.11

Isolation of antibacterial peptides from rabbit milk casein

M. Baranyi[1], U. Thomas[2] and A. Pellegrini[2], [1]Agricultural Biotechnology Center, Institute of Animal Biology, P.O.Box 411, H-2100 Godollo, Hungary, [2]University of Zurich, Institute of Veterinary Physiology, Winterthurerstrasse 260, CH-8057 Zurich, Switzerland*

It is well known that proteolytical digestion of bovine casein yields several bioactive peptide fragments. To look for possible antibacterial peptides encrypted in the rabbit casein structure, the acid-precipitated fraction of rabbit milk, the "whole casein" was enzymatically digested by trypsin. For the separation of peptide fragments reversed-phase chromatography was used. The collected fractions were pooled and their antimicrobial properties investigated against *Esch. coli*, *B. subtilis* and *Staph. lentus*. Two pools derived from the trypsin digestion inhibited the growth of *Staph. lentus* moderately. Three pools were found to strongly inhibit the growth of *B. subtilis*, whereas two pools inhibited this bacterial strain moderately. Three bactericidal peptide fragments present in two pools were isolated and identified. Their sequences were FHLGHLK (residues 19-25 of α_{S1}-casein), HVEQLLR (residues 50-56 of β-casein), and ILPFIQSLFPFAER (residues 64-77 of β-casein). These peptides were synthesised, and after purification their bactericidal activity was investigated up to a concentration of 6.0 mM against Gram-positive and Gram-negative bacteria. The peptides were active against the Gram-positive bacteria only. Our results suggest a possible antimicrobial function of rabbit caseins. It is conceivable that bactericidal peptides generated by endopeptidases of the mammalian gastro-intestinal tract possibly provide protection for newborn rabbits against aggression of microorganisms.

Theatre G2.1

Genomics for disease resistance: example of search and validation of QTLs for resistance to chicken coccidiosis and need for a european research network (EADGENE)

M-H Pinard-van der Laan, Animal Genetics Department, 78352 Jouy-en-Josas, France

There is an increasing concern in Europe to produce food of high quality and safety, while respecting animal welfare and the environment Genomics provides a valuable tool to develop improved disease control in animals. One approach includes mapping quantitative trait loci (QTL) linked to the control of disease resistance. An example of such a strategy applied to the control of coccidiosis in chicken is presented.

Two chicken lines were identified as showing extreme resistance phenotype to E*imeria. tenella*. An F2 cross was produced and challenged for *E. tenella* (n=860) and the technique of selective genotyping was applied within families by typing for 139 microsatellite markers the 15 % most resistant and susceptible birds. From QTLs linked to resistance traits, 8 were chosen for validation in two commercial « label » lines. If applicable, interest of including these markers in breeding schemes will be investigated.

But the applications of genomics for disease control remain too scarce, due particularly to fragmentation of research across Europe and the different fields. Hence, a European Network of Excellence, EADGENE (European Animal Disease Genomics Network of Excellence for Animal Health and Food Safety) will aim at bringing sufficient excellence and resources, co-ordinate research and apply genomics as a tool to improve animal health, food quality and safety by linking other players, in research and in industry.

 Theatre G2.2

Implementation of selection for carcass and meat quality in a pig breeding program in the Netherlands

E.F. Knol and D.T. Prins, IPG, Institute for Pig Genetics, Beuningen, Schoenaker 6, 6641 SZ, The Netherlands*

Classification and payment of pig carcasses is changing. Slaughter plants base their payment on the estimated value of primals and sub-primals using e.g. evaluation systems like AutoFOM and Vision. Weights and quality of (sub) primals often have an optimum. Size of bellies should fit boxes and marbling optimum. We participated in an experiment involving 2555 dissections in the USA and we entertain a carcass dissection program in Canada and one in the Netherlands. At present data of 7658 dissected carcasses were available for analysis. Genetic parameters, both heritabilities and genetic correlations were estimated (ASREML) for, among others, de-boned loin and ham yields, HGP -lean, -backfat and -loin depth, ALOKA loin depth and Renco backfat, marbling in the loin and ham, drip loss, shelf life and Japanese colour score. Different payment systems were analysed and different selection scenarios were evaluated to arrive at a selection program that favours boneless primal yields, drip-loss and Japanese colour score. Current research program, however, is only a first step to a more sophisticated selection program with distinct sire lines and refined selection indices per line which will try to optimise the total carcass value for different markets. The presentation will focus on the predicted results of different selection strategies using the estimated genetic correlation matrices.

Theatre G2.3

Selective breeding provides a method to control clinical disease in pigs

M. Henryon, P. Berg and H.R. Juul-Madsen, Danish Institute of Agricultural Sciences, Research Centre Foulum, P.O. Box 50, 8830 Tjele, Denmark*

The objective of this presentation is to demonstrate that selective breeding for resistance provides a method to control clinical disease in pigs. Current research findings indicate that selective breeding for resistance can be successful as additive genetic variation for resistance to clinical disease exits in pigs. However, the level of success depends largely upon the reliability with which the breeding values for resistance can be predicted. If we use the incidence of disease (i.e., health records) as the only source of information by which to estimate the breeding values, these estimates are generally unreliable unless records are available from large numbers of relatives. Consequently, selective breeding would be more successful if other traits, which reflect the resistance of the pigs, could be identified and used as additional sources of information. Three traits that may reflect resistance include immunological parameters, post-mortem lesions, and genetic markers linked to quantitative trait loci. These traits may be potentially useful as additional sources of information as there is some evidence to suggest that they are genetically correlated to resistance to clinical disease. These findings are encouraging for pig breeders. The additive genetic variation that exists for resistance, together with the potential for additional sources of information becoming available, demonstrates that selective breeding for resistance would provide a method to control clinical disease in pigs.

Theatre G2.4

Effects of genotype by environment interaction on genetic gain in breeding programs: selection candidates located in one environment

H.A. Mulder and P. Bijma, Animal Breeding and Genetics Group, Wageningen University, PO Box 338, 6700 AH Wageningen, The Netherlands*

Genotype by environment interaction (G×E) is becoming increasingly important, because breeding programs tend to be more internationally oriented and breeding goals put increasing emphasis on functional traits. The aim of this study was to investigate the mechanisms involved in the effects of G×E on genetic gain in breeding programs. Two environments were considered: a selection environment (SE) with all selection candidates and a production environment (PE). The breeding goal contained only performance in PE. A pseudo-BLUP selection index was used to predict genetic gain.
If no information of relatives in PE was available, genetic gain in the breeding goal was proportional to the genetic correlation (correlated response). If information of relatives in PE was available, genetic gain had a curvilinear relationship with the genetic correlation (symmetric around zero) and resulted in a much lower loss in genetic gain due to G×E. Progeny testing schemes had a lower loss in genetic gain than sib testing schemes. Increasing heritability or number of progeny per dam resulted in a higher loss in genetic gain, whereas increasing the number of progeny per sire in PE resulted in the opposite. It was concluded that recording of relatives in PE could minimize loss in genetic gain due to G×E. Progeny testing schemes were most robust for G×E.

Theatre G2.5

First results on modelling level of infection to detect a polygenic effect associated to susceptibility to scrapie in a Romanov flock

C. Díaz[1], Z. G. Vitezica[2], R. Rupp[2], O. Andréoletti[3] and J.M. Elsen[2], [1]Depto Mejora Genética Animal. INIA. Apdo 8111. 28080 Madrid, Spain.[2]INRA. Station d'Amélioration Génétique des Animaux, Castanet Tolosan, France, [3]Ecole Vétérinaire de Toulouse, Interactions Hotes-Agents Pathogènes, 31076 Toulouse, France*

The objective was to compare different measurements of infection and mechanisms of action to explain differences in risk of animals to show scrapie signs. Data consisted of 4049 Romanov animals at Langlade, 447 died of scrapie. Several measurements of infection were developed for each period of time. Measurements were a function of placenta positive units, number or proportion of scrapie animals and proximity to death of scrapie. Several mechanisms of infection were also compared. Firstly, animals were assumed to be affected just by the initial load; or be independently affected for successive doses of infection. Secondly, animals were assumed to be infected by a cumulative doses in the animal. For each case, doses were defined by each measurement, period of time and management group. Models were compared according to Akaike (AIC) and Bayesian Information Criteria (BIC). According to both criteria cumulative models performed better than models considering just one infection dose at first exposure or independent doses. With the available information, level of infection seemed to be better described by measurements involving number of scrapie animals and distances to death.

Model for developing breeding objectives for beef cattle used in different production systems

M. Wolfová[1], J. Wolf[1], J. Přibyl[1], R. Zahrádková[1] and J. Kica[2], [1]Research Institute of Animal Production, P.O.Box 1, CZ 10401 Prague-Uhříněves, Czech Republic, [2]Research Institute of Animal Production, SK Nitra, Slovak Republic*

A bio-economic model for a wide range of cattle production systems in which beef bulls are involved was developed. The model simulates the life-cycle production of both beef and dairy cow herds with an integrated feedlot system. The Markov chain approach is used to simulate herd dynamics. The herd is described in terms of states animals can be in and possible transitions between these states. The structures of the production systems in their stationary states are calculated. The economic efficiency of each system is expressed as a function of biological traits of animals and of management and economic parameters. The model allows for estimating marginal economic values for 15 traits in each system separately. The economic weight for each trait or trait component (direct, maternal) and for the selection group and breed of interest is then calculated as the weighted sum of the marginal values for the trait in all production systems in which the selection group will operate. The weighting factors for each system are given by the product of the number of discounted expressions for direct and maternal trait components transmitted in this system by the selection group and the proportion of cows in each system.

Theatre G2.7

Analysis of associations between the prion protein genotype and production traits in East Friesian milk sheep

F. de Vries[1], H. Hamann [1], C. Drögemüller[1], M. Ganter[2] and O. Distl[1], [1]Institute for Animal Breeding and Genetics, School of Veterinary Medicine Hannover, Germany, [2]Clinic for pigs and small ruminants, forensic medicine and ambulatory service, School of Veterinary Medicine Hannover, Germany*

The objective of this study was to analyse associations between scrapie ovine prion protein (PrP) genotypes and production, type, as well as reproduction traits in East Friesian milk sheep. Linear animal models were employed for the analysis of the PrP-genotype-effect. The data set included 643 genotyped sheep among 3306 recorded sheep for the traits scores for muscle mass, wool quality and type, 431 genotyped sheep among 2097 recorded sheep for reproduction traits, 258 genotyped sheep among 915 recorded sheep for milk production traits and 440 genotyped animals with body measurements. The genotyped milk sheep showed significantly superior performance in score for muscle mass, type, wool quality and fat yield in comparison to the non-genotyped animals. This indicates a pre-selection of animals for genotyping.
Significant differences for the traits withers height and heart girth were found between the ARR/ARR and ARQ/ARQ genotypes. No significant associations were found between the prion protein genotypes and the other performance traits investigated. The results from this study indicated that selection towards scrapie resistance mainly based on ARR homozygous animals might lead to a population of smaller sized East Friesian milk sheep.

Optimization of the Bavarian Pig testing and breeding scheme

D. Habier[1,2] and L. Dempfle[1], [1]Technical University of Munich-Weihenstephan, Department of Animal Science, Alte Akademie 12, 85354 Freising, Germany, [2]Bavarian Institute of Agriculture, Institute of Animal Breeding, Prof.- Dürrwaechter-Platz 1, 85586 Poing-Grub, Germany

The present optimization of performance testing and of the breeding scheme was stimulated by the implementation of a combined breeding value estimation utilizing purebred and crossbred information. Also the breeding objective was changed considering new market conditions. Another reason is the new testing situation, caused by the change of the traditional pens for two pigs to sized pens with automatic feeding system. The objective is to maximize profit from breeding by harmonizing the performance testing with the breeding scheme.
Performance testing was simulated with APL and the breeding process was modelled by a three stage selection. A first analysis should answer how many test boars, how many progenies per sire, how many full sibs per test group and how many purebred and crossbred progenies are optimal within a given station testing capacity and size of the breeding population.
The results show that a young sire design based on own performance and sib information combined with a strict selection of about 2% at the second stage is superior to the current progeny testing design. The within family breeding value (mendelian sampling) of these selected animals is then estimated by progeny testing to provide sib information for the next generation. To ensure that all young breeding sows participate with two piglets per litter about 25-30% of the testing should be done with purebred animals.

Theatre G2.9

Prediction of response to long-term selection considering skewness of distribution and curvilinear relationships between traits caused by previous selection

A. Nishida, H. Ogasawara, T. Konari, H. Shinohara, Y. Ohtomo and K. Suzuki, Graduate School of Agricultural Science, Faculty of Agriculture, Tohoku University, Sendai 981-8555, Japan*

Current theory to predict response to repeated selection is based on normal distribution of data with constant variance and based on linear relationships between traits and between breeding values(G) and phenotypes. However, it has been shown that genetic variance curvilinearly decreases with generation of selection, and skewness of distribution caused by truncation selection brings curvilinear relationships between variables. Further, it is becoming clear recently that the region in genome called intron plays an important role to control the expression of exon. The clarifying role of intron suggests the adequacy of the treatment of the result of QTL analysis not only as a point of peak in LOD score which possibly means exon but a region which significantly affects the quantitative trait through interacting exon and intron. Based on these facts, a continuous distribution model, not a discrete gene model, for the prediction of response to long-term single trait selection was developed by improving the previous model. The degrees of curvilinearity in selection response, in decrease of genetic variance and in increase of skewness of the distribution in G were larger under higher initial heritability and stronger selection intensity. The possibility to extend the model for index selection for multiple traits was discussed.

Theatre G2.10

Impact of selection environment on the evolution of environmental sensitivity using a resource allocation approach, illustrated by reaction norms

E.H. van der Waaij[1,2] and P. Bijma[2], [1]Department of Farm Animal Health, Veterinary Faculty, University of Utrecht, PO Box 80127 3508 TC Utrecht, The Netherlands, [2]Animal Breeding and Genetics Group, Wageningen University and Research Centre, PO Box 338, 6700 AH Wageningen, The Netherlands*

Consequences of selection for productivity in a particular environment on performance in other environments have been investigated using a resource allocation approach. Environments were defined by the percentage of energy intake required for fitness. Insufficient energy for fitness reduced fertility and increased mortality. Animals that died had zero production. Resource intake, production potential and allocation of resources were assumed to be heritable and genetically uncorrelated, whereas fitness and (observed) production are resulting parameters. A base population was placed in a poor, intermediate, or good environment and selected for 50 generations using stochastic simulation. Resource intake was either unlimited or limited to 150% of the initial population mean. Every fifth generation, production was recorded in all environments and reaction norms were derived accordingly. Results show that selection in a good environment causes increased environmental sensitivity, as illustrated by an increased slope of the population average reaction norm, occurring especially when resource intake was limited. These results are supported by analysis of real data available in literature.

Theatre G2.11

Somatic cell (neutrophil) counts in the war against Staphylococcus aureus: predator-prey models at the rescue

J. Detilleux, Faculty of Veterinary Medicine, University of Liège , Liège, 4000 Belgium

To address the question of whether a minimum concentration of blood neutrophils is necessary to decrease *Staphylococcus aureus* concentration in mastitic milk, literature was searched for studies in which neutrophils were incubated with *S. aureus*. Different mathematical models that describe the changes in *S. aureus* population as a function of neutrophilic concentrations were applied to the collected data. The best fitted model established (1) that the rate of bacterial killing depended on the ratio of neutrophils to bacteria with neutrophilic attack rate accelerating at first before decelerating as the ratio increases and (2) that neutrophil concentration should be within a limited range to trigger a decline in the bacterial population. Outcomes of this model are supported by what is known about neutrophilic functions and laboratory findings in bovine and human neutrophils. These results may be of assistance in setting selection goals for a better resilience to *S. aureus* mastitis in dairy cattle. Indeed, an optimal neutrophilic concentration appears to exist for successful clearance of *S. aureus* infection, which is neither the lowest nor the highest one.

Theatre G2.12

Genetic relationship between carcass and calving traits in beef cattle

S. Eriksson, A. Näsholm, K. Johansson and J. Philipsson, Swedish University of Agricultural Sciences, Dept. of Animal Breeding and Genetics, P.O. Box 7023, SE-75007 Uppsala, Sweden*

Swedish beef breeders aim at improving growth- and carcass traits and minimise calving difficulties. Both calf and dam sizes are of importance for the calving process, and may be influenced by selection for growth or muscularity. The aim of this study was to estimate genetic correlations between carcass and calving traits for beef cattle in Sweden. Field data on altogether 59,182 Charolais and 27,051 Hereford calvings, and commercially recorded carcass weight, fatness and fleshiness for 5,260 Charolais and 1,232 Hereford bulls were included in the analyses. Bivariate linear animal models were used. First and later parity calvings were treated as separate traits. Carcass weight was positively genetically correlated (0.11-0.53) with both direct and maternal effects on birth weight. Carcass weight was also positively (unfavourably) correlated with direct effect on calving difficulty score (0.12-0.40). The genetic correlation between carcass weight and maternal effect on calving difficulty was weak or negative (favourable). Carcass fatness was negatively genetically correlated with birth weight (-0.38 – -0.05) and in most cases also with calving difficulty. Genetic relationships between carcass fleshiness and calving traits were variable. Certain antagonistic genetic relationships between calving and carcass traits were thus found in this study. It is therefore important to evaluate and select for calving traits in addition to production traits in beef cattle.

Theatre GM2.1

Genetic parameters of calving traits at first and second calving in Danish Holsteins

M. Hansen[123], M. S. Lund[1], J. Pedersen[2], and L. G. Christensen[3], [1]Danish Institute of Agricultural Sciences, Dept. of Animal Breeding and Genetics, Research Centre Foulum, P.O. Box 50, DK-8830 Tjele, Denmark, [2]Danish Cattle Federation, Udkærsvej 15, Skejby, 8200 Aarhus N, Denmark, [3]The Royal Veterinary and Agricultural University, Dept. of Animal Science and Animal Health, Grønnegårdsvej 2, DK-1870, Frederiksberg C, Denmark*

The aim of this study was to estimate heritabilities and genetic correlations for stillbirth, calving difficulty, and calf size at second calving and to estimate genetic correlations between these calving traits at first and second calving. Data originated from Danish Holstein cows calving from 1988 to 2002. Bivariate threshold models using Gibbs sampling were fitted for estimation of heritabilities and genetic correlations between the traits at second calving. For estimation of the genetic correlation between calving traits at first and second calving bivariate linear models using REML were fitted. The direct and maternal heritabilities at second calving were 0.05 and 0.005 for stillbirth, 0.08 and 0.04 for calving difficulty, and 0.16 and 0.05 for calf size. The genetic correlations between these traits at second calving were high (0.65 to 0.99) between the direct effects and moderate to high (0.44 to 0.92) between the maternal effects. The genetic correlation of each calving trait at first and at second calving ranged from 0.89 to 0.99 for the direct effects and from 0.74 to 0.88 for the maternal effects.

Theatre GM2.2

New breeding tools for improving mastitis resistance in European dairy cattle

J. Vilkki[1], N. Schulman[1], L. Andersson-Eklund[2], A. Fernandez[2], H. Viinalass[3], A.M. Sabry[4], J.L. Williams[5], N. Hastings[5], R.F. Veerkamp[6], H.J. Eding[6] and M. Sando Lund[4], [1]MTT Agrifood Research Finland, Finland, [2]Swedish University of Agricultural Sciences, Sweden, [3]Estonian Agricultural University, Estonia, [4]Danish Institute of Agricultural Sciences, Denmark, [5]Roslin Institute, Edinburgh, Scotland, UK, and [6]Institute for Animal Science and Health, ID-Lelystad, The Netherlands

Mastitis is the most common and costly disease in dairy cattle. The heritability of mastitis resistance is low and hence genetic improvement based on phenotypic selection is difficult. The extensive recording of health traits in the Nordic countries provides a unique resource for developing of tools for marker-assisted selection. Genome scans show that QTL for mastitis resistance segregate in Nordic cattle populations. The EC funded project (QLK5-CT-2002-01186, MASTITIS RESISTANCE) uses these resources to confirm, fine-map and identify genetic loci underlying variation in mastitis resistance. During the first year a powerful granddaughter design material has been collected. It consists of DNA samples and comparable phenotypes from 78 half-sib families, including 4000 bulls representing Nordic Red and Holstein-Friesian breeds. Six chromosome regions with the best support for the presence of QTL and largest total number of segregating sires were selected for further analysis to refine the QTL-intervals. Additional markers were genotyped in the selected regions and meta-analyses across countries were performed, including linear regression, variance components methods, and combined linkage/linkage disequilibrium mapping.

Theatre GM2.3

Assessment of genetic change for clinical mastitis in Norwegian Dairy Cattle with a multivariate threshold model

B. Heringstad[1], Y. M. Chang[2], D. Gianola[1,2] and G. Klemetsdal[1], [1]Department of Animal and Aquacultural Sciences, Agricultural University of Norway, P.O. Box 5025, N-1432 Ås, Norway, [2]Department of Dairy Science, University of Wisconsin, Madison*

A Bayesian multivariate threshold model was used to infer genetic change for clinical mastitis (CM). Records on 372,227 daughters of 2411 Norwegian Dairy Cattle (NRF) sires were analyzed. Each of the first three lactations was divided into 4 intervals: (-30, 0), (1, 30), (31, 120), and (121, 300) days after calving. Absence or presence of CM was scored as "0" or "1" within each interval. CM was assumed to be a different trait in each of the 12 intervals. Posterior mean of heritability of liability to CM was 0.09 in the first interval, 0.04 in the last interval, and between 0.06 and 0.07 for other intervals. Posterior means of the genetic correlations between intervals ranged from 0.32 to 0.67. The trend of mean sire posterior means by birth year of daughters was used to assess genetic change. The 12 traits showed similar trends, with essentially no genetic change from 1976 to 1986, and genetic improvement in resistance to mastitis thereafter. Annual genetic change was larger for intervals in 1st than in 2nd or 3rd lactations. Within lactation, genetic change was larger for intervals early in lactation, and more so in the 1st lactation. This reflects that selection against mastitis in NRF has emphasized mainly CM in early 1st lactation, and points towards favorable correlated selection responses in 2nd and 3rd lactations.

Theatre GM2.4

Genetic evaluation for mastitis using survival analysis

E. Carlén, M. del P. Schneider and E. Strandberg, Department of Animal Breeding and Genetics, Swedish University of Agricultural Sciences, P.O. Box 7023, 750 07 Uppsala, Sweden*

Clinical mastitis was analysed by use of a linear model and survival analysis, using data from the first three lactations of > 200,000 Swedish Holstein cows with first calvings between 1995 and 2000. For the mixed linear model clinical mastitis was defined as a binary trait measured from 10 d before to 150 d after calving. For the survival analysis clinical mastitis was defined either as the time period from 10 days before calving to the day of first treatment or culling due to mastitis (uncensored record), or from 10 days before to the day of next calving, culling for other reasons than mastitis, movement to a new herd or to a period of 250 days (censored record). For a better comparison the model for both methods included the same fixed and random effects.
In agreement with our hypothesis, the heritability estimates from survival analysis (0.12 to 0.23) were distinctly higher than those estimated by the linear model (0.01 to 0.03). The resulting accuracies were also higher for the trait analysed with survival analysis, especially for later lactations. However, the differences between accuracies were smaller due to the large number of censored daughters in the survival analysis. Although more studies are needed, this study reveals a potential use of analysing clinical mastitis data with survival analysis.

Theatre GM2.5

Across country evaluation for clinical mastitis and somatic cell count by multitrait model

E. Negussie, M. Koivula and E.A. Mäntysaari, MTT Agrifood Research Finland, Animal Production Research, 31600 Jokioinen, Finland*

Clinical mastitis (CM) is recorded nationwide only in Nordic countries, whilst in most countries selection is limited to indirect means by using only somatic cell count (SCC). Owing to high genetic correlation between the two traits and availability of records across Nordic countries, opportunities exist to combine all these sources of information to increase the efficiency of selection. So far models for genetic evaluation have considered data only from first lactation and in most cases fitting a univariate model. In this study a multitrait model including SCC and CM over the first three lactations was developed and the effect of combining information from both traits in joint genetic evaluation of Nordic bulls was assessed. Data on SCC and CM were obtained from 1.2 million Finnish Ayrshire and 1.0 million Swedish Red cows. Variance and co-variance components were estimated from a sample of the joint data set. The heritabilties for SCC and CM in the first, second and third lactations range from 0.119 to 0.137 and 0.020 to 0.023, respectively, with a genetic correlation of 0.70 within lactations. The correlation of EBVs between univariate and bivariate model range from 0.98 to 0.99 for SCC and from 0.86 to 0.92 for CM indicating that CM benefited more from the combined evaluation. The observed genetic trends and correlations of EBVs ranging from 0.97 to 0.99 showed good agreement between national and across country evaluations.

 Theatre GM2.6

QTL for natural antibodies in two populations of laying hens

M. Siwek[1], A.J. Buitenhuis[3], S.J.B. Cornelissen[1], M.G.B. Nieuwland[2], H. Bovenhuis[1], H.K. Parmentier[2], R.P.M. Crooijmans[1], M.A.M. Groenen[1] and J.J. van der Poel[1], [1]Animal Breeding and Genetics Group, [2] Adaptation Physiology Group, Wageningen Institute of Animal Sciences, Wageningen University, P.O. Box 338, 6700 AH Wageningen, The Netherlands, [3]Danish Institute of Agricultural Sciences, Department of Animal Breeding and Genetics, Foulum P.O. Box 50, DK-8830 Tjele Denmark*

Natural Antibodies in cooperation with a.o. complement system, might act as a first mechanism to protect an animal from disease. Lipopolisacharide (LPS) and lipoteicholic acid (LTA) represent 'homotopes' antigenic determinants of gram negative and gram positive bacteria. The aim of this study was to identify QTL associated with levels of natural antibodies in two population of laying hens. The H/L F2 population originated from a cross between two divergently selected lines for either high or low primary antibody response to sheep red blood cells (SRBC) at 5 weeks of age. The FP F2 population was created from a cross of laying hens differing for behavioural traits. A genome scan has been performed using 174 microsatellite markers, equally distributed over the chicken genome. Total antibody titres to LPS and LTA were determined by ELISA. A paternal half-sib and a line-cross regression model were used to detect QTL involved in response to these two 'homotopes'. Overall more QTL for LPS and LTA were detected in the FP population. There is little overlap in QTL between two populations and both 'homotopes'.

Theatre GM2.7

Assessing the influence of dam genotypes on QTL mapping in a variance component model, based on the analysis of data from the German national MAS project

J. Szyda[1,2], Z. Liu[1], H. Thomsen[1], F. Reinhardt[1] and R. Reents[1], [1]VIT, Heideweg 1, 27-283 Verden, Germany, [2]Department of Animal Genetics, Agricultural University of Wroclaw, Kozuchowska 7, 51-631 Wroclaw, Poland*

Within the framework of the German national MAS project, data has been collected since 1995, which cover 14 markers on BTA6, BTA14, and BTA18 genotyped for 7624 individuals till November 2003. Among them there were 4717 bulls and 2907 cows or heifers. The practice of acquiring female genotypes has begun only recently so that among animals born before the start of female genotyping are only known for males.

The goal of our study is to assess the impact of female genotypes on QTL mapping based on a variance component model, i.e. a model with random polygenic and QTL effects. Practically, data sets stemming from the German MAS project are compared, which differ in the amount of female genotype information available. Completeness of marker genotypes on both sire and dam side will be quantified. The impact of dam genotypes, observed or derived, on the calculation of identity by descent will be assessed. The comparison involves estimates of QTL parameters and their standard errors, likelihood curvature in the vicinity of the most probable QTL position and the composition of the QTL (co)variance matrices.

 Theatre GM2.8

The genetics of indicators for host resistance to gastrointestinal parasites in Merino sheep

G.E. Pollott[1], F.C. Greeff[2], L.J.E. Karlsson[2] and S.J. Eady[3], [1]Imperial College London, Department of Agricultural Sciences, Wye, Ashford, Kent, TN25 5AH, UK., [2]Great Southern Agricultural Research Institute, 10 Dore Street, Katanning, WA 6317, [3]Australia, CSIRO McMaster laboratory, Armidale, NSW 2350, Australia*

Faecal egg count (FEC) has been widely used as an indicator trait for host resistance to gastrointestinal parasites. In this study two further indicators were investigated; dag score (DS) and faecal consistency score (FCS). These three traits and liveweight (BWT) were studied to see how their heritabilities and genetic correlations varied over the first 1.5y of life using random regression of sire breeding value on age. The overall estimate of FEC h^2 was 0.28±0.072 but varied from 0.2 at weaning (4mo) to 0.6 at 400d. Dag score had a low overall heritability (0.11±0.036) but it ranged form 0.02 at weaning to 0.25 at 280d. The heritability of FCS was also low (0.12±0.036) but fell from 0.2 at weaning to 0.1 at 300d and rose to 0.2 by 400d. Dag score and FCS were highly correlated (0.63±0.14) at all ages. However the genetic correlation between FEC and DS decreased with age (+0.1 to -0.1) and FEC and FCS increased with age (-0.05 to +0.25). This study indicates that FEC measures a different aspect of parasite resistance from DS and FCS and shows how all three measures change with age of the lamb.

Theatre GM2.9

Genetic relationships between indicator traits and parasitic nematode infection in sheep

G. Davies[1,2], M.J. Stear[2] and S.C. Bishop[1], [1]Roslin Institute, Midlothian EH25 9PS, UK, [2]Dept. Veterinary Clinical Studies, Glasgow University Veterinary School, Glasgow G61 1QH, UK*

Gastrointestinal nematodes are a major cause of economic loss to the UK sheep industry. The development of anthelmintic resistance is now a worldwide problem and alternative strategies are required. This study aims to investigate the potential of selective breeding for increased resistance to *Teladorsagia circumcincta*, the predominant nematode in the UK. 1000 Scottish Blackface lambs were continually exposed to a mixed nematode infection by grazing; anthelmintic treatment was administered at 28-day intervals. Faecal egg counts and blood samples were collected at 28-day intervals. Indicator traits representing the immune response and the consequence of infection were studied; IgA, eosinophil, pepsinogen, fructosamine and faecal egg count. The animals were necropsied at 6 months of age and parasitic traits recorded; worm length, worm burden, worm fecundity, L4 burden and L5 burden. ASREML was used to calculate heritabilities and correlations. Highly heritable were worm length (h^2=0.53) and fecundity (h^2=0.50) as were the indicator traits. Strong genetic correlations were observed at 5-6 months of age between indicator traits and parasitic traits, with IgA, eosinophil and pepsinogen negatively correlated (Rg c. -0.5) and fructosamine positively correlated. In contrast, to genetic correlations, the environmental correlations with IgA were positive. In conclusion the traits studied could aid selection for increased resistance to gastrointestinal nematodes.

Theatre GM2.10

Effect of herd environment on phenotypic and genetic levels of health and fertility

M.P.L. Calus, J.J. Windig and R.F. Veerkamp, Animal Sciences Group, Division Animal Resources Development, P. O. Box 65, 8200 AB Lelystad, The Netherlands*

The objective of this paper was to estimate the association of several descriptors of herd environment with 1) levels of fertility and health and 2) environmental sensitivity of breeding values for fertility and health. Fourteen environmental parameters were defined, that described the average of a certain characteristic of the available animals in the herd, such as average protein production, age at calving, number of animals or calving interval. The considered traits were days to first service, days to last service, days first to last service, calving interval, number of inseminations, first service conception, non-return at 56 days, survival, occurrence of mastitis, somatic cell score and body condition score. Phenotypic levels of traits were estimated by a fixed polynomial regression on environmental parameters. Associations between levels of the traits and values of the EP were expressed as the relative change of the fixed polynomial regression on EP between 25th and 75th percentiles of the data. Differences up to 35% were observed for the means. Variance components and breeding values of sires and maternal grandsires were estimated for all traits by applying a random regression on environmental parameters that represented the herd environment of their (maternal grand)daughters. Environmental sensitivity was expressed by the genetic correlation of a trait and the change in genetic variance between 25th and 75th percentiles of the data. Results showed genetic correlations between .76 and 1.00 and changes in genetic variance up to 83%.

Poster GM2.11

Improvements in genetic evaluation for longevity for the Italian Holstein

F. Canavesi, S. Biffani, and A.B. Samorè, Associazione Nazionale Allevatori Frisona Italiana, Via Bergamo 292, 26100 Cremona, Italy*

Transmitting abilities for longevity have been officially published for the Italian Holstein Friesian using survival analysis since August 2001. Productive life is calculated from national milk recording information as the number of days between first calving and the last milk recorded test day. The hazard function applied is explained by a baseline hazard function, only depending on time and equal for all animals, and fixed and random effects, possibly time dependent. Fixed effects included are stage of lactation by lactation number, age at first calving, annual change in herd size, herd-year-season (random), year-season (fixed), production level for milk, fat and protein, and random effect of the sire. This model, routinely used for the official sire proofs estimation for direct longevity, was recently improved: the sire model was modified into a sire maternal-grand-sire model and the milk production level was included in the model using a better measure of 305 days production and a higher number of production classes than before. The improvement in the adjustment for milk production led to an estimated heritability of 0.09. The introduction of the maternal grand sire increased the estimated heritability to 0.097 improving the overall accuracy of the system. Published longevity sire proofs are the result of the combination of direct longevity and type traits. The correlation with the official genetic sire evaluation was around 0.94.

Poster GM2.12

Investigating genotype x environment interactions for indicators of host resistance to gastrointestinal parasites in Merino sheep

G.E. Pollott[1], F.C. Greeff[2], L.J.E. Karlsson[2] and S.J. Eady[3], [1]Imperial College London, Department of Agricultural Sciences, Wye, Ashford, Kent, TN25 5AH, UK., [2]Great Southern Agricultural Research Institute, 10 Dore Street, Katanning, WA 6317, Australia, [3]CSIRO McMaster laboratory, Armidale, NSW 2350, Australia*

Breeding for reduced faecal egg count (FEC) as an indictor of host resistance to gastrointestinal parasites has become widespread. It is important to know if genes for such resistance are effective in all environments or whether site-specific genotypes are required. This trial was set up to investigate two specific resistant lines, bred in very contrasting environments, and to see how well their offspring performed in the reciprocal environment. Control groups of sheep from each source were kept alongside the resistant genotype in each environment. Two further indicators of parasite resistance were measured, dag score (DS) and faecal consistency score (FCS). All measurements were taken eight times, at monthly intervals from weaning (4 months of age) to about 400d. Genotype by environment (GxE) interactions were thus measured eight times for each trait as a site-by-genotype interaction in a general linear model analysis of the dataset. There was little evidence of a GxE effect for FEC or liveweight. Some evidence of a GxE effect was found for DS and FCS at the oldest three ages but this was mainly due to the control groups.

Poster GM2.13

Opportunities for breeding for disease resistance in UK sheep

G.J. Nieuwhof[1] and S.C. Bishop[2], [1]MLC, PO Box 44, Milton Keynes MK6 1AX, U.K., [2]Roslin Institute, Roslin EH25 9PS, U.K.*

This study aims to evaluate sheep diseases in terms of the opportunities for improving host resistance through genetic selection. For each disease, cost impacts and known genetic factors were identified and used to model breeding programmes. In a 1997 survey, gastro-intestinal (GI) parasites, footrot and sheep scab were identified as the three most important diseases in British sheep production, but actual costs were not quantified. Recent economic analyses estimated costs of Enzootic Abortion in Ewes (EAE) and Toxoplasmosis in Great Britain at £25 million and £18M per annum, respectively. In this study, using the same methodology, costs were estimated at £35M for footrot, £107M for GI parasites and £12M for sheep scab. Our models show that, within current breeding programmes and with moderate genetic parameters (as documented for GI parasites and footrot), selection on traits measured on lambs (like GI parasites and footrot) can lead to significant reduction in disease impact, while there is less scope for diseases expressed in adult females only (e.g. EAE). Breeding programmes for resistance to GI parasites exist in Britain and elsewhere. Currently, little is known about resistance to footrot in UK sheep populations and environments. Research is required on the genetics of footrot resistance in the UK, including epidemiological aspects of the disease in order to ascertain the full potential of selection.

Poster GM2.14

Genetic resistance to *Eimeria* infections in merinoland sheep

M. Gauly[1], J. Reeg[3], C. Bauer[2], H. Brandt[3], Christina Mertens[2], H. Zahner[2] and G. Erhardt[3], Institute of Animal Breeding and Genetics [1,3], Institute of Parasitology[2], University of Göttingen[1] and Giessen [2,3], Albrecht-Thaer-Weg 3, 37075 Göttingen, Germany*

Genetic parameters in resistance to *Eimeria* infections were studied in Merinoland lambs (n = 222) descending from 10 rams under natural challenge conditions in Germany, using *Eimeria* oocyst counts (oocysts per gram of faeces, OpG) and *E. ovinoidalis* antibody level index (Ab) as marker traits.

Faecal samples were taken in all lambs beginning at an age of 17 days until an age of 40 days every third day and afterwards every sixth day, respectively. Blood samples were taken from all lambs at an age of 7, 40 and 80 days to measure an *E. ovinoidalis* antibody level. *E. ovinoidalis* was the predominant species in the study.

Heritabilities of $\log_{10}$ OpG were between 0.00 and 0.79 (s.e. 0.27). The estimated values for log Ab were between 0.00 and 0.02 (s.e. 0.06), respectively. Phenotypic correlations among $\log_{10}$ OpG and body weight ranged between $r = -0.07$ ($p > 0.05$) and $r = -0.25$ ($p < 0.01$), respectively. The corresponding correlations between log Ab at various ages and body weight averaged between $r = -0.09$ ($p > 0.05$) and $r = 0.27$ ($p < 0.05$), respectively.

The results confirm in principal the feasibility of genetic selection in Merinoland sheep for resistance to *Eimeria*.

Poster GM2.15

Resistance parameters to gastrointestinal parasitosis in Zerasca breed: a Tuscany indigenous line

N. Benvenuti, J. Goracci, L. Giuliotti and P. Verità, University of Pisa, Veterinary medicine Faculty, Animal Production Department, Italy*

Zerasca is an indigenous sheep breed in endangered status. That breed carries out a primary position in the safeguard of Tuscany biodiversity, typical production improvement and agricultural district protection. It is characterized by a quite big size and a white medium open fleece. The study took on natural rangelands around Zeri in Massa Carrara region, placed in the north part of Tuscany at an altitude of 600-1200 m above sea level. Rough countries with self-shown sloping pastures typify that area. The Zerasca positive resistance to adverse climate conditions underlines a fine adaptation to the whole environment. However, common shepherds' habit is to treat animals twice a year with antiparasitarian drugs, mainly in consequence of the presumed loss of productions. Starting from these observations, we analyzed worm challenge in three Zerasca breeding under natural pasture, through faecal egg count, packed cell volume and classification of parasite species. The identification of animals resistant to gastrointestinal parasitosis and the recognition of relationships between sheeps pointed out the possibilities of evaluating genetic resistance. The purpose is to reduce antihelmintics, drugs' residual in the soil and pharmacological resistance, so that the treatments limitated to the non-resistance subjects move significant economical and ecological implications in breeding management.

Poster GM2.16

Breeding for scrapie resistance in France

I. Palhière[1] and M. Brochard[2], [1]INRA Station d'Amélioration Génétique des Animaux, BP 27, 31326 Castanet Tolosan, France, [2]France UPRA Sélection, 149 rue de Bercy, 75 595 Paris Cedex 12, France*

Numerous studies have shown that, in sheep, genetic susceptibility to scrapie is modulated by allelic variation in the *PrP* gene. They revealed the resistance of homozygous ARR/ARR animals and the high risk of VRQ/VRQ and VRQ/ARQ animals in scrapie affected flocks. Thus, the selection of ARR/ARR genotypes may be a strategy to control scrapie either at the flock level or at the population level. In France, the Ministry of Agriculture launched a global scrapie eradication programme which aims to improve resistance to scrapie through a selection on *PrP* genotype both in affected flocks and in the whole population. The selection programme for scrapie resistance had been implemented in all the French sheep breeds and benefited from the existing national scheme of livestock selection. To reach the general objectives, specific selection schemes were defined for each breed, considering their own characteristics, such as initial allele frequency. In 2002 and 2003, about 90,000 *PrP* analyses were realised in selected flocks. All the selection programmes strongly increased the ARR allele frequency while they efficiently eliminated the VRQ allele at the young males. Moreover, the selection on the production traits and the genetic variability seem to be maintained despite the effort on the *PrP* gene.

Poster GM2.17

PRNP polymorphisms at codons 136, 154 and 171 in ten sheep breeds of Greece

G. Lühken[1], S. Lipsky[1], C. Ligda[2], A. Georgoudis[2] and G. Erhardt[1], [1]Justus-Liebig University, Department of Animal Breeding and Genetics, Ludwigstrasse 21B, 35390 Giessen, Germany, [2]Aristotle University, Department of Animal Production, 54006 Thessaloniki, Greece*

In Greece, Scrapie is an epidemic disease in sheep and goats. Studies of natural and experimentally induced scrapie in sheep have shown that genetic susceptibility to the disease is mainly modulated by allelic variation in the three codons 136, 154 and 171 in the prion protein (PrP) gene (*PRNP*). Therefore, new strategies for the eradication of scrapie in sheep are now based on the selection for and against distinct PrP genotypes in many European countries.

Here we report the determination of *PRNP* polymorphisms at codons 136, 154 and 171 in the Greek sheep breeds Orino, Sfakia, Anogeiano, Kalarritiko, Pilioritiko, Kefalleneas, Karagonniko, Lesvos, Kymi and Skopelos. For haplotype analysis of 23 to 31 DNA samples per breed, a combined PCR-RFLP method was used.

In all breeds, the three PrP haplotypes (alleles) ARR, ARQ and AHQ have been found, while the VRQ allele occured in six and the ARH allele in three breeds only. The number of observed genotypes differed from five to eight between the breeds. The less scrapie susceptible genotype ARR/ARR occured in samples of eight breeds with frequencies up to 36% (Anogeiano and Sfakia). No sample was homozygous for the VRQ allele, which is presumable most susceptible to scrapie.

Poster GM2.18

Estimation of risk factors using genotype probabilities: an scrapie case

Z.G. Vitezica[1], C. Díaz[2], R..Rupp[1], J.M. Elsen[1], [1]INRA, Station d'Amélioration Génétique des Animaux, 31326 Castanet Tolosan, France, [2]INIA, Depto. Mejora Genética Animal. Ctra. De la Coruña km 7,5 28040 Madrid, Spain

Susceptibility to scrapie is mainly controlled by polymorphisms at PrP gene. In some sheep breeds genotyping for the PrP gene is not extensively used therefore the information on PrP genotypes is incomplete. The objective of this work was to evaluate the potential use of genotype probabilities to handle records of non-genotyped animals in the estimation of risk associated to PrP genotypes. Original data consisted of 4049 Romanov sheep. Out of those, 447 animals died of scrapie. Data were analyzed using survival analysis techniques. Three models differing in the way that PrP genotype information was handled were tested. Firstly, records of untyped individuals were discarded (P1); secondly those animals were assigned to an unknown group (P2); thirdly probabilities of genotypes were assigned (P3). In addition, different missing genotype information patterns were simulated for uncensored and censored data. When unknown genotypes were not a negligible part of uncensored records, the exclusion of these individuals from the analysis affect the accuracy of model estimates. Probabilities of genotypes allow to rank genotypes as expected. The use of genotype probabilities is a useful technique to deal with records of individuals with unknown genotype to estimate risk associated to PrP genotypes. However it is required to have available genotypes and informative pedigrees in the population.

Poster GM2.19

Prion protein gene polymorphisms in purebred sheep breeds of Nova Scotia

A. Farid[1], S.D.M. Llewellyn[1], G.G. Finley[2] and S. Nadin Davis[3], [1]Department of Plant and Animal Sciences, Nova Scotia Agricultural College, Truro, NS, B2N 5E3, Canada, [2]Nova Scotia Department of Agriculture and Fisheries, Truro, NS, B2N 5E3, Canada, [3]Ottawa Laboratory (Fallowfield), Canadian Food Inspection Agency, Ottawa, Ontario, K2H 8P9, Canada

The objective of this study was to determine genotypes of the entire purebred sheep population in Nova Scotia (Canada) at codons 136, 154 and 171 of the PrP gene. Genotypes of 1151 registered sheep of both sexes from 10 breeds (Border Leicester, Canadian, Clun Forest, Dorset, Finnish Landrace, Karakul, NC Cheviot, Rideau, Suffolk, Texel) on 26 farms were determined using RFLP-PCR. H_{171} was not identified and was combined with Q_{171}. All the four possible allele combinations were detected, but only $A_{136}R_{154}Q_{171}$ was present in all the 10 breeds. The most resistant allele ($A_{136}R_{154}R_{171}$) was observed in all the breeds, except Finnish Landrace, in which all animals had $AA_{136}RR_{154}QQ_{171}$ genotype. The frequency of the most resistant allele was high (0.57 to 0.69) in Border Leicester, Canadian, Dorset and Karakul, and low in Rideau (0.07) and Suffolk (0.24). Rideau, NC Cheviot and Texel selected sires had a higher proportion of susceptible AV_{136} or QQ_{171} genotypes than the ewes, which may suggest the presence of a negative association between resistant alleles and traits which breeders consider important.

Poster GM2.20

Mapping QTL affecting somatic cell count (SCC) using a longitudinal model in Danish dairy cattle

A. Sabry, M.S.Lund, B. Guldbrandtsen, B. Thomsen and P.Madsen, Dept. Anim. Breed. & Genet., DIAS, P.O.50, Tjele-8830, Denmark*

In many breeding programs SCC is an important source of information for genetic improvement of mastitis resistance. This is because most countries don't have recording systems for mastitis and many studies have shown high genetic correlation between mastitis and SCC. Usually, SCC is recorded repeatedly during lactation but in prediction of breeding values a form lactation average is used. However, using a lactation average of SCC to map QTL affecting SCC is only appropriate if the effect is constant during lactation. Incidence of mastitis is the major source to variation in SCC, and because mastitis primarily occurs in the beginning of lactation, we expected that individual QTL have variable genetic influences over time. This is supported by polygenic studies, which have shown that SCC exhibits heterogeneous additive genetic variance over lactation (Live. Prod. Sci. 2003: 79,239-247). In this setting using random regression models (RRM) has been shown to provide two advantages over traditional univariate models (7WCGALP. CD-ROM Communication N° 21-28). First, use of a RRM to map QTL increases the statistical power to detect stage-specific QTL. Second, a RRM improves the biological interpretation of the inferred QTL. Univariate and longitudinal models were employed to detect QTL affecting SCC. This comparison was carried out using data from the Danish Holstein granddaughter design with 20 families with an average of 68 sons per family.

Poster GM2.21

Genetic parameters for electrical conductivity of milk

E. Norberg[1], G. Rogers[2] and P. Madsen[1], [1]Danish institute of Agricultural Sciences, Department of Animal Breeding and Genetics, P.O. Box 50, Tjele, Denmark, [2]University of Tennessee, Department of Dairy Science, 201D McCord Hall, 2640 Morgan Circle, Knoxville, TN 37996-4500, USA

Electrical conductivity (EC) of milk has been introduced as an indicator trait for mastitis during the last few decades. The association of EC and mastitis together with new techniques for easy and cheap recording are properties that make EC a good indicator for both management and breeding purposes. Data used in this study were daily measurements of EC from 2101 first parity Holstein cows in 8 herds in the United States. Data were analyzed with an animal model that included herd-test-day, age at calving and days in milk as fixed effects, and random additive genetic and permanent environmental effects. A random regression model with a 4th order Legendre polynomial for both additive genetic and permanent environmental effects gave the best fit, and heritability estimates varied from 0.26 to 0.36 during lactation. Previous studies have shown that the genetic correlation between lactation mean of EC from first lactation cows and mastitis is about 0.60. Genetic correlation between test-day EC and mastitis will be estimated as well in this study. Due to the relatively high heritability and genetic correlation between EC and mastitis, EC might be a potential indicator trait to use in a breeding program designed to reduce the incidence of mastitis.

 Poster GM2.22

Trajectories and variance component estimation for economically important disease categories

D. Hinrichs, E. Stamer, W. Junge and E. Kalm, Institute of Animal Breeding and Husbandry, Christian-Albrechts-University, D-24098 Kiel

In the present study trajectories for the disease categories all diseases, fertility diseases, ovarian problems, udder diseases, and metabolic diseases are described. The data were collected on three commercial dairy farms with an overall total of 3200 German Holstein cows. Information from 10,521 cows, sired by 803 bulls were available resulting in 17,183 lactation with 4,361,909 different lactation days. All medical treatment (75,595) were recorded in the years 1998 to 2002. Variance components were estimated for all diseases and udder diseases, for the first 50, 100, and 300 days of lactation. Analyses were carried out with lactation threshold models (LTM) and test day threshold models (TDTM). Furthermore, the impact of the number of daughters per sire was analysed with udder disease information from the first 50 days of lactation.
For all diseases the frequencies varied between 62% and 77% in the LTM, and they were between 6% and 15% in the test TDTM. Udder disease frequencies ranged from 30% to 46% in the LTM, and varied between 3% and 7% in the TDTM.
The resulting heritabilities were low with values wthin the interval of 0.03 and 0.05. Higher heritabilities were estimated for udder diseases, where the estimates varied between 0.07 and 0.08 in the LTM, and they were significantly higher in the TDTM with values ranging from 0.12 and 0.25.

Poster GM2.23

Genetic parameters for claw disorders in Dutch dairy cattle and some relationships with conformation traits

E.H. van der Waaij[1,2], G. de Jong[3], M. Holzhauer[4], C. Kamphuis[2], and E. Ellen[2], [1]Department of Farm Animal Health, Veterinary Faculty, University of Utrecht, POBox 80127 3508 TC Utrecht, The Netherlands, [2]Animal Breeding and Genetics Group, Wageningen University and Research Centre, POBox 338, 6700 AH Wageningen, The Netherlands, [3]NRS, POBox 454, 6800 AL Arnhem, The Netherlands, [4]Animal Health Service, POBox 9, 7400 AA Deventer, The Netherlands*

Disturbed claw health is one of the major problems causing production loss and reduced animal welfare. In this study, claws of >75% of the lactating cows and heifers in 366 herds were trimmed by hooftrimmers and the claw health status of the hind claws were recorded, resulting in records on 18,637 animals. Eight claw disorders could be distinguished, varying in prevalence from ~1% (interdigital phlegmona) to ~40% (sole heamorrhages). In total, 70% of the animals had at least one claw disorder. Conformation traits (incl. locomotion) were recorded during the animal's first lactation. Heritabilities were estimated using logistic regression and a sire model, and ranged from 0.01 (interdigital phlegmona) to 0.13 (digital dermatitis), and from 0.12 (locomotion) to 0.24 (rear leg side view). Genetic correlations between claw disorders and locomotion were high, ranging from 0.19 (white line) to -0.93 (interdigital fibroma). Correlations with the rear leg conformation traits were lower, ranging from -0.05 (acute laminitis with rear leg hind view) to -0.69 (interdigital phlegmona with rear leg side view).

Poster GM2.24

Genetic analysis of osteochondral lesions, meat quantity and quality, growth and feed conversion traits in station-tested pigs

H.N. Kadarmideen[1,], D. Schwörer[2], H. Ilahi[1], M. Malek[1] and A. Hofer[2], [1]Statistical Animal Genetics Group, Institute of Animal Science, Swiss Federal Institute of Technology, Zurich CH 8092, Switzerland, [2]SUISAG, Allmend, CH-6204 Sempach, Switzerland*

Multiple-trait genetic analysis of osteochondrosis (OC) in station-tested pigs with proportion of premium cuts (PPC), 4 meat quality (MQ) traits, daily weight gain (DWG) and feed conversion ratio (FCR) by linear mixed models (LMM) and threshold or generalized linear mixed models (GLMM) were performed using 2710 records (6438 animals in pedigree), of which 1291 had OC lesions. GLMM was fitted to OC traits only, with logit, probit and poisson functions. For OC traits, estimates of heritabilities (h^2) were low for LMM (0.06-0.16) and moderate to high for GLMM (0.08-0.61), which suggests that selection against OC may be possible. Estimated h^2 for PPC was 0.60 and for MQ traits ranged from 0.12 to 0.66, with 0.28 for DWG and 0.42 for FCR. Genetic correlations (r_g) were generally unfavourable between OC and PPC (0.21 - 0.30), while MQ, DWG and FCR traits had both favorable and unfavorable r_g with OC (-0.54 to 0.58 for MQ traits; -0.44 to 0.31 for DWG; -0.23 to 0.40 for FCR). Among OC traits, r_g was mostly negative (-0.40 to 0.69). There were high standard errors for most r_g estimates, limiting us to draw conclusions. However, these results on possible genetic relationships will be useful to pig breeders.

Poster GM2.25

Evaluating disease resistance of pigs using an in-vitro test

P.K. Mathur[1], J. Phipps[2], D. Hurnik[3], L. Maignel[1] and B.P. Sullivan[1], [1]Canadian Centre for Swine Improvement, Central Experimental Farm, Bldg54 Maple Drive, Ottawa, Ontario K1A0C6. [2]PharmaGap Inc., 100 Sussex Dr., Ottawa, Ontario, K1A0R6, [3]Atlantic Swine Research Partnership Inc., Suite212, 420 University Avenue, Charlottetown, PEI, C1A7Z5, Canada*

Disease resistance of animals is often evaluated as innate immune response to specific antigens. This involves the inoculation of live animals and the estimation of the response in terms of skin thickness or haematological parameters. The method is usually effective but involves risk of a strong reaction and possible secondary infection that may adversely affect growth and productivity. As an alternative, a multi parameter assay has been developed to assess the general immune capacity using a small sample of blood. The lymphocytes of the peripheral blood are exposed to several chemicals known to trigger a blastogenic response. The proliferation of the various blood cell subsets is then measured using flow cytometry analysis, and reflects the potential of the animals to adoptive viral and bacterial disease resistance. The assay was applied to a group of pigs also tested for the effect of an immuno-modulator. It was effective at measuring the differences between animals, as well as the dose response to the modulator.

Poster GM2.27

Selection strategies to decrease ascites susceptibility in broilers

A. Pakdel, J.A.M. Van Arendonk, and H. Bovenhuis, Animal Breeding and Genetics Group, Wageningen Institute of Animal Sciences, PO Box 338, 6700 AH Wageningen, The Netherlands.*

Ascites syndrome is a metabolic disorder in broilers. Mortality due to ascites especially in later ages results in significant economic losses and has a negative impact on animal welfare. The aim of the present study was to evaluate the consequences of alternative selection strategies where selection on BW as well as ascites susceptibility traits carried out. Five different schemes were compared by deterministic simulation. Traits investigated in the index as indicator traits for ascites were a non-destructive indicator trait like hematocrit value (HCT) and a destructive indicator trait like ratio of right ventricle to the total ventricular of the heart (RV:TV). Further it was studied if testing the birds under cold conditions, i.e. conditions that trigger ascites, could improve the selection response. The evaluation of the consequences of alternative selection strategies indicated that when ascites susceptibility trait was ignored, response for ascites susceptibility 0.025 unit increased whereas response for BW was 130 gr. However by including ascites susceptibility in the breeding goal and by using the information of HCT and RV:TV in the index, response for BW slightly decreased whereas ascites susceptibility were fixed at zero level. The response for BW was 108 gr. when HCT and RV:TV measured under normal conditions and it was 111 gr. when those trait measured under cold conditions.

Poster GM2.28

Estimates of environmental and genetics effects on hip dysplasia in Italian German Shepherd dog population

E. Sturaro[1], K. Mäki[2], M. Ojala[2], G. Bittante[1], P. Carnier[1] and L. Gallo[1], [1]Dept. of Animal Science, University of Padova, Italy. [2]Department of Animal Science, University of Helsinki, Finland*

This study aimed to investigate sources of variation and genetic parameters of hip scores (FCI method) for the evaluation of hip dysplasia (HD) in German Shepherd dog in Italy. Screening data for hip dysplasia of 18212 X-rayed dogs (provided by CeLeMaSche) were analyzed. Frequency of dogs not affected was 76.4% and the frequency of animals with mild, moderate and severe hip dysplasia was 18.0, 4.9 and 0.7% respectively. Data were analyzed with a Restricted Maximum Likelihood procedure using a mixed linear model to obtain variance component estimates for hip score. The pedigree file consisted of 26704 animals and contained ancestors for up to eight generations. Environmental effects influencing hip dysplasia were sex of the dog, age at screening, year and period of birth and experience of X-ray veterinarian. Recent years of birth showed a decreasing hip score, and this result was confirmed from genetic trend. Estimate of heritability was relatively low, ranging from 0.14 ± 0.01 to 0.17 ± 0.01. Maternal effect and random breeder effect had a small effect on the hip score. However, some breeders may have selected the most healthy animals for official screening, and this aspect should be taken into account because it might have caused bias in the results.

Poster GM2.29

A screening program for subaortic and pulmonic stenosis in the Italian Boxer dog population: prevalence and preliminary analysis

L. Menegazzo, C. Bussatori, D. Chiavegato, P. Piccinini, E. Sturaro, L. Gallo and P. Carnier, Department of Animal Science, University of Padua, viale dell'Università 16, 35020 Legnaro, Italy*

Subaortic (SAS) and pulmonic (PS) stenosis are congenital heart defects commonly affecting some dog breeds. The aim of this study was to assess the prevalence of SAS and PS in the Italian Boxer dog population, to investigate variation of the cardiac murmur grade and echocardiographic measures, and to perform a risk analysis of heart defects through logistic regression. Cardiovascular examination was carried out by 7 different veterinarians on 999 random dogs. Female dogs were 53.85% of the sample and the same proportion resulted in the sub-sample of affected dogs. The diagnosis was based on clinical and echocardiographic findings (peak velocity, flow type, obstruction severity). Based on value of the peak pressure gradient, SAS and PS were classified into 3 categories: mild, moderate and severe. For 68% of the dogs the cardiac murmur was absent or physiological, for 23% mild, and for 9% pathological. In agreement with other european studies, the prevalence of SAS was 7.8% whereas that of PS was 3.2%. The linear model used to analyse echocardiographic dimensional measures explained 20-38% of variation whereas the logistic regression analysis showed the role of the murmur grade as a risk factor for stenosis occurence.

Theatre G3.1

Managing genetic variability in dairy cattle and pig breeding schemes: from research to practice

J.J. Colleau[1], S.Moureaux[1,2], M. Briend[1] and T.Tribout[1], [1]Station de Génétique Quantitative et Appliquée, INRA, 78352 Jouy-en-Josas, France,[2]Institut de l'Elevage, 75595 Paris Cedex 12, France

Due to intensive selection, dairy cattle and pig populations are increasingly challenged by multiple needs: saving genetic variability, avoiding inbreeding depression, containing spread and expression of genetic defects. Since more than 15 years, Research has been very active for testing many alternative solutions to theses issues. Some of them have turned out to be very efficient, albeit quite demanding from practitioners involved in cattle and pig breeding. The diversity of solutions and the corresponding efficiencies as found by the current research were briefly summarized. The potential gap between optimised procedures proposed by research and current real practices was analyzed on some real populations, based on retrospective analyses of past selection decisions. In these populations, the order of magnitude of the efficiency loss from practice was 20%, as compared with optimization. For helping breeders to identify the origin of this gap and to reduce it in the future, the quantitative impact of some factors potentially responsible for this situation (multiple threshold selection, ignorance of remote relationships) was evaluated.

Theatre G3.2

Optimum contribution selection conserves genetic diversity better than random selection in small populations with overlapping generations

K. Stachowicz[1,2], A. C. Sørensen[2,3] and P. Berg[3], [1]Department of Animal Science, Warsaw Agricultural University, ul. Ciszewskiego 8, 02-786 Warszawa, Poland, [2]Department of Production Animals and Horses, Royal Veterinary and Agricultural University, Gronnegardsvej 2, 1870 Frederiksberg C, Denmark, [3]Department of Animal Breeding and Genetics, Danish Institute of Agricultural Science, P.O. Box 50, 8830 Tjele, Denmark*

Small, closed populations are at risk, because of higher loss of genetic diversity and increased rates of inbreeding. Optimum contribution selection (OCS) limits inbreeding by controlling the increase of average relationship between individuals.

Stochastic computer simulations have been used to investigate how much benefit OCS is giving compared with random selection in different breeding scenarios.

This study showed that OCS results in lower average rates of inbreeding per generation, reduced by 3,5-90 % depending on the scenario. Generation intervals are about 20% longer. Less alleles are lost from populations, in scenarios with ten offspring per female per time step OCS maintains 5-6 times more alleles than random selection. Also effective number of founders is higher. OCS is more beneficial in smaller populations (50 breeding females) than in larger populations (200 breeding females). It is the most advantageous in scenarios with ten offspring per female per time step and fewer breeding males than females. In scenarios with one offspring per female per time step OCS gives more benefit in situations with equal number of breeding males and females.

Theatre G3.3

SAUVAGE, a software to manage a population with few pedigrees

J. Raoul[1], C. Danchin-Burge[2], H. de Rochambeau[3] and E. Verrier[1], [1]Institut National Agronomique Paris-Grignon, Animal Sciences Department, 16 rue Claude Bernard, 75331 Paris Cedex 05, France, [2]Institut de l'Elevage, Genetics Department, 149 rue de Bercy, 75595 Paris Cedex 12, France, [3]INRA SAGA, Auzeville BP 27, 31326 Castanet Tolosan cedex, France*

There are numerous ways to manage rare breeds genetic variability, but one of the main difficulty is to find one that the breeders and/or the rare breeds' managers will actually comply with. The purpose of this study was to realize a software to help farmers to choose their most appropriate breeders (from a genetic variability point of view) among all the available ones.

The starting point for this work was a program developed by H. de Rochambeau for two rare breeds. This tool's main strong point is that it works whatever the level of pedigree information in the breed (including the case where no pedigree information exists). However, this program is appropriate only for breeds without a selection program because it doesn't take into account any selection criteria (such as EBVs) to dispatch the males. The software was tested successfully in three different breeds. It is now fully operational but it needs to be transferred on a more user friendly platform before it can be transferred to rare breeds' managers.

Theatre G3.4

Characterisation and conservation of genetic diversity between breeds

M. Toro[1] and A. Caballero[2], [1]Instituto Nacional de Investigación y Tecnología Agraria y Alimentaria, Departamento de Mejora Genética, Carretera La Coruña km.7, 28040-Madrid, Spain, [2]Facultad de Ciencias, Departamento de Bioquímica, Genética e Inmunología, Universidad de Vigo, 36200 Vigo, Spain*

In domesticated animals breeds are considered the units of conservation. Genetic distances estimated from molecular markers have been used to assess the genetic diversity of livestock breeds. Thaon d'Arnoldi et al. (1998) proposed the Weitzman approach to measure global diversity and loss of diversity attached to each breed, and to determine conservation priorities according to this methodology. However, the approach has been critizised because genetic variation within groups is ignored. Here, we summarise the classical population genetics tools to analise between and within genetic variability. When both approaches are applied to some case studies they produce not only different but sometimes opposite results. We also review several practical approaches that have been proposed that weight the between and within breeds depending on the imagined scenario for the long term use of genetic diversity.
Some other important topics are reviewed and summarised: the use of other genetic criteria such as allelic diverstiy or mitochondrial DNA and the importance of neutral vs adaptive variation.

Theatre G3.5

Investigations on the selection of breeds for conservation of genetic variance

J. Bennewitz[1] and T. H. E. Meuwissen[2], [1]Institute of Animal Breeding and Husbandry, Christian-Albrechts-Universität, D-24098 Kiel, [2]Department of Animal Science, Agriculture University of Norway, Box 5052, 1432 Aas, Norway*

In making decision plans for conservation of farm animal genetic resources, a key question is the choice of breeds to be conserved. In this study a core set of breeds is defined in that the additive genetic variance of a hypothetical quantitative trait is maximised. This variance can be found within and between breeds. The breeds are ranked for conservation priority by their relative contribution to the core set. Breeds show a comparably high contribution if they are inbred and not related with other breeds. Alternatively, recent studies suggested to define a core set of breeds in that overlap genetic is as much as possible avoided and aim to recover a hypothetical founder breed as fully as possible. Both core set methods are based on the mean kinships between and within breeds, these can be estimated using molecular genetic marker data.
Bootstrap based methods for the estimation of the breeds contribution to the core sets are presented. By simulation it was shown that these methods improve the accuracy of the contribution estimates substantially.
The similarities and the differences of the two core set methods are illustrated by simulations and by an application of the methods to a field data set consisting of 10 cattle breeds genotyped for 11 microsatellite markers.

Theatre G3.6

Epistatic kinship - A new method to estimate the genetic similarity between individuals and between populations

C. Flury and H. Simianer, University of Göttingen, Institute of Animal Breeding and Genetics, Albrecht-Thaer-Weg 3, 37075 Göttingen, Germany*

The kinship coefficient (Malécot, 1948) focuses on single locus only. The extension to chromosomal segments leads to a new similarity index called ‚Epistatic Kinship', which basically is the probability that chromosome segments of a given length are identical by descent. This parameter reflects the number of meioses separating individuals or populations. Thus, it might be used as measure to quantify the genetic distance of sub-populations that have been separated only few generations ago, which may proof especially useful in studies with farm or experimental animals. The properties of the approach are investigated in a Monte Carlo simulation. In this study algorithms for Epistatic Kinship and Epistatic Inbreeding are presented. Further a test based on the mean Epistatic Kinship is demonstrated for the assignment of animals to their population. Based on the results the optimum design of SNP haplotypes will be derived and the respective genotyping will be done for the three subpopulations of the Goettingen Minipig. The typing results will be used to calculate the Epistatic Kinship, Epistatic Inbreeding and the diversity between populations based on the average Epistatic Kinship. We expect the project to provide a novel approach to the estimation of genetic similarity of closely related individuals or populations.

Theatre G3.7

A world wide emergency program for the creation of national genebanks of endangered breeds

E. Groeneveld, Institute for Animal Breeding, Mariensee, Federal Agricultural Research Center (FAL), 31535 Neustadt, Germany

As a response to the rapid loss of animal genetic resources, a world wide emergency program is proposed to create national genebanks on the basis of somatic cells. Contrary to other procedures, like storing deep frozen semen or embryos, collection and storage of somatic cells can be done cheaply and rapidly on any species. Only in this way, an emergency depot of animal genetic resources can be created fast enough to forestall the ongoing rapid erosion of animal biodiversity. While already now animals can be produced from somatic cells for 10 species, this number will continue to rise in the future. A layered strategy making use of the infrastructure of National Coordinators created by the FAO Animal Genetic Resources group is proposed which is based on within country collection and storage, thus leaving execution of the program, responsibility and ownership of the cryobanks with the countries. After a pilot study, more country gene banks could be created as funds become available. With a limited effort, the creation of a network of country genebanks of last resort are a realistic option.

Theatre G3.8

An *in situ* markers assisted conservation scheme of 11 Italian avian breeds

M. Cassandro[1], M. De Marchi[1], C. Targhetta[1], C. Dalvit[1], M. Ramanzin[1] and M. Baruchello[2], [1]University of Padova, Department of Animal Science, Viale dell'Università, 16, Agripolis, 35020, Legnaro, Padova, Italy, [2]Veneto Agricoltura, Agripolis, 35020, Legnaro, Padova, Italy

An *in situ* marker assisted conservation scheme (CoVa project) of 11 local pure avian breeds of 4 different species (poultry, duck, helmeted guinea fowl and turkey) is currently running, since 2000, in the Veneto region . Objectives of CoVa project are: to preserve local traditions and rural culture of avian productions, to maintain these sources of genetic biodiversity and to minimize average kinship coefficient within breed. A minimum number of 54 mature animals per breed (20 males and 34 females) is guaranteed within each of the 4 nucleus farms located in 4 different environments. A molecular markers information are used to monitor heterozigosity and genetic similarities among and within breeds and to select sires and dams. Production and reproduction performances are also recorded and a biannual complete substitution of sires and dams is applied. Individual identification is based on wing tagging at hatching. A pre-selection of males and females at about 4 months of age is applied using a threshold index based on family attribution, standard phenotype and production and reproduction performances. The future of this local avian breeds is based on their genetic conservation and on the development of niche production, possibly including crossbreeding with more productive commercial genotypes.

Theatre G3.9

Inbreeding in Danish dairy cattle populations

A.C. Sørensen[1,2], M.K. Sørensen[1] and P. Berg[1], [1]Department of Animal Breeding and Genetics, Danish Institute of Agricultural Science, P.O. Box 50, DK-8830 Tjele, Denmark, [2]Department of Production Animals and Horses, Royal Veterinary and Agricultural University, Grønnegårdsvej 2, DK-1870 Frederiksberg C, Denmark*

Inbreeding depression poses a potential threat to dairy cattle breeding. Therefore, inbreeding should be monitored. Calves born from 1999 until 2003 and registered as Danish Holstein (DH; 2,238,349) or Danish Red (DR; 292,395) were reference populations in this study. For calves born in 2003, average inbreeding was 3.4% for DH and 1.3% for DR. Since 1983, the effective population sizes has been 70 for DH and 274 for DR.

For DH, 8 ancestors contributed 50% to the reference population. The most important ancestors were Elevation (13.5%), Chief (10.7%), and Bell (8.3%). The genetic basis is very narrow as illustrated by an effective number of founders of 73. The effective number of ancestors is 22. The difference in numbers reflects unequal use of ancestors between founders and reference population.

For DR, 13 ancestors contributed 50% to the reference population. The 4 most important ancestors contributed 26%. They belong to 4 different breeds: Red Holstein Friesian, Swedish Red and White, Brown Swiss, and original DR. This highlights the synthetic character of the DR population. The effective number of founders in DR is 214 and the effective number of ancestors is 35.

These data provide material for estimation of inbreeding depression in dairy cattle.

Theatre G3.10

Economics of breeding programs with indigenous and adapted cattle in Africa and the effect on conservation

S.B. Reist-Marti[1], H. Simianer[2] and A. Abdulai[1], [1]Institute of Agricultural Economics, Swiss Federal Institute of Technology, CH-8092 Zurich, Switzerland, [2]Institute of Animal Breeding and Genetics, Georg-August-University Goettingen, Albrecht Thaer-Weg 3, D-37075 Goettingen, Germany*

The present study compiled and analysed unique data on costs and benefits of cattle breeding programs in Africa and showed problems and trends. Total average costs were mainly influenced by high salary (fixed) costs and high treatment (variable) costs. Costs and returns were only weakly correlated with the size of a breeding program.

Returns varied among programs, but covered for most programs the fixed costs. The production oriented and privately financed programs were profitable (in terms of money), whereas the non-private and improving/conservation programs broke even or showed loss.

The indigenous breeds were superior to other breeds farmed in the same region for climate adaptation, tick resistance, trypanotolerance and ceremonial use, but inferior in appearance. Effects on a breed's endangerment were mainly positive for all programs.

There is little information available about monetary and non-monetary costs and benefits of breeding programs. The results will help to better assess profitability, management and conservation of Africa's cattle diversity.

Poster G3.11

Genetic variability in sheep and goat breeds evaluated by DNA fingerprinting

Yakovlev A. ,V. Terletski, V. Tyschenko, All Russian Research Institute of Animal Genetics & Breeding, St.Petersburg-Pushkin, Moscowskoye Sh.55a, 196600, Russia

We have investigated RFLP data for 13 sheep and 4 goat populations, many of which are classified as endangered in some countries. This survey used a multilocus DNA fingerprinting technique based on the oligonucleotide probes, $(GTG)_5$ and $(GT)_8$. The RFLPscan™ program was used for band detection and for creating a fingerprint database. Several population genetic parameters were calculated from this data with the GELSTATS© program including average heterozygosity and band sharing. Interestingly, average heterozygosity (H) values for most critically endangered sheep breeds examined were higher (H = 0.83 in Stone sheep, H = 0.84 in Bavarian Forest sheep) than those obtained for several commercial sheep breeds (H = 0.65 for Texel sheep), a result of commercial pressure to breed from a small number of males with the highest genetic merit. Bandsharing (BS) values between the sheep breeds studied were relatively low (ranging from 0.18 to 0.33). Within the goat breeds studied, H values were lower (e.g. H = 0.40 for Swiss Toggenburg goats) and between goat breeds, BS values were higher (ranging from 0.37 to 0.48) than for sheep. The use of these population genetic parameters and their derivatives is discussed in relation to the objectives of breeding organizations engaged in the preservation of animal genetic resources.

Poster G3.12

Preliminary analysis on genetic variability and relationships among five native Italian ovine breeds (Appenninica, Garfagnina Bianca, Massese, Pomarancina and Zerasca)

R. Bozzi[1], P. Degl'Innocenti[1], P. Diaz Rivera[1], C. Sargentini[1], C. Vettori[2], D. Paffetti[3], and A. Giorgetti[1], [1]Dipartimento Scienze Zootecniche, Via Cascine 5, 50144 Firenze, Italy, [2]IGV - CNR, Via Madonna del Piano, Sesto F.no (FI), Italy, [3]DISTAF, Via S. Bonaventura, 11, 50145 Firenze, Italy

Genetic variability and relationships among five Italian ovine breeds (Appenninica, Garfagnina Bianca, Massese, Pomarancina and Zerasca) and one foreign breed (Pelibuey) were investigated. Genetic characterization were performed by six microsatellite loci analysed in 42 subjects for each breed. (MCMA8, MCMA11, OARAE119, OARCP49, OARFCB4 and MAF70). The following statistics were calculated: allele frequencies; mean heterozygosity values; average expected heterozygosity; gene diversity; coefficient of gene differentiation. Genetic distances and phylogenetic trees among populations were obtained. The analysis of microsatellite loci revealed observed heterozygosities in the range of 0.21 to 0.83. The expected heterozygosity over all the populations varied between 0.776 and 0.836. Almost all the diversity was retained between breeds as indicated by the low G_{ST} value (8.89%). Significant deviations from Hardy-Weinberg equilibrium were shown for all the locus-population combinations. The constructed dendrogram, using the NJ method, showed that Garfagnina Bianca and Appenninica resulted the genetically closest breeds forming a cluster with Pomarancina whereas Massese and Zerasca pertain to another cluster. Our study showed that a good level of genetic variability still exist even in the endangered breeds but further investigations are in progress to confirm this hypothesis.

Poster G3.13

Genetic diversity of original pig breed in Croatia

M. Bradic[1], M. Uremovic[1], Z. Uremovic[1], M. Konjacic[1] and T.Safner[2], [1]Department of Animal production, [2]Department of plant breeding, genetics and biometrics, University of Zagreb, Faculty of Agriculture, Svetosimunska25, 10000 Zagreb, Croatia

The number of original pig breeds and their genetic diversity has been decreasing during a long period of selection and by the increase of very similar breeds. The Croatian crna slavonska breed, which dates from 19th century, also belongs to this group. Up to the middle of the last century it was the dominant pig breed in Croatia. Today there is just a small number of this individuals (about 400 sows) in the extensive condition of productions. Therefore the aim of this research was to determine the genetic profile and variability of this breed. In order to see the genetic diversity between individuals 8 microsatellite loci were analised on Spredex gels EL 400, by which the difference between individuals was determined. Population genetic parameters were evaluated on the basis of 42 individuals from 6 husbandry. As expected, this research proved that there was not a big difference between individuals that confirm individual differentiation (F=0.216), Nei' s (1987) unibiased gene diversity (Hz=0.44), and the clustering based on the genetic distances between individuals, which did not group in different clusters. This results can be explained by very small population and extensive breeding at which number of boars in mating is small.

Poster G3.14

Conservation of genetic diversity in German draught horse breeds using DNA markers

K. Aberle, H. Hamann, C. Drögemüller and O. Distl, School of Veterinary Medicine Hannover, Institute of Animal Breeding and Genetics, Bünteweg 17p, 30559 Hannover, Germany*

All German draught horse breeds experienced severe decreases in population numbers in the 1960s. Most of these breeds are now considered as endangered or critical. We determined genetic diversity and relationships among the German heavy draught horse breeds using 31 microsatellites and sequence of the d-loop mt-DNA. The breeds included besides reference breeds were South German, Rhenish German, Mecklenburg, Saxon Thuringa, Black Forest, and Schleswig coldblood. The average observed heterozygosity differed little among the heavy horse breeds (0.64 to 0.71) but was negatively correlated to inbreeding coefficients. The mean number of alleles decreased with declining population size (5.2 to 6.3). Genetic variation among the heavy horse breeds was 11.6%. Heavy horse breeds formed a separate cluster from the reference populations, in which the Rhenish German Draught Horse, Saxon Thuringa and Mecklenburg Coldbloods were the most robust groups, the Schleswig Draught Horse was the most distinct breed and South German as well as Black Forest horses could be separated as own groups. Using a minimum-spanning network of mt-DNA haplotypes, we could find a common root of mt-DNA haplotype shared by all domestic horses as far as known. Breeding programmes could make use of this network to conserve horses closely related to the most ancient mt-DNA haplotype or along the branches of further domestication.

Poster G3.15

Genetic variability of caseins in goat populations bred in Lombardy

E. Budelli[1], S. Chessa[2], P. Bolla[2], A. Stella[1] and A. Caroli[3], [1]CERSA, Via Fratelli Cervi 9, 20090 Segrate, Milano, Italy, [2]Department VSA, Milan, Italy, [3]Department SBA, Bari, Italy

Goat milk has been recently revaluated by nutritionists for human consumption, although results about its use as a substitute for cows milk are still conflicting. A research was carried out on the genetic variability of caseins in goat populations bred in Lombardy (Italy), mainly focusing on the possibility of identifying null and/or faint alleles at α_{s1}- (CSN1S1), α_{s2}- (CSN1S2) and β-casein (CSN2), which could be usefully exploited for the production of milk with low casein content.
More than 400 individual milk samples from Saanen, Camosciata, Orobica, and Verzaschese breeds, were screened at the protein level by isoelectrofocusing, and compared for casein structure. Moreover, the presence of faint and null alleles at the calcium sensitive caseins was evaluated by DNA techniques in a sub-total of 145 samples. Very low frequencies were found for CSN1S1*0 (0.014) and CSN1S2*0 (0.003), while no CSN2*0 allele was observed. A high frequency of the CSN1S1*F allele, linked to a reduced level of the specific casein fraction, could be exploited for breeding purposes in selecting for lowering milk casein content. However, the F protein variant is associated with the typical caprine taste, which could involve organoleptic problems in milk consumption. The establishment of specific goat lines producing milk for nutritional and clinical trials could require the use of crossing between different goat populations.

Poster G3.16

Analyses of genetic relationships between various populations of domestic and commercial poultry breeds using AFLP markers

M. De Marchi[1], C. Targhetta[1], C. Dalvit[1], B. Contiero[1], G. Barcaccia[2] and M. Cassandro[1], [1]Department of Animal Science and [2]Department of Crop Science, University of Padova, Viale dell'Università, 16, Agripolis, 35020, Legnaro, Padova, Italy,*

Aim of this study was to assess the genetic distances between some Veneto local chicken breeds and a commercial broiler line, using AFLP fingerprinting technique. AFLP markers are known as an important tool to enrich existing genetic maps of plants, bacteria and animal genomes. Four Veneto local chicken breeds (Padovana, Pépoi, Robusta and Ermellinata of Rovigo) are compared with a broiler line (Golden). Genomic DNA was extracted from whole blood and AFLP polymorphism analysis was carried out according to the protocol described in Barcaccia *et al.* 1999, using three primer combinations (E32/T35, E45/T32 and E45/T33). Heterozigosity and genetic similarity estimates were calculated and an additional factorial analysis allowed a clear distinction between commercial broiler and local breeds. This study proved the applicability of AFLP markers for genetic characterization of the chicken genotypes examined and showed the possibility of defining a genetic breed-traceability system. Further studies will be needed in order to improve the use of AFLP markers and to reduce time and costs of analyses.

Poster G3.17

The endangered Algarvia cattle breed: 1 Morphological relationships with four geographical related Portuguese cattle breeds

M.F. Sobral[1], D. Navas[2], C. Roberto[3], I. Palmilha[4], J. Chumbinho[4], M.B. Lima[5] and A.Cravador[6], [1]DSRA, Rua Elias Garcia 30, Venda Nova, 2704-507 Amadora, Portugal, [2]ENMP, Estrada de Gil Vaz, Apartado 6, 7350-951 Elvas, Portugal, [3]EZN, Fonte Boa, 2000-763 Vale De Santarém, Portugal, [4]DRAAlg, Apartado 282, Patacão, 8000 FARO, Portugal, [5]EAN, 2784-505 Oeiras, Portugal. [6]Universidade do Algarve, UCTA, Campus de Gambelas, 8000-117 Faro, Portugal*

The Algarvia cattle is an indigenous breed originated in the southern region of Portugal. The breed has suffered a serious decline in the middle 20th century and was even considered as extinct. In order to evaluate the actual situation of this population, the morphological diversity of 257 females and 4 males of five cattle breeds from southern region of Portugal were analysed. The morphological characterisation was based on the analysis of 183 characters for the females and 170 for the males and the phenetic relationships among animals and among breeds were calculated by average taxonomic distance coefficients. Cluster analysis based on the similarity matrix using UPGMA (Unweighted Pair-Group Method using Arithmetic averages) indicates that Preta is the most distant population, while Alentejana, Algarvia, Garvanesa and Mertolenga appear as highly related populations. Principal co-ordinates analysis confirms these observations, what was expected on the basis of their geographical location. The results show that all the animals clustered according to their own breed and that Algarvia seems to be a distinct group.

Poster G3.18

Effect of inbreding restrictions on selection response for a family index

J.L. Campo, S.G. Dávila and I Peña, Instituto Nacional de Investigación Agraria, Departamento de Mejora Genética Animal, Apartado 8111, 28080 Madrid, Spain*

Selection responses for a selection index of two traits (weights at 21 and 28 days) were compared in two experiments with *Tribolium* under different inbreeding restrictions. There were four replicates and four generations of selection. Genetic change was calculated from the breeding values of individual animals, which were estimated using an animal model incorporating the best linear unbiased prediction. Experiment 1 involved two lines: IFH, selected for a family index using individual, full sibs, and half sibs information; RIFH, selected as line IFH but choosing one male from each sire's progeny and one female from each dam's progeny. Experiment 2 involved two lines: IF, selected for a family index using individual and full sibs information; RIF, selected as line IF but using a reduced weight on full sibs information. The selected proportion was 25%, except in males from the IFH and RIFH lines (5%). Results show that line RIFH gave a significant smaller selection response (1.99 ± 0.17) than line IFH (2.83 ± 0.08). The N_e in the IFH line (26) represented a strong reduction of the breeding animals in the base population (48), the value in the RIFH line (41) being similar to that in the base population. The line RIF did not differ from the line IF, selection responses being 2.20 ± 0.07 and 2.40 ± 0.14. Values of N_e in both lines (52 and 49) represented a strong reduction of the breeding animals in the base population (80).

Poster G3.19

A new method to estimate relatedness from molecular markers

J. Fernández and M. A. Toro, Instituto Nacional de Investigación y Tecnología Agraria y Alimentaria, Ctra. Coruña Km. 7,5, 28040 Madrid, Spain*

Three major problems affect the efficiency of methods developed to estimate relatedness between individuals from the information of molecular markers: (i) they are dependent on the knowledge of the real allelic frequencies in the base population; (ii) pairwise methods can lead to incongruencies because they take into account only two individuals at a time; (iii) they are usually constructed for particular structured populations (only consider a few relationship classes, e.g. full-sibs or unrelated).

We have developed a new approach to estimate relatedness that is free from the limitations above. The method uses a 'blind search algorithm' (actually *simulated annealing*) to find the genealogy that yield a coancestry matrix with the highest correlation with the molecular coancestry matrix calculated using the markers. Thus, (i) it makes no direct assumptions about allelic frequencies; (ii) it always provide congruent relationships, as it considers all individuals at a time; (iii) degrees of relatedness can be as complex as desired just increasing the 'depth' (i.e. number of generations) of the proposed genealogies. Computer simulations have shown that the accuracy of this new approach is comparable with other proposed methods in those situation they were developed for, but it is more flexible and can cope with more complex situations.

Poster G3.20

Establishing a national cryo bank for ovine breeds with semen collected post mortem

T.A. Schmidt, U. Baulain, C. Ehling, E. Groeneveld, M. Henning, D. Rath, S. Schwarz, S. Weigend, Institute for Animal Breeding (FAL), Höltystr. 10, 31535 Neustadt-Mariensee, Germany*

Genetic based resistance against TSE is a major selection criterion for the favourite PrP genotype in all ovine breeds in the EU. *ARR*-homozygous animals are most resistant, while *VRQ* is the most susceptible out of five known alleles. Despite the very different frequencies among breeds, strong selection for *ARR* is carried out. Allele frequencies in German breeds vary from 6.2 % for *ARR* and 17.8 % for *VRQ* in the Bentheim breed as the worst case up to more than 70 % *ARR* and no *VRQ* in the German Whitehead and Blackhead. 11 breeds have frequencies of *ARR* around or below 25 % that will result in strong bottle necks, if high selection pressure is put on *ARR*.
An efficient semen preservation method to maintain the current genetic diversity had previously been developed in Mariensee, employing deep frozen epididymal semen. Semen was successfully collected from 84 rams of 16 German breeds after slaughter. 400 straws were frozen per ram with a sperm concentration of 50 millions per straw. Less than 10 % of all rams could not be used for semen collection or missed the semen quality requirements. Endoscopical inseminations resulted in a pregnancy rate of 75 %. This pilot project was supported by the Federal Ministry (BMVEL) and the Sheep Breeds Associations (VDL), initiated by the Advisory Board within the National Gene Preservation Programme.

Poster G3.21

Evaluation of crossbreeding strategy for dairy cattle in Pakistan: two breed rotational crossing

M. Ahmad, Livestock Production Research Institute, Bahadurnagar, Okara, Pakistan

Such a system of crossbreeding in which a different sire breed is used in each generation. The dams are chosen from within the system. In this case two breeds (Friesian and Sahiwal) are used for rotational crossing, so the heterosis will be 67 percent after n generations. The main objective of this study was to increase the productivity of local cattle through crossbreeding by getting desired characteristics of both breeds. This simulation study was based on the data set including 1113 Sahiwal and 745 crossbred animals of Sahiwal with Friesian cattle maintained at Livestock Production Research Institute, Bahadurnagar, Okara (Pakistan) during 1973-2003. Three breeding objective traits i.e. lactation milk production, age at first calving and calving interval were included in this study. The means for lactation yield in pure bred Friesian and Sahiwal cows were 3000 and 1700 kg, respectively. The age at first calving and calving interval averaged 913 and 418 days for Friesian and 1218 and 482 days for Sahiwal cattle, respectively. The results were derived by the Genup program version 4.6 (Kinghorn, 1998) for Friesian and Sahiwal breeds and their crossbreds. The mean of rotational crosses for lactation milk yield, age at first calving and calving interval is 2701 kg, 962 days and 399 days, respectively. The overall genetic merit over different traits is considerably higher as compared to purebred Sahiwal cattle.

Poster G3.22

Population and pedigree analysis of indigenous South African beef breeds

R.R. van der Westhuizen[1] and E. Groeneveld[2], [1]ARC-Animal Improvement Institute, P. Bag X2, Irene, 0062, South Africa, [2]Institute for Animal Breeding, Mariensee, Federal Agricultural Research Center (FAL), Höltystr. 10, D-31535 Neustadt, Germany

Reports for the assessment of a population under genetic resources aspects can be generated from any APIIS database. After loading the four native South African breeds: Afrikaner, Bonsmara, Drakensberger and Nguni into an APIIS database population and pedigree information reports were generated for them with the APIIS population report package. Population structure and pedigree information obtained from these reports were the population size of animals in reproduction per year, average family sizes for all animals and for animals selected as parents, generation intervals, effective population sizes and inbreeding. The numbers of reproductive animals decreased over years for all breeds except for the Nguni breed. The generation interval of the Bonsmara was the shortest varying between 5.4 and 4.9 years and, on average, a year shorter than the Afrikaner breed. The generation interval for the Drakensberger has shortened by 1.2 years over the period 1981 to 1997 while the interval for the Nguni breed is increasing. The average inbreeding values for offspring are increasing for all breeds; however, they do fall within a low and acceptable rate of inbreeding. The highest rate of inbreeding was 0.08% per year for the Afrikaner followed by the 0.05% for the Drakensberger and Nguni and the 0.04% per year for the Bonsmara.

Poster G3.23

Cluster analysis of kinship in the Iceland Dog as a small pedigreed population

P.A. Oliehoek[1], G.J. Ubbink[2] and R.F. Hoekstra[3], [1]Wageningen-UR, Animal Sciences Group, Marijkeweg 40, 6709 PG Wageningen, The Netherlands, [2]University of Utrecht, Department of Clinical Sciences of Companion Animals, Yalelaan 8, 3584 CM Utrecht, The Netherlands, [3]Wageningen-UR, Laboratory of Genetics, Arboretumlaan 4, 6703 BD Wageningen, The Netherlands

Cluster analysis of kinship can elucidate the population structure, since this method divides the cohort in clusters of individuals in a dendrogram. Previous research shows that the incidences of breed-specific diseases are bound to specific clusters.
Research has been carried out on the entire Iceland Dog population, a sheep-herding breed. When kinships were calculated up to seven generations, as has been done in previous research, the cohort split up in 5 clusters, which is much lower than other dog populations. Next cluster analysis was based on kinships calculated up to the founder-population. The results demonstrate that the cluster-analysis dendrogram is rather different when based on all generations instead of seven. This contradicts predictions of previous research. Furthermore, the results suggest that kinship-based clustering reveals animals with low mean kinship (animals that are genetically important for genetic diversity). The research shows that despite increasing population size, the Iceland Dog population lost genetic variation. Average mean kinship of the population was 0.22. The number of puppies per litter decreased significantly with 0.22 per 10% increase of inbreeding coefficient. Cluster analysis of kinship coefficient provides better insight in the potential genetic variation.

Poster G3.24

Inbreeding levels in the Italian Heavy Draught Horse population

R. Mantovani[1], G. Pigozzi[2] and G. Bittante[1], [1]Dept. of Animal Science, Agripolis, 35020 Legnaro (PD), Italy, [2]Italian Heavy Draught Horse Breeders Association, Via Belgio, 10, 37135 Verona, Italy*

Inbreeding coefficients (IC) of about 38000 Italian Heavy Draft Horse (IHDH) mares and stallions were computed. Relative to the 1982 population the IC ranged from 0.6 to 30.0%. The inbreeding level within the period of study was upwards, although the annual increase in the IC was generally low (+0.03%/year) and only 1.6% of live animals (4900) had an IC higher than 12.5%. The trends in the level of inbreeding showed a cyclic pattern over years, due to the allowed use of stallions imported from the French Breton population (FB). Indeed, stallions obtained by mating IHDH mares with FB stallions showed a lower mean IC than those obtained within the IHDH population (2.90 vs. 4.35%), although the FB stallions imported were more related within group than the IHDH stallions. Despite the widespread of the IHDH population over all the country and a different use of natural mating (NM), inbreeding levels were very similar from the north to the centre of the country. However, a higher number of highly inbred animals (IC>12.5%) were detected in regions located in the centre of Italy, where NM is more used. Within the IHDH mares no relationship was found between the length of life and the mean no. of abortion/year of life with different levels of inbreeding.

Poster G3.25

Use of experimental dataset of chickens to validate two methods for predicting the evolution of inbreeding in small populations under selection

V. Loywyck[1], P. Bijma[2], M.H. Pinard-Van der Laan[1], J. Van Arendonk[2] and E. Verrier[1], [1]UMR Génétique et diversité animales, Institut national de la recherche agronomique / Institut national agronomique Paris-Grignon, 78352 Jouy-en-Josas cedex, France, [2]Animal breeding and genetics group, Wageningen University, PO Box 338, 6700 AH Wageningen, The Netherlands*

Different theories about the evolution of inbreeding within a small population have been developed but have been mainly been validated through Monte Carlo simulation studies. The purpose of this study is to validate theory with experimental data with different selection schemes. Comparisons were done by analysing inbreeding of two sets of experimental lines of chickens. Selection chicken experiments are powerful tools of investigation on a more or less long term because of the completeness knowledge of the pedigree, generations are not overlapping and generation interval is small. This study has confirmed, on an experimental basis, that modelling is an efficient approach to make useful predictions of the evolution of selected populations despite that basic assumptions considered in the models (polygenic additive model, normality of the distribution, base population at the equilibrium, etc.) are not met in the reality. Especially, the theory developed by Woolliams and Bijma (2000), based on genetic long-term contributions was validated on real data, whereas it has been previously validated only by comparison with simulated data.

Theatre G4.1

Study of simplified designs in French dairy goat milk recording

V. Clément[1], A. Piacère[1], E. Manfredi[2] and F. Barillet[2], [1]Institut de l'Elevage, France, [2]INRA-SAGA, France*

In recent years, the number of French dairy goat flocks over 200 animals has more than doubled. In 2000, the average size was 83 goats per flock, versus 61 in 1995. A consequence of this growing size is that milk recording becomes more and more difficult to implement for technical and economic reasons. Therefore, we started a study in order to simplify the current milk sampling designs while preserving the main objective, the genetic evaluation for dairy traits. Currently, two official designs are used, based either on monthly recordings of the two daily milkings (A4), or on monthly recordings of one of the two daily milkings alternatively in the morning and in the evening (AT4). In a first step, genetic parameters for test-days records have been estimated for the dairy traits : milk, fat and protein yields, fat and protein contents. Results showed that the traits recorded in the beginning of the lactation were less heritable and less representative of the total lactation than those recorded in the middle of the lactation (between 85 and 155 days). In a second step, several simplified designs, based on part-lactation sampling for fat and protein contents in the middle of the lactation were simulated on an on-farm milk recording sample. To compare these simplified designs, the loss of accuracy compared to A4 or AT4 samplings was computed.

Theatre G4.2

Large scale discovery of Single Nucleotide Polymorphism (SNP) markers in Atlantic Salmon (*Salmo salar*)

B. Hayes[1,2], A. Adzhubei[3], J. Lærdahl[3] and B. Høyheim[4,5], [1]AKVAFORSK, P.O. 5010, 1432 Ås, Norway, [2]Centre for Integrative Genetics, Ås, Norway, [3]Biotechnology Centre of Oslo, University of Oslo, Norway, [4]Norwegian School of Veterinary Science, BasAM-Genetics, Oslo, Norway [5]National Hospital, Dept of Dermatology, Oslo, Norway

Single nucleotide polymorphism (SNP) markers are variations among animals in the genetic code at specific sites in the genome. If in aquaculture species this variation is associated with variation in economic traits this information can be used in breeding programs to improve genetic gains. Our goal was to find SNP markers in the Atlantic salmon genome. We took 34,815 expressed sequence tags (ESTs) sequences from the Salmon Genome Project database. These ESTs were fragments of DNA expressed in a variety of tissues. Using the *phred* and *phrap* programs, the sequences were assembled into 5,543 overlapping clusters, or 'contigs'. By comparing sequences within a contig, we were able to detect a large number of base substitutions. The *PolyBayes* program was used to predict whether a base substitution in the aligned sequences was sequencing error, or a true SNP. We detected 458 base substitutions with very high probability of being a true SNP. As the SNPs are within the expressed part of the genome, we will be able to position the genes containing the SNPs on the Atlantic salmon linkage map. This will be a valuable resource for the genetic improvement of Atlantic salmon.

Theatre G4.3

Genetic analysis of mice lines long-term selected on high body weight

U. Renne[1], L. Bünger[2], V. Guiard[1], S. Kuhla[1], M. Langhammer[1], K. Nürnberg[1], Ch. Rehfeldt[1] and G. Dietl[1], [1]Research Institute for the Biology of Farm Animals, Wilhelm-Stahl-Allee 2, 18196 Dummerstorf, Germany, [2] SAC, Bush Estate, Penicuik, EH26OPH, UK*

Crossing experiments, particularly if implemented as diallel, can provide information about the magnitude and kind of genetic causes of population differences and is an important basis for choosing lines for a follow-up linkage mapping study.
Seven lines of mice long-term selected for high growth in several laboratories around the world were used for a complete diallel cross. The extreme selected lines differ substantially in such measured traits as body mass, body dimensions, body composition and in several morphological, physiological and biochemical characteristics. As phenotypic differences are an uncertain criterion for the selection of lines for a mapping study, this diallel, conducted in 4 replicates with overall 3746 animals, was used to estimate F1 and F2 variances for 42 traits, and maternal, heterosomal, autosomal and mitochondrial effects.
Because of recombination an increase of the variance within the F2 compared to the F1-offsprings can be expected. From the relationship of both variances the suitability of special line crosses for a QTL linkage analysis can be quantified, but will be specific for the different traits. For example, the maxima of line differences in direct genetic effects were found between the lines from Roslin and Dummerstorf in 9 different growth traits.

Theatre G4.4

Marker assisted introgression of a sex linked major gene into a mouse line with extreme growth

L. Bünger[1§], F. Oliver[1], U. Renne[2], P.D. Keightley[1] and W. G. Hill[1], [1]School of Biological Sciences, University of Edinburgh, EH93JT, UK, [2]Research Institute for Biology of Farm Animals, , D-18196 Dummerstorf, Germany, [§]Current address: SAC, Bush Estate, Penicuik, EH260PH, UK*

In earlier studies a substantial sex-linked effect accounting for >20% of the divergence between two mouse lines divergently selected for body weight was found to be due a single QTL for body weight at 70d (BW70) at about 23cM from the proximal end of the X-chromosome. The additive genotypic effects in both males (half the difference between hemizygotes) and females (difference between homozygotes) were c2.6g. The average male BW70 in the inbred high line (EDHi) is c47g. Another long term growth selected and inbred line (DUHi) with BW70 of nearly 80g, derived from the heaviest known mouse line, was used to test how much its growth could be further increased by introgressing the X-linked high QTL, and if there were pleiotropic effects on fitness. A reciprocal F1 cross showed that the DUHi carries a low X chromosome relative to the EDHi line. Males hemizygous for the high X-QTL were 5.4, 9.5 and 11.9g heavier than the low QTL carriers at 42, 70 and 84d, respectively. Replicated marker assisted selection was undertaken to backcross (> 2 and 7 generations) the region of the X-QTL in EDHi into DUHi. In this background, the effects of the QTL were non-significant in females and small in males.

Theatre G4.5

Genetic trends in hunting behaviour in the Finnish Hound

A.E. Liinamo, Wageningen University, Animal Breeding and Genetics Group, P.O.Box 338, 6700 AH Wageningen, The Netherlands

Finnish Hound is the most popular dog breed in Finland, with 2 000 - 3 000 pups registered annually in the last years. The great majority of the Finnish Hounds are still used actively as working dogs, so the main breeding goal of the breed is improving its behavioural abilities with respect to hunting traits. Hunting ability of individual dogs is judged in field trials, and the selection of breeding animals is based mainly on dogs' own test results or those of their progeny. From 1996 onwards the breed club has also estimated BLUP breeding values for the most important hunting traits based on the field trial results.
This paper studied the genetic trends in the most important traits evaluated in the hunting trials of the Finnish Hounds in the past 20 years. The data included 85 048 trial records from 12 243 dogs collected between 1987 - 2001. The studied traits included measures related to search, pursuit, tongue and ghost trailing. Despite of continuous selection on these traits, very little genetic improvement was observed prior to the implementation of the BLUP evaluation. Most of the traits have very low heritabilities and are difficult to evaluate reliably. More active use of the BLUP breeding values by the breeders should improve the accuracy of selection and genetic progress in the future.

Theatre G4.6

Genetic and non-genetic factors influencing fibre quality of Bolivian llamas

M. Wurzinger[1], J. Delgado[2], M. Nürnberg[2], A. Valle Zárate[2], A. Stemmer[3], G. Ugarte[4] and J. Sölkner[1]. [1]Department of Sustainable Agricultural Systems, BOKU - University of Natural Resources and Applied Life Sciences Vienna, Gregor-Mendel-Strasse 33, A-1180 Vienna, Austria, [2]Institute of Livestock Production in the Tropics and Subtropics, University Hohenheim, Garbenstrasse 17, D-70599 Stuttgart, Germany, [3]Universidad Mayor de San Simón, Cochabamba, Bolivia, [4]Asociación de Servicios Artesanales y Rurales ASAR, Cochabamba, Bolivia*

Llamas display a great variability of fibre traits that determine the quality of the fleece as raw material for textiles. Little research has been conducted on the extent of this variability, although it is important for optimal use of natural resources in the Andean region. Fibre samples of 1869 llamas were analysed with the optical fibre diameter analyser (OFDA). The following traits were considered: Mean fibre diameter (MFD), standard deviation (SD), diameter of fine fibre (DFF), proportion of fine fibre (PFF), proportion of kemp (PK) and proportion of medullated fibre (PMF). The effects of type of llama, age, sex, coat colour and shearing interval were studied. The type of llama influences all traits showing that Th´ampulli (fibre type) is better than Kh`ara (meat type). With increasing age of the animal MFD, SD, DFF and PK increased whereas PFF decreased. Comparing the two sexes, females showed better fibre quality. Light coat colours tended to be of better quality than darker ones.
Heritabilities and genetic correlations were estimated using animal model procedures where all information came from mother-offspring relationships. Heritability estimates were 0.33, 0.28, 0.36, 0.32, and 0.25 for MFD, SD, DFF, PFF and PK, indicating potential for genetic selection.

 Theatre G4.7

Variance components for litter-size and survival in Danish Landrace and Yorkshire pigs

G. Su, M. Sandø Lund and D. Sorensen, Danish Institute of Agricultural Sciences, Department of Animal Breeding and Genetics, DK-8830, Tjele, Denmark*

Selection for total number of piglets born (TNB) since 1992 has led to a significant increase in this trait in Danish Landrace and Yorkshire, but also an increase in mortality. The objective of the study was to estimate genetic and phenotypic parameters for litter-size and survival in order to find alternative selection criteria to improve litter-size at weaning. Data from Landrace (5178 litters) and Yorkshire (3938 litters) were analysed using REML based on additive genetic model including sow and service-sire. The estimate of heritability (based on sow component) for TNB, number born alive (NBA), number alive at 5 days after birth (N5d) and at 3 weeks (N3w) was 0.081, 0.096, 0.109 and 0.111 in Landrace and 0.065, 0.080, 0.106 and 0.105 in Yorkshire. Heritability for survival ratio per litter at birth, from birth to 5 days and from 5 days to 3 weeks was 0.127, 0.127 and 0.016 in Landrace, and 0.073, 0.043 and 0.009 in Yorkshire. Genetic correlation between TNB and N3w was 0.39 in Landrace and 0.72 in Yorkshire, but between N5d and N3w was 0.99 in both lines. Variance components for survival were also estimated from individual record of piglet. These results suggested that selection for N5d or a combination of TNB and survival would be interesting alternatives to improve litter-size at weaning.

Theatre G4.8

Genetic analysis of body condition in the sow during lactation, and its relation to piglet survival and growth

K. Grandinson, L. Rydhmer, E. Strandberg and F.X. Solanes, Department of Animal Breeding and Genetics, Swedish University of Agricultural Sciences, Box 7023, S-750 07 Uppsala, Sweden*

The aim of this study was to estimate genetic parameters for measures of sow body condition around farrowing, and their genetic correlations to piglet survival and early growth rate. Records were available from 24549 Swedish Yorkshire piglets born in 2198 litters. Sows had records on weight and backfat depth at farrowing and at weaning. Piglets had individual records on birth weight and weight at weaning, and cause of death for those who did not survive throughout lactation. Mixed linear bivariate models were used to estimate correlations between traits. Estimated heritabilities for sow weight and backfat at farrowing and loss of weight and backfat during lactation were low to moderate (0.10-0.47). We found strong genetic correlations between loss of weight and backfat during lactation and piglet survival and growth, indicating that sows with the genetic capacity for a high early piglet growth and survival rate will lose more body reserves during lactation. We conclude that in a selection program aiming at improving piglet survival and growth, attention should be paid to the sow's body condition during lactation. A high enough level of body reserves has to be maintained in the sow, in order to not increase the incidence of reproductive problems and involuntary culling.

Theatre G4.9

A Bayes factor analysis for testing genetic determinism in the rate of hoof growth in pigs

R. Quintanilla, L. Varona and J.L. Noguera, Àrea de Producció Animal, Centre UdL-IRTA, Rovira Roure 191, 25198 Lleida, Spain*

The excessive growth and the consequent deformity of hoofs is a frequent disorder in some purebred pig populations. A genetic determinism could be suspected as one of main causes for this phenomenon. The presence of genetic variability in hoofs growth has been tested using a Bayes Factor (BF). Available data proceeded from individuals of three selection lines (561, 225 and 183 records respectively). Animals were scored in four categories according its rate of hoofs growth. A bayesian analysis was performed through a threshold model with a probit approach. In addition, the Bayes Factors were computed to contrast between models with and without additive genetic effects. Results showed that models with genetic variance were highly more probable than models without genetic variance (BF values of 321, 46 and 16 respectively - posterior probabilities of 0.99,0.98 and 0.94). Posterior mean estimators for heritability took values medium to high (0.24, 0.38 and 0.41 respectively), with posterior standard deviations of 0.08 0.13 and 0.16. These results allow us to conclude that an important genetic determinism of rate of hoof growth in pigs exists.

Theatre G4.10

PigAce: an integrated pig genome database

Jan W.M. Merks[1], Tony J.A. van Kampen[2], Rik van Wijk[1], Barbara Harlizius[1], Annemieke Rattink[3], Gerard Albers[3] and Martien A.M. Groenen[2], [1]IPG, Institute for Pig Genetics BV, P.O.Box 43, 6640 AA Beuningen, The Netherlands, [2]Wageningen University and Research Centre, Department of Animal Sciences - Animal Breeding and Genetics Group, P.O.Box 338, 6700 AH Wageningen, The Netherlands, [3]Nutreco Breeding Research Centre, P.O. Box 220, 5830 AE Boxmeer, The Netherlands

Knowledge of farm animal genomes has increased enormously over the last decade. A large part if this information is publicly available for a variety of species and through specific databases such as for pigs; PiGBASE for mapping data, Pig EST Database, TIGR SsGI for genes and data on their expression patterns and the INRA Comparative and Cytogenetic mapping home pages. Potentially these databases provide comprehensive public repositories for genome research. However, these data are difficult to combine from the different sources or with private data, but also with genome data of model organisms. This strongly hinders comparative mapping and positional fine-mapping.

A new pig genome database - PigAce was set up in the Netherlands to enable integration of data from the different sources. For this, the widespread database system of AceDB has been adapted and links with existing farm animal databases but also databases like LocusLink, Genbank, MGI, GeneCards are included to facilitate an efficient comparative mapping with human and mouse. In addition published information on porcine QTL has been included. This database with more than 5000 genetic markers and loci and about 500 QTL's will be available publicly from July 2004.

Poster G4.11

Effect of age and housing system on genetic parameters for broiler carcass traits

S. Zerehdaran [1], A.L.J. Vereijken[2], J.A.M. van Arendonk[1], and E.H. van der Waaij[13], [1]Wageningen University, PO Box 338, 6700 AH, The Netherlands; [2]Nutreco Breeding research Center, P.O. Box 220, 5830 AE, Boxmeer, The Netherlands; [3]Veterinary Faculty, University of Utrecht, Yalelaan 7, 3584 CL Utrecht. The Netherlands*

Objective of this study was to estimate genetic parameters for BW and carcass traits at different ages and housing systems. Birds were divided into three age groups (2000 per group), raised in free housing until either 48 or 70 days or in individual cages until 63 days of age. BW (in individual cages also at 48d), carcass, breast muscle (BMP), abdominal fat (AFP) and back half percentage (BHP) were recorded at slaughter age. Heritability for BW was lower in older birds (0.34, 0.25, and 0.21 at 48, 63, and 70d respectively). Positive genetic correlations (r_g) occurred at 48 days between BW and BMP (0.35), or BHP (0.30), whereas, negative r_g were found at 70 days between BW and BMP (-0.12), or BHP (-0.27). However BW and AFP showed a higher r_g at 70d (0.37) than 48d (0.28). Current results indicate that growth at 48d is accompanied by increase in valuable parts whereas at 70 d it is accompanied by increase in abdominal fat. The genetic correlation between BW at 48 days in different housing systems (0.81) indicated the presence of G*E interaction.

Poster G4.12

Population genetic parameters in selected chicken breeds

V.I. Tycshenko, O.V. Mitrofanova, N.V. Dementieva, V.P. Terletski and A.F. Yakovlev, All Russian Research Institute for Animal Genetics & Breeding, 196625 St.Petersburg-Pushkin

The main goal of our research has been to evaluate genetic diversity in chicken breeds (Black Australorp, Andalusian, White Leghorn, Russian White) and several experimental highly selected populations. Two approaches has been used - DNA fingerprinting with digoxigenated oligonucleotide probe (GTG)5 and RAPD analysis. Similarity index (BS) varied 0.82 to 0.92 (RAPD) and 0.43 to 0.53 (DNA fingerprinting). Both methods revealed similar genetic relationships between breeds and populations. For example, low values of genetic distances were observed between Black Pied Australorp and experimental population "Aurora" (D=0.025 and D=0.060 for RAPD and DNA fingerprinting, correspondingly) which were the lowest values for genetic distances calculated. Selection in many generations led to reduction in intra-population genetic variability evaluated by criteria of average heterozygosity. This was confirmed in two experimental populations by both methods. In general, average heterozygosity in all breeds and populations varied 0.050 to 0.112 (RAPD) and 0.56 to 0.062 (DNA fingerprinting).

Poster G4.13

Researches about traits linkage to a paternal chicken line

I. Custura, I. Van, St. Popescu Vifor, E. Popescu Miclosanu and M. Tudorache, University of Agronomical Sciences and Veterinary Medicine Bucharest, Faculty of Animal Sciences, 59 Marasti Blvd., 1 sector, Romania*

Study of linked characters concerning relationship between them is especially important in breeding science and technology because it make possible to establish causes of relationship and so the measure in which improvement of a trait would change other traits as well. The goal of this work is to establish phenotypic, genotipic and environmental connections between characters: body weight, body shape, feed conversion and fertility at a paternal Cornish line.

Study was made during a two years interval, with generations F and G, based on production and breeding control results of 60 chicken families from which male parents of next generation were chosen.

Study results have shown, based on average flock performances, a medium genetically causality, expressed by intermediate probability values, different from one generation to another. There are also differences about size and direction of the interdependence between studied characters. Phenotypic relationship between pairs of analysed traits has values between -0.203 and +0.263 in generation F and between -0,153 and +0,148 in generation G, with respectively three and four negative values on generation. Genetic relationship between the four studied traits had have bigger values than these of the phenotypic characters, between -0.367 and +0.434 in generation F and between -0,481 and +0,456 in generation G, with respectively three and four negative values on generation.

Poster G4.14

Mapping of growth quantitative trait loci in a cross of chickens divergently selected for body weight

D. Tercic, P. Dovc, A. Holcman and S. Horvat, University of Ljubljana, Biotechnical Faculty, Zootechnical Department, Groblje 3, 1230 Domzale, Slovenia*

We previously developed high body (H) weight and low body (L) weight chicken lines by divergent selection for over 25 generations. To identify quantitative trait loci (QTL) for growth, a mapping cross using a full-sib intercross line (FSIL) design was initiated. A single H male was crossed with an L female producing 19 F_1 progeny. One F_1 sire was then mated to 9 F_1 dams producing 135 F_2 progeny. Five randomly chosen F_2 sires were mated with 12 randomly assigned F_2 females resulting in 506 F_3 offspring. In the F_3 population, body weights at 1, 21, 42 and 55 day and various carcass traits at 8 weeks of age were recorded. So far, 120 microsatellites covering 22 linkage groups were screened for informativeness in the F_0. Among these, 68 informative markers were used for the bulked segregant analysis of two DNA pools consisting of extreme 10 % individuals from the F_3 phenotypic distribution for the body weight at slaughter. Markers showing differences in allelic composition between the pools were analysed further by analysing F_3 pools within the F_3 families and eventually the whole of F_3 population. Analysis of additional markers is currently underway, and an up-to-date status of marker and QTL analyses will be presented.

Poster G4.15

Chromosome and gene mapping homology between river buffalo, cattle and sheep using molecular markers

Othman El Mahdy Othman, Cell Biology Department - National Research Center - Dokki - Egypt*

Chromosomal localization of sixteen bovine microsatellites in river buffalo has been assigned in this work, using polymerase chain reaction and buffalo-hamster somatic cell hybrids. These tested microsatellites were previously assigned to sheep chromosomes. This study also aimed to confirm that the chromosome band identity between river buffalo, cattle and sheep is a good indicator of genetic homology between these closely related species.
The correlation coefficients between these tested microsatellites and other markers-representing syntenic groups and chromosomes in river buffalo- were calculated to assign these microsatellites to river buffalo chromosomes (BBU). The results showed that BM719 is assigned to BBU5q, BM827 to BBU12, BM1818 to BBU2p, BM1824 to BBU1q, BM2113 to BBU2q, CSSM015 to BBU3p, CSSM034 to BBU4q, CSSM037 to BBU3q, CSSM043 to BBU1p, CSSM058 to BBU21, ILSTS011 to BBU15, ILSTS013 to BBU10, ILSTS019 to BBU5p, ILSTS029 to BBU6, ILSTS054 to BBU20 and ILSTS060 to BBU4p.
The results also showed that 16 tested microsatellites are localized in river buffalo, cattle and sheep on the equivalent chromosomes, which have chromosome-band homology in these closely related species belonging to Bovidae family, which has a high degree of chromosome conservation between its members and where the bi-armed autosomes are formed by centric fusion of acrocentric autosomes.

Poster G4.16

Goat milk somatic cell count is a heritable trait

R. Rupp[1], V. Clément[2], A. Piacere[2] and E. Manfredi[1], [1] INRA-SAGA, France, [2]Institut de l'Elevage, France*

Goat milk somatic cell counts have been collected for several years in France by the national milk recording organisations. Information is used for mammary health management. Moreover, French dairy industries apply a quality payment system, including penalties on somatic cell counts, following official recommendations of the French association of goat's milk producers and processors. This paper reports a first genetic analysis of somatic cell counts data from the French national data base. Genetic parameters were estimated for 59203 primiparous goats of the Saanen breed in 857 herds. Goats were sired by 3500 bucks, 173 of them insuring connection amongst herds. Estimates were obtained by REML with an animal model. The trait considered was the arithmetic mean of test-day scores, i.e. log transformed cell counts, recorded in first lactation. The heritability estimate was 0.19 in primiparous goats, i.e. a somewhat higher than for French dairy cattle and sheep. Relationships with milk production traits were additionally estimated through multitrait analyses. Heritability estimates of lactation milk yield, fat and protein content, and fat and protein yield, were, 0.32, 0.60, 0.50, 0.40 and 0.34, respectively. Genetic correlation of somatic cell scores with milk production traits were closes to zero, ranging from -0.03 to -0.10. Corresponding environmental correlations were negative for milk yield, protein and fat content, -0.22, -0.18 and -0.25.

Poster G4.17

Genetic variability and divergence of some salmon breeds

N.V. Dementieva[1], V.P. Terletski[1], V.I. Tycshenko[1], D.E. Belash[1], V.M. Golod[2], E.G Terentieva[2] and A.F. Yakovlev[1], [1]All Russian Research Institute for Animal Genetics & Breeding; [2]Federal Genetic and Breeding Center of Fisheries

Variation of polymorphic DNA sequences in two new breeds Rofor and Rostalj and other salmon fish Steelhead Trout, Baltic, Onezhskij, and Caspian trout was evaluated using minisatellite DNA probe $(GGAT)_4$. The large observed genetic distance between Onezhskij and Caspian trout (D=0,150) is explained by natural geographic isolation and limited exchange of genetic resources between these populations. The same true Baltic and Caspian Trout (D=0,130). In contrast genetic relatedness of Rofor, Rostalj and Steelhead Trout was clearly confirmed by small values of genetic distances (0,040-0,085). Intensive selection and breeding aimed to achieve a genetically homogeneous population in Rostalj breed resulted in the lowest value of observed heterozygosity (H=0,41). In conclusion, analysis of band distribution in DNA patterns provides a good estimate of the current status of fish breeds and populations.

Poster G4.18

Genetic parameters among body condition score, fertility, type and production traits in Italian Brown Swiss dairy cows

R. Dal Zotto[1], M. Cassandro[2], G. Bittante[2], [1]Superbrown consortium of Bolzano-Bozen and Trento, Italy, [2]University of Padova, Department of Animal Science, Viale dell'Università, 16, Agripolis, 35020, Legnaro, Padova, Italy*

Aim of this study was to estimate genetic parameters for body condition score (BCS), calving interval (CI), final score and milk yield and to investigate the genetic relationship between BCS and CI. Data set consisted of 5,617 test day records of BCS and milk yield, of 1,291 Brown Swiss lactation cows with registered calving interval, reared in 57 herds of the "Superbrown Nucleus". In addition 1,176 individual final scores and pedigree data (4,916 animals) were provided by National Breeders Association of Brown Swiss. Heritabilities for average of BCS, recorded during the whole lactation and during the first 150 days in milk were around 30% and 20%, respectively. Heritabilities for milk yield, type traits and CI were 30%, 20% and 5%, respectively. Genetic correlations of CI with milk yield and final score were high and unfavourable: 66% and 56%, respectively. Genetic correlations between CI and BCS were favourable: -70% and -58% with average BCS overall lactation and within first 150 days in milk. In conclusion, the selection for milk yield and type seems to have a strong negative effect on reproduction ability in Brown cattle population, but this preliminary results indicate that BCS, recorded during lactation, might be used as an indirect selection tool for cows' fertility.

Poster G4.19

Identifying QTL for growth rate in the Hungarian Merino sheep

I. Komlósi[1], I. Anton[2], A. Zsolnai[2], M. Árnyasi[1], L. Fésüs[2], [1]Univesrity of Debrecen, Department of Animal Breeding and Nutrition, H-4015 Debrecen,P.O. Box. 36, Hungary, [2]Research Institute for Animal Breeding and Nutrition, Gesztenyés u. 1. 2053 Herceghalom, Hungary*

The objective was to search QTL affecting growth rate before and after weaning in the Hungarian Merino sheep. 280 mixed sex lambs of 5 sires were blood sampled after a fattening performance test. Panels of bovine and sheep microsatellites were chosen and the lambs and sires were genotyped for 15 markers altogether. For the PCR reaction a Perkin Elmer (DNA Thermal Cycler) equipment was used. The PCR products were identified by ABI PRSIM 310 Genetic Analyser. The effect of a loci was analysed within sire (half-sib families) using the sex, birth type and the genotype as fixed effects in a GLM analysis. As the preliminary results show, the SPS115 loci had a significant effect on before weaning within one sire progeny group. A QTL affecting growth rate before weaning linked to markers CSRD0247 and HSC is probable ($P<0.05$). The extreme genotypes had 284 g/day and 460 g/day growth rate, respectively. The lambs of 269-284 genotype for the HSC loci significantly differed from the 273-279 genotypes. No effects were observed in the case of the following locus: ETH225, OarA129, OarCP49 and MAF214.

Poster G4.20

Creating germ line chimeric rabbits by blatomer injection

Sz. Bodó[1], E. Gócza[1], T. Révay[2], L. Hiripi[1], B. Carstea[1], B. Bender[1], A. Kovács[3], L. Bodrogi[1] and Zs. Bösze[1], [1]Department of Animal Biology, Agricultural Biotechnology Center, P.O. Box 411, 2100 Gödöllö, Hungar,y [2]Institute for Small Animal Research, P.O. Box 417, 2100 Gödöllö Hungary, [3]Research Institute for Animal Breeding and Nutrition, Gesztenyés str.1, 2053 Herceghalom, Hungary*

Single blastomers from a precompacted human factor VIII (hFVIII) transgenic embryo derived from wild type donor does mated with a hFVIII homozygote buck was microinjected into the perivitellin space of 10-16 cell stage wild type embryos in order to produce chimeras. Chimeric embryos were transfered to recipient females by laparoscopy. HFVIII was used as a marker gene to detect chimerism and for quantitative analysis. The chimeric offsprings were detected by presence of the hFVIII transgene in their tissue samples. Among the four chimeric animals one was identified as a chromosomal intersex. The sexually matured chimeras were mated with wild type rabbits and their litters analysed for the presence of hFVIII. Two adult animals were identified as germline chimeras. The degree of chimerism in different tissues of the chimeric animals was estimated by real-time PCR and was found to be in the range of 0.1 and 42%. We took the advantage of a rabbit transgenic line and the efficient microinjection method of preimplantation stage embryo reconstruction which combined with the improved reproductive technologies enabled us to create germ line chimera rabbits.

Poster G4.21

Estimation of genetic and phenotypic parameters for some growth and carcass traits of Japanese Black calves using a multi-trait animal model

M.A. Aziz[1], S. Nishida[2], K. Suzuki[1] and A. Nishida[1], [1]Graduate School of Agricultural Science, Faculty of Agriculture, Tohoku University, Sendai 981-8555, Japan, [2]Miyagi Animal Research Station, Iwadeyama Miyagi 989-6445, Japan*

Data on 1170 records of fattening calves were collected on some growth and carcass traits from a herd located in Miyagi prefecture, Japan. The objective was to determine direct and maternal heritabilities, direct and maternal genetic correlations, phenotypic correlations and direct-maternal correlations between initial weight at fattening (INWT), final weight after fattening (FNWT), carcass weight (CAWT), daily body weight gain during the fattening period (DBWG), rib eye area (REA), rib thickness (RTH), backfat thickness (BFTH), cutability (CUT) and beef marbling score (BMS). Direct heritability estimates of BFTH (0.16) and BMS (0.07) were low, while those of the other traits were medium to high and ranged between 0.44 (REA) and 0.78 (CAWT). Direct genetic correlations were all positive, except those between INWT and each of BFTH and CUT (-0.49 and -0.14, respectively). The lowest positive genetic correlation was between INWT and BMS (0.04) and the highest was between FNWT and CAWT (0.99). The phenotypic correlation coefficients ranged between -0.41 and 0.96. Maternal heritability estimates were generally low and ranged between 0.003 and 0.08. The permanent maternal environmental variance ranged between 0.01 (BMS) and 0.46 (DBWG). Based on the above mentioned results, genetic improvement based on a selection index combining information on growth and carcass traits is suggested.

Poster G4.22

Studies concerning the effect of the inbreeding on some physiological characteristics (bombyx mori l.)

G. Dinita, University of Agronomical Sciences and Veterinary Medicine Bucharest, Faculty of Animal Sciences, Technology Department, 59 Marasti Blvd., 1 sector, 011464, Romania

Thirty silkworm inbreed lines were studied during six generations and the effect of the inbreeding on some physiological characteristics.
The prolificity ranged between 632 - 494 eggs/lay inside the White Baneasa lines and 748 - 522 eggs/lay in Baneasa 75 lines. The number of eggs/lay was in I_6 10.39 % lower in Baneasa 75 lines and 11.41 % in White Baneasa lines by comparison with I_0.
Hatching percentage was affected by inbreeding depression beginning with I_3 in both lines groups. This was lower with 6.10 - 15.66 percentages in White Baneasa lines and 6.80 - 15.90 percentages in Baneasa 75.
The survival rate of larvae proved a gradual diminution together with the increase of inbreeding coefficient and became significant in I_3. In I_6 the survival rate of larvae ranged in the limits 64.40 - 80.16 %.
The pupation rate in I_6 was 78.16 - 84.20 % (White Baneasa lines) and 76.10 - 80.86 % (White Baneasa lines).
The inbreeding had a negative influence on the resistance of lines to the induced polyhedrosis. The survival percentage of larvae, after their maintenance in low temperature was 4 - 32.11 % in I_6. The control larvae showed 48.00 - 50.67 % survival rate in the same experimental conditions.

Poster G4.25

Research on obtaining inter specific goose hybrids for ecological rearing

E. Popescu-Miclosanu, I. Van, I. Custura, M. Tudorache and C. Roibu, University of Agronomical Sciences and Veterinary Medicine, Faculty of Animal Science, 59 Marasti Blvd., 1 Sector, Bucharest, 71331, Romania*

In order to obtain goose hybrids for ecologically rearing, crossings were made between Toulouse and Embden breeds (Anser cinereus) with the Chinese goose (Cygnopsis cygnoides), with knob and recognized for her resistance on keeping conditions and illness. Were obtained 4 hybrids by interbreeding crosses between 2 and 3 breeds (112 individuals for every combination). The growing performances were compared with the same number of goslings from every pure breed. All birds were kept in houses with free outside access, with a ration of mixed feed having exclusively vegetal composition and alfalfa from 2 weeks old. The growth intensity of the simple reciprocal hybrids of Embden (E) and Chinese (C) breeds was significantly greater than that of the Chinese pure breed and closed enough of the Embden breed, comparatively with they attained 87.3% (EC) and 91.6% (CE). Back crossing these hybrids towards Embden breed (EEC) lead to an average weight of 3995 g at 8 weeks of age, greater than that of both parental forms (with 10.4% compared with Embden and 18.2% with EC). Hybridization of EC geese with Toulouse (T) ganders ensure the bigger weight of 3911 g, better than that of the maternal hybrid with 17.5%. The differences between the performances of EEC and TEC are not statistically assured.

Poster G4.26

Using cluster analyses in genotype x environment interaction studies in beef cattle

F.W.C. Neser and G.J. Erasmus, Dept. of Animal, Wildlife and Grassland Sciences, University of the Free State, PO Box 339, Bloemfontein, 9300, Republic of South Africa*

A cluster analysis was performed on weaning weight records of 116435 Bonsmara calves, the progeny of 2218 sires in 69 herds in four regions. The following environmental factors were used to classify herds into clusters: solution for herd effects (indicative of the management level in a herd), herd size and average temperature and - rainfall. Bivariate animal models were used to obtain correlation estimates between the four clusters for weaning weight. The direct genetic correlations between the clusters varied between 0.47 and 0.88 while the maternal genetic correlations varied between 0.39 and 0.86. The low correlation estimates between some of the clusters indicate a possible genotype x environment interaction. However, further research is needed to identify and prioritize variables that can describe the genetics, management and climate of each herd more accurately.

Poster G4.27

Research about heritability of some traits in Frasinet, Ineu and Ropsa carp breeds

C. Nicolae and L.D. Urdes, University of Agronomical Sciences and Veterinary Medicine, Faculty of Animal Sciences, Marasti Bvd, no 59, sector 1, Bucharest, Romania

For any characterization of an animal population to be as complete as possible, the overall study of the particular population by analyzing the quantitative traits with biometric genetics, is essential. The knowledge of heritability represents an important factor in all the improvement programmes; this importance derives from the fact the this parameter explains the correlation between the phenotype and improving value. With that point of view, the present work is to estimate the heritability for more traits into a pattern of three different breeds of carp individuals, Frasinet, Ineu and Ropsa. The studied traits was represented by: body weight, length, utmost height, H/l proportion and the rapidity of the rising process.

Poster G4.28

Inbreeding in a closed sheep flock

J.B. van Wyk and M.D. Fair , Department Animal, Wildlife and Grassland Sciences, University of the Free State, P.O. Box 339, Bloemfontein, 9300, Republic of South Africa*

Data of the Elsenburg Dormer sheep flock, which was kept closed since its inception, were collected over a period of 62 years (1941-2002). These data were analysed to quantify the increase in actual level of inbreeding and to investigate the effect of inbreeding on phenotypic values, genetic parameters and estimated breeding values. After editing 11954 pedigree, 11721 birth weight (BW) and survival, 9205 weaning weight (WW) and 7504 reproduction records were available for analysis. Mean, minimum and maximum level of inbreeding (F) for the lambs in 1997 (when 3 rams from outside were introduced) were respectively 22%, 21% and 24%. Estimates of inbreeding depression of individual inbreeding were -0.006kg for BW and -0.093 kg for WW respectively. These were the only estimates that were significantly ($P<0.01$) different from zero. There were virtually no difference in the genetic parameters estimated fitting the two models (inclusion or exclusion of inbreeding coefficients as covariates). Ranking of animals on breeding values for WW obtained from the two models were compared. The high correlation coefficients (0.990) indicated that inclusion of inbreeding coefficients did not cause important changes in ranking of animals and sires. It was concluded that slow inbreeding (rate of inbreeding of approximately 1.53 % per generation over 19 generations) allows selection to operate and to remove the less fit animals.

Poster G4.29

Preliminary evaluation of the effect of the *Compact* mutant *myostatin* allele: *Mstn*$^{Cmpt\text{-}dl1Abc}$ on growth in the mouse line of origin

W. Schlote[1], T. Hardge[2], M. Reissmann[1], S. Lutz[1] and F. Major[3], [1]Humboldt-Universität zu Berlin, Institut für Nutztierwissenschaften, Unter den Linden 6, 10099 Berlin, Germany, [2]Boehringer-Ingelheim Animal Health GmbH, Binger Strasse 173, 55216 Ingelheim am Rhein, Germany, [3]Backa Topola, Pere Segedinca 6, Serbia and Montenegro*

The compact allele mutant *Mstn*$^{Cmpt\text{-}dl1Abc}$ has been shown to affect muscle size and muscle cell number in specific mouse lines. A total of 1569 animals of genotypes homozygous compact (C/C), heterozygous (C/+; +/C) and homozygous wild type (+/+) were used to study the effect of the compact allele on growth traits in animals of the original selection line. The animals were tested under standard conditions over four generations without selection. The model for weekly individual body weights from birth to 63 days of age included fixed effects generation, sex, genotype and sex-genotype interaction. Average weights at different ages were 2.0 (0d), 14.7 (21d), 34.2 (42d) and 40.1g (63d). Significant sex effects ($P<.01$) were found for animals older than 21 days, sex-genotype interactions were significant ($P<.01$) at 42 days and at higher age. Weights of myostatin genotypes were significantly different ($P<.01$) for all weights after weaning, the difference between homozygotes compact (C/C) and wild type (+/+) amounting to 0.7, 1.3 and 1.5 standard deviations for 21, 42 and 63 day weights, respectively. Thus, the compact allele caused quite a strong increase in weight and growth.

Poster G4.30

Molecular analysis of Water buffalo (*Bubalus Bubalis*) genes involved in muscle development

M. Strazzullo[1,2], C. Campanile[3], M. D'Esposito[1] and L. Ferrara[3], [1]Institute of Genetics and Biophysics "A.Buzzati Traverso" CNR, Naples Italy, [2]Biogem-SCARL, Italy, [3]ISPAAM-CNR, Naples, Italy

Water buffalo (Bubalus bubalis) represent an important specie for Italian food industry, especially in Southern Italy. In spite of these considerations, it lacks a systematic approach to the understanding of the water buffalo genome, a condition which is indispensable for its genetic improvement. We are taking advantage of technologies we already set up for the structural and functional analysis of the Human Genome (Roberts, R.M. et *al.* 2001). Our final goal is to construct tissue specific cDNA libraries from muscle, brain and mammary gland of Bubalus bubalis. Here we present the analysis of the Bubalus bubalis myostatin gene (Kambadur, et *al.* 1997) , identified by comparative genomics.
After extraction of total RNAs from targeted tissues, checking of the quality, we used gene specific primers to amplify myostatin gene, involved, in mammals, in muscle development. We cloned and sequenced the whole coding region of the gene, of roughly 1100 bp. We confirm by RT-PCR the high selectivity of the gene expression. Northern analysis has beeen performed to reveal possible alternative mRNAs. A search for DNA polymorphisms is currently undergoing by using different water buffalo breeds.

Poster G4.31

Genetic parameters of weight productivity traits of D'Man ewes in Tunisia

R. Aloulou[1], H. Hentati[2], M. Rekik[3] and M. Ben Hamouda[4], [1]E.S.H.E., Chott Mariem, B.P. 47, 4042 Chott-Mariem, Sousse, Tunisia, [2]Faculté des Sciences de Tunis, Tunisia, [3]E.N.M.V., 2020 Sidi Thabet, Tunisia, I.R.E.S.A., 30 Rue Alain Savary, 1002 Tunis, Tunisia

Data on weight productivity of 1023 ewes of the D'Man breed calculated using an original file on growth performances of 2026 lambs of the same breed were used in this study. All the lambs were born and raised in Tunisia between 1994 and 1999 in the north of the country (Mateur, Kef) and in the south under an oasian environment (Gabès). The studied variables included weights of the litter at 10, 30, 70 and 90 days of age and respectively designated LW_{10}, LW_{30}, LW_{70} and LW_{90}. Average calculated figures were 8.22 kg (±2.91), 13.72 kg (±4.64), 24.51 kg (±8.79) and 29.46 kg (±11.11) respectively for LW_{10}, LW_{30}, LW_{70} and LW_{90}. Analysis of the non genetic sources of variation revealed a highly significant effects ($P<0.001$) of the factors rank of lambing within station, age of the dam, and type of lambing-rearing on all the studied traits. Heritability coefficients estimated using the REML method of variance analysis, were 0.17, 0.23, 0.08 and 0.08 for respectively LW10, LW30, LW70 and LW90. Genetic and Phenotypic correlations between all the studied traits were high and positive respectively varying between 0.81 and 0.94 and between 0.82 and 0.96.

Poster G4.32

Expression of endothelial and inducible nitric oxide synthase genes in different stages of in vitro produced preimplantation cattle embryos

A.K. Kadanga, D. Tesfaye, S. Ponsuksili, K. Wimmers, M. Gilles and K. Schellander, Institute of Animal Breeding Science, University of Bonn, Endenicher Allee 15, 53115 Bonn, Germany*

Nitric oxide (NO) is known to regulate various reproduction processes, such as stereogenesis, pregnancy, folliculogenesis and tissue remodelling. NO is synthesized from L-arginine by nitric oxide synthase (NOS). In the present study we have investigated the expression pattern of bovine endothelial and inducible NOS genes (eNOS & iNOS) during preimplantation development. For this mRNA isolated from pools of immature and mature oocytes, 2-cell, 4-cell, 8-cell embryo, morulae and blastocyst were reverse transcribed and subjected to real time PCR using specific primers in ABI-PRISM 7000 SDS instrument (Applied biosystems). The effect of NOS inhibitor (N-nitro-L-arginine methyl ester: L-NAME from Calbiochem) on development was examined by adding different doses of the inhibitor (1mM, 10mM and 20mM) at maturation, fertilisation and culture periods. The iNOS was detected at higher level in immature oocyte and 2-cell stages and down-regulated in the later developmental stages. The eNOS was abundant from immature oocyte up to 4-cell stage and further down-regulated up to the blastocyst stage. Addition of L-NAME at maturation and fertilisation significantly affected embryo development at 10 and 20 mM level. In conclusion, bovine oocytes and preimplantation embryos express both eNOS and iNOS transcripts in stage specific manner. NO inhibition affects normal preimplantation development in vitro.

Poster G4.33

The polymorphism of kappa-casein gene in Russian Red- Step cows

A.V. Barshinova, N. Yukhmanova, All-Russian Institute of Animal Breeding, Lesnye Polyany, Pushkin district, 141212 Moscow Region, Russia

The Red-Step dairy cattle bred was founded by crossing of Simmental cows with Red-White Holstein sires. The identification of the kappa-casein genotype in Red-White dairy cattle was carried out by PCR-RFLP methods. Animals are rearing in regions of European Russian areas with different climatic conditions (Central-Chernozem zone and Volga region - temperate and continental climate) and in Eastern Siberia - with sharply-continental climate. The findings are presented in table.

Areas of	Number	Frequency of genotype,%			Frequency of alleles	
breeding	of animals	AA	AB	BB	A	B
Central-Chernozem	162	48	36	16	0,66	0,34
Volga region	170	50	32	18	0,66	0,34
Eastern Siberia	105	42	55	3	0,70	0,30
On average	Total - 437	47	39	14	0,67	0,33

As a whole for the breed the frequency of allele A kappa-casein gene ranges from 0,66 to 0,70 and the frequency of allele A ranges from 0,30 to 0,34. The frequency of genotypes AA, AB and BB are greately differing in the areas of breeding. The highest frequency of geterozygous kappa-casein genotype AB and the lowest frequency gomozygous genotype BB are observed in Eastern Siberia.

Poster G4.34

Meiotic chromosomes of cattle oocytes in relation to seasonal variation

Karima Gh. M. Mahmoud[1] and N. T. Eashra[2], [1]Dept. Anim. Reprod. & A.I National Research Center, Egypt, [2]Dept. Field Inspection, Anim. Reprod. Research Institute, Egypt

The present work aimed to study the effect of seasonal changes on the recovery rate, oocyte quality and meiotic chromosomes of cattle oocytes in vitro. The follicular oocytes were collected from ovarian follicles (2-5 mm in diameter) of slaughtered cattle by aspiration method. The recovered oocytes were classified morphologically into three categories (A,B &C). The selected recovered oocytes (class A & B) were matured in tissue culture medium-199 supplemented with 10% fetal calf serum and antibiotics at 39°C, 5% CO_2 and high humidity for 22-24 hrs. Cultured oocytes were analyzed for meiotic chromosomes during the four seasons. The results indicated that, the average number of cattle oocytes was significantly increased ($P <0.05$) in spring season than other seasons. A higher percentage ($P<0.05$) of high quality oocytes with lower percentage ($P<0.01$) of poor quality oocytes were obtained in winter and spring than autumn and summer seasons. Moreover, the percentage of oocytes suitable for maturation was significantly decreased ($P<0.05$) in hot seasons than cooled seasons. The oocytes meiotic chromosomes as represented by telophase I and metaphase II were significantly ($P<0.01$) decrease in summer and autumn than winter and spring. The highest incidence of diploid oocytes was recorded at spring and lowest incidence at autumn. It is concluded that the number and quality of cattle oocytes and also meiotic chromosomes were affected by the seasonal variations, they increased at cool seasons and decreased at hot seasons.

Poster G4.35

Genetic traceability of meat of Italian cattle breeds

L. Orrù, F. Napolitano, G. Catillo, M. Iacurto and B. Moioli, Animal Science Research Institute, via Salaria 31, 00016 Monterotondo, Italy

Consumers have become more and more demanding as regards to the origin of the beef they buy, asking for the cut of a specific animal of which they know the herd and the production system where it was fattened. In order to prevent frauds and to offer reliable and low-cost tools to ascertain the individual origin of meat, it is sufficient to collect, freeze and store a meat sample at the slaughtering, so to perform the genotyping of both the stored sample and the sample at the shop after request of the consumer. Among the various types of genetic markers, microsatellites are preferred because of their high polymorphic information content, ease of interpretation and the possibility to amplify more markers together in one multiplex PCR reation. Purpose of this work was to define the most efficient five-microsatellite multiplex in the common Italian breeds: Friesian, Chianina, Piedmontese and Italian Simmental. The multiplex includes: TGLA227; TGLA122; TGLA126; BM2113; SPS113, that were selected among 13 microsatellites, as they showed the highest polymorphic information content in the considered breeds. Eighty non-related animals for each breed were genotyped. The probability of declaring that two meat samples taken from two different animals come from the same animal, calculated through the maximum likelihood approach, ranged from a maximum of 7 out of 100 million (Simmental) to 3 out of 10 billion (Piedmontese). The multiplex was further tested on three groups of half-sibs, and gave maximum probabilities of 3 out of 100,000. Considering the above probabilities almost close to null, this multiplex is recommended for the genetic traceability of beef in Italy.

Poster G4.36

Impact of the *ESR* gene on litter size and litter weight in Czech Large White pigs

E. Goliásová and J. Wolf, Research Institute of Animal Production, P.O.Box 1, CZ 10401 Prague-Uhříneves, Czech Republic*

The effect of the *Pvu*II polymorphism of the oestrogen receptor gene on reproductive traits was evaluated in a Czech Large White population (1250 sows, 3600 litters). Data were analysed with a four-trait repeatability animal model. The traits were number of piglets born, number of piglets born alive, number of piglets weaned and litter weight at weaning. The animal model was modified by considering the herd-year-season effect as random or fixed and by including or not including the heterotic effect. The *ESR* gene showed a mostly significant effect ($P < 0.05$) on litter size traits in favour of allele *A*. For all three litter size traits, the difference between *AA* and *BB* sows was approximately 0.3 piglets ($P < 0.05$). No significant heterotic effect was found for litter size traits. For litter weight at weaning no significant additive effect was observed, but a significant negative heterotic effect (1.3 to 1.5 kg) was estimated.

Poster G4.37

Nonconstitutional karyotypical variability in ontogenesis at Black and White cattle

F.R. Bakai, Department of Genetics and Animal Breeding, Moscow State Academy of Veterinary Medicine and Biotechnology, 109472, Academician Scryabin st., 23, Russia

The parameters describing aneoploids' anomalies of karyotype during postnatal ontogenesis, show the tendency to reduction, but appreciably grow during achievement of sexual maturity, then again are reduced.

The similar increase to nine-monthly age can be connected with reorganization of hormonal status of animal, which can provoke occurrence of karyotypical anomalies and decrease of natural resistance. The share of polyploidical cells with age is increased, reaching a maximum by 12 months, however, by 18 months - is reduced.

Cells with the large number of aberrations with age become a little less. First of all, it is connected to increase, during of ontogenesis, system of reparations. The share of cells, in which the associations of chromosomes are found out, is increased with 52,5 up to 87,9 % (from 2 to 12 months) further their stability is kept.

The positive correlation with age of heifers such parameters as share of hyperploidical cells, level of chromosomes unsplitting and their associative ability is established. The share hyperploidical and aneoploidical cells, low level of cells' osmoresistance and number of aberrations has negative correlation with age.

Poster G4.38

Preliminary study on the polymorphism in beta-defensin 4 gene and its association with dairy production traits and somatic cell count in Black-and-White cows

Z. Ryniewicz, L. Zwierzchowski, E. Bagnicka, K. Flisikowski, A. Maj, J. Krzyzewski and N. Strzalkowska, Institute of Genetics and Animal Breeding, Polish Academy of Sciences, Jastrzebiec, 05-552 Wólka Kosowska, Poland

Because of the antimicrobial role of defensins, their genes might be considered as molecular markers of genetically determined resistance of the mammary gland to *mastitis*. A total of 1428 records were collected of daily milk, fat and protein yield, fat, protein and lactose content, and somatic cell count (SCC-log) of 88 cows. Basing on the sequence of the beta4-defensin the following primers were designed for the amplification of 924-bp: forward - 5' GAGGATGCGGAGACTGAGAC-3'; reverse - 5'-ACGGCACAAGAACGGAATAC-3'. PCR reactions were performed (DNA denaturation - 94°C; annealing - 63.5°C, elongation - 72°C; 34 cycles) and SSCP analysis was made. DNA samples representing different SSCP variants were sequenced. New nucleotide sequence polymorphism was found in the intron of bovine beta-defensin 4 gene - substitution G→A in position 2238. This mutation creates additional *Nla*III restriction site, which was found in 33% of investigated DNA samples. Clear associations between beta-defensin4 genotype and dairy traits were shown. The genotypes had significant effect on almost all production traits studied - except milk and protein yield, and protein content. This may lead to the use of defensins as genetic marker(s) in the cattle breeding programmes aiming at selecting highly productive dairy cattle with increased resistance to udder infections but further investigations are needed.

Poster G4.39

Effect of three polymorphisms in the estrogen receptor gene on litter size in Slovak Landrace pigs

R. Omelka[1], L. Hetényi[2], D. Peskovicová[2], M. Bauerová[1], M. Martiniaková[1] and M. Bauer[2], [1]Constantine the Philosopher University in Nitra, Trieda A. Hlinku 1, 949 01 Slovakia, [2]Research Institute of Animal Production Nitra, Hlohovská 2, 949 92 Slovakia*

The identification of genes or genetic markers associated with reproductive traits in pigs could have a great economic impact on pork production. We investigated the effect of Pvu II, Ava I and MspA1 I polymorphisms in the estrogen receptor gene on total number of born (TNB), number of born alive (NBA) and number of weaned (NW) piglets. We analyzed 162 pigs (535 litters) of Landrace breed from two Slovak breeding farms. The genetic polymorphisms were detected by PCR-RFLP method. Associations between the polymorphisms and reproductive traits were evaluated by linear model which included fixed and random effects. The frequencies of B (Pvu II), D (Ava I) and F (MspA1 I) alleles were 0.08, 0.12 and 0.07, respectively. We found out significant additive effects of B allele (Pvu II) of +0.62±0.18 (TNB, P≤0.01), +0.65±0.18 (NBA, P≤0.01) and +0.51±0.16 (NW, P≤0.05) piglets/litter. Significant differences (P≤0.05) between CD and CC genotypes (Ava I) were also identified in TNB (+0.41±0.16) and NBA (+0.37±0.16). MspA1 I polymorphism showed the smallest effect on the traits. A positive association of F allele with TNB, NBA and NW was found but the differences were not confirmed statistically.

Poster G4.40

Polymorphism PCR-RFLP in an intron of the bovine μ-calpain (CAPN 1) gene and its association with meat quality trait

Edyta Juszczuk-Kubiak[1], Stanislaw Józef Rosochacki[1,2], Krystyna Wicinska[1], [1]Molecular Cytogenetics Department, Institute of Genetics & Animal Breeding, Jastrzebiec, Poland, [2]Bialystok Technical University, Chair of Sanitary Biology and Biotechnology, 45e Wiejska str, 15-351 Bialystok, Poland

The calpain system orginally comprised molecules: two Ca^{2+}-dependent proteases, μ-calpain and m-calpain, and a third polypeptide, calpastatin, whose only known fuction is to inhibit the two calpains. This proteolytic system plays a key role in the tenderisation process that occurs during *post-mortem* storage of meat under refrigerated conditioning.
Their polymorphism is examined from the point of view of their effect on corresponding production traits. The calpain genes are investigated as a potential candidate gene for quantitative trait locus (QTL) affecting meat tenderness.
In this study the new nucleotide polymorphism was found within the intron 13 of the bovine CAPN1 gene-substitution C→T at position 4686 bp (consensus sequence - GenBank no. AF 248054). As this mutation creates new *Fok*I restriction site detected with PCR-RFLP analysis. The RFLP-*Fok*I polymorphism was studied in cattle (n=123) belonging to six breeds including one considered Polish native. The overall frequencies of alleles A and B were 0.30 and 0.70, respectively. We observed association between the calpain gene polymorphism and organoleptic valuation of meat, where BB genotype was associated with highest fat content, but AB genotype have best consistency of meat.

Poster G4.41

Single nucleotide polymorphism (SNP) in the 5' region of the bovine growth hormone receptor gene and its association with dairy production traits

N. Strzalkowska, A. Maj, J. Krzyzewski, J. Oprzadek and L. Zwierzchowski, Institute of Genetics and Animal Breeding, Polish Academy of Sciences, Jastrzeebiec, ul. Postepu 1, 05-552 Wólka Kosowska, Poland

The effects of cow's genotype for growth hormone receptor (*GHR*) *locus* were determined on milk production traits of the Polish Black-and White (BW) cattle. It was shown, that GHR genotypes significantly influenced most of the dairy traits studied. Cows of the RFLP-NsiI -/- and +/- genotypes of *GHR* produced more milk with higher content of most milk components, including fat, protein, and lactose then of those with +/+genotype. The RFLP *Fnu*4HI had little effect on the milk yield and composition traits. The RFLP-*Alu*I +/+ genotype appeared favourable for most of the traits. The heterozygous +/- genotype at RFLP-*Acc*I *locus* appeared superior with respect to several milk yield and composition parameters - fat, gross energy, % of total solids, fat, protein, and lactose. The combined GHR genotype (CGG) - *Alu*I, +/- *Acc*I, +/-; -*Fnu*4HI, +/+; *Nsi*I, /- was clearly superior for most traits under study. Cows carrying this genotype combination produced daily more milk, fat, protein, lactose more then other genotypes, and their milk contained significantly more total solids and fat. The contribution of the single and combined GHR genotypes to the pleiotropic variation in the milk composition yield and traits was significant and in some cases exceeded 10% of all variation sources tested.

Poster G4.42

LINE-1 element insertion in 5'-noncoding region of *bovidae* growth hormone receptor gene

A. Maj and L. Zwierzchowski, Institute of Genetics and Animal Breeding, Polish Academy of Sciences, Jastrzebiec, ul. Postepu 1, 05-552 Wólka Kosowska, Poland

Structure and nucleotide sequence was studied in the 5' noncoding region of the GHR gene of cattle (*Bos taurus* and *Bos indicus*), yak (*Bos mutus*), European bison (aurochs; *Bison bonasus*), sheep (*Ovis aries*), moufflon (*Ovis musimon*), goat (*Capra hircus*) and markhor (*Capra falconeri*). In all species studied a consensus sequence recognised by the endonuclease of LINE-1 element was found upstream to P1 promoter for exon 1A of the GHR gene. Insertion of 1,206 kb LINE-1 was found typical for taurine cattle, but five cows were identified - 3 Polish Red and 2 Bialogrzbietka - with the heterozygous deletion of the LINE-1 element from the GHR gene 5' noncoding region. A 317-bp LINE-1 element was detected in the goat and markhor GHR gene. No LINE-1 element insertion was detected in moufflon, sheep, yak and aurochs GHR gene. The structure and sequence of 5'GHR gene region was compared in two cattle subspecies - the European cattle (*B. taurus*) and of *B. indicus*, as well as of the another representative of the family *Bovidae* - European bison (*Bison bonasus*). Alignment of the *B. bonasus* GHR sequence to the sequences of the relevant fragment of the bovine gene showed 99.0% similarity.

Poster G4.43

Kappa-casein polymorphism in Russian Red-White dairy cattle

L.A. Kalashnikova, N.A. Yukhmanova and A.V. Barshinova, All-Russian Institute of Animal Breeding, Lesnye Polyany, Pushkin district, 141212 Moscow Region, Russia

The κ-Cn genotypes have been detected by PCR method in Red-White cows (n=437) breeding in regions with different climate (Central-Chernozem, Volga region and Eastern Siberia - temperate continental, continental and sharply-continental climate respectively). All animals were crosses between Simmental and Red-White Holsteins; the mean fraction of Holstein genes was 0.75. The highest frequency of geterozygous genotype was observed in Siberia.

Areas of breeding	Number of animals	Frequency of genotype,%			Frequency of alleles	
		AA	AB	BB	A	B
Central-Chernozem	162	48	36	16	0,66	0,34
Volga region	170	50	32	18	0,66	0,34
Eastern Siberia	105	42	55	3	0,70	0,30

The association between κ-Cn genetic variants and milk yield was not found. The B allele had significant effects on protein and casein percent. At the beginning of first lactation (2-3 months) estimates for κ-Cn BB in comparison with AA were +0.12% protein (p<0.05), +0.09% casein (p<0.05), for AB were +0.05% protein, +0.03% casein. In the middle of lactation (5-6 months) estimates for κ-Cn BB were +0.30% protein (p<0.01), +0.24% casein (p<0.01), for AB were +0.17% protein, +0.14% casein. Towards the end of lactation (8-9 months) estimates for κ-Cn BB were +0.32% protein (p<0.01), +0.25% casein (p<0.01), for AB were +0.19% protein, +0.15% casein.

Poster G4.44

No detectable association of the LEP Hinf I polymorphism with leptin mRNA levels in the White Improved Pig breed

M. Bauer[1], D. Vasícek[1], A. Bábelová[2], R. Omelka[2], M. Bauerová[2] and L. Hetényi[1], [1]Research Institute of Animal Production, Hlohovská 2, 949 92 Nitra, Slovakia, [2]Constantine the Philosopher University, Tr. A. Hlinku 1, 949 74 Nitra, Slovakia*

Much effort has focused recently on understanding the role of leptin in regulating feed intake, energy balance, reproduction, immune response and stress in domestic animals. Leptin is a small, 16 kDa protein, secreted predominantly by white adipocytes. It acts on the central nervous system to regulate body weight and fat deposition through the control of appetite and energy expenditure. Several genetic polymorphisms have been detected in pig leptin gene. Possible association between the polymorphism at 3469 (C→T) and backfat thickness in the Large White breed has been suggested. Although the mutation is not predicted to affect protein structure, effects on transcription and/or transcript stability cannot be ruled out.

The 96 animals of the White Improved pig breed were screened for Hinf I polymorphism (C→T transition at 3469) in the leptin gene by PCR-RFLP. Two groups of unrelated animals homozygous for C and T allele respectively were selected (8 CC and 21 TT) and RT-PCR was developed to determine leptin mRNA levels in white adipose tissue collected from the middle layer of subcutaneous neck fat. Statistical analysis was done using an association methods.We found no detectable association of the leptin mRNA levels with C/T polymorphism at 3469 of leptin gene (P>0.1).

Poster G4.45

Polymorphism in kappa-casein gene in Russian Black-White cattle

A.S. Tinaev, E.A. Denisenko and L.A. Kalashnikova, All-Russian Institute of Animal Breeding, Lesnye Polyany, Pushkin district, 141212 Moscow Region, Russia

DNA from 222 Black-White (BW) heifers has been analysed by PCR-RFLP method. All heifers were crossbreds of original BW cows and Holstein sires; the mean fraction of Holstein genes of heifers was 0.83. For the BW cattle in Moscow region (n=63) the genotype frequency for κ-Cn AA was 0.30, for κ-Cn AB was 0.59, for κ-Cn BB was 0.11, the allele frequency for κ-Cn A was 0.60, for κ-Cn B was 0.40. For the same breed in Eastern Siberia (n=159) the genotype frequencies were 0.40, 0.52 and 0.08 respectively, the frequency for A allele was 0.66, for B allele was 0.34.

The κ-Cn genotypes was associated with 305-day lactation milk yield, fat yield, protein percent and protein yield. In Moscow region analyses of the officially first lactation records showed cows carrying κ-Cn BB to produce +352 kg milk ($p<0.05$), +10 kg fat, +0.09% protein, +15 kg protein than those with κ-Cn AA. In Eastern Siberia cows with κ-Cn BB produced +678 kg milk ($p<0.05$), +18 kg fat, +0.15% protein, +30 kg protein. Test the following number lactation records in Eastern Siberia showed the tend to strengthening this association. The analyses of third lactation records demonstrated cows with κ-Cn BB to produce +915 kg milk ($p<0.001$), +57 kg fat, +0.37% protein ($p<0.001$), +57 kg protein ($p<0.05$) than cows carrying genotype AA. The cows with κ-Cn AB had intermediate values for milk yield, fat yield, protein percent and protein yield.

Poster G4.46

Detection of the ryanodine receptor mutation in pig breeds in Russia

N.V. Rydzova, L.A. Kalashnikova and I.J. Pavlova, All-Russian Institute of Animal Breeding, Lesnye Polyany, Pushkin district, 141212 Moscow Region, Russia

The mutation in the ryanodine receptor (RYR1) gene having predisposition to stress was typed by PCR on a total of 594 breeding pigs. The frequency of n allele for the Early Meat (EM) was 0,042 as a whole (n=83), 0,063 in boars (n=32), 0,023 in sows (n=22), 0,042 in suckling piglets (n=29). The frequency of n allele for the meat type of Large White (n=198, involving 25 boars, 149 sows and 25 suckling piglets) from central European part of Russia was 0,04. The n allele had the frequency 0,060 in boars, 0,045 in sows and 0,042 in piglets. For other unrelated population of meat type Large White (n=83) from East European part of Russia the frequency of n allele was 0,012. The frequency of n allele was zero for LW pigs from Siberia (n=85) and for Cyvilskaya pigs (n=50). The frequency of n allele in Landrace from East Siberia (n=20) and from Central European part of Russia (n=75) was determined to be 0,075 and 0,025, respectively. Homozygous nn- animals were not found in populations whatever. There are no significant negative effects of RYR1 Nn- genotype on productive traits (total number born in litter, number born alive, test average daily gain, individual birth weight, individual weight at weaning) for boars and sows of meat type Large White.

Poster G4.47

Genetic variability of falcon microsatellite DNA markers

L. Putnová[1], J. Pokorádi[2], K. Civánová[1], T. Urban[1], I. Krízanová[1], A. Kúbek[2] and J. Dvorák[1], [1]Mendel University of Agriculture and Forestry Brno, Department of Genetics, Zemedelská 1, 613 00 Brno, Czech Republic, [2]Slovak Agriculture University, Department of Genetics and Breeding Biology, Trieda A. Hlinku 2, 949 76 Nitra, Slovakia*

The variability based on microsatellite DNA markers is using for the identification of species, populations, individuals and parent-offspring relationships. All species of Falco family (Falconidae sp.) are protected by CITES and EU laws. Consequently we want to contribute to the legal breeding of endangered species, wildlife protection, crime suppression and also to determination of populations in Slovakia and the Czech Republic. We studied the genetic variability of five falcon microsatellite markers (*NVHfp13, NVHfp31, NVHfp79-4, NVHfp92-1* and *NVHfp89*) analysed by multiplex PCR reaction and fragment analysis. The investigated population consisted of *Falco biarmicus, Falco cherrurg, Falco peregrinus, Falco tinnunculus* and their interspecific hybrids: *Falco cherrug*x*Falco rusticolus, Falco peregrinus*x*Falco cherrug, Falco rusticolus*x*Falco cherrug* and *Falco rusticolus*x*Falco peregrinus*. All microsatellites show polymorphic variability across the study samples. A total number of 44 alleles was obtained. The number of alleles at individual loci ranged from 4 (*NVHfp13*) to 16 (*NVHfp79-4*). The highest heterozygosity and polymorphism information content (over 70%) was observed for locus *NVHfp79-4, NVHfp31* and *NVHfp92-1*. The probabilities of paternity exclusion/one parental genotype unavailable/and parentage exclusion were for this panel 98.20%/90.92%/99.91%, respectively. Supported by Ministry of Education, Youth and Sports of the Czech Republic (Project No. MSM 432100001).

Poster G4.48

Prediction of breeding values of Czech Pied young bulls from pedigree information

D. Rehák[*], M. Stípková, J. Volek and L. Barton, Research Institute of Animal Production, Prátelství 815, 104 00 Prague, Czech Republic

The objective of this study was to predict breeding values (BV) of Czech Pied young bulls from pedigree information. Recent BV and pedigree information of 218 bulls entering artificial insemination from 1996 to 1997 were included into the analysis. Only 5.6 % of the overall variability ($R^2 = 0.056$) were explained by regression of BV of evaluated bulls on BV of sires at the beginning of evaluation ($b_s = 0.421$). Inclusion of BV of maternal grand sire and BV of sire at the beginning of evaluation resulted in 1.9 % increase in the accuracy of prediction of BV determined by coefficient of determination ($R^2 = 0.075$) and a higher estimate of the regression coefficient on BV of sire ($b_s = 0.439$). The estimated regression coefficient of BV of maternal grand sire was $b_m = 0.273$. The results revealed by regression analyses are also confirmed by Spearman rank-order correlation coefficients. A significant rank-order correlation ($P<0.001$) was determined between BV of evaluated bulls and BV of their sires at time of their sons' selection ($r = 0.22$). It is concluded that the above mentioned results indicate a low reliability of BV prediction in young bulls based on BV of sires at the beginning of evaluation. It is only slightly improved by use of information on BV of maternal grand sires. The study was supported by the project NAZV QD0176.

Poster G4.49

Study of melanocortin receptor 1 (*MC1R*) gene polymorphisms in some Italian dairy cattle breeds and their possible use for the traceability of milk and milk products

V. Russo, L. Fontanesi, E. Scotti, M. Tazzoli, S. Dall'Olio and R. Davoli, DIPROVAL, Sezione di Allevamenti Zootecnici, University of Bologna, Via F.lli Rosselli 107, 42100 Reggio Emilia, Italy*

Coat colour is a distinctive trait of the breeds. Several genes that affect coat colour have been isolated in cattle. The melanocortin receptor 1 (*MC1R*) gene has a major role in the regulation of black/brown versus red/yellow pigment synthesis and four main alleles at this locus have been identified in bovine: E^+, E^D, *E1* and *e*. With the aim to identify DNA markers useful to differentiate and trace the milk products obtained from different breeds we studied the frequency of these alleles in 934 animals belonging to five Italian dairy cattle breeds (Italian Holstein-Friesian, Italian Brown, Italian Simmental, Reggiana and Modenese) with diverse coat colour. E^D was identified only in Italian Holstein-Friesian with frequency of 0.915, *E1* was present only in Italian Brown (0.392), E^+ was detected in Italian Brown, Modenese and Italian Simmental (0.587, 0.958 and 0.046 respectively), *e* resulted fixed in Reggiana breed and was identified also in Italian Simmental (0.958) and at low frequency in Italian Holstein-Friesian, Italian Brown and Modenese. These data may be used to distinguish or to exclude, at least in some cases, the dairy products obtained from some of the analysed breeds.

Poster G4.50

Sexing of roe deer (Capreolus capreolus L.) by PCR amplification reaction

J. Kobolák[1], I. Majzinger[2], M. Szabari[3], E. Gócza[1], Sz. Bodó[1], Gy. Bicsérdy[2], G. Palotás[2] and Zs. Bösze[1], [1]Department of Animal Biology, Agricultural Biotechnology Center, PO.Box 411, H-2100 Gödöllö, Hungary, [2]University of Szeged, College of Agriculture Hódmezövásárhely Andrássy str. 15, Hungary, [3]University of Kaposvár Faculty of Animal Science H-7400 Kaposvár, Guba Sándor str. 40, Hungary*

Polymerase Chain Reaction enables us to study the genes of a given species even from a small amount of tissue sample. In a given animal, a PCR based method can help in identification of the gender or a trait of economic importance.

Sex-specific sequences of related species published earlier were analysed in the hope that, due to interspecies relations, similar sequences could be obtained in Roe deer (Capreolus capreolus L.). Sequences of several primers applied in bovine were analysed, but the primers used in Sika deer by Takashaki et al. (J.Vet.Med.Sci 60(6):713-716, 1998) were found to be the most suitable. Using those primers under modified reaction conditions, roes could be sexed successfully. The amplified Y-specific sequence which size was different from that of the Sika deer fragment is supposedly the Sry gene. The PCR product was cloned and the sequence is being analysed. Hopefully, when the sequence analysis of the Sry gene will be completed, Roe deer specific primers could be planned for use in one-cell PCR for embryo sexing.

Poster G4.51

Promoter variants of the bovine *STAT5A* affect the expression level in liver tissue.

K. Flisikowski, R. R. Starzynski, A. Maj and L. Zwierzchowski, Institute of Genetics and Animal Breeding, Jastrzebiec, 05-552 Wólka Kosowska, Poland

Signal transducers and activators of transcription (STATs) are a family of transcription factors. STAT5A, previously known as mammary gland factor (MGF), transduces prolactin signals to the milk protein genes. Here we describe the detection of nucleotide sequence polymorphism in the promoter region of the bovine STAT5A gene. Using PCR-HD and sequencing technique A/G substitution at position -488 of the promoter region of STAT5A gene was found. In a group of 162 young Fresian bulls thee genotypes were identified. TESS analysis revealed that the polymorphic sequence is located in a putative binding site for transcription factor HNF-3 (hepatocyte nuclear factor-3) and may thus influence gene expression. To evaluate that, total RNA was isolated from liver tissues of cattle with different STAT5A genotypes and used for RT-PCR and Real-Time PCR analyses. Both methods showed significantly higher STAT5A mRNA levels in AA then GG genotypes. Additionally, Western-Blotting analysis of the nuclear extracts revealed higher STAT5A protein level in AA as compared with GG genotypes. EMSA showed that A→G mutation at position -488 increased the STAT5A gene promoter binding capacity to liver nuclear protein (possibly HNF-3). Thus, our analyses showed a possible effect of the A/G substitution at putative HNF-3 transcription factor site on the expression of STAT5A protein in bovine liver tissue and a possible role of this region on the expression of STAT5A protein.

Poster G4.52

Characterisation of polymorphisms in the porcine MC3R gene

K. Civánová, Mendel University of Agriculture and Forestry Brno, Department of Genetics, Zemedelská 1, 613 00 Brno, Czech republic

The central melanocortin system is critical for the long-term regulation of energy homeostasis. Important member of this gene family is MC1R, having an important role in pigmentation regulation. MC4R and MC5R, expressed in brain areas, play a key role in complex control of appetite and body weight. Disfunction or overexpression cause obesity, disturbances of food intake and energy imbalance in pigs. However, little is known about the function, localisation and structure of another porcine central melanocortin receptor, the MC3R.

To obtain a specific PCR fragments of MC3R gene, several primers were designed using comparison of a human MC3R cds sequence and partial porcine sequences of MC3R gene. Identity of PCR fragments was confirmed by sequencing. The 311 bp exone fragment was used for PCR-RFLP analysis. In this fragment two biallelic polymorphisms were found (*MnlI, DdeI*). Codominant Mendelian inheritance and allele frequencies were confirmed by analysing 78 animals of five pig breeds (Landrase, Large White, Duroc, Hampshire, Pietrain). Results show that *MnlI* polymorphic alelle B was represented by the 149 bp fragment, while in alelle A this restriction site absents (179 bp). Polymorphic restriction site *DdeI* (alelle B-197 bp and 114 bp) was present only in Pietrain breed (P). *DdeI* alelle A, with polymorphic restriction site absent, was epitomized by the 311 bp uncut amplimer.

Work supported by IGA MZLU in Brno (2481/2003)

Poster G4.53

ECOWEIGHT - a C program for modelling the economic efficiency of cattle production systems

J. Wolf[1], M. Wolfová[1], E. Krupa[2] and D. Peskovicová[2], [1]Research Institute of Animal Production, Department of Genetics and Biometrics, P.O.Box 1, CZ 10400 Praha Uhlínves, Czech Republic, [2]Research Institute of Animal Production, Hlohovská 2, SK 94992 Nitra, Slovak Republic*

A program in C was written on the basis of a bio-economic model for a wide range of cattle production systems. The model simulates the life-cycle production of a cow herd and the growth performance of offspring born in the herd. An integrated cow-calf and feedlot system was assumed. The Markov chain approach was used to simulate herd dynamics. The herd was described in terms of states animals can be in and the probabilities of transitions between these states. The program calculates the structure of the integrated production system in its stationary state, the economic efficiency of the system expressed as a function of biological traits of animals and of management and economic parameters, the number of discounted expressions for direct and maternal traits transmitted by breeding animals and the economic weights for 15 (beef cattle) or 18 (dairy cattle) economically important traits. The program runs under LINUX and Windows and possibly under further operational systems and is freely available on request with a detailed manual in pdf format.

Poster G4.54

The genetic structure for the polymorphous proteins of the different Large White pigs populations

E.D. Ambrosieva, All-Russian Institute of Animal Breeding, Lesnye Polyany, Pushkin district, 141212 Moscow Region, Russia

The genetic structure for five polymorphous protein locus in seven population of Large White pigs (n=3768) been bred in 6 Russian geographical regions (Yaroslavskaya, Lipezkaya, Kaluzchskaya, Volgogradskaya provinces, Republics Chuvaschiya and Mariy El) was studied.
The findings allow to describe the averaged model of given breed's genetic structure: Pi-1 F - 0,600; Tf A - 0,270; Po-2 F - 0.400; Ptf A - 0,250; Hpx 2 - 0,010; Hpx 3 - 0,100. At the same time some populations have typical differences from the averaged model for either loci: Yaroslavskaya Pi-1 F - 0,730; Republic Chuvaschiya Pi-1 F - 0.480. The populations did not differed for the polyallelic locus Hpx. The new type of Large White breed "Svobodovskiy" had been reared in Republic Chuvaschiya differs from the averaged model in four from five researched loci (n=2039): Pi-1 F - 0,482; Tf A - 0,427; Po-2 f F - 0,521; Ptf A - 0,365. The analysis of genetic structures variability for some temporal period reveals that the genetical structure for loci Pi-1 and Hpx did not change during late two decades of years. But Tf A allele's frequency tends to decrease about by 25 percents for locus Tf in pig's population reared in Russia.

Poster G4.55

Hereditability of plasma vitamin C level in pigs

E.V. Kamaldinov, Research Institute of Veterinary Genetics and Selection of Novosibirsk State Agrarian University

AA plasma concentration in all animals was determined right after blood taking. The plasma vitamin C content was normally distributed with slight positive asymmetry.

In the investigated population the coefficient of repeatability was 0.424 ($P<0.01$). The differences in plasma ascorbic acid level in LW and ERM piglets are revealed ($P<0.05$). The great phenotypical variety of the AA plasma level of some West Siberia pig breeds was identified 2 weeks after weaning.

We found that the level of vitamin C variability in the blood plasma of the LW boars' descendants was lower than that in the descendants of ERM boars. The extreme values of the AA level were nearly 2 times different. It means that the ERM sires have a big phenotypical variability of the studied quantitative trait in plasma. It may be explained by the fact that the ERM breed was registered as late as 1993. Hence, the ERM breed is insufficiently consolidated because a certain time is needed to achieve this goal. We suppose that the variability mentioned above will decrease. Thus, it has been established that the sow and sire genotype influences the level of vitamin C biosynthesis in LW and ERM pigs. The high phenotypical plasma AA content variability was found in the sires' and sows' descendants of the ERM animals in comparison with the LW animals.

Poster G4.56

Estimation of genetic (co)variance components for weaning weight and preweaning daily gain of Hungarian Charolais population

Z. Lengyel[1], Z. Domokos[2], I. Komlósi[3] and F. Szabó[1], [1]University of Veszprém, Georgikon Faculty of Agricultural Science, H-8360 Keszthely, Deák F. str. 16. [2]Association of Hungarian Charolais Breeders, H-3525 Miskolc, Vologda str. 3. [3]Debrecen University, Centre for Agricultural Sciences, H-4032 Debrecen, Böszörményi str.138.

Weaning performance of 10808 purebred offspring (4944 male and 5817 female) of 75 sire were analised with animal model in 27 farms. Heritability, breeding value, (co)variance components of weaning weight (WW), preweaning daily gain (PDG), 205-day weight (CWW) were calculated. Farm, number of calving, year of birth, season of birth, sex, were treated as fixed, and the maternal permanent environment was treated as random effect. In case of WW and PDG, the age of the calves at weaning was fitted as a covariant. Data were analyzed with MTDFREML (1993) pogram. The overall mean value and standard deviation of WW, PDG and CWW were 221±47 kg, 1.111±0.21 kg/day and 226±42 kg, respectively. The age of the calves at weaning was 202 days. The direct heritability of WW and PDG was 0.57±0.08, 0.49±0.07 and CWW was 0.44±0.06, respectively. The maternal heritability of these traits was 0.32±0.09, 0.33±0.10 and 0.33±0.09, respectively. The direct-maternal correltaions were strong and negative.

Poster G4.57

Dry-cured ham production: correlation between breed and quality of raw meat

G.L. Restelli[1] and G. Pagnacco[1], [1]Dipartimento di Scienze e Tecnologie Veterinarie per la Sicurezza Alimentare (VSA), Via Grasselli 7, 20137 Milano, Italy*

Two hundred forty pigs of four different genotypes, Duroc × (Landrace × Large White), Duroc × Large White, Duroc × Landrace and Landrace × Large White were studied to investigate the effect of breed on raw meat quality.
In the slaughterhouse the data regarding carcass weight, trimmed weight of green hams, pH at 45' and 24h post mortem were recorded. Quality of green hams were evaluated using a linear score for fatness, meat colour, meat tenderness, absence of defects. Ham weight losses were recorded at different stages of processing.
The multivariate analysis shows a positive relationship between carcass weight and trimmed weight of green hams, also the weight losses at different stages of processing are positive related. The principal component analysis shows clear distinctions between genetics types. Green hams coming from Duroc × Landrace cross have the best score for quality traits and shows the lower weight losses during ham processing. Duroc × (Landrace × Large White) and Duroc × Large White crosses are quite similar for quality and weight losses. Landrace × Large White cross have the lower score for quality traits.

Poster G4.58

Variability in candidate genes for meat performance traits in pigs

R. Mikolasova, T. Urban and K. Civanova, Mendel University of Agriculture and Forestry Brno, Department of Genetics, Zemedelska 1, 613 00 Brno, Czech Republic*

The genes for growth hormone *GH*, heart fatty acid-binding protein *H-FABP*, leptin *LEP*, leptin receptor *LEPR* and *MYC* protooncogene are potential quantitative trait loci for growth, body composition, meat quality and feed conversion ratio in pigs. We analyzed a set of 181 pigs of Large white (LW) and 120 Landrace (LA) breeds. Genotyping was performed using PCR-RFLP methods according to information from the literature. Calculated population parameters were: frequencies of genotypes and alleles, Hardy-Weinberg equilibrium, observed and expected heterozygosity; further Nei's genetic identity and genetic distance through program Popgene version 1.31 (1999) were established. Both of the breeds were in genetic equilibrium in the tested loci, except the breed LA in *LEP* and *GH1* loci. LW breed was more heterozygous than Landrace in accordance with the mean of the observation heterozygosity. The value of the genetic identity of breeds in the testing was approximately 97%.
This study is supported by researche order MSM 432100001

Poster G4.59

Response to selection of weaning weight based on maternal effect in Syrian hamsters: Results from fifteen generations

K. Ishii[*1], M. Satoh[2], O. Sasaki[1], H. Takeda[1] and T. Furukawa[1] and Y. Nagamine[1], [1]National Institute of Livestock and Grassland Science, Tsukuba-shi 305-0901, [2]National Institute of Agrobiological Sciences, Tsukuba-shi 305-8602, Japan*

Fifteen generations of selection were conducted to study the response for weaning weight (WW) of standardized litter size in Syrian hamsters. The experiment involved four lines: selection on an estimate of direct genetic effect of WW (line A); selection on an estimate of maternal genetic effect of WW (line M); selection on estimate of an aggregate genetic value of direct and maternal effects of WW (line B); a randomly selected control (line C). Direct and maternal effects were estimated from an animal model BLUP. Significant differences between C and each selected line for WW were found at after generation 5; linear regressions of WW in selected lines as a deviation from C on generation number were all significantly positive. Direct and maternal heritability estimates for WW were 0.15 and 0.18, respectively, and correlation between direct and maternal genetic effects was 0.09. Mean estimates of direct and maternal genetic effects for WW increased linearly with generation in each selected line: estimated direct effects in generation 15 were 8.9 g, 9.7 g, and 9.2 g for A, M, and B, respectively; estimated maternal effects were 4.4 g, 7.1 g, and 5.1 g for A, M, and B, respectively. Average inbreeding coefficients in generation 15 were approximately 31 % in all selected lines. Selection using maternal effect had beneficial to genetic improvement for WW.

Poster G4.60

The evaluation of the effect of the season, different weather and parity on the milk, fat percentage and calving interval of the Iranian buffaloes

H.R.Nader Fard, shahin ataee sheiq, Ardabil province

There are about 500000 heads of river buffaloes in the I.R. of IRAN which during centuries the y have maintained a fine natural adaptation with the annual climatic cycles of temperature, humidity, rainfall, sunshine, etc, in the provinces of: KHUZESTAN (warm and humid located in the south east of Iran),West Azarbaijan, East Azarbaijan and ARDABIL (cold located in the north west of Iran),GILAN and MAZANDARAN (rainy and humid located in the south of Caspian Sea north of Iran). Effect of climate seems to be significant on the natural productive and reproductive performance of the buffaloes. By producing milk and different dairy products buffaloes have a very important economic role in the rural areas of the above mentioned provinces. Lactations from 1992 up to now were used to evaluate the effect of season on the milk, fat and calving interval. the number of records were 64000 with 4687 animals in seven provinces. the different fixed effects included in the model were: Four level of season (spring, summer, autumn and winter) Seven levels of provinces .1308 herds nested in the provinces.12 levels of year.3 levels of milking times per day10 levels of parity. the model was: Yijklmne = mean + provi + herd (provi) + parity + times milked + year + season + error. Analysis was done by the system of SAS. The procedure of GLM. The results showed that the total milk yield, peak yield, fat percentage and calving interval were significantly ($p<0.01$)influenced by year, season .province and parity.

Poster G4.61

Genetic parameters of production traits in automatic and conventional milking systems

H. Täubert[1], W. Brade[2] and H. Simianer[1], [1]Institute of Animal Breeding and Genetics, University of Göttingen, Albrecht-Thaer-Weg 3, D-37075 Göttingen, [2]Landwirtschaftskammer Hannover, Johannsenstr. 10, D-30159 Hannover*

Milk recordings of 1297 animals from 23 farms with automatic (AMS) and 20555 animals from 616 farms with conventional milking systems (CMS) were analysed. The data was collected within the years 2001 - 2003. Recordings were available for milk-kg, fat and protein content from the first three test days of each cow after calving. Genetic parameters between AMS and CMS recordings for each single test day and with a fixed regression test day model were calculated.
Heritabilities of the three test days were in an expected range from .182 and .418. Genetic correlations were high with values near 1 for milk-kg and fat-%. Values for protein-% ranged between .760 and .881.
Heritabilities for the fixed regression test day model with all three test days as repeated measures were for milk-kg .242 (AMS) and .201 (CMS), for fat-% .316 (AMS) and .254 (CMS) and for protein-% .202 (AMS) and .337 (CMS). Genetic correlations between AMS and CMS were 1.00 (milk-kg), .999 (fat-%) and .994 (protein-%).
It can be concluded, that under this model no genetic difference can be found between AMS and CMS. Therefore genotype by environment interaction between automatic and conventional milk recording systems is irrelevant for milk yield and fat percentage and small for protein percentage.

Poster G4.62

Genetic and phenotypic parameters of milk yield and lactation curve parameters estimated by the Gamma function in Egyptian buffaloes

M.A. Aziz[1], O.M. El-Shafie[2] and A. Nishida[1], [1]Graduate School of Agricultural Science, Faculty of Agriculture, Tohoku University, Sendai 981-8555, Japan, [2]Animal Research Centre, Ministry of Agriculture, Egypt*

Data on 2617 records of weekly milk yield representing the first ten lactations of 1080 Egyptian buffaloes were collected from a herd of the Animal Research Centre, Egypt. A multi-trait animal model was used to estimate heritability, genetic and phenotypic correlations and the permanent environmental variance of milk yield and lactation curve parameters calculated by the Gamma function. Heritability estimate of total milk yield (TMY) was 0.31. Heritability estimates of the initial milk yield (log a) and the rate of increase to peak production (b) of the linear form (0.25 & 0.33) were higher than those of the non-linear form (0.20 & 0.28), while those of the rate of decrease after peak (c) were equal (0.54). The highest genetic correlation was between b and c (0.97), and it was equal in both forms, while the lowest was between TMY and b (-0.84 and -0.82, respectively). The phenotypic correlations ranged between -0.88 and -0.84 (b and c) and 0.05 and 0.14 (TMY and a) for the two forms, respectively. The permanent environmental variance components of all traits in both forms of the function were approximately equal and the highest and the lowest values were observed for TMY and a (0.12 and 0.001), respectively.

Poster G4.63

Factors affecting length of productive life in Swedish dairy cattle

M. del P. Schneider[1], E. Strandberg[1] and A. Roth[2], [1]Department of Animal Breeding and Genetics, Swedish University of Agricultural Sciences, P.O. Box 7023, SE-75007 Uppsala, Sweden, [2]Swedish Dairy Association, Box 1146, SE-63180 Eskilstuna, Sweden

The length of productive life of Swedish Dairy Cattle has been analyzed with Survival Analysis. After editing, the data set had information on 1.019.562 cows calving from 1988 to 1996 from the Swedish Official Milk Recording System and AI scheme. Four breeds: Swedish Red and White (SRB), Swedish Friesian (SLB), Swedish Polled Breed, Jersey, and crossbreds (SRB x SLB), were included. Length of productive life was defined as days from first calving to culling (uncensored) or end of data collection (censored). A preliminary analysis included the effects: stage of lactation by parity number, peak yield deviation within herd-year, age at first calving, year, season, region, breed, and the random effect of herd. The proportion of censored records was 36%. All the effects were significant at $P < 0.001$. The effects with the greatest contribution to the likelihood function were stage of lactation by parity and peak yield deviation. The risk of culling increased within lactation and with parity number. Low producing cows were ten times more likely to be culled than average producing cows.

Poster G4.64

Genetic Evaluation of Fertility Using Direct and Correlated Traits in Italian Holstein - Pilot study

S. Biffani1, F. Canavesi and A.B. Samorè, Associazione Nazionale Allevatori Frisona Italiana (ANAFI), Via Bergamo 292, 26100 Cremona, Italy*

Genetic parameters and breeding values for dairy cow fertility were estimated from 7780 cows. Multiple-trait analysis of calving interval (CI), days to first service (DTFS), Body Condition Score (BCS), Angularity (ANG) and milk yield (ME305) were investigated as a method to estimate fertility breeding values. For fertility traits, estimates of heritability was 0.013 to 0.076 and genetic correlations were moderate (.32). Genetic correlations of fertility with ME305 were unfavourable (range .34 to .61). Heritability estimates for BCS and ANG were moderate (0.22 and 0.24) and their genetic correlation was high (.74). Between BCS and CI, the genetic correlation was -0.29, whereas it was -0.35 between BCS and DTFS. The genetic correlation between ANG and CI was 0.31, and between ANG and DTFS was .27. Genetic correlations of BCS and ANG with ME305 were -.45 and .58, respectively. Single and multiple trait BLUP Animal model were used to predict sire breeding values for DTFS. There were differences between EBVs and rankings for DTFS from single versus multi-trait analyses. The range for rank correlation was 0.90-0.98 for sires with more than 25 daughters. Milk yield as a correlated trait seems to reduce the effects of culling / selective treatment bias. Using or not BCS when ANG observations are available is not very clear and future studies are needed.

Poster G4.65

Estimation of genetic parameters for milk production of Czech Spotted cattle

E. Nemcová, L.Dedková and J. Wolf, Research Institute of Animal Production, P.O.Box 1, CZ 10401 Prague-Uhríneves, Czech Republic

Genetic parameters for milk production traits of Czech Spotted cattle were estimated with the aim of their future use in genetic evaluation. About 150,000 test-day records from approximately 10,000 animals were taken from the official milk recording database, the year of test being in the range between 1995 and 2002. Model equations included the fixed effect of herd-test-day and fixed regression on the stage of lactation (within subgroups of cows) and the permanent environmental and animal effects modeled by random regression. Third order Legendre polynomials (with four coefficients) were used for all regressions. The models employed differed in the definition of the subgroups for fixed regression on stage of lactation, or (alternatively) in definition of the residual effect. Gibbs sampling and REML were used for the parameter estimation. The heritabilities were found in range from 0.18 to 0.25 in the first, from 0.19 to 0.27 in the second and from 0.16 to 0.38 in the third lactation. The genetic correlations between lactations varied from 0.76 to 0.82 between the first and second; from 0.63 to 0.68 between the first and third; and from 0.71 to 0.82 between the second and third lactations.

Poster G4.66

Genetic diversity of mitochondrial DNA in Zemaitukai Horse

R. Juras[1,3], E.G. Cothran[2] and V. Macijauskiene[1] [1]Lithuanian Institute of Animal Science, Zebenkos g. 12, LT-5125 Baisogala, Lithuania, [2]University of Kentucky, Department of Veterinary Science, Lexington, 40506-0099, KY, US, [3]Siauliai University, Visinskio 25, LT-5400 Siauliai, Lithuania

Program for the preservation of the genetic resources of different farm animals was developed in Lithuania. In this study, genetic variation in Zemaitukai horses was investigated using mitochondrial DNA (mtDNA) sequencing. The study was performed on 421 bp of the mitochondrial DNA control region, which is known to be more variable than other sequencies. Samples from each remaining mare family lines of Zemaitukai horses and three random samples for other Lithuanian (Lithuanian Heavy Drought, Zemaitkai large type) and ten European horse breeds were selected. Five distinct haplotypes were obtained for five Zemaitukai mares' families supporting the pedigree data. A total of 20 nucleotide differences compared to the reference sequence were found in Lithuanian horse breed. All detected mutations were transitions. Total of 37 haplotypes were found out of 13 breeds investigated. Phylogenetic analysis of mtDNA haplotypes was performed in order to derive genetic relationships between Lithuanian and other horse breeds.

Poster G4.67

Variability and heritability of some chosen milk performance traits in cows

B. Sitkowska and S. Mroczkowski, University of Technology and Agriculture, Genetics and Animal Breeding, Zootechnical Department, Mazowiecka 28, 85-084 Bydgoszcz, Poland

The research was carried out basing on breeding documentation which covered 1676 cows of the leading milk cattle breeding centre. There were estimated the coefficients of variability and of heritability of daily milk performance, chemical composition and the level of somatic cells in milk.

The greatest variability in the population researched was observed for milk performance, which is linked to various lactation stages. The heritability of milk traits was evaluated with the REML method from the paternal component. The values of the heritability coefficient obtained ranged from 0.136 to 0.372, for the level of somatic cells heritability was 0.209. The content of the basic milk components is more genetically conditioned than milk performance, which suggests that the selection towards a daily content of milk components can be more effective than towards milk performance.

Poster G4.68

Note about increasing the economical efficiency of the Romanian Hucul breed

M. Maftei, D. Sandulescu and R. Popa, University of Agricultural Science and Veterinary Medicine of Bucharest, Faculty of Animal Science

By reason of hipometric body sizes we can say that the Hucul horse is on the last place regarding the Romanian horse market request. The Romanian Hucul breed has a lot of characteristics whence recommend it as if the best horse for working in the mountain area. We remind here: great resistance at morbidity, great stamina, cheap maintenance, great reproductive coefficients and a good traction power related with his body weight and body sizes. Unfortunately, the romanian horse owner from the mountain area is interest now by the heavy horses who have a expensiv maintenance, are impressibles and diseaseds, and who are not adapted at the enviroment condition from the mountain area. Till 1992, the romanian army was a great customer of the Hucul horse and along with the cessation of the buying activity was decreasing the number of the brood mares & sires from the Lucina stoodfarm. That makes from romanian Hucul horse a breed in danger concerning to economical efficiency. Anyway, because of the new activities who are occurred in activity with horses, we find a few ways to help the Romanian Hucul horse concerning increasing the economical efficiency of this breed.

Poster G4.69

Quality of heritability estimates as affected by heritability level, progeny number per sire, type of algorithm, type of model and type of trait

*R**. Elsaid[1], M. Elsayed[2], S. Galal[2] and H. Mansour[2], [1]Animal Production, Desert Environment Research Institute, Menoufia University, Sadat City, Egypt, [2]Animal Production, Faculty of Agriculture, Ain Shams University, Cairo, Egypt*

The study objectives were to investigate the effect of heritability level, progeny number per sire, type of algorithm, type of model and type of trait on the quality of heritability estimates. Two simulation programs were used, one to simulate a continuous trait and another to modify it into a binary trait. Twelve populations were created (three heritability levels * four levels of progeny number per sire), each with three parities and twenty replicates. Each replicate was analyzed by sire and animal models, using two algorithms (MTDFREML or Gibbs Sampling, GS). Bias and mean squared errors (MSE) of heritability estimates were used to assess the quality of heritability estimates. The effect of heritability level, progeny number per sire, type of algorithm, type of model, type of trait and the interactions on bias and MSE were examined. For estimating variance components, for continuous trait, the animal model was the best in the case of using MTDFREML and GS. For binary trait, within GS, the sire model was the best at heritability equals to 0.1 with progeny number more than 5 whereas, at heritability equals to 0.25 or 0.5 with 20 progeny, the use of animal and sire models were equivalent.

Poster G4.70

Differences in gene expression in myostatin-mutant Belgian White-Blue satellite cell cultures during differentiation

J. Kobolák[1], B. Görhöny[1], I. Király[2], K. Kiss[2], K. Bölcskey[3], E. Gócza[1] and Zs. Bösze[1], [1]Institute of Animal Sciences, Agricultural Biotechnology Center, H-2100 Gödöllö, Hungary, [2]Petöfi Mgtsz H-9512 Ostffyasszonyfa, Hungary, [3]Research Institute for Animal Breeding and Nutrition, H-2053 Herceghalom, Hungary*

The myostatin protein is a regulator factor in normal muscle that determines the maximum amount of muscle mass. If the myostatin gene is mutant, its negative regulating function does not work resulting in muscle hypertrophy and hyperplasia. In the view of quality meat production, this is an outstanding trait. This phenomenon occurs in Belgian White-Blue breed, where 'double-muscled' phenotype is common due to successful selection. Crossing with Belgian White-Blue shows that, although, the gene is recessive and monofactorial, its effect is apparent even in heterozygotes due to its partial dominance: the meat: bone ratio and meat yield is better than in the other breed.

Muscle stem cell (satellite cell) cultures were isolated from biopsy of hind limb muscle (m.gluteus medius) of one-year-old Belgian White-Blue bulls. For controls, muscle biopsies and cell cultures obtained from Holstein-Friesian bulls of the same age were used. Expression of 17 genes playing role in muscle differentiation was examined. In mutant cells, a significant increase was observed in the level of MRFs, Igf-1, Cdk2, and a decrease was observed in p21. The difference detected in early stages of differentiation supposes that myostatin mutation may affect muscle differentiation causing hypertrophy.

Poster G4.71

Comparison of direct and maternal genetic effect on growth and carcass trait between two breeds of Japanese beef cattle

T. Yamazaki[1], T. Sasaki[1], K. Suzuki[1], S. Nishida[2], S. Takahashi[3], T. Kondo[3], S. Komatsu[4], K. Suzuki[4], J. Yasuda[4] and A. Nishida[1], [1]Graduate School of Agric. Sci. Tohoku University, 981-8555, Sendai, Japan, [2]Miyagi Prefectural Livestock Experiment Section, 989-6445, Miyagi, Japan, [3]National Agricultural Research Center for Tohoku Region, 020-0198, Morioka, Iwate, Japan, [4]Iwate Animal Industry Research Institute, 020-8570, Iwate, Japan

Japanese Shorthorn has shown high growth rate and nursing ability, and has been kept on pasture in summer. Japanese Black has shown high marbling score, and has been kept in cattle shed all the year around. On the breeds, genetic parameters of direct and maternal effect were estimated for carcass and growth traits.
The heritability estimate of marbling score (BMS) was remarkably lower (0.05) in Japanese Shorthorn than that in Japanese Black (0.36). The estimate of maternal heritability for daily gain until calves were shipped to a calf market (DG1) in Japanese Shorthorn was high (0.50), while very low value (0.04) was estimated in Japanese Black.
The estimate of genetic correlation between the direct effects on BMS and DG1 was negative (-0.34) in Japanese Shorthorn, but in Japanese Black, positive value (0.54) was estimated. In both breeds, direct genetic effect on BMS and maternal genetic effect on DG1 were closely and negatively correlated (-0.90 and -0.53, respectively). Using these results, genetic improvement in consideration of the genetic characteristic of each breed is possible.

Poster G4.72

A comparison of selection procedures with constraints based on restricted best linear unbiased prediction of breeding values

M. Satoh and M. Takeya, National Institute of Agrobiological Sciences, Tsukuba-shi 305-8602, Japan*

Two restricted selection schemes were compared under various combinations of genetic parameters using Monte Carlo computer simulation. The selection schemes are based on restricted best linear unbiased prediction (RBLUP) of breeding values for only selected animals (RBLUP-S) or all animals (RBLUP-A) in a population. Three closed breeding herds of 50, 100, and 400 females per generation and mating ratio of 1:10 were considered. From each litter, about one male and two females were reared to weaning for breeding stock. Two-trait selection was assumed, in which animals were selected to maximize the genetic gain in trait 1 under no change in trait 2. The heritabilities of the traits were all combinations of 0.1, 0.4, and 0.7, and the five different genetic correlations (-0.6, -0.2, 0.0, 0.2, or 0.6) were examined for each set of heritabilities. All animals in each generation were mated at random. Selection was carried out during 10 generations without overlapping. When genetic correlation was close to zero, the average RBLUP of true breeding values of animals in the last generation using RBLUP-S was similar to that using RBLUP-A. However, selection on RBLUP-S resulted in higher response than that on RBLUP-A under high genetic correlation between the traits. The population size and heritability of the traits did not affect the difference of genetic gain between selection schemes.

Poster G4.73

Used of sexed semen in Italian Holstein Friesian cows: a field trial

D. Marcomin, P. Povinelli, M. Cassandro, L. Gallo and G. Bittante, University of Padova, Department of Animal Science, Viale dell'Università, 16, Agripolis, 35020, Legnaro, Padova, Italy*

Aim of this study was to examine the performance of sexed semen in Holstein Friesian cattle in field condition. A total of 1,282 insemination (98% of heifers) were recorded in 92 Italian commercial dairy herds, from September 2002 to December 2003. Four selected Holstein Friesian dairy bulls with sexed semen, produced by flow citometry method, were used. Results are currently recording and they will be discussed. However, a preliminary subset of the first 164 artificial inseminations, using ultrasound, showed a pregnancy rate of 50% with around 90% of females embryos. However, a large variability was showed on these results due to herd, region, sire and their interactions. An economic simulation study, was also conducted in order to verify the most important factors to modify the economical benefits on use of sexed semen. The two most important factors were: price of crossing calves and percentage of sexed semen purity. In conclusion a field application of sexed semen respect to the use of conventional sperm seems to be useful but more data are needed before to clear its economical and technical impacts on animal production sector.

Poster G4.74

Effect of AI-sires on connectedness and accuracy of breeding values in Norwegian beef cattle breeding: a simulation study

S.U. Narvestad[1], E. Sehested[1,2] and G. Klemetsdal[1], [1]Department of Animal and Aquacultural Sciences, Agricultural University of Norway, P.O. Box 5003, N-1432 Aas, Norway, [2]Geno Breeding and AI Association, 2326 Hamar, Norway*

The objective of this study was to investigate the effect of artificial insemination (AI) on connectedness and accuracy of predicted breeding values using stochastic simulation. Data from the Norwegian Hereford population was analyzed to find population structure parameters like reproductive life for males and females, number and size of herds, exchange of animals between herds, use of AI sires etc. Beed cattle production in Norway is characterized by several breeds in few and small herds. AI-sires continue as natural mating-sires after a short semen collection period. Use of AI was varied from zero to all matings in the herds. Number of herds (125), herd size (average 11) and population size (1370 calves born each year) were kept constant throughout the simulation. To evaluate the effect of accuracy only, no selection was performed at any stage. No import was considered. The model used was a single-trait animal model for yearling weight with sex as fixed effect and herd-year as random effect. Different alternatives were run for 20 years and with 20 replicates. Correlation between true genetic values and predicted breeding values were obtained for each situation.

Poster G4.75

Effects of outliers on parameter estimates and robust statistical modelling in animal breeding

M.Z. Firat, Akdeniz University, Faculty of Agriculture, Department of Animal Science, 07059 Antalya, Turkey

Statistical inference based on the normal distribution is known to be vulnerable to outliers that may have a large impact on inferences about parameters of interest in a statistical model. Outlying milk yield records may arise due to preferential treatment of some cows, which is a likely practice in dairy cattle breeding. Some of the preferential treatments include separate housing, hormonal treatment, feeding according to production and longer milking intervals on test day. These outlying observations may then be influential on the estimates of genetic parameters. The adoption of robust estimation procedures is required to analyze data having outliers in animal breeding. This can be achieved by assigning to the random terms of the model, the residuals in particular, probability distributions that have thicker tails than the normal distribution, such as the *t*-distribution. In this paper, robust statistical modelling based on the scale mixture distributions was developed to analyse data having outlying observations in mixed linear models in animal breeding and the effect of such observations on the estimates of parameters was investigated.

Poster G4.76

Validation of region effects in the random regression model for Simmental cattle in Germany and Austria

R. Emmerling and K.-U. Götz, Bavarian State Research Centre for Agriculture, Institute of Animal Breeding, Prof.-Dürrwaechter-Platz 1, 85586 Poing-Grub, Germany*

In November 2002 the joint random regression test day model for Germany and Austria was introduced for milk production traits in Simmental. These countries comprise very different regions due to differences in climate and topography as well as in herd management. Therefore, separate effects for lactation stage, pregnancy, calving-year-season and production-year-month have been defined within regions in the fixed model. The herd specific environment is accounted for by a herd test day effect. Also heterogeneous variances are considered within region-production-year-month-lactation subclasses and herd test days (Lidauer et al., 2002). To validate the effect of the region in comparison to the effect of the herd itself, four herds were treated as if they were displaced to regions with different average production intensities in Bavaria, a state in Southern Germany. A separate evaluation for protein yield was performed and compared to the routine evaluation with the herds in their original region. Results show only minor changes in average EBVs of animals in displaced herds. The averaged changes were between -0.16 to +0.11 kg in EBV for protein yield. The observed differences have shown that the defined regions have only small effect on EBVs. The influence of the herd specific environment covered by the herd test day effect is much more important in comparison to the average effect of the defined regions. Regions are nevertheless useful for small herds and the estimation of lactation curves.

Poster G4.77

Advantages and disadvantages of delayed first breeding of dairy goats

F. Panzitta[1], A. Stella[1] and P. Boettcher[2], [1]Parco Tecnologico Padano and [2]IBBA-CNR, Via Fratelli Cervi 93, 20090 Segrate, Italy*

Most goats mature quickly enough to support reproduction at one year of age. However, milk production increases as they mature physically. For that reason, some goat breeders in Italy choose to delay the time of first kidding until two years of age. This practice is expected to yield greater production during the first lactation, but effects on lifetime production and profit are not known. The objective of this study was to compare performance of goats bred to kid at one year of age versus those for which the first breeding was delayed by one year. The data used were records of milk production for 3881 goats calving in Italy during 2001 and 2002. For each parity, average production was compared for goats with standard (1-year) and delayed times of first kidding. As expected, goats kidding first at two years of age produced more milk in their first lactation than did those kidding at one year of age. However, the difference was less than 10 kg per lactation and not statistically significant. In addition, this advantage disappeared in later lactations, as the late-bred goats tended to produce less. Moreover, goats with delayed first-breeding had much lower rates of survival to subsequent lactations. Considering these factors, delayed first-breeding is expected to be associated with decreased lifetime production and profit.

Poster G4.78

Estimation of genetic and phenotypic parameters of some production traits in Japanese quail

Ensaf El Full and G. Shaaban, Department of poultry, Faculty of Agriculture, Cairo University, Fayoum branch, Egypt*

Genetic parameters represent an important step in a breeding program design. The purpose of this paper is to estimate inheritance, heritability with different approaches and genetic and phenotypic correlations of live body weight and shank length at different ages, growth rate and egg characteristics to obtain additional information concerning productive traits of quail to be used in different selection programs. Heritability of body weight at 28 days of age based on paternal-half sib correlations which considered a better estimator of heritability in the narrow sense was higher magnitude than heritability of body weight at 14 and 42 days of age. It can be concluded that body weight at 28 days of age can be used as a selection criteria to improve body weight in quail. Live body weight at 28 days of age is the most appropriate criterion for selection. This age not only one of the peaks at which the variability rises to a maximum, but it is also has the advantage of a relative earliness. In addition to the significantly high correlations found between live body weight and most of the studied traits.

Poster G4.79

Effects of sex and different slaughter ages on live body weight and carcass traits of Japanese quail

Ensaf El Full and G. Shaaban, Department of poultry, Faculty of Agriculture, Cairo University, Fayoum branch, Egypt*

In Egypt production of animal protein has been recently initiated through commercial rearing of Japanese quail. The aim of this study was to investigate the effects of sex and different slaughter ages on live body weight and carcass traits to derive equations from which weight and part yield can be estimated without slaughtering the bird. Slaughter test was performed using males and females which were randomly chosen at 35, 49, 63, 77, 91 and 105 days of age. Live body weight, carcass traits and body measurements were recorded. The results showed that sex and slaughter age significantly affected all studied traits. Sex had significant effect by first order polynomial equations on most of studied traits. First and second order polynomial coefficients were found to be significant for age on most of studied traits. Our data indicated that slaughtering at earlier ages better than other ages.

Poster G4.80

The aggregate genotype based on the additive and epistatic effects

C.I. Draganescu, Research Institute for Cattle Production Balotesti, Romania

In a panmictic population, the contribution of the parent to the progeny genetic value is a linear combination of its breeding value $a=a^{(1)}$ and epistatic effects, i.e. $a^{(2)}=aa$, $a^{(3)}=aaa$, ..., $a^{(Q)}=aaa...a$. The transmitting ability TA=a/2 may be extended as Total Transmitting Ability $TTA=\Sigma_q(1/2)^q \cdot a^{(q)}$, and as the Total Transmitting Ability in the kth generation $TTA^{(k)}=\Sigma_q(1/2)^{kq} \cdot a^{(q)}$ (k=1,2,...). On this basis, it is possible to define $g^{(k)}=TTA^{(k)}/(1/2)^k$ as the aggregate genotype of the additive and epistatic effects based on the size of genetic effects transmitted toward the kth generation. By consequence, $g^{(\infty)}=a$.

Using the interest rate C and generation length L, an aggregate genotype based on the genetic effects transmitted toward all generations was created as a sum of all ratios $TTA^{(k)}/(1+C)^{kL}$, divided by the coefficient of additive value in this sum. This aggregate genotype is a function $G(C,L)=\Sigma c_q(C,L) \cdot a^{(q)}$ with $c_1(C,L)=1$.

It results that $G(\infty,L)=2 \cdot TTA^{(1)}=g^{(1)}$. This result suggests that selection for the first generation puts an exaggerate emphasis on economic. Because $g^{(\infty)}=a \notin \{G(C,L)|C \geq 0\}$, when epistatic effects are evaluated also, the selection on the breeding value cannot be optimal from economic point of view.

The established aggregate genotypes may be extended to an arbitrary number of traits.

Poster G4.81

Relations between rearing parameters and polymorphic form of chosen hormones of calves Charolaise and Limousine breeds

A. Nowopolska-Szczyglewska and A. Dobicki, Institute of Animal Breeding, Agricultural University of Wroclaw, Chelmonskiego str. 38 C, 51 630 Wroclaw, Poland

In order to determine the relations between rearing parameters and meat performance compared to polymorphic forms of the cattle growth hormone (bGH), *Pit1*, leptin (Lep.) and myostatin (Mio.) DNA was isolated from the blood of 212 Limousine (LMS) and 200 Charolaise (CHL) head of cattle. The PCR products were separated electrophoretically in agar gel. Relations between the selected physiological indicators and muscle deposition were determined on the basis of the level of urea (ur), phosphatase (AP), alanine aminotransferase (AlAT), asparagine aminotransferase (AspAT), creatinine (Cr), cholesterol (Chol) and glucose (Glu).
A determination was made of the polymorphic genotypes of *Pit1*, MSTN, LEP and bGH in CHL and LMS cattle breeds. Calves with genotype MSTN-AB showed a statistically higher body weight at the age of 210 days (compared to genotype MSTN-AA), while animals with genotype bGH-LV - a higher live body weight and daily weight gains (compared to genotype bGH-LL). Polymorphic genotype LEP-AB proved to be correlated with a higher live body weight at birth and higher daily weight gains (by 100 g/day when compared to genotype LEP-AA). No statistically significant differences were observed between the physiological indicators analysed and muscle deposition in the animals studied.

Poster G4.82

Analysis of the variability at mitochondrial DNA in Spanish cattle breeds

S. Pedrosa, J.J. Arranz, Y. Bayón and F. San Primitivo, Departamento de Producción Animal, Facultad de Veterinaria, Universidad de León, Spain*

The common origin for all cattle was estimated about 10000 years BP and it was one of the most significant achievements of Neolithic peoples in different regions of Eurasia. From this zone, several strains emerged to Asia, Africa and Europe by subsequent movements of people in their migrations during the Pleistocene. The origin of present-day Iberian cattle is controversial, but the analysis of morphological evidences indicates that all entrances of cattle were from the *Bos Taurus* lineage. A more recent and sporadic introduction of cattle from North Africa may have occurred during the Moorish occupation. All these historical facts highlight the importance of the geographical position of the Iberian Peninsula, which has played a key role in human movements between Europe and Africa. In this study we use 637 bp of the mitochondrial displacement loop sequences from 10 Spanish autochthonous cattle breeds belonging to three groups defined by their geographical position, related origin, and morphological characteristics: the *Iberian*, the *Blond-Brown Cantábrico* and the *Turdetano*. Total DNA was extracted from blood samples and was amplified through PCR and subsequently cycle-sequenced using an ABI PRISM377 DNA sequencer. A neighbour-joining tree constructed from pairwise distances provides evidence of high variability and narrow relationships between breeds of three geographical groups. We analyze the introgression of African mtDNA haplotypes in the genetic pool of the Spanish breeds.

Poster G4.83

Immunogenetic characteristic Ig G of Kemerovskaya pig breed

A.I. Zheltikov, V.V. Gart, V.L. Petukhov and O.S. Korotkevich, Research Institute of Veterinary Genetics and Selection of Novosibirsk State Agrarian University

Kemerovskaya pig breed is one of the two produced in Siberia. Kemerovskaya animals are characterized by high productivity and adaptivity to extreme conditions of environment. Investigations were carried out in Kemerovskaya pigs raised on the breeding farm "Yurginsky" Kemerovskaya region. All animals were titrated with the bank of reagents for pig serum protein allotypes, the bank was setup in Research Institute of Cytology and Genetics (Siberian Division of Russian Academy of Sciences). We studied two allotypes of class G immunoglobulins. Two allotypes belonging to different subclasses Ig G (Ig G1a, Ig G 2b) were revealed with antiserums available. Ig G^{1a+} allele frequency was equal to 67.2 % and Ig G^{2b+} was 89.6 %. Immunoglobulin molecules are responsible for mammals humoral immunity, and their intra-species diversity has been found for immune genetic types (allotypes). Individuals with 1a+/1a+ genotype, heterozygous and 1a-/1a homozygous were 45.2, 44.1 and 10.7 %, respectively. Analyses of another Ig G2b allotype showed that in this case the frequency of 2b+/2b+ homozygous animals reaches 80.3 % that is 4.3 times higher than that of heterozygous individuals.

Poster G4.84

A comparison of different selection indices for genetic improvement of some milk traits by using two sets of relative economic values in Friesian cattle in Egypt

*M.N. El -Arian[1], F.H. Farrag[1], E.A. Omer[2], A.M. Hussien[2] and A.S. Khattab[*3]. [1]University of Mansoura, Faculty of Agriculture, Egypt, [2]Animal Production Research Institute, Ministry of Agriculture, Egypt and [3] University of Tanta, Tanta, Faculty of Agriculture, Egypt*

Milk records on 2181 Friesian cows kept at Sakha, Experimental Station, belonging to the Animal Production Research Institute, Ministry of Agriculture, Egypt, between 1996 and 2002 were used. Traits studied were 305 day milk yield (MY), 305 day fat yield (FY), 305 day protein yield (PY), age at first calving (AFC) and calving interval (CI). Variance components, heritability (h^2), repeatability (t) and genetic (r_g), permanent (r_{pe}), environmental (r_e) and phenotypic (r_p) correlations for different traits were estimated using MTDFREML procedure of multi trait analysis. In addition, twenty six of selection indices were constructed using all combination among five traits studied, for two sets of relative economic values (actual economic values (set 1) and one standard deviation (set 2)).

Estimates of h^2 for MY, FY, PY, AFC and CI were, 0.29, 0.24, 0.29, 0.49 and 0.12,respectively. Estimates of t were 0.37, 0.44, 0.51 and 0.12 for MY, FY, PY and CI, respectively. The second index (I_2) , incorporating MY, FY, PY and AFC was the best for two sets of relative economic values (R_{IH} = 0.85 and 0.75), for set 1 and set 2, respectively.

Poster G4.85

Estimation of genetic parameters and breeding values for milk yield and milk composition in Friesian cattle in Egypt through MTDFREML program

A.S. Khattab[*1], M.N. El-Arian[2], F.H .Farrag[2], E.A. Omer[3] and A.M. Hussein[3], [1]University of Tanta, Faculty of Agriculture, Tanta, Egypt, [2]University of Mansoura, Faculty of Agriculture, Egypt, [3]Animal Production Research Institute, Ministry of Agriculture, Dokki, Cairo, Egypt*

A total of 632 normal first lactation records of Friesian cattle Kept at Sakha Experimental Station, belonging to Animal Production Research Institute, Ministry of Agriculture, Egypt, during 1996 to 2002 were used to estimate the parameters and breeding values through sire, cow and dam for 305 day milk yield (MY), 305 day fat yield (FY) and 305 day protein yield (PY). Data were analyzed by (MTDFREML) computer program. The model included the fixed effects of month and year of calving and age at first calving as a covariate and the random effects of individual, permanent environmental and errors.
Estimates of heritability were 0.36, 0.28 and 0.37 for MY, FY and PY, respectively. Genetic correlations among all traits were equal unity. Predicted breeding values of sire ranged from -464 to 888 kg for MY, from - 19.72 to 37.78 kg for FY and from - 16.81 to 32.41 kg for PY. Also, predicted breeding values of cows were between -882 to 1104 kg , -37.65 and 47.66 kg , and - 25.10 and 31.92 kg for the same traits, respectively. Similarly, predicted breeding values of dams, being from - 551 to 628 kg, -26.18 to 26.23 kg and - 22.75 to 23.35 kg for the above mentioned traits, respectively. Product moment correlations between all traits were positive and highly significant, (0.94 to 0.99).

Poster G4.86

The genetic potential of pigs raised in Lithuania

R.Klimas, A. Klimiene, S. Rimkevicius, Siauliai University, P. Visinskio 25, 76285 Siauliai, Lithuania.*

Pig breeding in Lithuania is a traditional branch of animal husbandry. The major pig breed in our country is Lithuanian White. Of late years, pigs of Yorkshire (Swedish, Danish, Norwegian, Dutch selection), Large White (German, English, Dutch selection), Landrace (German, Danish, Finnish, Swedish, Norwegian, Dutch selection), Pietrain, Hampshire and Duroc breeds have been imported for crossbreeding. Also, Lithuanian native pigs are raised and preserved as a gene pool. In Lithuania, pig breeding is based on the pyramidal principle. The breeding progeny is raised and distributed by 49 pig breeding centres. About 9-10% of all pigs bred is in them. Pig selection carried out in pig breeding centres is directed towards higher litter size and milk yield and improvement of fattening and carcass traits. In 2003, the average litter size of the main sows for all breeds was 10.6 piglets. The milk yield (litter weight at 21 days of age) of the main sows of various breeds ranged from 34.7 to 70.1 kg ($P<0.001$). According to the control fattening data by January 1, 2003, all purebred pigs and their crossbreds reached 100 kg weight in 180 days with the daily gain of 800 g and compound feed consumption of 3.00 kg per kg gain. Their average muscularity reached 54.6 % (*Piglog 105*). The backfat thickness behind the last rib was 20 mm in average for all slaughtered pigs (n=1932).

Poster G4.87

Influence of some genetic factors on the birth weight of piglets

A. Klimiene, Siauliai University, P. Visinskio 25, 76285 Siauliai, Lithuania.

We investigated weight of piglets of purebreds Lithuanian White (n=347), Yorkshire (n=383), Landrace (n=147), Hampshire (n=20), Duroc (n=26) and crossbreds of these breeds (n=557) in time of birth. The study was carried out in 4 pig breeding centres.
Referring to the investigation data, the average birth weight of piglets of Lithuanian White breed in separate breeding centres was 1.27 - 1.30 kg, when of imported breeds it was 1.39 -1.48 kg (P<0.001). The biggest were piglets of Duroc breed (1.48 kg). When coupling Lithuanian Whites with the boars of Yorkshire, Landrace, Hampshire and Duroc breeds, derived crossbreds of the first generation, according to this indicator, occupied intermediate place (1.35-1.40 kg). Comparing with purebred Lithuanian Whites, difference is statistically reliable (P<0.05-0.001). It was determined that newborn purebred and crossbred boars were more bigger than gilts. Variation coefficient (Cv) of this parameter was ranged from 12.54 to 23.44 for the groups of different breeds. Higher variability of birth weight of piglets in the litter manifested in the sows distinguished for higher prolificacy. Besides that, litter size of the sows in most cases had direct influence on development of newborn piglets: the higher litter size, the lower weight of piglets. Therefore the birth weight of piglets is depending on some genetic factors: breed, sex, choosing of parents, individual inherited features and litter size of sows.

Poster G4.90

A genetic study of the lactation curve parameters in Iranian Holstein primiparous

H. Farhangfar, Animal Science Department, Agriculture Faculty, Birjand University, Birjand, P.O. Box 97175-331, Iran

Genetic aspects of the lactation curve of Iranian Holstein heifers were studied with the use of Wood's incomplete gamma function ($y_t=at^b(exp)^{-ct}$). Data comprised 179610 monthly test day milk records from 17961 Iranian Holstein dairy heifers calving between 1986 and 2001 and distributed in 287 herds of Iran. Wood's incomplete gamma function parameters (a,b and c) were initially estimated for individual cows using **SAS** programme and subsequently along with some production characteristics including peak time (b/c), peak yield ($a(b/c)^b(exp)^{-b}$), persistency (-(b+1)Log_ec) and 305-day lactation milk yield were subjected to a multivariate **REML** animal model to estimate genetic parameters of seven traits using **MTC** programme. Based on the results, it was revealed that heritability estimates for the parameters of Wood's incomplete gamma function (0.1, 0.03 and 0.05 for a, b and c respectively), peak time (0.1) and persistency (0.08) were low indicating that these traits were substantially influenced by the environmental factors and as a result additive genetic variation between animals for these traits was not significant suggesting that direct selection could not meaningfully change the shape of the lactation curve. In this study, the heritability estimates of peak yield (0.28) and 305-day milk yield (0.29) were found to be approximately the same.

Identification of a gene causing complex vertebral malformation (CVM) in cattle

C. Bendixen[1], P. Horn[1], F. Panitz[1], E. Bendixen[1], A.H. Petersen[1], L-E. Holm[1], V.H. Nielsen[1], J.S. Agerholm[2] and B. Thomsen[1], [1]Danish Institute of Agricultural Sciences, Department of Animal Breeding and Genetics, Research Center Foulum, DK-8830 Tjele, Denmark. Royal Veterinary and Agricultural University, Frederiksberg, Copenhagen, Denmark

In cattle breeding, the extensive use of relatively few sires can generate very large families counting hundreds or even thousands of descendants. This inbreeding presents a major concern since recessive disease-causing alleles are rapidly transmitted to a large number of offspring. A new disease called Complex Vertebral Malformation (CVM) was in 2000 observed with high frequency in the Holstein-Friesian breed. Pedigree analysis revealed that all diagnosed cases of CVM were genetically related to the US bull Carlin-M Ivanhoe Bell, which has been used extensively world wide, CVM thus presents a significant problem in Holstein-Friesian breeds all over the world. A genome-scan of families of affected animals using a panel of microsatellite markers was used to map the disease to a narrow region on bovine chromosome 3 (BTA3). A bovine BAC contig covering the region was established and candidate genes predicted from homologies to the human genome were evaluated by sequencing both healthy and diseased animals. The region covers approximately 3 million base pairs and contains about 25 known and predicted genes. Sequencing analysis revealed a mutation in a candidate gene encoding a nucleotide-sugar transporter. Functional analysis confirmed that the mutation disrupts the molecular function of the protein, demonstrating that the detected mutation is the direct cause of CVM.

Theatre G5.2

Implementation of the French bovine genetic disease programme

A. Ducos[1], A. Eggen[2] and D. Boichard[3], [1]ENVT-INRA, UMR cytogénétique, 23, chemin des Capelles, 31076 Toulouse cedex 3, France, [2]INRA, LGBC, 78352 Jouy-en-Josas cedex, France, [3]INRA Département Génétique animale, 78352 Jouy-en-Josas cedex, France*

During the last 15 years, bovine selection organizations have been confronted with the appearance and the massive diffusion of genetic mutations responsible for various defects, especially in the Holstein-Friesian breed (BLAD, bulldog, CVM). In France, the so-called "bovine genetic disease observatory" programme (BGDO) aiming at allowing an early detection of such defects has been built up two years ago. This programme, coordinated by the animal genetics department of INRA, takes all representative professional organizations involved in cattle production and breeding (including veterinarians) as partners. A new form allowing an accurate description of most congenital defects has been created and widely distributed. The corresponding information is centralized and analysed at the INRA level. Thorough clinical analyses, as well as collection of biological samples obtained from interesting families are also organized within the BGDO. More than 400 individual descriptions have been received during the last two calving campaigns. The incidence of two different defects was rather high : *atresia coli*, a long known defect in Holstein-Friesian cattle (about 25% of the declarations in this breed), and a new anomaly ("generalized hypoplasia", also called "sheep head" syndrome) in the Montbéliarde breed. For the latter, DNA samples for more than 50 calf-dam pairs (+sires) have already been collected for further investigations.

Theatre G5.3

Genetic relationships between speed of greying, melanoma and vitiligo prevalence in grey horses

J. Sölkner[1], M. Seltenhammer[2], I. Curik[3] and G. Niebauer[2], [1]BOKU-University of Natural Resources and Applied Life Sciences, Gregor-Mendel-Str. 33, A-1180 Vienna, Austria, [2]University of Veterinary Medicine Vienna, Clinic for Surgery and Ophtalmology, Veterinärplatz 1, A-1210 Vienna, Austria, [3]University of Zagreb, Faculty of Agriculture, Animal Science Department, Svetosimunska 25, 10000 Zagreb, Croatia*

The quantitative inheritance of grey level, vitiligo and melanoma prevalence was analyzed on 725 Lipizzan horses born in five state studs (Austria, Croatia, Hungary, Slovakia and Slovenia). Melanoma status (levels 0-5) was recorded for 5 years, greying level (L* parameter of the CIE L*a*b system) for 4 years and vitiligo status (levels 0-3) for 2 years. Heritabilities and genetic correlations were estimated by multivariate animal model REML. As differences in speed of greying are expressed at young ages whereas melanoma and vitiligo tend to develop at later ages, data for level of greying (573 obs.) were restricted to ages of 1-6 years, whereas data on melanoma (700 obs.) and vitiligo (379 obs.) were only included for horses older than 6 years. Estimated heritabilities were 0.76±0.07 for level of greying, 0.32±0.10 for melanoma status and 0.33±0.07 for vitiligo status. The genetic correlation between speed of greying and vitiligo was high (0.66±0.10), whereas the others were very close to zero. A separate run including grey levels of old horses (> 6 years, 646 obs.) indicated a positive genetic correlation between grey level of adult horses and melanoma (0.42±0.12).

Theatre G5.4

Fine mapping of the syndactyly locus in the Holstein population

A. Duchesne[1], C. Grohs[1], A. Ducos[2] and A. Eggen[1], [1]INRA, LGBC, 78352 Jouy-en-Josas cedex, France, [2]ENVT-INRA, UMR cytogénétique, 23, chemin des Capelles, 31076 Toulouse cedex 3, France

Bovine syndactyly, also referred to as "mulefoot", is a hereditary condition that has been shown to segregate as an autosomal recessive trait with incomplete penetrance. The main feature is the fusion of functional digits exhibiting variable expressivity and a right-left / front-rear gradient. Syndactyly has been described in a number of cattle breeds, but occurs with the highest frequency in the Holstein population. Previously mapped to bovine chromosome 15 by identity-by-descent mapping, we constructed a BAC contig spanning a large portion of the region of interest in order to start positional cloning of the underlying gene(s) and we identified several new genetic markers. The genotyping of over 250 animals allowed us to determine the disease associated haplotype and therefore to validate a marker-based genetic test for this disorder. Several "Mulefoot" affected embryos are currently being produced from an affected cow and homozygous or heterozygous sires and collected at different gestation stages. These biological samples will allow us to undertake sequence comparison and expression studies for several candidate genes in the chromosomal region. Identification of the causal mutation will facilitate eradication of this disorder in the Holstein population and help to verify if, in other breeds, the same mutation is responsible for this phenotype.

Gene expression profiles in inbred and non-inbred *Drosophila melanogaster*

T.N. Kristensen[1,2], P. Sørensen[2(*)], M. Kruhøffer[3], K.S. Pedersen[1], and V. Loeschcke[1], [1]Aarhus Centre for Environmental Stress Research (ACES), Department of Ecology and Genetics, University of Aarhus, Building 540, Ny Munkegade, DK-8000 Aarhus C, Denmark, [2]Department of Animal Breeding and Genetics, Danish Institute of Agricultural Sciences, P.O. Box 50, DK-8830 Tjele, Denmark, [3]Aarhus University Hospital, Molecular Diagnostic Laboratory, Brendstrupgaardsvej DK-8200 Aarhus N, Denmark

The negative consequences of inbreeding on fitness related characters, termed inbreeding depression, have been known for a very long time. Empirical and theoretical investigations have led to an understanding of the phenomenon in great detail. In animal breeding the use of reproductive technologies and statistical tools have made directional selection very efficient. However, the use of these technologies have also contributed to the observed decrease in effective population sizes leading to inbreeding within some breeds. In spite of the tremendous number of papers published on inbreeding little is known about how inbreeding affects gene expression and physiology of populations. We compared gene expression profiles of non-inbred control populations of *D. melanogaster* and populations being inbred at different rates. Affymetrix *Drosophila* gene chips were used to reveal changes in gene expression that are associated with inbreeding. Data will be presented and its implications discussed.

Poster G5.6

Genetics of displaced abomasum in German Holstein cows

H. Hamann, M. Ricken and O. Distl, School of Veterinary Medicine Hannover, Institute of Animal Breeding and Genetics, Bünteweg 17p, 30559 Hannover, Germany*

The objectives of this study were to analyse the importance of genetic factors responsible for abomasal displacement (DA) in German Holstein cows. The data set included 4090 lactations of 3706 cows from 50 farms. The cows calved between July 2001 and January 2003. The lactational incidence of DA was 3.59%. Left DA was much more prevalent than right DA (74 vs. 26%). A threshold model yielded heritability estimates of $h^2 = 0.20$ for all cases of DA and $h^2 = 0.12$ for left DA. In order to estimate correlations of DA with milk production traits, we used milk recordings of the lactation prior to DA. Genetic correlations between all cases of DA and milk yield traits were close to zero. However, there were highly positive genetic correlations between left DA and milk yield traits of $r_g = 0.68$ (milk yield), $r_g = 0.60$ (fat yield) and $r_g = 0.65$ (protein yield). The genetic correlations of left DA were negative with fat and protein content and their ratio, with $r_g = -0.77$, $r_g = -0.64$ and $r_g = -0.59$, respectively. Our results indicated that cows sensitive to metabolic stress measured via milk protein and fat are genetically predisposed to left DA. Prediction of breeding values using fat and protein content proved useful to reduce the prevalence of left DA.

Poster G5.7

Prevalence and preliminary analysis of severe udder edema occurrence in Italian Friesian cows

A. Comin, F. Cesarini, L. Gallo, P. Carnier and M. Cassandro, University of Padova, Department of Animal Science, Agripolis, 35020 Legnaro (PD), Italy*

Udder edema is caused by an excessive accumulation of lymphatic fluid in the interstitial spaces of udder, leading to alterations of appearance, consistency and flexibility and increasing the risk of physical damage. The present study aimed to investigate the prevalence of severe udder edema (SUE) in Italian Friesian cows and to analyse the relevance of some possible risk factors of SUE occurrence. Data were from a program currently running in cooperation with some breeder associations and aimed to control dairy cows for functional and health traits. A total of 8,734 udder edema observations was evaluated after calving on 4,666 cows in 46 herds of northern Italy as alternatively severe (all quarters and navel and extending to the upper half of udder over the midline) or not severe. Overall prevalence of SUE approached 9,1% and appeared higher in primiparous than in multiparous cows (11,4 vs 8,5%, respectively). Risk of occurrence of SUE was investigated using logistic regression analysis and considering the role of herd, calving year and season, parity, cow genetic merit and the concurrent occurrence of retained placenta. To be in the first parity and to have a high genetic merit for milk yield appeared associated with an increased risk of SUE. Further research is needed for better understanding the genetic aspects of such disorder.

Poster G5.8

Chromosome abnormalities and infertility in cattle

I. Nicolae and D.-M. Hateg, Research and Development Institute for Bovine, Balotesti, Sos.Buc.-Ploiesti, km.21, 077015, Romania

As it became known, the aberrations such as chromosomal breaks and small deletions could have an important ethiological role on infertility. The presence of strructural chromosome abnormalities is very important because the carriers have a normal phenotype but most of them have a reduced fertility due to the production of abnormal gametes, always associated with embryonic losses. In this connection, our cytogenetic analysis was carried out in a group of eight Romanian Black Spotted cows because of their reproductive problems, expressed by repeated inseminations. One of them showed, not only the instability of heterosomes determined by the monocromatidic Xq breaks, but also, a frequent presence of a middle shape autosome with unequal chromatids . It could be a deletion or a duplication of a chromatidic segment, but this will be elucidate by further investigations. Any way, the heterosomes instability associated with a structural chromosomal abnormality have a common effect of the reproductive performances. For this reason it is stressed to keep breeding animals under cytogenetic control, especially in animal populations applying artificial inseminations.

Theatre G6.1

Incorporating molecular markers into genetic evaluation

R.L. Fernando, Iowa State University, Department of Animal Science, 225 Kildee Hall, Ames, Iowa 50011, USA

Molecular markers can be broadly classified into two types: I) those that have a direct effect on a trait, and II) those that do not have a direct effect on any trait but are linked to a trait locus. A marker of type I can be incorporated into genetic evaluation by including it as a fixed effect in the model used for genetic evaluation. Even in this ideal situation, genetic evaluation may not be straightforward when marker genotypes are missing on a significant proportion of the pedigree. Markers of type II can be further classified into two types: IIa) markers that are in linkage disequilibrium with the trait locus, and IIb) markers that are in linkage equilibrium with the trait locus. When disequilibrium is strong, a marker of type IIa can be treated as a type I marker. If disequilibrium between the marker and the trait locus is weak, the marker will have little effect on the trait mean. When the marker locus and trait locus are in linkage equilibrium, the allele states at the two loci are independent, and thus, the maker has no effect on the trait means. But, even in this situation, marker information can be used to model trait covariances by treating marker within animal as a random effect. Strategies for including these types of markers in genetic evaluation will be discussed.

Theatre G6.2

The efficiency of mapping quantitative trait loci using cofactor analysis

G. Sahana, B. Guldbrandtsen, P. Sorensen and M.S.Lund, Department of Animal Breeding and Genetics, Danish Institute of Agricultural Sciences, Foulum, 8830, Tjele, Denmark

Detection of QTL in outbred half-sib family structures has mainly been based on interval mapping of single QTL on individual chromosomes. Cofactor analysis builds a single model by combining information from individual analysis after which trait scores for a specific linkage group are adjusted for identified QTL at other linkage groups. It has been reported in the literature that cofactor analysis identifies more QTL than individual chromosome analyses. The aim of the study was to compare analyses with and without cofactors with respect to power of detection and false discovery rate. In a simulation study, different scenarios of a half sib design with 15 chromosomes and multiple QTL were simulated. The scenarios differed in family size (large vs. small), heritability of the trait (0.30 vs. 0.05), marker density (20cM vs. 5 cM), and a few QTL with large effect versus more QTL with small effects. Different levels of threshold at entry level for cofactor and QTL were also investigated.

Theatre G6.3

A new method for fine mapping quantitative trait loci using linkage disequilibrium

H. Gilbert, M.Z. Firat, L.R. Totir, J.C.M. Dekkers and R.L. Fernando, Department of Animal Science, ISU, Ames, IA 50011, USA

A new approach was developed to fine map a biallelic QTL using linkage disequilibrium , relating means and covariances of the QTL gametic values to the QTL allele effect and their frequencies among the founders' marker haplotypes. This reduces the number of parameters compared to usual models. MCMC was used to derive the conditional probabilities of inheriting maternal and paternal QTL alleles. A residual maximum likelihood method was implemented to map the QTL, using a Newton-Raphson algorithm to jointly estimate QTL position and effect of the QTL, QTL allele frequencies of the founders' marker haplotypes carrying the mutant QTL allele, and polygenic and residual variance components for each interval. Simulated populations were analyzed to compare its ability to fine map a QTL to two others methods: one based on identity by descent QTL covariances, and the other modeling independently the QTL effect means and covariances of the QTL gametic values.

Theatre G6.4

Fine mapping of QTL using combined linkage disequilibrium and linkage analysis for non-normal distributed traits

R. Labouriau and M.S. Lund, Department of Animal Breeding and Genetics, Danish Institute of Agricultural Sciences, P.O. box 50, DK-8830 Tjele, Denmark*

Fine mapping of QTL regions using linkage disequilibrium and linkage analyses can be accomplished by using appropriate multivariate normal linear mixed models (Genetics 163: 405-410, 2003). We present extensions of these models based on the analogous Generalized Linear Mixed Models. In the new class of models introduced, traits are assumed distributed according to an Exponential Dispersion Model (J.R. Statist. Soc. B 58: 573-592, 1996), which includes classic families of distributions such as the normal, gamma, binomial and the Poisson distributions. Here an interesting case is the compound Poisson that is a family of positive valued distributions that present positive probability of taking the value zero. This can be used to represent a QTL situated in a switch-regulated chromosomal region or a QTL related to genes with very low penetrance with their effects sometimes being below a minimum action threshold.
We present simulated examples where some SNP markers typed in a real population of about 12,000 pigs are replaced by simulated QTLs.

Theatre G6.5

Interval mapping methods to detect QTL on survival data

C.R.. Moreno-Romieux[1], J.M. Elsen[1], P. Le Roy[2], V. Ducrocq[2], [1]INRA, Station d'Amélioration Génétique des Animaux, 31326 CastanetTolosan Cedex, France, [2]INRA, Station de Génétique Quantitative et Appliquée, 78352 Jouy-en-Josas, France

Quantitative Trait Loci (QTL) are usually looked for using classical interval mapping methods which assume that the trait follows a normal distribution. However, these methods cannot take into account the characteristics most of survival data such as non normal distribution and presence of censored data. In this paper, we propose two new QTL detection approaches which allow to consider censored data. One interval mapping method uses a Weibull model (IMWM) which is popular to model survival trait and the other uses a Cox model (IMCM) which avoids to make any assumption on the trait distribution. Using simulated data, we compare IMWM, IMCM and a classical interval mapping method using a Gaussian model (IMGM). An adequate mathematical transformation was used for both parametric methods (IMGM and IMWM). When no data were censored, the three methods gave similar results. But as soon as some data were censored, the IMGM had a power of QTL detection but also an accuracy of QTL location and of QTL effects, which decreased considerably with censoring, particularly when censoring is at a fixed date. The results obtained with IMWM and IMCM were little affected by censoring.

Theatre G6.6

A count-location model for the chiasma process allowing detection of differing location distributions for single and multiple chiasmata

F Teuscher and Volker Guiard, Research Institute for the Biology of Farm Animals Dummerstorf, Wilhelm Stahl Allee 2, 18196 Dummerstorf, Germany

All models of chiasma processes developed and applied so far assume a unique distribution of chiasma locations, independent of the actual number of appearing chiasmata. Direct observations of chiasmata led to the hypothesis that this cannot be assumed in any case. A count-location model was developed here to infer such events. It takes into account suppression interference and allows the determination of the chiasma number distribution and the distribution of the chiasma locations for each number of appearing chiasmata. The usefulness of the novel model was shown with an example.

Theatre G6.7

Mixture models - use and applications within the field of animal breeding

J. Ødegård[1], G. Klemetsdal[1], J. Jensen[2] and P. Madsen[2], Agricultural University of Norway, [1]Department of Animal and Aquacultural Sciences,P.O. Box 5003, N-1432 Ås, Norway, [2]Department of Animal Breeding and Genetics, Danish Institute of Agricultural Sciences, Research Centre Foulum, P.O. Box 50, DK-8830 Tjele, Denmark

Quantitative traits with unknown underlying group structure, caused by e.g. QTL or disease, may distributed as a mixture, consisting of separate, but mutually overlapping distributions with different means and, possibly, different variances. Mixture models identify the underlying group state for each observation with a certain probability. For example, SCC data can be used to estimate probability of mastitis at a given test-day. This information may be used for herd managament decisions, or for prediction of breeding values for liability of putative mastitis, assuming individual *a priori* probabilities of mastitis with a genetic component. In addition, the model may allow estimation of genetic variance in baseline SCC, as well as SCC response to infection. Further, the model can be expanded to allow for several mixture traits, assuming the same underlying group structure (e.g. SCC and electrical conductivity in milk). Alternatively, multivariate analysis consisting of both mixture and non-mixture traits, e.g. SCC and libility to observed clinical mastitis, may be carried out. Another application of the mixture model is to adjust for unobserved underlying group effects might biasing genetic evaluation, e.g. preferential treatments.

Theatre G6.8

Structural equation models for quantitative genetic analysis

D. Gianola[1,2] and D. Sorensen[3], [1]Department of Animal Sciences, University of Wisconsin-Madison, WI 53706, USA, [2]Department of Animal and Aquacultural Sciences, Agricultural University of Norway, P.O. Box 5025, N-1432 Ås, Norway, [3]Department of Animal Breeding and Genetics, Danish Institute of Agricultural Sciences, P.O. Box 50, 8830 Tjele, Denmark*

Multivariate models are of great importance in theoretical and applied quantitative genetics. We extend quantitative genetic theory to accommodate situations in which there is linear feedback or recursiveness between the phenotypes involved in a multivariate system, assuming an infinitesimal, additive, model of inheritance. It is shown that structural parameters defining a simultaneous or recursive system have a bearing on the interpretation of quantitative genetic parameter estimates (e.g., heritability, offspring-parent regression, genetic correlation) when such features are ignored. Matrix representations are given for treating a plethora of feedback-recursive situations. The likelihood function is derived, assuming multivariate normality, and results from econometric theory for parameter identification are adapted to a quantitative genetic setting. A Bayesian treatment with a Markov chain Monte Carlo implementation is suggested for inference and developed. When the system is fully recursive, all conditional posterior distributions are in closed form, so Gibbs sampling is straightforward. If there is feedback, a Metropolis step may be embedded for sampling the structural parameters, since their conditional distributions are unknown. Extensions of the model to discrete random variables and to non-linear relationships between phenotypes are discussed.

Theatre G6.9

Estimating genetic principal components for multi-trait and random regression models directly using restricted maximum likelihood

K. Meyer[1] and M. Kirkpatrick[2], [1]Animal Genetics and Breeding Unit, University of New England, Armidale NSW 2351, Australia, and [2]Section of Integrative Biology, University of Texas, Texas 78712, USA*

Principal component analysis is a widely used `dimension reduction´ technique. In analyses estimating multiple, correlated genetic effects per individual, either effects for individual traits or random regression coefficients to model trajectories, we generally estimate the corresponding covariance matrices assuming they have full rank. If correlations between effects are strong, however, most genetic variation is explained by few linear combinations of the original effects. The loadings in these principal components (PC) are given by the eigenvectors corresponding to the largest eigenvalues of the genetic covariance matrix. It is shown that genetic PCs can be estimated directly through a simple reparameterisation of the usual linear mixed model, and that this parameterisation can be used to estimate reduced rank, smoothed covariance matrices by restricted maximum likelihood, employing derivatives of the likelihood. This can result in considerable reductions in computational resources required : the number of parameters to be estimated is decreased from $k(k+1)/2$ to $m(2k-m+1)/2$ for k traits and m PCs, the number of effects in the model is reduced by k-m genetic effects, and the work per likelihood evaluation, in part proportional to m^2, can be reduced dramatically, making routine multivariate analyses involving more than 3 or 4 traits feasible.

Theatre G6.10

Comparison of two methods to approximate reliabilities of breeding values from test day model

I. Strandén, E.A. Mäntysaari, Agrifood Research Finland, Animal Production Research, 31 600 Jokioinen Finland*

Calculation of accuracies or reliabilities of breeding values is computationally more challenging than calculation of estimated breeding values itself. Commonly used methods rely on two steps. First, reliabilities accounting for data design are calculated. Second, the relationship information is accounted. Misztal and Wiggans (MW), and Jamrozik and Schaeffer (JS) have presented methods for approximating reliability that account for relationship information. In this study, reliabilities of 305 day breeding values were calculated by exact method and by the above mentioned approximations. First lactation test day model for milk, fat and protein using sample of the Finnish Ayrshire data with 3479 animals was used in the analysis. Differences between exact and approximated reliabilities were small. Correlations of the MW and JS methods to the exact reliabilities of milk 305 day breeding values were 99.96% (0.3%) and 99.92% (-0.2%), respectively (mean difference between approximation and exact reliability in parenthesis). Values for protein and fat were similar. The MW method gave always greater or equal reliabilities to those of the JS method. The JS method seemed to slightly underestimate reliabilities while the MW method slightly overestimated them. In particular, the JS method tended to underestimate low reliabilities while the MW method tended to overestimate large reliabilities.

Theatre G6.11

An improved model for the French genetic evaluation of dairy bulls on length of productive life of their daughters

V. Ducrocq, Station de Génétique Quantitative et Appliquée, Institut National de la Recherche Agronomique (INRA), 78352 Jouy-en-Josas, France

Functional longevity of dairy cows has been routinely evaluated in France since 1997 using a survival analysis model. Recently, we proposed a genetic trend validation test that could be used before including national data in an international evaluation for longevity. Its application to the French Holstein data revealed a large overestimation of the genetic trend. It was found that the bias is the result of a change of the baseline hazard rate over time. A new proportional hazard model is proposed that accounts for this change. In the new sire-maternal grandsire model, the baseline is described as a stratified, piecewise Weibull hazard function within lactation, i.e., a function of the number of days since last calving. Stratification is within year-season and lactation number. Different Weibull hazard functions are used over 3 periods: 0-270 days of lactation, 270-380 and 380-day when dried. The non genetic effects included in the model are an age at first calving effect; the interaction effects between the within herd-year class of milk production and a) lactation number x stage of lactation b) year-season ; the within herd classes of fat and protein percent effects; a class of variation herd size effect and a random herd-year-season effect. The new genetic trend is almost flat in the Holstein breed.

Poster G6.12

Estimates of genetic principal components and reduced rank covariance matrices for 'live' carcass traits recorded by ultra-sound scanning of Australian Angus cattle

K. Meyer, Animal Genetics and Breeding Unit, University of New England, Armidale NSW 2351, Australia

Records for eye muscle area, fat depth at the 12/13th rib and the P8 site, and percentage intra-muscular fat measured by live ultra-sound scanning were extracted for 36 herds of Angus cattle with at least 1000 animals measured. Records on heifers or steers and bulls were treated as different traits. There were a total of 262,862 records on 74,268 animals, and including pedigree information yielded 103,467 animals in the analysis. Data were analysed by restricted maximum likelihood considering all 8 traits simultaneously, but estimating only the first 3, 4, 5 and 6 genetic principal components (PC), respectively. In addition, 28 bivariate analyses were carried out to estimate all pairwise genetic correlations. Results showed good agreement in estimates of eigenvalues and -vectors from PC and pooled bivariate analyses, with the first three PCs explaining most genetic variation and two eigenvalues not significantly different from zero. Heritabilities and genetic correlations derived from reduced rank covariance matrices corresponded to literature values. Implications for a multi-trait genetic evaluation for the traits considered are discussed.

Poster G6.13

Studies on the allometric growth of muscles in young Romanian Spotted bulls

S. Acatincai, G. Stanciu, L.T. Cziszter, L. Keresztesi, I. Tripon. Banat University of Agricultural Sciences, Calea Aradului 119, 300645 Timisoara, Romania*

Experiments were carried out on 80 Romanian Spotted bulls, fattened in intensive system and slaughtered at 6, 9, 12, 15, and 18 months of age. Live weight at slaughter, carcasses weight and killing out percentages was measured. Five muscles were cut out from the carcasses after refrigeration, that is *Longissiums dorsi*, *Biceps brachii*, *Biceps femoris*, *Semimembranosus*, and *Semitendinsous*. Allometric coefficients for each muscle were calculated during different age intervals. *L. dorsi* has an isometric relative growth to that of the whole body during the studied period, except the fourth age interval (15-18 months). *B. brachii* has a positive allometric growth during the whole studied period, though the allometric coefficients decreased with age. *B femoris* showed an increasing allometric growth with age, being nearly isometric at the beginning of the studied period. *Semimembranosus* has a positive allometric growth in all age intervals, except for the 9-12 age interval when its relative growth was lower than the growth of the whole body. *Semitendinosus* showed a similar allometric growth pattern with *Semimembranosus*, except that the lower relative growth was encountered during the age interval 12 to 15 months. Overall, the studied muscles showed a different pattern of allometric growth, but had a positive allometry compared to body weight.

Poster G6.14

Comparing alternative definitions of the contemporary group effect in Avileña Negra Ibérica cattle in a Bayesian analysis

M.J. Carabaño[1], A. Moreno[1], P. López-Romero[2] and C. Díaz[1], [1]Depto Mejora Genética Animal, INIA, Apdo 8111, 28080 Madrid, [2]Ud. de BioinformáticaCBMSO-CSIC 28049 Madrid, Spain*

Data on weaning weight from 12,740 animals were used to compare different contemporary group (CG) definitions to be applied in the current genetic evaluation in the Avileña Negra Ibérica beef cattle breed. Several classical 'ad-hoc' statistics and criteria for statistical model comparison within a Bayesian analysis were used. Six alternative definitions for the CG effect were compared: Herd- year-season of calving (HYS), with seasons defined according to the four natural seasons; herd-year-month of calving (HYM); herd clusters with length of 30 days (HC30-30) or 90 days (HC90-90), and, adaptive clusters with two time limits, 30 and 90 days (HC30-90), and 30 and 180 days (HC30-180). Classical statistics pointed at HYM and HC30-30 as the models showing smaller within CG variation and smaller residual variance, with slightly worse performance in terms of accuracy of prediction for breeding values for direct genetic effects. Bayes factors and cross-validation predictive densities, allowed for a better discrimination among models. Models including CG spanning 30 days were more plausible and showed better predicting ability than models spanning 90 days. Adaptive CG showed intermediate results. Model definition had a relatively large impact on the variance components estimates. Overall, HYM showed best results but implied the largest loss of data (14%). HC30-90 might represent a compromising solution for this population.

Computerized morphological description of the Ragusana donkey

L. Liotta, B. Chiofalo and L. Chiofalo, Sect. Animal Production, Dept. MOBIFIPA, University of Messina, Italy

The morphological evaluation of 70 Ragusana donkeys were investigated with the aim of identifying the type of this Sicilian (Italy) race. The study was carried out measuring the morphological traits on each animal from a side, front and rear view using a computerised system of image analysis (Image Pro Plus) with a digital camera. The most significant morphometric traits were measured, and the most significant bio-metric indices (*i*): format, construction, constitution, volume were calculated, so as to try and define the morphological type. The mean values of these indices show diametrical proportions of the mesodolichomorphic type (height-length proportion *i*: 99; thoracic *i*: 47; body proportion *i*: 90; *i* of the pelvis: 24; chest length *i*: 78; dactylo-thoracic *i*: 11), with a morphological variability among the measured animals. In fact, the typical diametrical proportion of the dolichomorphic type (height-length proportion *i*: 101; thoracic *i*: 51; body proportion *i*: 108; *i* of the pelvis: 19; chest length *i*: 95; dactylo-thoracic *i*: 10) and those of the brachymorphic type (height-length proportion *i*: 89; thoracic *i*: 36; body proportion *i*: 74; *i* of the pelvis: 36; chest length *i*: 67; dactylo-thoracic *i*: 12) were observed. Therefore, it seems important, considering the correlation between the morphological type and the productive aptitude, to verify whether the two morphological types identified in this asinine race are also different as regards productions, e.g. the milk production.

Theatre N4.1

Effect of feeding a mixture of maize silage and alfalfa silage ensiled separately or together on feed intake and milk production in dairy cows

U. Meyer[1], G. Pahlow[2] and G. Flachowsky[1], Federal Agricultural Research Centre, [1]Institute of Animal Nutrition, [2]Institute of Crop and Grassland Science, Bundesallee 50, 38116 Braunschweig, Germany*

The aerobic stability of maize silage is usually low if compared to alfalfa silage. It can be expected that co-ensiling of alfalfa silage and freshly harvested whole plant maize can improve the aerobic stability of the mixture during feed-out. The aim of the experiment was to investigate the influence of feeding a freshly prepared mixture of alfalfa silage and maize silage ensiled separately (contr.) or a mixture of alfalfa silage and maize ensiled together (exp.), on feed intake and milk production.

Forty six Holstein cows were assigned in one of two groups. During a period of 117 days the animals received silage mixtures (contr. or exp.) which were composed of alfalfa and maize at a ratio of 35:65 on dry matter base. Concentrates were added to the rations according to milk yield in order to meet the energy and protein requirements. Individual feed intake and milk yield were recorded daily.

The feed intake was significantly different (exp. vs. contr.): silage mixture 10.1 vs. 12.7kg dry matter, concentrates 10.6 vs. 9.7kg DM. Cows fed the exp. mixture showed a significantly increased milk yield (34.2 vs. 32.0kg/d) and a decreased fat content (3.92 vs. 4.06%), while there were no significant differences in the amount of FCM (32.1 vs. 33.7kg/d).

Theatre N4.2

The effect of silage microbial inoculant with and without additional preservatives on the aerobic stability of maize silage

S. Hall[1], P. Moscardo Morales[1], J.K. Margerison[1], D. Wilde[2], P. Light[2], M. Smith[2] and N. Adams[2], [1]University of Plymouth, Seale-Hayne faculty, Newton Abbot, Devon, UK, [2]Alltech UK Ltd, Alltech House, Ryhall Road, Stamford, Lincs. PE9 1TZ, UK

The aim of this experiment was to measure the effect of inoculating maize silage with Maize-all GS (inoculant) and Sil-all Fireguard (inoculant and preservative). Forage maize (DM 29 (±1.29) had; No additional additive with 200 ml of water / 100 kg FM (0), Sil-All fireguard 0.5g/100kg FM (SAFS), Maize All GS 1g/100kg FM (MAS) applied (3 replicates of each treatment) applied, stored (17 to 20°C) and opened after 30 days. Containers were punctured with 20, 0.5-cm diameter holes, insulated using 5cm polystyrene foam, stored (17 and 20 °C) and tested for lactic, acetic acid levels, temperature and pH. Lactic acid levels (g/kg) 48 to 68 h, 0, -37.2^{b}, SAFS -78^{a}, MAS -52.5^{a,b} (11.90), Acetic acid levels (g/kg DM) 0 and 48 h, 0, -8.2^{c}, SAFS, 0.6^{a}, MAS -2.2^{b} (2.60), 0 and 168 h, 0 -14.6^{b}, SAFS -23.8^{a}, MAS -18.4^{b}(2.67). Maximum pH, 0, 7.3^{a}, SAFS 6.7^{b}, MAS 7.4^{a} (0.22), pH change over 48-168 hours, 0, 3.3^{b}, SAFS 3.0^{c}, MAS 3.6^{a} (0.17), time to increase temp. 2 °C (h), 0, 41.1^{b}, SAFS 76.5^{a}, MAS 56.4^{a,b} (10.30). Silage additives MAS and SAFS increased both lactic acid levels and silage stability as indicated by an increased amount of time to reach maximum temperature and reduction in maximum and silage pH change.

Theatre N4.3

The performance of finishing cattle offered high or low cereal-based concentrate diets with or without mycosorb

R. J. Fallon and P.O'Kiely, Teagasc, Grange Research Centre, Dunsany, Co. Meath, Ireland.*

The objective was to determine the effects of the inclusion of Mycosorb (a yeast derived glucomannan) at 1.0 kg/tonne in high and low cereal-based concentrate diets offered *ad libitum* on the productivity of finishing bulls. Eighty, 20-month old Friesian bulls with an initial liveweight of 585 kg were allocated to the following diets (150 g protein/kgDM) for a 91 day period: High cereal (CB), High cereal plus Mycosorb (MB), Low cereal (CP) and Low cereal plus Mycosorb (MP). The bulls were accommodated on concrete slats (4 pens of 5 bulls/treatment) with *ad libitum* access to barley straw and water. The mycotoxin content was greatest in the low cereal diet. Carcass weights were 394, 393, 386 and 386 (sem 3.4) kg for treatments CB, MB, CP and MP respectively. The corresponding values for liveweight gain (g/d) were 1790, 1740, 1600 and 1660 (sem 62) for mean daily concentrate DM intake (kg/d) were 12.3, 11.9, 12.2 and 12.0 (sem 0.20), for carcass gain (g/d) were 1105, 1103, 1014 and 1032 (sem 33), for carcass fat score was 3.51, 3.53, 3.29 and 3.45 (sem 0.061) and for conformation score were 2.08, 2.50, 2.35 and 2.30 (sem 0.070). It was concluded that 1.0 kg of Mycosorb inclusion per tonne of concentrate did not effect productivity of bulls offered either a high or low cereal-based diets.

Theatre N4.4

Comparative evaluation of grass silage, fermented whole crop wheat silage, urea-treated processed whole crop wheat and maize silage in the diet of early lactation cows

J.J. Murphy[1], S. Kavanagh[2] and J. J. Fitzgerald[1], [1]Teagasc, Dairy Production Research Centre, Moorepark, Fermoy Co Cork, Ireland, [2]Teagasc, Kildalton College, Piltown, Co Kilkenny, Ireland*

The objective of these experiments was to compare the effects of including fermented whole crop wheat (WCW), urea-treated processed WCW or maize silage with grass silage on intake and production of dairy cows. Experiment 1 was a Latin Square change-over design (20 cows) and experiment 2 was a 10-week randomised block design (56cows). In both experiments the forage treatments were *ad libitum* (i) grass silage, (ii) grass silage: fermented WCW, (iii) grass silage:urea-treated processed WCW and (iv) grass silage: maize silage. In both experiments the starchy forages were included at a level of 670 g/kg of DM with grass silage. In experiment 1 and 2 concentrate supplements were fed at 10 and 8 kg/cow per day, respectively. In both experiments forage and total DM intakes were increased ($P<0.001$) by all the forage mixtures compared to grass silage alone and urea-treated processed WCW resulted in higher intakes ($P<0.05$) than the other two forage mixtures. Forage mixtures resulted in higher milk ($P<0.05$) and fat + protein ($P<0.01$) yields and protein contents (<0.01) than silage alone. The results indicate that fermented WCW, urea-treated processed WCW and maize silage will increase intake and production when included at 0.67 of the forage mixture with grass silage.

Theatre N4.5

Effects of washing procedure, grain type and particle size on the size of non-washout and insoluble washout fractions in concentrate ingredients

A. Azarfar, S. Tamminga and H. Boer, Wageningen University, Animal Nutrition Group, P. O. Box 338, 6700AA Wageningen, The Netherlands*

It assumes rumen degradation of material washed out of nylon bags to be instantaneous and complete. Verifying this assumption requires a standardised laboratory procedure that mimics the results of washing in the *in situ* technique. In a 2x3x6 factorial design of treatments we compared the effects on the size of the non-washout (NWF) and the insoluble washout fraction (ISWF) of three washing procedures (Y=Yang, M=Melin, I=*In situ*) in six grains (maize, barley, milo, peas, lupins and faba beans), ground at two particle sizes (1 and 3 mm). Differences between methods were washing in a washing machine (I), repeated plunging of nylon bags in a beaker (Y) and filtration according to the procedure of Melin et al (2001) (M).
The effects on NWF and ISWF of grain, washing method and particles size were significant ($P<0.05$). NWF in method Y was smaller than in the other methods (53.4, 58.0and 60.4% in Y, M and I). Method M was more similar to method I. NWF in maize, barley, milo, lupins, faba beans and peas was 77.5, 68.1, 61.3, 65.9, 37.8 and 33.5%. ISWF in peas, faba beans, milo, barley, maize and lupins was 40.9, 37.3, 32.7, 24.8, 15.5 and 13%. Increasing the particle size increased NWF whereas ISWF was decreased.

Theatre N4.6

The influence of mineral mixture with buffering activity on milk production, metabolic profile and rumen fluid parameters in cows

M. Adamovic[1], J. Lemic[1], M. Jovicin[2], G. Grubic[3], O. Adamovic[3], B. Stojanovic[3] and M. Radivojevic[4], [1]ITNMS, P.O. Box. 390, 11000 Belgrade, [2]NIV NS, Rumenacki put 20, 21000 Novi Sad [3]Faculty of Agriculture, Nemanjina 6, 11080 Zemun, [4]Institute Agroekonomik, 11213 Padinska Skela, Serbia

The influence of mineral mixture with buffering activity (bentonite, organo-zeolite, magnesium oxide and sodium bicarbonate) on milk production and composition, metabolic profile, blood, urine and feces pH, and rumen fluid parameters. The mixture named »Mix Plus« was produced in ITNMS, Belgrade.
The experiment was done on two groups of Holstein dairy cows (control - K and experimental - O), lasting 30 days (June-July). The average temperature in the stable, (fixed system), was 32.6 °C. Diet for control and experimrntal group was the same, formulated on the basis of lucerne hay, maize plant silage, concentrate mixture (18% TP), and extruded soybean grain (NEL MJ 146.35:150.47; TP 16.55:16.80%; ADF 20.79:21.16%, NDF 34.05:34.55%). Experimental group had received (for diference from control) concentrate mixture with inclusive 1% »Mix Plus«. The experimental treatment had positive effects on 4% FCM production (K: O = 26.97:27.40 kg/day), milk fat (3.29:3.58%) and dry matter (11.62:11.99%), ($p >0.05$). The content of glucose, protein, urea, Ca and P in blood was within normal physiological values. The pH values were: rumen 6.64:6.50, blood serum 7.38:7.60, urine, 7.42:7.77, feces, 6.74:6.88 ($p >0.05$).The ratio of acetic an other acids was 1.65:1.62.

Theatre N4.7

Effect of starch content of maize silage based diets for dairy cattle

D.L. De Brabander, N.E. Geerts, S. De Campeneere and J.M. Vanacker, Agricultural Research Centre, Department Animal Nutrition and Husbandry, Scheldeweg 68, 9090 Melle, Belgium*

Assuming that the effect of starch content could depend on milk yield level and diet composition, 2 trials were carried out in early lactation (T1, T2) and 2 in mid lactation (T3, T4). Maize silage (MS) was given as the sole roughage in T1, or with prewilted grass silage (PGS) in T2. In T1 and T2, three starch (S) levels, originating from the MS-cultivar and the concentrate (C), were compared in a latin square design with 18 Holstein cows: MSmCm, MShCm, MShCh (m, h: moderate, high S-content). In T3 only S-content of MS was compared: MSmCm, MShCm, whereas in T4 the effect of PGS addition to MSh was studied. T3 and T4 were cross-over trials with 16 Holstein cows.

Milk yield amounted to 27.8 , 26.7 and 27.4 kg in T1, and to 31.4 , 30.5 and 30.7 kg in T2, for MSmCm, MShCm and MShCh, respectively. Differences were not significant ($P > 0.05$). In T3 milk yield was significantly lower for the diet with MSh (19.2 vs. 20.2 kg). Adding PGS to MSh (T4) significantly increased milk yield (19.1 vs. 18.4 kg). Milk composition was almost unaffected.

Generally, differences in milk yield were better related to differences in dry matter and net energy intake, than to the starch content of the diet.

Theatre N4.8

Bioactive natural compounds from herbs as new feed additives

D. Tedesco and S. Galletti, Department of Veterinary Sciences and Technologies for Food Safety, Via Celoria 10, 20133 Milano, Italy*

Impact of animal feed including functional additives on animal and human health, and safe and environmentally friendly production process for healthy food have been outlined as objectives of the Food Quality and Safety Priority of FP 6. Use of subtherapeutic doses of antibiotics is the major cause of cross-resistance in pathogenic bacteria. Increasing consumer pressure to reduce antibiotics in feed has led U.E. to ban antibiotics used as feed additives from 2006. Due to their antimicrobial, antioxidant, immunostimulant properties, the use of natural bioactive compounds as additives in livestock nutrition instead of other chemical compounds is becoming widespread. There is an increasing number of herbal feed additives on the market as e.g. essential oils and herbal mixtures, often lacking a scientific background and information as to safety and efficacy. To assure batch-to-batch conformity of herbal plant preparations, efficacy, standardisation and characterisation are necessary. Safety standards used for food additives and feed ingredients could be used for establish safety of these substances. Considering that their legal status is not clear (feed additives or medicinal products?), approaches for the evaluation of safety and efficacy of natural compounds will be discussed.

Theatre N4.9

The influence of meat on oxidative status in pigs

V. Rezar, T. Pajk, A. Levart, K. Salobir and J. Salobir, University of Ljubljana, Biotechnical Faculty, Zootechnical Department, Groblje 3, 1230 Domzale, Slovenia

To determine the importance of lean meat and fruit and vegetables for the etiology of oxidative stress formation in humans the effect of lean and fat meat was studied in diets with or without fruit and vegetables in pigs as a model for humans. After an adaptation period the animals were divided in five groups receiving isocaloric diets with or without lean or fat meat, fruit and vegetable. The oxidative stress was evaluated by measuring malondialdehyde (MDA) in blood plasma, the urine MDA excretion, the degree of the leukocyte nuclear DNA damage (Comet Assay), concentration of tocopherols in blood plasma and total antioxidant status (TAS) of the blood plasma. The results of experiment show that in a diet with fruit and vegetables, the substitution of lean meat for fat meat only modestly increased oxidative stress and that diet without fruit and vegetable significantly increased oxidative stress. It could be concluded that a substitution of lean meat for fat meat in otherwise balanced diet is much less detrimental for the oxidative stress formation than a diet without fruit and vegetables. The double lean meat intake in an otherwise balanced diet has no effect on oxidative stress formation. The results of the experiment confirmed the crucial role of fruit and vegetable in the balance diet.

Poster N4.10

A new rumen bacteria species?

A. Chiariotti, S. Lucioli, F. Grandoni, and S. Puppo, Instituto Sperimentale per la Zootecnia, Via Salaria, 31. 00016 Monterotondo scalo. Roma, Italy*

Highly preserved sequences of 16S rDNA in bacteria were used to identify bacteria species in rumen fluid collected from fistulated buffalo. A rapid rDNA sequencing method, using crude lysate from a single colony of rumen bacteria demonstrated improved efficiency and reduced time for rumen bacteria identification in comparison with traditional microbial techniques.

Single colonies were PCR amplified after beadbeater treatment with two different steps. In the first P3mod and PC5 were used as primers (fragment from 787 to 1507 of 16S *Escherichia coli* rDNA).

Sequencing condition were also optimized and allowed the identification of *Prevotella ruminicola* (98%, gi2073136), *Butyrivibrio fibrisolvens* (97%, gi4098353) and *uncultured fiber-attaching rumen bacteria* (99%, gi13429881).

In the second step, the unknown colony was analized by changing the primers to 27F and 1492R covering the entire gene sequence. This did not result in a high omology with any other classified rumen bacteria sequence reported in BLAST databases. Further research is necessary to confirm the identification of this potential new rumen bacteria species.

Poster N4.11

Prediction of *in vivo* organic matter digestibility of corn silage by means of the gas production technique

A.Garcia-Rodriguez[1], L. Lorenzo[1], A. Igarzabal[1], A. Legarra[1], L.M. Oregui[1] and G. Flores[2], [1]NEIKER, Apdo 46, 01080 Vitoria-Gasteiz, Spain, [2]Centro de Investigacions Agrarias de Mabegondo (CIAM). Apdo 10, 15080 A Coruña, Spain*

In the Basque country and Galicia conditions corn silage accounts for a great part of cows winter rations. Robust techniques that predict the quality of corn silage for developing improved feeding strategies and screening large number of samples are needed. The objective of the current work was to assess in vivo organic matter digestibility (OMD) by means of the gas production technique. Sixty seven samples of known OMD were chosen on the basis of their range of variation (mean=669 g kg^{-1}, range=161 g kg^{-1}). Samples were analysed for gas production using a pressure transducer. Stepwise regression analysis related cumulative gas production and OMD. The obtained equations were validated using the same pool of samples (cross-validation). Stepwise regressions analysis revealed significant relationships between *in vitro* parameters and OMD. After validation gas production alone allowed moderate results (R^2=0.39, rsd=24.41). However, the inclusion of chemical parameters improved the variance expressed by the model and the prediction accuracy (R^2=0.549, rsd=21.18). The best results were achieved when the pH of the silage was included in addition to gas production and analytical parameters (R^2=0.574, rsd=20.63). Therefore, it can be concluded that although the results are promising more research is needed.

Poster N4.12

Nylon bag degradability and mobile nylon bag digestibility of crude protein in two variety of black oats (Avesta and Noirine)

J. Harazim[1] and P. Homolka[2]. [1]Central Institute for supervising and testing in Agriculture, Opava, [2]Research Institute of Animal Production, Uhríneves, Czech Republic*

Samples of two variety of black oats (Avesta and Noirine) were incubated in situ to evaluate rumen effective degradability of crude protein (CP). Animal diet was composed of maize silage (10 kg), clover silage (7 kg), meadow hay (5 kg) and mixture. Nylon bags containing tested samples were suspended in the rumen of three cannulated steers for 2, 4, 8, 16, 24 and 48 h. In each steer, two repetitions were done in each time interval (the nylon bags were made of nylon with pore size 42 μm). The rumen effective degradability was calculated at rumen outflow rate 5 % and no correction for microbial contamination was included. The contents of CP in sample of Avesta was 113.0 and Noirine was 106.0 g per kg DM. Effective degradability values for the same feeds were 89.2 and 84.7 %, respectively.
The mobile bag technique with dry cows fitted with the large ruminal cannulas and the T-piece cannulas in the proximal duodenum was used to measure the intestinal digestibility.
Intestinal digestibility of rumen undegraded protein was 57.1 % for Avesta and 63.1 % for Noirine. The study was supported by the the Ministry of Agriculture of the Czech Republic (NAZV, project No. QD 0211).

Poster N4.13

Characterization and nutritive value of a new forage sorghum variety "Pnina" recently bred in Israel

A. Carmi, N. Umiel, A. Hagiladi, E. Joseph, D. Ben-Ghedalia and J. Miron, ARO, Bet-Dagan, Israel*

A new sorghum variety, Pnina, was recently bred in Israel. Its performance was studied compared to the widely used commercial sorghum hybrids, FS-5 and BMR-Nutriplus. Plants grew in summer irrigated with 190 mm and harvested at early milk or dough stages. Pnina was characterized by low height (135 cm) showing absolute resistance to lodging. The other strains were taller (>235) and suffered from high lodging (46% in FS-5 and 73% in Nutriplus). Pnina was more leafy, having 0.79 ratio of leaves: stems, compared with 0.34 in Nutriplus and 0.36 in FS-5. A dry matter (DM) content of < 30% may cause drainage and silage spoiling in the silo. In dough stage, DM content of Pnina, Nutriplus and FS-5 plants were 36.9%, 27.1% and 26.4%, respectively. Yield of these varieties in dough stage were similar, (13.3 to 14.7 tons DM/hectare). In vitro DM digestibility of silages made from the varieties was similar (62.5 to 67.1%). NDF content of Pnina silage was higher than that of Nutriplus and FS-5 (61.9 *vs* 56.9 and 52.7%, respectively). Yields of digested silage/Hectare in dough stage were similar in the varieties. Thus, the Pnina strain was superior to the other two varieties with respect to the absolute resistance to lodging, high DM content and leafy morphology.

Poster N4.14

Enhancement of fiber digestion and cellulase activity by the yeast culture of *Saccharomyces cerevisiae*

A.M. El-Waziry[1], H.E.M. Kamel[1] and H.R. Ibrahim[2], [1]Dept. Animal Prod., Fac. Agriculture, Alexandria University, Alexandria, Egypt, [2]Dept. Biochem. Biotechnol., Fac. Agriculture, Kagoshima University, Japan*

Dietary supplementation with dry yeast cultures has long been practiced to improve the nutrient digestibility by increasing the rumen bacterial. However, no study was undertaken to elucidate the effect of yeast on cellulases and fiber digestibility so far. In this study, we examined the effect of yeast culture of *Saccharomyces cerevisiae* (YC) on fiber digestion and cellulase activity. Three groups of ruminally cannulated sheep were used; (1) berseem hay (basal diet), (2) basal diet plus 0.75% YC of feed intake, and (3) basal diet plus 1.5% YC of feed intake. Digestibility of neutral detergent fiber (NDF) and acid detergent fiber (ADF) was increased ($P<0.05$) when the supplementation level was 1.5%. Degradation rate of NDF and ADF was enhanced by the addition of YC at both levels of supplementation. Reduction of lag time was noticed in the digestion of NDF and ADF at both levels of YC supplementation. The *in vitro* activity of cellulase was enhanced by the addition of YC, whereas the maximum cellulase activity was observed with YC from 6 days yeast culture. Interestingly, the cellulase activity was increased with an increase in the added dose of YC. In conclusion, YC enhanced fiber degradation and reduction of lag time as a result of its direct enhancement of cellulase activity.

Poster N4.15

The effect of addition *Saccharomyces cerevisiae* SC-47 on ruminal fermentation in cow´s

P. Dolezal, L. Zeman and J. Dolezal, Department of Animal Nutrition, Mendel University of Agriculture and Forestry Brno, Zemedelska 1, 613 00 Brno, Czech Republic

Twelve dairy cows were used in a experiment at which studied the effect of *Saccharomyces cerevisiae* SC-47 addition on the rumen fermentation. Animals were divided in to experimental and control groups, each of them 6 cows. The feeding ration consisted of maize silage with higher dry matter (16 kg), clover-grass silage (16 kg), meadow hay (3 kg), and feeding mixture (7,5 kg). Mean value of pH rumen fluid reduced in experimental group (6.11 ± 0.008), vs. control (6.30±0.100).

The most intensive increase ($P<0.05$) of VFA (1.158±0.013 g/100 g) content in rumen fluid of experimental group was observed. Statistical differences ($P<0.05$) were also found in rumen fluid in the experimental group of cow´s in the acetic acid, propionic, and butyric acid content. Statistical significant ($P<0.05$) differences were ascernaited by the content of infusoria value in experimental group (359.2±1.304 thousand/ml vs 302.0±12.349 thousand/ml of fluid).

The addition of Saccharomyces cerevisiae SC-47 (6 g/day) reduced the value of ammonia content in rumen fluid (8.68±0.084 vs. 9.06±0.207 mmol/L).

This study was supported by Project MSM 432100001.

Poster N4.16

Intensity of microbial proteosynthesis in the rumen of calves that were fed different feeding rations up to their weaning

Z. Mudrik[1], B. Hucko[1], A. Kodes[1] , V. Christodoulou[2], V. Kudrna[3] and P. Dolezal[1], [1]Czech University of Agriculture, Depoartment of Animal Nutrition, Prague 6 - Suchdol, 165 21, Czech Republic, [2]Research Institute of Animal Production, Giannitsa, Greece, [3]Research Institute of Animal Production,Prague - Uhríneves, Czech Republic

In our tests we have found, that the development of micro organisms, which cause the breaking down of cellulose (we're watching cellulolytic activity such as the decrease of cellulose in reaction to the rumen liquid, and using accurately weighted crushed cellulose), was stimulated in calves that were fed rations without hay (only grain and milk drink). We performed series of tests aimed to determine the proteosynthetic intensity of micro-organisms in the calves' rumen. We were watching the rumen proteosynthesis utilizing an indirect method: we established purin derivates of nucleic acids (allantoin, uric acid, hypoxantine, xantine) in the calves' urea. The supposition was, that purin derivates in the urine are related to the quantity of microbial purin synthesized in the calves' rumen.

The content of purin derivates (allantoin) in the urine of calves that were fed rations containing hay = 4228 µmol (s = 813,08) and without hay = 4834 µmol (862,41).

Proteosynthetic activity in calves fed rations containing hay = 921,704 g . den^{-1} , without hay = 1053,812 g . den^{-1}. The tests support our initial expectations that the calves, which were fed no hay, show higher proteosynthetic activity.This proteosynthetic activity as well as the development of microorganism in the calves' rumen is favourably influenced by the composition of VFA.

Poster N4.17

Influencing the cellulolytic activity of microorganisms in the rumens of calves that were fed different rations up to their weaning

B. Hucko[1], Z. Mudrík[1], A. Kodes[1], V. Christodoulou[2], V. Kudrna[3] and P. Dolezal[1], [1]Czech University of Agriculture,Agronomic Faculty, Depoartment of Animal Nutrition, Prague 6 - Suchdol, 165 21, Czech Republic, [2]Research Institute of Animal Production, Giannitsa, Greece, [3]Research Institute of Animal Production, Prague - Uhríneves, Czech Republic

According to our earlier tests with different feeding rations in calves we found out, that calves, which were fed rations without hay into their weaning, had larger rumen and more developed papillas on the rumen mucous membrane, different content of VFA in the rumen liquid, but also markedly increased acidity of the rumen content. The higher contents of milk acid caused a state of rumen acidosis and an indication of necrotic mucous membrane. We found references in literature, which support our findings concerning possible stimulation of cellulolytic microorganisms.
We performed a series of tests with the aim to determine cellulolytic activity of microorganisms in the rumen of calves fed rations with hay and without hay. The liquid from the rumen (ventral bag) of calves slaughtered at 73 + 2 days was extracted and pure crushed cellulose was then added into it. Cellulolytic activity was determined as the decrease in weight of the cellulose.
Cellulolytic activity of rumen microorganisms in the calves fed rations with hay were significantly lower statistically (about 23.89 %).
This supports our presupposition that the production of VFA with longer carbon chain stimulates microorganisms, which break down cellulose.

Poster N4.18

Prediction of protein degradability of grassland forage by near-infrared spectroscopy (NIRS)

T. Znidarsic, J. Verbic and D. Babnik, Agricultural institute of Slovenia, Hacquetova 17, 1000 Ljubljana, Slovenia*

The aim of the present work was to examine the potential of NIRS to predict the ruminal protein degradability of grassland forage. Protein degradabilities of 29 grass silage samples, 19 green forage samples and 1 hay sample were determined by means of nylon bag technique. Effective protein degradabilities (EPD) in the rumen were calculated on the basis of degradation characteristics and theoretical outflow rate 0.05 h^{-1}. All samples were scanned using the NIRSystems (Silver Springs, MD) Model 6500 NIR monochromator operating with software equipment (WinISI II - version 1.50). Calibration equations for the estimation of EPD were developed for the wavelength range 1100-2500 nm. The concentration of crude protein and crude fibre ranged from 77.8 to 218.1 g kg^{-1} and from 187.2 to 396.8 g kg^{-1} dry matter, respectively. EPD varied from 582 to 847 g kg^{-1}. Coefficient of determination (R^2) and standard error of cross validation (SECV) used for the prediction of EPD were only slightly lower (0.94 and 3.7 %) than values for crude protein (0.99 and 4.1 %) and crude fibre (0.98 and 4.5 %). The expansion of wavelength range (400-2500 nm) did not improve the results of calibration equation for the estimation of EPD. It was concluded that NIRS could predict the protein degradability of grassland forage with high degree of accuracy.

 Poster N4.19

In sacco degradability of crude protein in soybean, extruded soybean and soybean meal

P. Homolka, Research Institute of Animal Production, Department of Animal Nutrition, Prátelství 815, 104 00 Prague, Czech Republic

In this study, nutritive value of soybean, extruded soybean and soybean meal were compared. The nitrogen degradability experiments were performed using *in sacco* method in three dry cows (Black Pied) with a large ruminal cannula (120 mm internal diameter). The cows were fed twice a day (at 6 a.m. and 4 p.m.) and their daily rations consisted of 4 kg alfalfa hay, 10 kg maize silage and 1 kg barley meal with a vitamin and mineral supplement. The bags (nylon bags of pore size 42 µm - Uhelon 130 T, Silk & Progress Moravská Chrastová) containing feed samples were attached to a cylindrical carrier. They were incubated in the rumen for 2, 4, 8, 16, 24 and 48 hours and the estimated ruminal nitrogen degradation was: 63, 67, 79, 97, 99 and 99 % in soybean, respectively, 6, 10, 20, 47, 68 and 97 % in extruded soybean, respectively, 21, 26, 50, 94, 98 and 99 % in soybean meal. Nitrogen effective rumen degradability (rumen outflow rate 6 $\%.h^{-1}$) was determined 88.0 % for soybean, 46.5 % for extruded soybean and 72.9 % for soybean meal. This research was supported by the Grant Agency of the Czech Republic (Grant No.523/02/0164) and Ministry of Agriculture of the Czech Republic (NAZV, project No. QD 0176).

Poster N4.20

Comparative digestibility and prediction of intake of a low quality grass hay by two breeds of cattle using internal and external markers

L.M.M. Ferreira, A. Dias-da-Silva and M.A.M. Rodrigues, CECAV-UTAD, Apartado 1013, 5001 Vila Real Codex, Portugal*

The digestive ability of the Portuguese native breed Barrosão and the Holstein breed was measured with meadow hay (72 - 74 g CP and 641 - 671 g NDF per kg DM) offered either alone or with 15% soya bean meal (DM basis), using four mature cows of each breed. Average live weight (LW) was 457 and 635 kg for Barrosão and Holstein cows, respectively. Animals were fed close to energy maintenance level. Acid insoluble ash (AIA) was used as internal marker and chromium sesquioxide (Cr_2O_3) administrated via controlled release capsules (CRC) was used as external marker to estimate digestibility and intake of hay when offered as the sole component of the diet.

Breed had no effect on DMD of hay offered alone and prediction of DMD was similar by the two markers: 607 and 610 for Holstein and 582 and 583 g per kg DM for Barrosão when AIA or Cr were used, respectively.

Dry matter digestibility (DMD) of the hay was unaffected by supplementation (P>0.05). No interactions between breed and diets were found. Intake was accurately predicted using Cr/AIA.

The results of this experiment suggest that other mechanisms than digestive ability of the native breed may explain its adaptation to their natural environment.

Poster N4.21

Palatability and intake of four indigenous shrubs

D. Camacho-Morfín[1], C.A. Sandoval Castro[2]; J.A. Ayala Burgos[2] and L. Morfín-Loyden[1], [1]Department of Animal Sciences. Faculty of Superior Studies Cuautitlán. UNAM. Campo 4. km 2.5 Carretera Cuautitlán-Teoloyucan, Edo. de Méx. México. [2]Faculty of Veterinary Medicine and Animal Science, University of Yucatán. km 15.5 carret, Xmatkuil, Apdo. 4-116 Itzmná, Mérida, Yucatán, México*

The aim of research was to evaluate the palatability and voluntary intake of *Mimosa biuncifera* Benth (MB), *Acacia farnesiana (*L.) Willd. Sp. (AF), *Eysenhardtia polystachya* (Ortega) Sarg. Silva (EP) and *Buddleia cordata* (H.B.K.) (BC) on sheeps, and their interactions with chemical composition, digestibility, *in vitro* gas production and *in situ* rumen degradability. Palatability of the shrubs was determined by a series of four 4x4 latins squares, each square utilized one sheep. Intake was determined by 4x4 latin square design. The shrubs were offer in fresh and *ad libitum* every morning in both cases. The dry matter degradation was determined using *in saco* method in three cows with a rumminal cannula, Gas production from the tree fodder's fermentation was measured at 3, 6, 9, 12, 15, 18, 21, 24, 30, 36, 48, 60, 72, 96 and 120 h. Correlation analysis was used to test relationships. According the palatability test, sheeps prefered in decreasing order: EP, BC, AF and MB. The intakes data were 76, 29, 11 and 4 g DM/day/kg $BW^{0.75}$. Intake was significantly different between the shrubs ($p < 0.01$). The gas *in vitro* productions between 15 and 24 h were correlated to intake.

Poster N4.22

Silage making from mango and citrus leaves for resource poor emerging livestock farmers in Southern Africa

I.B. Groenewald and B.D. Nkosi, Centre for Sustainable Agriculture, University of the Free State, P.O. Box 339, Bloemfontein, 9300, Republic of South Africa*

Extensive livestock production systems in this country are characterized by major seasonal variations in quality and quantity of natural pastures. This is more evident with resource poor emerging livestock farmers where biological performance is negatively impaired. Mankind and livestock compete for cereals and an alternative feedstuff must be considered.

Numerous agri-byproducts are available. However, such should be preserved for periods of scarcity. Ensiling appears to improve palatability, chemical composition and eliminate pathogens. Consequently, this study was performed where mango and citrus leaves were ensiled either as is or with whey and *L. buchneri* as silage inoculants.

L. buchneri inoculations increased DM yield significantly ($P<0.05$) whilst whey lowered DM yield ($P<0.05$) relative to the control treatment. Main effects showed no significant differences in crude protein and crude fibre. The pH levels declined rapidly to 4.7 with no significant ($P<0.05$) main effects. Whey inoculation produced higher ($P<0.05$) water-soluble carbohydrates than *L. buchneri* or control. Ammonia nitrogen concentrations were not altered ($P<0.05$) by inoculations. *L. buchneri* showed significant ($P<0.05$) higher acetic acid production than whey or the control. Both inoculants greatly decreased ($P<0.05$) CO2 production over the control with no difference between inoculants.

Further studies on lactic acid levels and palatability must be conducted.

Poster N4.23

Evaluation of the nutritive value of ensiled wheat straw and apple mixtures

A.L. Rodrigues, C.V.M. Guedes, L.M.M Ferreira and M.A.M. Rodrigues, CECAV-UTAD, PO Box 1013, 5000-911 Vila Real, Portugal

Four mixtures of wheat straw and apple (0:100, M_{100}, 15:85, M_{85}, 30:70, M_{70}; 50:50, M_{50}) were ensiled at and analysed after15, 30 and 45 days. Every treatment was performed in quadruplicate. The wheat straw IVOMD was 358 g kg MO $^{-1}$. After the different levels of apple were mixed the IVOMD increased to 457, 483 and 590 g kg $^{-1}$ MO for M_{50}, M_{70} and M_{85} respectively.
Non-fiber carbohydrates (NFC) decreased along the ensiling process for all treatments (P<0.001), excepting the M_{100}. The pH decreased with the increase of wheat straw incorporation (P<0.001), but increased along the ensiling process. The acetic acid concentration decreased (P<0.001) during the ensiling process.
The organic matter (OM) in vitro digestibility (IVOMD) of the silages increased with apple incorporation (P<0.001). However, IVOMD decreased along the ensiling process for all treatments (P<0.001), ranging from 770 to 702, 546 to 452, 435 to 385 and 404 to 371 g kg $^{-1}$ MO, for M_{100}, M_{85}, M_{70}, and M_{50}, respectively.
The mixture of wheat straw and apple has potential to be used in animal feeding. However, probably due to low homo-fermentative bacteria activity, the silages were not well preserved and the nutritive value declined markedly. The addition of silage additives should be utilised enabling the faster decrease of pH values and the increase of NFC concentrations.

Poster N4.24

Cell wall degradation kinetics of eight pasture forages in the rumen

P. Shawrang, A. Nikkhah and A.A. Sadeghi, Dep. of Animal Science, Faculty of Agriculture, Tehran University, Karadj, Iran*

This study was carried out to determine ruminal NDF degradation kinetics of eight pasture forages for using in models to estimate the physical fill of fibrous feeds in the rumen. Forage samples were collected by plotting at the preflowering stage in spring. Dried forage samples were incubated for 0, 12, 24, 48 and 72 h in the rumen of three Varamini wethers. An exponential model was used to describe the NDF degradation parameters. The various degradability parameters were analyzed by a variance analysis GLM procedure of SAS in a completely randomised design according to this model: $Y= \mu + Ti + Eij$, where μ is overall average, Ti is the feed effect and Eij is the residual error.
NDF content of Vicia villosa, Bromus tomentellus, Hordeum bulbusum , Festuca ovina, Agropyron tauri, Agropyron trichophorum, Prangus ferulacea and Ferula orientalis were 39.85, 45.1, 61.13, 62.66, 60.45, 63.82, 25.11 and 23.94 respectively. Results of *in situ* trial showed that the effective NDF degradability of these forages at rumen outflow rate 8%/h were significantly (p<0.05) different and were 43.3, 42.8, 32.0, 34.0, 47.7, 38.3, 61.8 and 39.2 percent, respectively. In conclusion, the rate and extent of cell wall degradation in the rumen is different between pasture forages and must be considered as a parameter in developing models for physical fill and intake capacity of sheep.

Poster N4.26

Mercaptoethanesulfonic acid as a new derivatization reagent in the amino acid analysis of food and feed

J. Csapó, G. Pohn, E. Varga-Visi, Zs. Csapó-Kiss and É. Terlaky-Balla, University of Kaposvár, Faculty of Animal Science, Institute of Chemistry, Department of Biochemistry and Food Chemistry. Guba S. u. 40, 7400 Kaposvár, Hungary*

Mercaptoethanesulfonic acid (MES-OH) can be applied for hydrolyzing proteins. The aim of our research was to examine if it can be used also as a derivatization reagent for fluorescence detection of amino acids together with OPA (o-phthaldialdehyde) instead of ME (mercaptoethanol) owning to its thiol-group. Corn, soybean and meatmeal samples were hydrolyzed with hydrochloric acid or MES-OH, derivatized with ninhydrin or OPA/ME or OPA/MES-OH and analysed with IEC or HPLC. There were no significant differences among the amino acid composition results of samples irrespective of choice of the hydrolysis or derivatization methods. MES-OH can be applied not only for hydrolysis but also for derivatization of amino acids. In case of samples with high protein content (>50%), due to the dilution after hydrolysis the MES-OH concentration could be insufficient for derivative formation, in these cases an extra MES-OH addition is required prior to derivatization.

Poster N4.27

Changes in D-amino acid content of corn grain during extrusion

É. Varga-Visi[1], P. Merész[2], É. Terlaky-Balla[1], J. Csapó[1], [1]University of Kaposvár, Faculty of Animal Science, Institute of Chemistry, Guba S. u. 40., H-7400 Kaposvár, Hungary, [2]Budapest University of Technology and Economics, Hungary*

Racemization of peptide-bonded amino acids results in a decrease of protein digestibility, and long-term negative health consequences cannot be excluded. D-amino acid formation may be significant during the processing of food and feed with different technologies. The aim of the research was to determine the amount of D-enantiomers and the level of racemization of amino acids with the highest rate of racemization and occurring in the largest quantities in corn grain extrudates treated with different heat effects (temperature and residence time combinations). Extrusion trials below 144°C did not induce significant ($P<0.05$) racemization. Treatment at 171°C and at 200°C induced significant racemization of aspartic acid (2.4% and 6.1%, respectively). In case of serine and glutamic acid the ratio of D-enantiomers increased significantly due to the treatments at 200°C (0.75% and 0.69%, respectively). With the consumption of 200 g product extruded at 200°C the intake of aspartic acid could be 120 mg and that of D-serin 7 mg. The L-aspartic acid and L-lyisine content of the products extruded at 200°C were significantly lower than in control and in products produced at lower temperatures. The primary cause of the loss of L-aspartic acid was D-aspartic acid formation. In contrast, racemization of lysine played a minor role in the decrease of L-lysine content.

Poster N4.28

Effect of the addition of L-ascorbic acid megadoses to the diet on concentration of 8-oxo-2-deoxyguanozyne and morphological changes in the liver of rats

T. Niemiec[1], E. Sawosz[1], T.Ostaszewska[2], R. Lechowski[3] and W. Bielecki[3], Warsaw Agricultural University, [1]Department of Animal Nutrition and Feed Science, [2]Division of Ichtiobiology and Fisheries, [3]Department of Clinical Science 02-786 Warsaw, Ciszewskiego 8.

The aim of the conducted studies was to determine the effect of megadoses of L-ascorbic acid, added to the diets for rats, on the concentration of 8-oxo-2-deoxyguanozyne (oxidatively modified DNA base products) and on the morphological status of liver. Forty male rats were divided into 4 groups and fed the mixtures which were supplemented with ascorbic acid in the groups (in % of the mixtures): I - 0,0; II - 0,3: III - 0,6 and IV - 0,9. The increase concentration of 8-oxo-2-deoxyguanozyne, determined by HPLC-EC, in liver of the rats received 0,9% vitamin C was observed. In the liver of the rats from group II and IV, the focal necrosis of hepatocytes is found. In the group IV, the smallest number of two-nuclei hepatocytes was observed. The least number of cells was found in the group IV and the greatest hepatocytes were stated in the group II.

The results obtained in our experiments showed that L-ascorbic acid supplemented to the mixtures (0.9%) caused a necrosis damage and reduction of the ability of the liver cells to regenerate as well as the increasing of the oxidatively degradation of DNA.

Poster N4.29

Assessment of neonatal proteinuria in kids using the urinary protein-osmolality ratios in the second-morning and an evening voided urine samples

D. Drzezdzon and W.F. Skrzypczak, Department of Animal Physiology, Agricultural University, Str. Doktora Judyma 6, 71-466 Szczecin, Poland

A preliminary study was conducted to estimate amount of proteinuria and to observe the dynamics of changes during maturation of kid kidneys for protein handling within the first month of postnatal life using the urinary protein-osmolality ratios (Uprot:Uosm). On each day, analyses were done in 12-hour intervals (at 7.00 a.m. and 7. p.m.). In the urine samples the osmolality was measured automatically using cryoscope (Knauer), while total protein concentration was established with Lowry's method.

Upr:Uosm determined in the morning increased on the 2nd day of age from 12.29 to 14.34 mg/l:mosmol/kg. Within the next day decreased nearly twice and the decreasing tendency persisted until the end of the 1st week of age. The slow development of proteinuria to the value of 8.43 mg/l:mosmol/kg was observed in the 2nd and 3rd weeks of age and after that time maintained stable in the range (8.51-9.12 mg/l:mosmol/kg) until the end of 1st month.

The index measured in the evening hours was found to decrease very rapidly to the 3rd day of age (17.46 to 4.95 mg/l:mosmol/kg) and remained stable until 12th day in the range (4.45-5.24 mg/l:mosmol/kg). From the end of the 2nd week the trends of changes in Upr:Uosm observed in the morning and evening were similar. The values of Upr:Uosm in the evening were lower than in the morning, which may indicate the higher tubular reabsorption of protein in the evening hours.

Poster N4.30

The effect of captopril on neonatal proteinuria in calves

M. Ozgo, W.F. Skrzypczak amd A. Dratwa, Department of Animal Physiology, Agricultural University, Str. Doktora Judyma 6, 71-466 Szczecin, Poland

The aim of the study was to explain the effect of angiotensin convertase inhibitor (captopril) on the amount of excreted proteins by calves during the first week of life.

The experiment was carried out on 10 Black&White calves. The calves were p.o. administered with 0.3 mg/kg b.w. of captopril (Sigma). The urine samples were collected before captopril administration, 0.5h and 6h after ACE- inhibitor administration. Total protein concentration was established with Lowry's method.

The results show that the amount of proteins excreted in urine before captopril administration ranged from 5.98 to 3.77 mg/min./m^2. The highest value was observed in the first day of age and the lowest one in the seventh day of life of calves.

The protein excretion in urine 0.5h after ACE- inhibitor administration was decreasing within the first week and ranged from 4.9 to 3.0 mg/min./m^2.

The decreasing tendency of proteinuria was also observed 6h after ACE-inhibitor administration and ranged 5.13 - 2.69 mg/min./m^2.

The inhibition of the angiotensin converting enzyme (by captopril) reduced proteins excretion in calves during the first week of life.

The study was financed by the Committee for Scientific Research (grant no. 3 P06D 02822

Poster N4.31

Detection and quantitation of cellulolytic bacteria *Fibrobacter succinogenes* and *Ruminococcus flavefaciens* in rumen of calves, using quantitative PCR technique

B. Saremi, A.A. Naserian, M. Nassiri, and A. Mohammadi, Animal Science Department, Agricultural college, Ferdowsi University of Mashhad, P.O.Box: 91775-1163, Iran

Eighteen preruminant Holstein calves were placed at three groups and received similar basal diet and different percentages of yeast (*Saccharomyces cerevisiea*) daily (0, 0.5 and 1%). Rumen samples derived at 30, 60 and 90 days old. Results showed that PCR could detect *Fibrobacter succinogenes* (ATCC 19169_T) up to two copy of Pure DNA. Also it had been shown that *Fibrobacter succinogenes* (ATCC 19169_T) and *Ruminococcus flavefaciens* (ATCC 19208_T) could be detected at rumen samples and introduced bands were similar to other studies. In addition Multiple-PCR could be used for detection of them. Results showed that quantitative PCR could be used for enumeration of *Fibrobacter succinogenes* (ATCC 19169_T) and showed that different levels of yeast had no significant effect on population of *Fibrobacter succinogenes* (ATCC 19169_T), but population of this bacterium was increased with growth of calves up to 90 days old. Generally, it seems that quantitative PCR technique could be used for enumeration of *Fibrobacter succinogenes* (ATCC 19169_T) in rumen samples.

Poster N4.32

Feeding liquid whey to holstein dairy cattle

*M. Tahmorespoor**, *R. Valizadeh, A.A. Naserian and B. Saremi, Ferdowsi University of Mashhad P.O. Box 91775-1163, Mashhad, Iran*

Eight Holstein dairy cows with mean milk over 25 Kg in a Latin-squaredesign were allocated to four dietary Treatments as follows; 1) basal diet + tap water (ad lib); 2) basal diet + tap water (ad lib) + Liquid whey (ad lib); 3) basal diet + Liquid whey (ad lib); 4) basal diet + a mixture of tap water and Liquid whey in a ratio of 1:1(ad lib). The animals were subjected to a collection period of 7 days following an adaptation period of 14 days. The dry matter intakes for treatments 1, 2, 3 and 4 were 24.7, 26.6, 27.5 and 25.7 kg/day respectively. The average dry matter provided from feeding whey to the cows in treatments 2, 3 and 4 was 4.6, 8.8 and 4.2 kg/day. Almost a similar trend to total dry matter intake was detected in daily milk production. Generally, milk constituents did not differed significantly, although fat content in treatments with whey was lower. The measured blood indicators and rumen pH were into the normal ranges for an appropriate fermentation. It was concluded that utilization of liquid whey in feeding of dairy cattle not only reduced its pollution problem but also helps to provide least-cost diets in Iranian dairy industry.

Poster N4.33

The performance of finishing bulls offered an all concentrate diet with or without yeasacc[1026]

R.J. Fallon and B. Earley, Teagasc, Grange Research Centre, Dunsany, Co. Meath, Ireland*

The objective was to determine the effects of the inclusion of YeaSacc[1026] (a viable yeast culture, 10^8 CFU/g) at 1.25 kg/tonne in a rolled barley (825 g/kg) and soyabean meal (150 g/kg) diet offered *ad libitum* on the productivity of young finishing bulls. Fifty, 6-month old Holstein/Friesian bulls with an initial weight of 215 kg were allocated following a 3 month pre-experimental acclimatisation period to the following treatments: Concentrate ration (150 g crude protein/kg DM) (C) or Concentrate ration plus YeaSacc[1026] (Y). The bulls were accommodated on concrete slats (3 pens of 6 and 1 pen of 7 bulls) with *ad libitum* access to barley straw and water throughout the 196 day experimental period (25 animals per treatment). Carcass weights were 230 and 245 (sem 6.5) kg for treatments C and Y respectively. The corresponding values for liveweight gain (g/d) were 1190 and 1320 (sem 0.041), for carcass gain (g/d) were 620 and 710 (sem 0.078), for carcass fat scores were 2.76 and 3.00 (sem 0.2), for carcass conformation scores were 1.77 and 2.05 (sem 0.07), and for concentrate dry matter intake (kg/d) were 6.75 and 7.05. It is concluded that the 1.25 kg of YeaSacc[1026] inclusion per tonne of concentrate increased liveweight gain by 25 kg, carcass weight by 15 kg and improved carcass conformation score of young bulls.

Poster N4.34

Effect of the incorporation of Yucca extract in the diet of beef calves

A.E.S. Borba, S. Vieira, C.F.M. Vouzela, O.A.Rego and A.F.R.S. Borba, Departamento Ciências Agrárias. Universidade Açores, 9700 Angra Heroísmo, Portugal*

The main purpose of this report was to study the viability of an additive, Yucca extract, on the productive performance of beef calves.

For this purpose 104 calves were divided in two groups, fed with Yucca extract (I) and without Yucca extract (II). In each group 10 animals were selected (5 males and 5 females, I' and II'), that were weighed and an individual blood sample was taken at each 10 days for both groups, for subsequent plasma urea nitrogen analysis.

Calves were suckled for 50 days And the Yucca extract was incorporated in the milk (3g/day/animal).

The values of weight variation (PV), plasma urea nitrogen levels (PUN) and mean daily gain (GMD) were compared by analysis of the variance.

In the case of groups I and II PV, there were significant differences ($p<0.05$). In groups I' and II', there were no significant differences.

There was a significant ($p<0.05$) sex effect, the males having the largest GMD. Within each Group I and II, the difference between both sexes was significant ($p<0.05$) in Group I for PV values, but not for GMD values. Within Group II, there were no significant differences, neither for PV nor for GMD. In relation to GMD of both groups, there were significant effects ($p<0.05$). No significant "Yucca effect" ($p<0.05$) on PUN levels could be seen.

Poster N4.35

Effects of liquid diet composition and feeding frequency on rumen fermentation and performance in calves

B. Niwinska, J.A. Strzetelski and K. Bilik, National Research Institute of Animal Production, 32-083 Balice, Poland*

Differences in rumen fermentation and performance in calves can be affected by liquid diet composition and frequency of feeding. The experiment was carried out with bull-calves, aged from 7 to 120 days, The calves were given to appetite a concentrate mixture and from 7 until 56 days of age were fed two types of isonitrogenous liquid feeds: cow's milk or milk replacer according to INRA (1989) requirements, once or three times a day. Crude protein of the milk replacer was composed of 40% soya concentrate and 60% whey. The liquid diet composition and feeding frequency did not influence the daily intake of concentrate mixture ($P>0.05$). Higher levels of butyric acid ($P\leq0.05$) and lower of propionic acid ($P\leq0.06$) in the sum of VFA in the rumen content of 3-, 6-, 9- and 13-week-old calves fed cow's milk were found in comparison with calves receiving the milk replacer. In calves receiving milk at 3, 6, 9 ($P\leq0.01$) and 13 weeks of age ($P\leq0.02$) higher daily weight gain and better liquid feed nutrient utilization were noted. The feeding frequency did not influence the VFA concentration and molar proportions of individual VFA in the rumen content, although higher daily intake of liquid feed ($P\leq0.05$) and higher daily weight gain ($P\leq0.09$) were noted in the 3- and 6-week-old calves fed three times a day.

Poster N4.36

Effect of Terramycin on growth Performance and feed Conversion efficiency of growing calves

A.A. ElShahat[1] and E.E. Raghab[2], [1]Anim. Prod. Dept. NRC, Dokki (12622), Cairo, Egypt, [2]Anim. Prod. Res. Institute, Dokki, Cairo, Egypt*

Terramycin, a composite antibiotic, is an antimicrobiol feed additive that is produced as a fermentation product of streptomyces. It has been reported that terramycin may improve average daily gain and/or feed conversion efficiency of dairy cattle. Therefore, the objective of the present trial was to evaluate the effect of terramycin on growth rate, digestibility coefficients and feed conversion of feed lot cattle. However, Brown Swiss bread heifer calves were fed either a control or terramycin supplemented ration. Live weights were measured at 14 day intervals. Routine vaccinations were administered in the receiving trial and heifers were treated for internal and external parasites. At the end of the experiment, the data were analyzed, and it was observed generally that daily gain and the digestion of dry matter, crude protein and nitrogen free extract were not affected by feeding treatment differences. On the other hand, feed to gain ratios were somewhat improved by terramycin supplementation. The results will be discussed on the scientific basis of animal nutrition and cattle production.

Poster N4.37

Description of a mechanistic model integrating preruminant calves growth

B. Saremi, A.A. Naserian, M. Bannayan and F. Shahriary, Animal Science Department, Agricultural college, Ferdowsi University of Mashhad, P.O.Box: 91775-1163, Iran

The objective of this study was to describe a mechanistic model, which stimulates preruminant calves growth based on three different indexes (Net energy, Apparently digestible protein and mean of them). Calves growth period was divided: 1) Milk fed calves 2) Milk+calf starter fed calves (Pre-weaning period) 3) Post-weaning period. Model contains six pools: 1) Daily dry matter intake stimulation 2) Feeds and their percent, Weather condition and calves birth weight 3) Stimulation of feed's nutrients and their efficiency for growth or maintenance 4) Simulation of protein requirement of calves 5) Simulation of energy requirements of calves 6) Simulation of weight gain, based on net energy and apparently digestible protein. The lowest estimation is based on net energy and highest one is based on apparently digestible protein in growth period, but simulation values based on net energy, apparently digestible protein and mean of them are in other studies reported ranges. Model behavior about daily weight gain is similar to body weight at different days. Model made good dry matter intake stimulation in 90 days, although it was really hard to stimulate the dropping dry matter intake at weaning.

Poster N4.38

Evaluation of a mechanistic model integrating prerumiant calves growth

B. Saremi, A.A. Naserian, M. Bannayan and F. Shahriary, Animal Science Department, Agricultural college, Ferdowsi University of Mashhad, P.O.Box: 91775-1163, Iran

A model was designed according to NRC 2001 equations, which simulates growth of preruminant calves based on three different indexes (Energy, Apparently digestible protein and mean of them). Results showed that the best estimation of calves weight in different ages was based on energy index (RMSD=7.251). Also model estimations of calves weight were generally greater than observations, so there were positive MBE values with RMSD=11.549, 16.705 and 7.251 respectively for mean index, Apparently digestible protein and energy. Daily weight gain estimation according to energy index was lower than observed data, so MBE value was negative (RMSD=0.250). About the other two indexes (Apparently digestible protein and mean index), results showed that model estimations were greater than observations with positive MBE values (RMSD=0.241 and 0.307 respectively). Model made an over estimation about dry matter intake with positive MBE value (RMSD=0.666). According to model estimation trends, this model is a start point for designing a complex model. With respect to correct determination of observations trend, it seems a suitable model.

Poster N4.39

Effects of starch level and roughage intake on animal performance, rumen wall characteristics and liver abscesses in intensively fed Friesian bulls

K.F. Jorgensen, S.B. Larsen, H.R. Andersen and M. Vestergaard, Danish Institute of Agricultural Sciences, P.O. Box 50, DK-8830 Tjele, Denmark*

In total, 41 Danish Friesian bulls (154±20 kg LW) were allocated into four treatment groups in a 2x2 factorial design fed either a high starch (N) or a high fibrous concentrate (F) and either chopped barley straw (S) or a mixture of 75% chopped barley straw and 25% sugar beet molasses (Sm). In the F ration, 25% of the wheat from N was replaced by dried sugar beet pulp and grass pellets giving a starch level of 24.3% and 39.1% (wt) and an energy content of 0.99 and 1.04 Scandinavian Feed Units/kg feed, respectively. The *ad libitum* fed bulls were slaughtered at 11.5±0.2 mo. weighing 429±41 kg LW on average. Feed intake was higher for F than N ($p=0.005$) (7.1 vs. 6.4 kg/d), whereas the addition of molasses increased straw intake from 0.5 to 0.9 kg/d ($p<0.001$). The higher intake of Sm compared with S reduced concentrate intake ($p<0.05$), but did not affect total net energy intake. The average daily gain (1370 g/d) did not differ between groups. The rumen wall was more severely damaged with N compared with F ($p<0.05$) and with S compared with Sm ($p<0.003$). The higher straw intake prevented rumen wall damage, but did not reduce the number of animals with liver abscesses, like the lower starch level did (1 vs. 9 animals, $p<0.005$).

Poster N4.40

Selected aspects of the mineral nutrition of the European bison

M. Dymnicka[1], M. Debska[1], W. Olech[2], Warsaw Agricultural University, The Faculty of Animal Sciences, [1]Department of Animal Nutrition and Feed Science, [2]Department of Genetics and Animal Breeding, 02-786 Warsaw, Ciszewskiego 8, Poland

The aim of the studies was to evaluate the content of nutrients in feeds, consumed by the bison, with the particular consideration of the mineral components, and to determine the selected biochemical parameters in tissues and in the hair in order to state the degree of the organism's supply of minerals, proteins and energy.
The practical purpose is to develop a formulation of mineral mixture and to correct the quantity and composition of the administrated concentrates (being additionally given to the bison), basing on the analysis of feeds and the selected biochemical parameters in tissues and hair of bison.
Based on previous conducted studies, i.e. the determined biochemical indices in blood (from 58 bison), the results show: mean glucose contents (14,3 mmol/l), exceeding many times the upper reference values for cattle (4,5 mmol/l) and values, obtained by other authors (for bison); levels of Cr and Co below the sensitivity of the method; very low levels of Se (0,15-0,20 μmol/l); low levels of Zn (12,3 μmol/l); low levels of Cu (9,7 μmol/l); high levels of Fe (35,8 μmol/l); high levels of K (9,1 μmol/l).The studies will be continued.

Poster N4.41

Estimation of rumen volume by different markers as reference to expected volume in Holstien bulls

S.M Salem, Anim. Prod.Dept., Fac.ofAgric.,Cairo University.Giza, Egypt

Twenty five Fresian bulls, weighing about 420 Kg±32 were used in an experiment under the same dietary regime , to determine the correlation between the different methods of rumen volume determination and live body weight and to postulate regression equations which help to determine rumen volume without using marker techniques,also to define an accurate method for rumen volume determination.
The rumen volume was estimated from the rate of decline in concenteration of a marker in the rumen fluid following a single ruminal injection. Rumen volumes as determined by different methods are significantly ($p<0.01$) correlated with live body wight. The correlation coefficient values ranged from 0.60 to 0.78.. The results of several determination methods of rumen volume using different materials (lithium sulphate, polyethelienglycol, Cr-EDTA and Cr_2O_3) as markers gave values with differences ranging from 6 to 18%. The direct measurements through substituting rumen content by water (physical volume) gave values ranged from 30 to 36% higher than estimated by using marker methodes. the accuracy of estimating rumen volume by using the method of PEG marker (solid state marker) was the highest accurecy ($r^2 = 0.94$) in comparison to physical rumen volume (true volume), meanwhile, the accuracy (r^2) of other markers to estimate rumen volume were 0.62, 0.45, 0.31 for lithium sulfate, Cr_2O_3, Cr-EDTA, respectively, in comparison to physical rumen volume.

Poster N4.42

Passage of nylon capsules and Cr-mordanted corn silage through the digestive tract of cows in lactation

J. Trinácty[1], M. Richter[1], P. Homolka[2], P. Vesely[3], [1]Research Institute for Animal Breeding, Ltd., Department Pohorelice, [2]Research Institute of Animal Production, Praha-Uhríneves, [3]Mendel Agricultural and Forestry University Brno, Czech republic*

The trial was performed with four dairy cows, producing in average 21.9 (s.e. 0.82) kilograms of milk per day. The animals consumed daily 18.3 kg of DM. The evaluation was performed in two periods lasting 5 days. In each of them 500 capsules were given to each experimental animal. The shape of capsules was lenticular with outside and inside diameters 10 and 8 mm. Capsules were made of nylon cloth with the mesh of 42 micrones. The capsules were applied orally in the form of a paper bolus. The total amount of Cr-mordanted corn silage applied (in diet) one cow was 100 g (3.04 % of Cr in DM). The recovery of nylon capsules and Cr during 120 hours was 94.7 and 91.6 %, respectively. The total mean retention time (TMRT) of nylon capsules and Cr was 36.2 and 44.8 h, respectively. Similarly delay time (the midpoint of the interval when the capsule or Cr appeared for the first time in the faeces) was 12.8 and 8.3 h, CMRTS (sum of two compartmental retention times) was 23.4 and 36.5 h, respectively. Calculated ruminal retention time was 22.4 and 33.0 h, and calculated outflouw rate from reticulo-rumen was 4.47 and 3.03 % ($P<0.05$) for capsules and Cr, respectively. This study was supported by GACR (523/02/0164).

Poster N4.43

The effect of feeding ration with protected methionine on milk yield and physiological parameters in dairy cows

V. Kudrna, P. Lang and P. Mlázovská, Research Institute of Animal Production, 10401, Prátelství 815, Prague-Uhríneves, The Czech Republic*

A supplement with protected methionine included in a feeding ration (FR) with 16.1 % of crude protein (CP) was verified in the experiment with 27 dairy cows, divided into 3 groups: C - control, with 18.2 % of CP in FR, E- experimental- 16.1 % of CP, and EM - experimental with methionine-16.1 % of CP +MET). FR of the EM-group was supplemented with 16 g/head/day of rumen protected methionine.The main goal of the experiment was to ascertain, whether the examined supplement, contained in the FR (formed above all by the large proportion of corn and alfalfa silages), may improve utilisation of CP and thus allow this lower dosage.The higest milk yield (39.88 kg/head/day) was achieved in the C-group with CP concentration of 18.2 %. The ascertained difference in milk perfomance compared with the groups with 16.1 % of CP was statisticaly significant ($P<0.05$). On the other hand the difference, which was found in milk protein concentration, was detrimental to 18.2 % CP concentration. Together with milk yield the CP concentration in the FR also statistically significantly influenced contents of urea in blood plasma and in milk. The supplement of rumen protected methionine also affected insemination interval and service-periode positively.
The project supported by NAZV (National Agency for Agricultural Research of The Czech Rep.) No. QD 0176.

Poster N4.44

Effect of dietary protein on rumen fiber kinetics and fiber digestion in dairy cows

C. Brøkner[1,2], P. Lund[1], M. R. Weisbjerg[1], T. Hvelplund[1] and P. Nørgaard[2], [1]Department of Animal Nutrition and Physiology, Danish Institute of Agricultural Sciences, Research Centre Foulum, P.O. Box 50, DK-8830 Tjele, Denmark, [2]Department of Animal Science and Animal Health, The Royal Veterinary and Agricultural University, Grønnegårdsvej 2, DK-1870 Frederiksberg C, Denmark*

Effect of insufficient supply of rumen degraded protein on rumen fiber kinetics was investigated in four rumen cannulated Holstein cows. The cows were fed ad libitum with the same total mixed ration low in degradable protein. Feed intake varied from 16 to 20 kg DM/d. Two cows were supplemented with urea (150 and 50 g/d, respectively), to increase rumen protein balance (PBV). Faecal excretion was calculated using feed indigestible NDF (INDF) as marker and rumen pools were determined from total evacuations. Obtained PBV-level was -13 g/SFU (Scandinavian Feed Unit) for the unsupplemented cows and -4 and +13 g/SFU for the supplemented cows; and mean retention time (MRT) of INDF and fractional rate of degradation (k_d) of digestible NDF were: MRT = 40 h, 37 h and k_d = 0.023 h^{-1}, 0.040 h^{-1} for the two supplemented cows; and MRT = 26 h, 25 h and k_d = 0.043 h^{-1}, 0.034 h^{-1} for the two unsupplemented cows. Based on these four observations, urea supplementation at the actual PBV levels seemed not to have a positive effect on fractional rate of fiber degradation, in contrast to earlier studies.

Poster N4.45

Effect of timing to feed corn silage-based supplement for grazing dairy cows on intake, milk production and nitrogen utilization

T. Mitani, M. Takahashi, K. Ueda, H. Nakatsuji and S. Kondo, Graduate school of Agriculture, Hokkaido University, Sapporo 060-8589, Japan

As herbage contains high ruminal degradable protein, the efficiency of nitrogen utilization is low in grazing dairy cows. Corn silage (CS) supplementation to grazing would be effective to reduce ruminal NH_3-N loss because its high contents of easily digestible carbohydrate (NFC) will improve nitrogen capture by ruminal microbes. The timing of NFC supplementation is importance to maximize microbial growth. Four Holstein dairy cows were used to determine intake, milk production and nitrogen utilization of grazing dairy cows supplemented CS-based ration 2 hours before grazing (Pre) or immediately after grazing (Post). Cows were grazed on pasture for 1700-1930 and 0530-0800. Herbage intake was estimated by the body-weight gain method. Total intake was not difference between Pre and Post as herbage intake was not different. Milk protein yield for Pre tended to be higher than for Post, while milk yield did not differ between Pre and Post, Total nitrogen intake for Pre tended to be higher than for Post. There was no difference in urinary nitrogen output between Pre and Post, but the proportion of urinary nitrogen output to total nitrogen intake for Pre was lower than for Post. Milk nitrogen output and nitrogen retention for Pre were higher than for Post ($P<0.06$). Therefore, CS-based supplementation 2 hours before grazing would improve nitrogen utilization.

Poster N4.46

Effect of energy and protein feeding levels in prepubertal and postpubertal Black-and-White X Holstein-Friesian heifers on productive traits and milk yield

K. Bilik, J.A. Strzetelski. and B. Niwinska, National Research Institute of Animal Production, 32-083 Balice, Poland*

Effects of feeding pre- and postpubertal heifers with different energy (UFL) and protein (PDI) levels on productive traits and milk yield were investigated. The studies were conducted with 43 Black-and-White heifers (77.5% Holstein-Friesian breeding) from 6 months of age to 100 days of lactation. From 6 do 11 months (period 1) and from 12 to 13 months of age (period 2) they were fed: in group K, the diets met 100% of the UFL and PDI requirement, for weight gains about 700 g/day. In other groups compared to group K, UFL and PDI in periods 1 and 2 were (%): 85/85 and 115/115 (group A); 85/115 and 100/100 (group B); 115/85 and 100/100 (group C); 115/115 and 115/115 (group D) respectively. From 14 months of age to 100 days of lactation the animals were fed similarly. For each age group and for pre- and post-calving periods, highly significant correlations were found between body condition scores on a scale of 0 to 5 and thickness of subcutaneous fat on the back. Increasing the dietary levels of UFL and PDI in pre- and postpubertal heifers by 15% in relation to those recommended for a weight gain of 700 g/day improved body condition and increased fatness of the heifers in the range considered optimal for this breed.

Poster N4.47

The incorporated model for the using of the grassland trough the pastured with cows for milk and the superior procession of milk

I. Scurtu, 5,Cucului Street, code 500128, Brasov, Romania

In the countries with an advanced zootechnology (France, Switzerland, Italy, Austria etc.) the production of milk per cow is superior to that obtained in our country. In the current study, 20 ha of grassland, which had been fenced, provided with water supply and improved through the application of chemical complex fertilizer (NPK) at times when the climatic conditions will allow this. The number of cows in the experiment is 124 cows of which 114 are in the milk period and 10 are dry waiting to gave birth for the next months. The average production of milk on head / animal is 16,5 l. The total production of milk from the herd is processed within the development where there exists a section for the industrial processing of milk. Through the procession of milk different milk products (cream cheese, green cheese, cream, whipped cream) are obtained. For the laboratory analysis regarding the quality of milk, the milk products as well as the blood analysis from the animals are examined.

Poster N4.48

Comparison of different processing method of barley for dairy cows

G. R. Ghorbani, H. Sadri, and M. Alikhani, University of Isfahan Technology, Animal Science Department, College of Agriculture, Isfahan, 84156 Iran*

A duplicated 4 × 4 Latin-square experiment (21- d periods) was conducted to compare differently processed barley in dairy cows diets and their effects on milk yield and ruminal fermentation characteristics. Eight multiparous lactating Holstein cows, 85 ± 15 d in milk, were given a total mixed ration of 43:57 forage:concentrate ratio (DM basis). Treatments diets were: 1) ground barley (with a mean particle size of about 1 mm), 2) steam-rolled barley (with PI = 68%), 3) dry-rolled barley (with PI = 72%), 4) dry-rolled barley (with PI = 81%). The processing index (PI) is a measure of weight of a given volume of barley after rolling as a percent of the weight of the whole barley. Treatment 2 and 3 in compared to treatment 4 showed higher daily milk yield (28.25, 28.82 vs. 26.59 kg/d), protein yield (0.84, 0.86 vs. 0.80 kg/d) and treatment 1 and 3 showed higher production of 4% FCM (27.42, 27.44 vs. 25.57 kg/d), as compared with treatment 4. However, milk fat and fat yield, milk protein, milk lactose, milk total solid, DMI, feed efficiency, apparent digestibilities of organic matter and dry matter, ruminal pH, urine pH, proportion of VFA, and acetate-to-propionate ratio were not affected by the treatments ($P > 0.05$). These results indicate that dairy cows fed TMR diets consisting of 25% ground barley grain (dry matter basis), and with adequate fiber, can be used effectively, without any negative effects on dry matter intake, milk production, milk fat and ruminal pH.

Poster N4.49

The effect of different forage type levels on ruminal pH and chewing activity in dairy cows

A.A. Naserian, A.R. Alizadeh, M. Sarei and B. Saremi, Animal Science Department, Agricultural college, Ferdowsi University of Mashhad, P.O.Box: 91775-1163, Iran*

The study was carried out to investigate effects of different levels and types of fiber source with constant level of whole cottonseed on ruminal pH and chewing activity in early lactation dairy cows. Eight first lactation dairy cows (60 days in milk) were assigned to four treatments in a Latin square design. The treatments were consisting of 36% alfalfa hay, 24%: 12% alfalfa hay: barely silage, 12%: 24% alfalfa hay: barely silage and 36% barely silage. The cows in different treatments were fed a supplementary 10% whole cottonseed. Diets were fed adlibitum as a total mixed ration with forage to concentrate ratio of 36:64. Diets were calculated to meet requirement of cows in early lactation according to NRC (2001). Ruminal pH was significantly different between treatments ($P<0.05$) and the highest was related to cows that barely silage is the major part of forage portion. Cows non-significantly spent most time chewing eating when alfalfa hay was as a major fiber source of diet, and spent most time rumination when fed 12% alfalfa hay: 24% barely silage. The results of this experiment showed that alfalfa hay can better stimulate physiological responses to effective fiber.

Poster N4.50

Research on on the influence of mineral pellets on the performance of the sheep in lactation

D. Dragotoiu, M. Marin, E. Pogurschi and C. Pana, University of Agronomical Sciences and Veterinary Medicine Bucharest, Romania, Faculty of Animal Sciences, Marasti 59, 70000 Bucharest, Romania

The scientific calculation of fodder rations allows their energetic and protein balancing by changing diet/forage. Howver, the mineral balance is, in most of the cases, difficult to achieve, as the mineral elements can be found in relatively small amounts.
Under these conditions, inorganic minerals should be supplemented in various ways according to the feeding characteristics of each animal species and category.
Starting from the macro and microelements needs of the milking sheep and taking into account the toxicity limits, three variants of mineral pellets were created, either by mineral elements sources of by the content of microelements.
The minerals added in the rations of the milking sheep, both by sources and by different levels of mineral elements, had a positive influence on milk production.
During the first 3 months of milking, the production was lower with the control, compared with the experimental batches (2.45-4.92% in the first month; 6.5-16.2% in the second, 21.2-29.2% in the third).

Poster N4.51

Subterranean clover on male lamb feeding: effect on carcass composition

D. Settineri, V. Pace, M. Iacurto, K. Carbone, F. Spirito and F. Vincenti, Instituto Sperimentale per la Zootecnia, via Salaria, 31 - (00016) Monterotondo (Roma)

Subterranean clover (**SC**) is well known for its detrimental effects on reproduction; otherwise there are only few studies on anabolic effects of phytoestrogens which are present in **SC**. The aim of this research was to investigate these effects on the carcass composition and dissection data of growing lambs. Two groups of 8 Comisana male lambs were fed, for about 185 days (I.A.L.W =29.7±6.16 kg), on subterranean clover (phytoestrogen content =0.2% on DM) or Italian ryegrass (**G**) *ad libitum* and supplemented with mais grains and sunflower meal to maintain adequate and similar energy and protein intakes. During the feeding period the ADG of **SC** group resulted 33.9 % higher than **G:** 154.25 *vs* 101.93g/d; the slaughtering weights (corrected for I.L.W.) were significantly ($P \leq 0.05$) different: 58.7**a** and 47.7**b**kg respectively (+ 19.4%) and the right half carcasses were noticeably heavier (14.66**a** *vs* 11.18**b**).
The main differences, recorded at leg dissection, were on weights (4.29**a** *vs* 3.42**b**kg); meat and subcutaneous fat percentages were very similar (65.21 *vs* 66.12%; 7.09 *vs* 6.53%), therefore the differences were consistent on total and internal fat (11.51**a** *vs* 9.85**b**%; 4.41**a** *vs* 3.31**b**%) and bone (20.84**b** *vs* 21.88**a**%).
These results seem to confirm the anabolic positive effect of SC administration that was previously noticed, by a sharp weight increase, on Friesian male calf growth.

Poster N4.52

Comparison of nutrient digestibility of sheep ration's differing in main protein and/or non-forage fiber source

Ch. Milis[1], D. Liamadis[2], M. Dasilas[2] and I. Prapas[2], [1]Ministry of Agriculture, [2]Department of Animal Nutrition, Aristotle University of Thessaloniki 540 06. Greece

An in vivo digestibility trial was conducted by the use of 4 castrated mature rams in a 4x4 Latin square experimental design. Significance level was determined at P<0.05. Four ration's differing in main protein source, cotton seed cake (CSC) vs corn gluten meal (CGM), and/or non-forage fiber source, wheat bran (WB) vs corn gluten feed (CGF), were used, as follows: A) CSC and WB (control), B) CSC and CGF, C) CGM and WB and D) CGM and CGF. The four diets were designed to provide equal amounts of energy, protein and fiber, and were covering maintenance requirements for energy and protein. Ration D had higher digestibility of DM (76.1 vs 73.9; 74.7; and 74.5%, for rations D, A, B, and C respectively), OM (80.2 vs 78.9; 78.8; and 78.6%), NDF (59.9 vs 56.8; 56.6; and 56.2%), ADF (60.9 vs 52.1; 48.3; and 52.2%) and cellulose (73.1 vs 65.7; 66.4; and 64.9%). Above results suggest that CGF has superior fiber fraction digestibility compared to WB. Moreover, hemicellulose digestibility of B ration (67.1%) was higher compared to A (62.4%), which is in line with former conclusion. CP digestibility of A ration was the lowest (62.9 vs 74.1; 74.0; and 73.4%, for A, B, C and D rations, respectively), suggesting that the combination of CSC and WB affect negatively ration's CP digestibility. This could be explained, probably, by a reduced MCP synthesis, due to inadequate FME of this ration, or lower RUP digestibility or both.

Poster N4.53

Concentrate level and Vitamin E supplementation do not interact on milk yield and fatty acid (FA) composition in goats receiving alfalfa hay and linseed oil

A. Ferlay[1], J. Rouel[1], P. Capitan[1], E. Bruneteau[2], P. Gaborit[3], L. Leloutre[4] and Y. Chilliard[1], [1]INRA-Theix, 63122, France, [2]INRA-Lusignan,[3]ITPLC-Surgères, [4]INZO, France*

Forty-eight multiparous mid-lactation goats received alfalfa hay *ad libitum* and 130 g/d (i.e. 4.5% of diet DM) of linseed oil, and were used in a 2x2 factorial design, including 2 levels of concentrate (medium=1.4kg/d, MC, vs low=.7kg/d, LC) and 2 supplements (0, or 1250 IU/d VitE), for 5 weeks. Total DM intake was 2.88 kg/d and did no differ across diets. Diets MC, compared to LC, tended to increase (P<.10) milk yield (+.28 kg/d), decreased (P<.05) milk fat % of branched-chain (BC) FA, 18:0 and 18:3n-3, increased 10:0 to 14:0 (21.4 *vs* 17.6%) and *trans*10-18:1 (1.03 vs .49%), and increased the normalized delta-9 desaturation ratios for *trans*11-18:1. VitE supplementations tended to increase (P<0.10) milk fat yield (+17 g/d), decreased milk fat % of 10:0 to 16:0 (36 vs 40%) and odd-numbered (ON) FA, increased BCFA, 18:0, *trans*11-18:1 (8.8 *vs* 7.6%), *cis*9,*trans*13-18:2 and *trans11,cis15*-18:2, and decreased the normalized delta-9 desaturation ratios for 14:0, 18:0 and *trans*11-18:1. A small significant concentrate%-VitE interaction was observed on ONFA only. In conclusion, increasing concentrate and starch levels, up to 52 and 25% respectively, did not change largely milk yield and composition in goats receiving a linseed oil-rich diet. Vit E supplementation increased several *trans* isomers (but not rumenic acid nor the *trans*-11/*trans*-10 18:1 ratio) and did not interact with concentrate level (*This work was funded by EU BIOCLA Project QLK1-2002-02362*).

Poster N4.54

^{32}P uptake by erythrocytes in sheep supplemented with phosphorus at pasture

D. A. Antunes[1], H. Louvandini[1], J. C. da Silva Filho[2], C. M. McManus[1], B. S. Dallago[1] and B.O. Machado[1], [1]University of Brazilia, Agronomy and Veterinary Medicine Faculty, C.P. 04508, Brazilia,DF, 70910-900 Brazil, [2]University of Federal of Lavras, Departament of Animal Science, C. P. 37, Lavras, MG,72.000-000, Brazil*

This experiment aimed to evaluate the effect of phosphorus supplementation on ^{32}P uptake by erythrocytes in sheep at pasture. Twenty sheep, with a live weight of 13.88 ± 2.51 kg, kept on *Andropogon gayanus* pasture were divided in two treatments of 10 animals each, one with supplementation of 3g/phosphorus/animal/day and the other without P supplementation. After 30 days, four fortnightly blood collections were carried out and ^{32} P was added to determine uptake, as well as P, calcium and glucose in the plasma. No significant differences were found for Ca and glucose levels between treatments. The level of P in the plasma was higher for animals receiving supplementation (6.75, 6.96, 6.83 and 6.64 mg/100ml) to those without supplementation (3.9, 4.5, 4.34 and 4.39 mg/100ml) for the fortnightly collections respectively ($p<0.05$). The ^{32}P uptake was greater in the treatment without supplementation (7.4, 8.4, 9.4 and 12.1%) and progressive with time compared to supplemented animals (4.2, 4.3, 3.3 and 4.2%) in respective collections ($p<0.05$). The ^{32}P uptake was shown to be an important additional tool to evaluate sub-clinical P deficiency in sheep.

Poster N4.55

Optimum level of concentrates for gaining rapid growth in lohi lamb

F. Ahmad, M. Afzal and K. Javed, Livestock Production Research Institute, Bahadurnagar, Okara Punjab, Pakistan

A feeding trial was conducted to determine the optimum level of concentrates for accelerating the growth rate in Lohi lambs. A ration containing 16% CP and 80% TDN was tested on three groups A.B & C having 26 animals in each group. Concentrate ration was given @ 1.2 and 3% of their body weights. The daily growth rate recorded was 105 127 & 132 gms and feed efficiency was 2.61, 4.41 & 6.41 in groups A, B & C respectively. The trial lasted for 79 days. The cost of feed for production of one kg live weight was Rs. 15.66, 26.46 and Rs.38.46 in groups A.B & C respectively. The cost of production was low in Group-A which was fed on 1 per cent ration of its body weight.

Poster N4.56

Performance of sheep supplemented with P using 12th rib analysis

H. Louvandini, D. A. Antunes, C. M. McManus, B. S. Dallago and B.O. Machado, University of Brazilia, Agronomy and Veterinary Medicine Faculty, C.P. 04508, Brazilia,DF, 70910-900 Brazil*

The aim of this study was to evaluate the performance of sheep supplemented with P using the analysis of 12th rib as an indicator. Twenty sheep weighing 13.88 ± 2.51 kg and kept on *Andropogon gayanus* pasture, divided in two treatments of 10 animals. In group 1 supplementation with 3g/phosphorus/animal /day was carried out and in group 2 no P supplementation was given. After 75 days the animals were slaughtered and carcass measurements taken. The 12th rib was removed for analysis. No significant differences in live weight, carcass kill out, or commercial cut weights were found between the two treatments, except for abdominal cavity organs which tended to be heavier in the non-supplemented group ($p=0.0954$). The 12th rib analysis showed that the supplemented group had a tendency for heavier rib and bone weight ($p<0.05$) as well as muscle weight ($p=0.0618$) when compared with animals not supplemented with P in the feed. Other measurements such as fat cover and rib eye area were not significantly affected by the treatments. It was shown that P supplementation improved performance in young sheep, including improved formation of bone and muscle tissue using the composition of the 12th rib as a reference.

Poster N4.57

Effect of Silymarin on plasma levels of retinol and α-tocopherol in periparturient goats

M.S. Spagnuolo[1], P.Abrescia[2], A.Carlucci[2], L. Cigliano[2], S.Galletti[3], D.Tedesco[3] and L. Ferrara[1], [1]ISPAAM-CNR, via Argine 1085, 80147 Napoli, Italia,[2]Dipartimento di Fisiologia Generale ed Ambientale-Università di Napoli Federico II, via Mezzocannone 8, 80134 Napoli, Italia[3]Dipartimento di Scienze e Tecnologie Veterinarie per la Sicurezza Alimentare, via Celoria 10, 20133 Milano, Italia*

The peripartum period is associated with an increase in metabolic processes, and in free radicals production. The imbalance between free radicals production and defense mechanisms could lead to "oxidative stress".It was reported that dietary supplementation of periparturient cows with antioxidants contributes to protect from the harmful effects of free radicals. Retinol and α-tocopherol enhance immunity and mantain the structural and functional integrity of cells. Silymarin was reported to possess antitoxic, antioxidant, and antiinflammatory properties. Twenty goats, in the dry period, were divided into two groups, according to health condition, and parity (=2). Starting from seven days before calving until 15 days after calving, ten goats received 1 g/d of silymarin, as a water suspension, by oral drench. Plasma samples were collected at d- 7, 0, 14, 21, and 28 from calving, and analysed for retinol and α-tocopherol concentrations. Plasma levels of retinol and α-tocopherol did not significantly differ between control and treated animals during peripartum. The concentrations of these antioxidants were found lower at calving than at d-7, and increased after parturition, in control goats, while did not change, in treated animals, in the peripartum. These results suggest that silymarin might reduce antioxidants consumption associated with calving.

Poster N4.58

Dietary linseed oil in lamb production. 1. Productive performances and carcass and meat quality

A. Caputi Jambrenghi, F. Giannico, R. Favia, F. Minuti, M. Ragni, A. Vicenti and G. Vonghia, University of Bari, Faculty of Agriculture, Department of Animal Production, Via G. Amendola 165/A, 70126 Bari, Italy*

The use of linseed oil containing C18:3ω-3 in feed could be a valid alternative to soybean oil containing C18:2ω-6. This study observed productive performances, carcass and meat quality of lambs whose feed contained linseed oil.

20 male Gentile di Puglia lambs were weaned at 45 days old and immediately divided into two homogeneous groups of 10. Control group lambs were fed on hay and commercial concentrated feed containing soybean oil (30 g kg^{-1}); experimental group lambs on hay and commercial concentrated feed containing linseed oil (30 g kg^{-1}). Energy, protein and fibre contents of both diets were the same. Lambs were slaughtered at 75 days. Productive performances of lambs were evaluated live and after slaughtering. Raw and cooked samples of *Longissimus lumborum* (Ll) and *Semimembranosus* (Sm) muscles were tested.

No effect of dietary treatment was observed regarding feed consumption, weight gain, final live weight, body and carcass form, dressing percentage, pH, chilling loss, half-carcass section, lumbar region dissection, chemical composition, colour and cooking loss of the meat. The linseed oil treatment differed from the soybean oil ($P<0.05$) in determining an increased bone percentage of pelvic limb (27.30 *vs* 24.51%) and an increased WBS of cooked Sm (4.57 *vs* 3.39 kg/cm^2).

Poster N4.59

Dietary linseed oil in lamb production. 2. Ω-3 PUFA of meat

A. Caputi Jambrenghi, F. Giannico, R. Favia, S. Sanapo, V. Megna, G. Marsico and G. Vonghia, University of Bari, Faculty of Agriculture, Department of Animal Production, Via G. Amendola 165/A, 70126 Bari, Italy*

Ruminant meats are a relatively good source of ω-3 PUFA due to the presence of C18:3 in grass. In Mediterranean areas where grazing lands are scarce and on farms where livestock is reared in sheds, it is possible to increase ω-3 PUFA contents of ruminant meats by feeding animals on grain-based diets which include whole linseed. The present study observed the effects on the fatty acid composition of lamb meat when linseed oil was used in feed.

The experimental procedure was the same as that described in Caputi Jambrenghi et al., Proceedings EAAP 2004a. Gas-chromatography was used to determine the fatty acid composition of intramuscular fat extracted from the raw and cooked *Longissimus lumborum* muscle.

When feed containing linseed oil was compared to feed containing soy oil, it increased the percentage of C18:3ω-3 in the cooked meat (1.24 *vs* 0.98%; $P<0.05$); this can be attributed to the higher content of this fatty acid in the raw meat, and to the smaller loss of the same following cooking. Linseed oil did not change the content of saturated, monounsaturated and polyunsaturated fatty acids, the indexes relating to atherogenicity and thrombogenicity, the ratio PCL/PCE, but improved the ω-6/ω-3 ratio in both raw and cooked meat, although this was not statistically significant.

Poster N4.60

Effects of previous diet and duration of soybean oil supplementation on conjugated linoleic acid and octadecenoic isomers in lamb meat

R.J.B. Bessa, M. Lourenço, P.V. Portugal and J. Santos-Silva, INIAP-Estação Zootécnica Nacional, 2005-048 Vale de Santarém, Portugal

Forty Merino Branco lambs were housed and randomly assigned to four groups. The trial duration was 42 days divided in 3 periods of 14 days. The diets used in the trial were commercial concentrate (C), dehydrated lucerne pellets (L), and pellets consisting in 90% lucerne meal and 10% soybean oil (O). Experimental groups were named as CCO, COO, LLO and LOO according to diet fed in each period (i.e. CCO group were fed C diet in the first 2 periods and O in the last). After 42 days the lambs were slaughter and *Longissimus thoracis* muscle were sampled. Lipids were extracted and fatty acid methyl esters prepared and analyzed by gas chromatography and expressed as % of total fatty acids. ANOVA was conducted using a model considering the effect of the initial diet (C vs. L), duration of lipid supplementation (14 days vs. 24 days) and their interaction.

The C lambs (CCO and COO) had higher C18:1 *trans*-10 and less C18:2 *cis*-9, *trans*-11 (0.98 vs. 1.38) than L lambs (LLO and LOO). The duration of lipid supplementation increased ($p<0.05$) C18:1 *cis*-12, C18: n-6, C18:3 n-3 only in C lambs whereas the concentration of C18:1 *trans*-11, C18:1 *trans*-12 and C18:2 *cis*-9, *trans*-11 (1.02 vs. 1.34) increased in both C and L lambs.

Poster N4.61

Cottonseed manufacturing process effect on nutrient digestibility and N balance of sheep rations

D. Liamadis[1], Ef. Lales[1], Ch. Milis[2] and Z. Abas[3], [1]Aristotle University of Thessaloniki, [2]Ministry of Agriculture, [3]Dimokritio University of Thrace, Greece

The high production of cotton in Greece results in a number of industrial cottonseed by-products that are available, mainly, as feedstuffs in ruminant rations. Nevertheless, in most cases, these by-products have not been nutritionally evaluated enough. Four rations containing one of four cotton seed by-products, whole cotton seed (WCS), cotton seed pellets (PCS), cotton seed meal (CSM) and cotton seed cake (CSC), where fed at four castrated mature rams, in an *in vivo* digestibility trial, using a 4x4 latin square experimental design. The four rations, A) WCS (control), B) PCS, C) CSM and D) CSC, additionally, contained equal amounts of alfalfa hay, and where covering maintenance requirements for energy. Rations A and B had higher ($P<0.05$) DM and OM digestibility in comparison to rations C and D (61.0; 61.0; 59.1 and 59.4%, for DM and 63.8; 64.1; 62.3 and 62.2%, for OM, respectively), suggesting that WCS and PCS have a superiority compared to CSM or CSC. Significant differences were not revealed on CP digestibility (75.0; 75.4; 75.4 and 74.3%, respectively) nor on Biological Value $(N_{intake}-N_{feacal}-N_{urine})/(N_{intake}-N_{feacal})$ (51.5; 51.0; 49.7 and 42.6%, respectively) or Net Protein Utilization $(N_{intake}-N_{feacal}-N_{urine})/N_{intake}$ (38.6; 38.4; 37.5 and 31.6%, respectively). Former result indicates that manufacturing process doesn't impair CP digestibility and N balance of cottonseed by-products.

Poster N4.62

Interactions between starchy concentrate and linseed oil supplementation on goat milk yield and composition, including *trans* and conjugated fatty acids (FA)

J. Rouel[1], A. Ferlay[1], E. Bruneteau[2], P. Capitan[1], K. Raynal-Ljutovac[3] and Y. Chilliard[1], [1]INRA-Theix, 63122, France, [2]INRA-Lusignan, [3]ITPLC-Surgères, France*

Forty-eight multiparous mid-lactation goats received alfalfa hay *ad libitum* and were used in a 2x2 factorial design, including 2 levels of concentrate (1.6kg/d, or medium, MC, vs .9kg/d, or low, LC) and 2 lipid intakes (0 g/d, or 130 g/d, i.e. 4.4% of diet DM, of linseed oil, LO) for 5 weeks. Diets MC (9% starch), compared to LC (30% starch), increased (P<.05) milk lactose and protein contents, decreased milk fat % of branched-chain (BC) FA, 18:0 and 18:3n-3, increased 10:0 to 14:0 and *trans*10-18:1 (.67 vs .28%), and increased the normalized delta-9 desaturation ratios for *trans*11-18:1. Diets LO increased milk fat (+5.6 g/kg) and lactose (+2.3 g/kg) content and milk fat yield (+22 g/d), decreased 10:0 to 16:0 (40 vs 58%), odd-numbered (ON) FA and 18:2n-6 and increased branched-chain (BC) FA, 18:0, *trans10-* and *trans11*-18:1, *cis*9,*trans*13-18:2, *trans*11,*cis*15-18:2 , *cis*9,*trans*11-18:2 (3.2 vs .5%) and 18:3n-3 (1.4 vs .6%). Diets LO decreased the normalized delta-9 desaturation ratios for 14:0, 17:0, 18:0 and *trans*11-18:1. Significant concentrate%-oil interactions were observed for almost all the FA%, particularly 16:0, and *trans*10-18:1. In conclusion, diets LO improved production performances, increased 18:0, 18:3n-3 and *trans*-FA, particularly *trans*11-18:1 and *cis*9,*trans*11-18:2, and decreased delta-9 desaturation ratios, and medium-chain FA. The effects of 54 *vs* 29% concentrate were slight, but interacted strongly with LO effects on FA% (*This work was funded by EU BIOCLA Project QLK1-2002-02362*).

Poster N4.63

Effect of the form of rapeseed and linseed in lamb diets on some health quality parameters of meat

B. Borys[1] and A. Borys[2], [1]National Research Institute of Animal Production Cracow, Experimental Station Koluda Wielka, 88-160 Janikowo, Poland, [2]Meat and Fat Research Institute, Jubilerska 4, 04-190 Warsaw, Poland*

Four groups of ram-lambs were fattened intensively with complete diets to 30-35 kg of body weight. Group I (control) received a standard diet. The experimental diets contained 10% rapeseed and linseed at the 2:1 ratio; in the form of whole seeds (group II), by 50% seeds and meal (III), and 100% meal (IV). No clear effects of the nutritional treatments on the growth rate and nutrient utilization of the feeds were found. Feeding the oilseed diets increased dressing percentage (by 3.8 p.u.), muscle tissue in leg (2.1 p.u.) and loin eye area (17.3%), with 22.8% greater external fatness. Ground oilseeds in the diet increased the content of intermuscular fat (IMF) by 4.4% and unground seeds caused a 6.6% decrease. Health parameters of meat based on the fatty acid profile tended to be less favourable in the groups of lambs fed with the oilseed diets than in the control group. However, oilseeds increased the content of CLA, with a clearly beneficial effect of grinding the seeds on the value of this parameter (increases of 13.8, 24.1 and 58.6% in groups I, II, and III). The use of oilseeds, whether ground or not, had a beneficial effect on decreasing the cholesterol content of lamb muscles by 7.4%.

Poster N4.64

Effect of feeding milked sheep with rapeseed on the yield and quality of milk and curd cheese

B. Borys[1] and A. Borys[2], [1]National Research Institute of Animal Production Cracow, Experimental Station Koluda Wielka, 88-160 Janikowo, Poland, [2]Meat and Fat Research Institute, Jubilerska 4, 04-190 Warsaw, Poland*

The effect of feeding rapeseed to milked Friesian×Polish Merino ewes on the quantity and quality of milk and curd cheese were investigated. Control [C] and experimental [R] group were fed during lactation with the same bulky feeds and with different concentrate (standard in group C; with rapeseed at 150 g/day/animal in R). Two series of observations were made on bulk milk obtained at 8. and 13. weeks of lactation. Analysis was made on milk and cheese yield and quality, with regard to selected parameters of health quality, yield of cheese mass, and organoleptic evaluation of cheese. Feeding the ewes with rapeseed compared to the control group was found to cause increases in body weight, daily milk production (7.7%) and fat percentage in milk and cheese (10.3 and 15.1%). The addition of rapeseed to the diet modified the fatty acid profile of milk and cheese, mainly by increasing the proportion of MUFA in milk and cheese (34%). The use of rapeseed had a favourable effect on increasing the CLA in milk and cheese (34.1 and 41.8%) and exerted an unfavourable effect on increasing the cholesterol (23.4 and 14.9%). There were no differences in the organoleptic characteristics of the cheese obtained from the milk of group C and R.

Poster N4.65

Yield and quality of culinary meat from fattened lambs as related to form of dietary rapeseed and linseed

J. Strzelecki[1], E. Grzeskowiak[1], B. Borys[2], A. Borys[1] and K. Borzuta[1], [2]Meat and Fat Research Institute, 4 Jubilerska St., 04-190 Warsaw, Poland, [2]National Research Institute of Animal Production in Krakow, Experimental Station Koluda Wielka, 88-160 Janikowo, Poland*

The effect of dietary form of rapeseed and linseed on the weight and percentage of culinary meat in half-carcasses of lambs fattened intensively to 30-35 kg of body weight was investigated. 24 ram-lambs were fattened in four groups. The control group [I] was fed a all-mush with no oilseeds and experimental groups [II-IV] received 10% rapeseed and linseed at the 2:1 ratio - group II whole seeds, III by 50% whole and kibbled seeds, IV only the kibbled. The half-carcasses were dissected according to the procedure developed at the Meat and Fat Research Institute. Muscles and some culinary products were evaluated organoleptically. Meat quality was analysed for basic chemical composition and physico-chemical properties. The studies showed a favourable effect of feeding with oilseeds on the weight of half-carcasses, cuts and culinary meat obtained. The use of 50% kibbled rapeseed and linseed proved the most beneficial, giving the highest rate of growth and dressing percentage in this group. No significant differences in percentage of cuts and saleable meat in half-carcasses were found in relation to the form of dietary oilseeds. The analysed parameters of meat quality confirmed no significant differences among the feeding groups of the lambs.

Poster N4.66

Investigations on the tryptophan requirement of lactating sows

D.A. Roth-Maier and B.R. Paulicks, Technical University Munich, Department of Animal Sciences, Division of Animal Nutrition, Hochfeldweg 6, D-85350 Freising-Weihenstephan, Germany*

The requirement of tryptophan (Trp) for lactating sows was examined in a total of 72 lactations of sows nursing 10-12 piglets. The sows were assigned to one of the six treatments (1-6; n=12), which differed only in the Trp concentration of the lactation feed (gTrp/kg feed: 1.2/1.5/1.8/2.4/3.0/4.2, resp.) derived by adding crystalline L-Trp to a basic feed (1.2g Trp, 0.8g apparent, 0.9g true ileal digestible Trp/kg). All the other parameters were kept constant. Dietary Trp contents of up to 2.4g/kg increased feed intake of the sows during lactation significantly by about 40% to 5.53kg/d. Further dietary Trp supplementations did not affect feed intake. Weight loss of the sows during lactation, milk yield, and piglet weaning weight reacted correspondingly. Contents of milk fat, protein, lactose, energy, dry matter and amino acids in milk protein were not affected. Due to the lower milk supply the piglets grew slower and could not compensate this delayed growth performance even by consuming an additional starter feed. Regression analysis of the present data resulted in a requirement of 2.3g Trp/kg feed for lactating sows corresponding to 1.9g apparent (2.0g true) ileal digestible Trp/kg feed to avoid impairments of feed intake and lactation performance. These values exceed the NRC recommendations by about 20% and suggest a Trp supplementation of practical diets based on corn, cereals and soybean meal.

Poster N4.67

Effect of organic acids on dietary self-selection by the piglet

F.X. Roth[1], K. Mentschel[1] and T. Ettle[1], [1]Technical University of Munich,Department of Animal Science, Division of Animal Nutrition and Production Physiology, Hochfeldweg 6, 85350 Freising-Weihenstephan, Germany*

Dietary additions of organic acids are often used to improve growth and health status in pig production. The objective of the present study was to investigate if there is a preference of piglets to diets supplemented either with potassium(K-)diformate, formic or sorbic acid compared with an unacidified diet. For this purpose, two growth experiments in weaned piglets were carried out with a choice-feeding equipment. In the first experiment, acidifying piglet diets with K-diformate tended to increase growth and feed intake compared to a reference group given exclusively a non-acidified diet. If given a choice, piglets consumed the unacidified diet or the alternative diet acidified with either 1.2% or 2.4% K-diformate at random throughout the experimental period. In the second experiment, acidifying the diets with either 1.2 % formic or sorbic acid increased feed intake and growth numerically compared to a reference group given exclusively an unacidified diet. In contrary to the K-diformate supplementation, acidifying diets with either formic or sorbic acid lead to the rejection of the acidified diets in a free choice situation. In each experimental week piglets consumed significantly more from the unacidified diet compared to diets acidified either with 1.2% formic or sorbic acid. Even if all tested organic acids have the potential to stimulate feed intake, formic and sorbic acid are avoided at a great extent in a choice feeding situation.

Poster N4.68

The influence of nutritive regime on the meat quality of pigs

S. Sevcíková, M. Koucky, Research Institute of Animal Production Prague - Uhríneves, Czech Republic

The objective of the experiment was to study of effect of a diet with lower energy level and higher fibre content on the content and composition of fatty acids in the adipose tissue of fattened pigs and to compare it with a commercial control diet. A fattening test was carried out in 2 groups of pigs (n=40) of the final hybrid combination (Norway Landrace x Landrace) x (Large White x Pietrain) in the finishing stage from 65 kg of live weight to average slaughter weight of ca. 110 kg. The experimental diet was fed for 45 feeding days and a representative number of individuals from each group (n=10) was slaughtered at the end of the experiment.
Fatty acids in the adipose tissue were determined by FAME gas chromatography; the contents of particular fatty acids were expressed as percentage proportions out of the sum of total fatty acids. Total SFA were reduced significantly ($P \leq 0.001$) while the effect on total MUFA was negligible. There was an increase in total PUFA ($P \leq 0.001$): in PUFAn-6 ($P \leq 0.001$) LA (C18:2), C20:2, AA (C20:4), C22:2 and C22:4 increased, and in PUFAn-3 ($P \leq 0.01$) LNA-α (C18:3) and C20:3 increased and DPA (C22:5) and DHA (C22:6) decreased. The content of PUFAn-6 was higher in the index by 50.1% and of PUFAn-3 by 23% compared to the control group.
This study was supported by the Ministry of Agriculture of the Czech Republic (project No. M02-99-04).

Poster N4.69

Effect of purified cellulose and pectins on digesta passage rate in growing pigs

F. Tagliapietra, S. Schiavon, L. Bailoni R. Mantovani and C. Ceolin, Dipartimento di Scienze Zootecniche, Viale dell'Università 16, 35020 Legnaro (PD), Italy

Two groups of 3 pigs each, housed in metabolism cages, received a commercial feed supplied with of 0, 5 and 10% of pure cotton cellulose or apples pectins, according to a two 3x3 latin square designs with periods of 5 weeks. IDF and SDF ranged from 14.9 to 23.3 and 1.2 to 7.2% of dry matter intake (DMI). During the last week of each period a pulse dose of 5 g of Cr_20_3 was supplied to each pig and faeces were collected from 4 to 96 hrs. after Cr_20_3 exposure, with intervals of 4 hrs.. From the kinetic of excretion the time of first appearance of marker in faeces (τ), the compartment turnover rate (τ), and the total mean retention time (MRT) were estimated. Data were subjected to ANOVA for the effects of period, pig and the percentage of IDF and SDF on DMI. Apparent digestibility of the added IDF and SDF averaged 7.7±7.7 and 98.2±2.0 %, respectively. MRT, τ and λ averaged 40 hrs., 27.9 hrs. and 0.177 hrs.$^{-1}$, respectively. The increase of 1% of IDF/DMI significantly reduced MRT (-0.85 hrs., $P<0.01$) due to significant increase of λ (+0.0109 hrs.$^{-1}$; $P<0.01$) and decrease of τ (-0.19 hrs.; $P<0.01$). SDF/DMI did not influence MRT because of opposite effects of λ (-0.0056 hrs.$^{-1}$; $P<0.01$) and τ (-0.65 h; $P<0.01$).

Poster N4.70

Nitrogen balance and protein utilization in growing Cinta Senese pigs

L. Pianaccioli, A. Acciaioli, O. Franci, C. Pugliese, G. Ania, Dipartimento di Scienze Zootecniche, Firenze, Italy

This work aimed to study the protein digestibility and the nitrogen balance in Cinta Senese in comparison with Large White pig, employing 4 diets with different crude protein levels (7, 10, 13, 16 %). Soybean meal was used as protein source.
Six Cinta Senese (CS) and 6 Large White (LW) barrows were submitted to digestibility trial using individual metabolic cages to collect faeces samples and total urine. Animals live weight ranged from 50 to 90 kg and diets were supplied on a basis of 90 g/kg of metabolic weight.
The endogenous faecal protein, estimated as intercept of the regression equation of total faecal protein on the diet CP, was employed to calculate the real digestibility of protein which was no different between breeds (88.4 vs 87.8 % in CS and LW respectively). With the 7% diet, CS pigs showed the highest value of retained nitrogen (19.4 vs 9.9 g/d) at 50 kg of l.w., but the lowest at 90 kg of l.w. (15.3 vs 77.8 g/d). With 16% diet no differences between breeds for retained nitrogen were found at 50 kg (65.8 vs 64.3 g/d for CS and LW respectively) while at 90 kg of l.w. CS pig showed the lowest value (86 vs 161 g/d).

Poster N4.71

Quantity of deposited fatty-acids in swine products depending on feeding time

T. Ács[1], Hermán Istvánné[1], Andrea Lugassi[2] and J. Gundel[1], [1]Research Institute of Animal Breeding and Nutrition, 2053 Herceghalom, Hungary, [2]National Institute of Food Hygiene and Nutrition, P.O. Box 52. 1097 Budapest, Hungary

The goal of our experiment was to present the effect of the feed with various fatty-acid contents on the fatty-acid composition of animal tissue.
We created two types of feeds, one with extracted soy(ES) and the other with full-fat soybean(FSB), and fed one after the other in 3 treatments for 25-75, 50-50 and 75-25 days before slaughter.
During the entire fattening process, the DG was the best on the group fed FSB for 75 days, but because of the high DFI, the FCR was poor.
We found that the quantity of SFA and MUFA in fatty acid composition of back-fat increased significantly, while the level of PUFA decreased significantly during treatments 1-3. The rate of linoleic acid(C18:2) decreased from 26,6% to 14,1% and 20,5%, and the linolenic acid(C18:3) from 2,9% to 2,6% and 2,2%.
There was no change in the composition of the limbs' SFA level. The level of MUFA increased significantly, while the level of PUFA decreased significantly in the treatments. The rate of C18:2 decreased from 19,5% to 18,2% and 14,2%, and the C18:3 from 2,9% to 2,6% and 2,2%.
We can conclude that the 75 day long intake of FSB increased significantly the C18:2 and C18:3 contents of the intra and extra muscular fat.

Poster N4.72

Different Cu- and Zn-Supplementations in nutrition of high performance pigs

A. Berk[1], G. Flachowsky[1], J. Fleckenstein[2] and U. Meyer[1], Federal Agricultural Research Centre, [1]Institute of Animal Nutrition, [2]Institute of Plant and Soil Science, Bundesallee 50, 38116 Braunschweig, Germany*

A three factorial experiment to investigate i) the level, ii) the influence of binding form of copper (Cu) and zinc (Zn) supply, where iii) half of the animals get feed without and a half with additional phytase, were arranged.
The 3 levels of trace element content were: 1. Native feed content (8 mg Cu, 36 mg Zn/).
2. Supplementation of 2 mg Cu, 35 mg Zn/kg and 3. Of 9 mg Cu, 95 mg Zn/kg feed. The elements were of inorganic ($CuSO_4$ and $ZnSO_4$) or of organic (amino acid-trace element-complex) sources. 5 groups get a phytase supplementation (750 U/kg feed) to the feed and 5 groups received only the native phytase content. That means 10 groups, 5 barrows and 5 gilts each, were tested while the whole fattening period from 25-115 kg LW.
Mean life weight gain (LWG) was 887 g/day. A variance analyse showed only a significant influence of the trace elements origin. The "organic" groups showed better performance data then the "inorganic", without a significant difference to the "native" groups. Comparing all 10 groups by a multiple t-test showed no significant differences.
It seems that the native Cu and Zn content in common pig feed is sufficient for high performances. Organic supplements results in slightly better performances.

Poster N4.73

Aktivität der selenabhängigen Glutathionsperoxydase (GSH-Px) im Ebersperma in verschiedenen Jahreszeiten

B. Lasota, B. Seremak, B. Blaszczyk, J. Udala, Landwirtschaftliche Universität Szczecin, Poland, 71-466 Szczecin, ul.Judyma 6, Poland

GSH-Px-Aktivität im frischen und konservierten Ebersperma in 4 Jahreszeiten (Frühling, Sommer, Herbst und Winter) wurde untersucht. Von 39 Besamungsebern wurden 87 Ejakulate entnommen. Der Se-Gesamtgehalt im für die Zuchteber üblichen Futter lag bei 0,43 µg/kg. Die Ejakulate wurden in Spermaplasma und Spermien getrennt In beiden Spermakomponenten wurde GSH-Px-Aktivität unter Anwendung von fertigen Bestecken bestimmt. Die Spermaqualität erfüllte die für Besamungseber gestellten Anforderungen. Die mittlere GSH-Px-Aktivität im frischen Spermaplasma für alle Jahreszeiten lag bei 165 U/l; die größte war im Sommer (205 U/l) und die geringste im Herbst (122 U/l). Unmittelbar nach der Verdünnung stieg die GSH-Px-Aktivität im Plasma - im Vergleich mit dem Frischsperma- rasch ab und verringerte sich bis zum 1. Aufbewahrungstag. Dann blieb sie auf etwa gleichem Niveau bis zum 2. Aufbewahrungstag unverändert. Am 3. Tag erfolgte ein nächster Abfall der GSH-Px-Aktivität auf etwa zweifach niedrigere Werte. Ähnliche Veränderungen der GSH-Px-Aktivität im verdünnten Sperma wurden in allen Jahreszeiten beobachtet. Die mittlere GSH-Px-Aktivität in Spermien lag bei 37 mU/ 10^9 Spermien; höher war sie im Sommer (41 mU/^{9} Spermien), am geringsten im Winter (28 mu/10^9 Spermien). Als Schlussfolgerung kann gesagt werden, dass die Beteiligung von GSH-Px am Antioxidant-Mechanismus im verdünnten Ebersperma beträchtlich geringer als im Frischsperma ist. Dies kann eine der Ursachen für ungenügenden Schutz der Spermien vor ROS sein. Die GSH-Px-Aktivität im Ebersperma, besonders im verdünnten Sperma, zeigte eine Tendenz zur jahreszeitlichen Abhängigkeit.

Effect of organic and inorganic selenium sources on selenium status of sows and their piglets

K. Kirov[1], M. Dimitrov[2], S. Denev[2], S. Yotov[2], [1]Central Laboratory of Veterinary Control and Ecology, National Veterinary Service, Sofia [2]Trakian University, Stara Zagora 6000, Bulgaria*

The objective of the study was to compare the efficacy of inorganic and organic selenium (Se) sources to sow diets during gestation and lactation on Se status of sows, on the transfer of Se into milk and subsequently to the nursing pigs. Clinically healthy (*Dunabe White* x *Landrace*) pregnant sows were allotted on the basis of parity and body weights within three groups: (I) Inorganic Se (sodium selenite), (II) Organic Se (Sel-Plex®, *Alltech Inc.*) and Negative control (III). The both Se sources were added to the diet at 0.3 ppm Se. A non-Se-fortified corn-soybean meal basal diet served as negative (untreated) control. Both sources of Se increased sow serum Se 20 day postpartum (P<0.001). The differences between the supplemented groups and the control group were also highly significant (P<0.001). Sel-Plex® in comparison with sodium selenite was the superior Se source not only during gestation, but also lactation. Se level in colostrum and milk was higher for the group given organic Se (P<0.001). Inorganic Se was less effectively incorporated into the milk of lactating sows. Organic Se improved Se status of sows, increased colostrum and milk Se content more than did inorganic Se and increased the nursing piglet serum Se. Organic Se significantly increased piglet serum Se at 20 (P<0.01) and 35 d postpartum (P<0.001). These results indicate a higher bioavailability of organic Se than sodium selenite.

Poster N4.76

Pea (*Pisum sativum*) and Faba bean (*Vicia faba*, variety minor) as an alternative protein source in growing-finishing pig diets

M. Morlacchini[2], A. Prandini[1], M. Moschini[1] and F. Masoero[1], [1]ISAN, Facoltà di Agraria, Piacenza-Italy, [2]CERZOO San Bonico, Piacenza-Italy

The use of alternative Mediterranean feed protein sources into pig diets was investigated. 96 male (38.2±6.9 live weight) and 72 female (39.1±5 live weight) Duroc × (Large white × Landrace) pigs were randomly assigned to four dietary treatments and raised in 28 pens (7 pens per treatment/6 animals per pen). Experimental treatments were identified by the protein source as: soybean meal (CTR), raw Pea (PPC), raw Faba bean and the CTR diet at lower protein content (14 vs. 16% as fed) plus synthetic amino acids (CTR-N). Diets were formulated according to the INRA requirements for the 40-80 kg, 80-120 kg and 120-160 kg growing live weight and fed at 9% of the pen metabolic body weight. Pig mortality was recorded daily and dead animals were removed and weight was recorded. The pen feed consumption and weight were recorded and the pen final weight (196d) was cleared from weight of removed animals. The pen average daily gain (ADG) was obtained and adjusted feed to gain ratio (F:G) was calculated. The ADG, average daily intake and F:G were not statistically affected by the treatment diets. Data suggest substitution of soybean with raw Pea or raw Faba bean had no effect on animal performance and could represent an alternative valuable protein source in swine diet formulation.

Poster N4.77

Hematological indices of Large White piglets

E.A.Tyan, Research Institute of Veterinary Genetics and Selection, Novosibirsk State Agrarian University, Russia

Morphological and physical-chemical bloods features are the result of prolong biological evolution. The blood flows through all organisms' cells and performs the nutritious, protective, excretory and other function. All organism changes are reflected by blood. We suppose that it's necessary to organize the comprehensive cytogenetic, hematological, biochemical, immunological, chemical and etc. monitoring of selected animals' populations.

We studied some hematological indices in Large White piglets aged 19-21 days of "Kudryashovskoye" pig-raising complex.

The average erythrocytes content in pigs was $7.46*10^{12}$/L, leucocytes - $10.60*10^{9}$/L which had the high phenotypical variability. The erythrocytes' and leucocytes' blood concentration were established within physiological levels (EFL). Hemoglobin level, one erythrocyte hemoglobin level and color index were slightly higher of EFL and were 130.25 g/L, 19.94 g/100 mL, 1.28 respectively. These indices were increased and characterized by hemoglobin saturation of erythrocytes. It's necessary to underline that all studied indices were normal distributed.

Thus, it's more simple and effective to study hematological indices by preliminary physiological organism state evaluation during the complex monitoring and make an appraisal of animal populations.

Poster N4.78

The alanine and aspartate aminotransferase activity of Large White piglets

E.A.Tyan, Research Institute of Veterinary Genetics and Selection, Novosibirsk State Agrarian University, Russia

The blood enzymes are more effective and sensitive indices that play the paramount role in organism's metabolism. They are used for productivity traits evaluation at young age.

The researches were carried out on Large White piglets aged 19-21 days in "Kudryashovskoye" pig-raising complex. The alanine and aspartate aminotransferase (ALT, AST) activity as well as the DERITIS coefficient were found. The levels of ALT and AST activity were 1.13 ± 0.04 and 1.15 ± 0.04 mumol/L*hour respectively. The level of transferase activity and DERITIS coefficient were within physiological limits for studied age bracket. It indicate the good physiological status of animals. We suppose that the obtained data can be used as population norm of biochemical indices of Large White piglets.

It is necessary to underline that the individual variability of studied biochemical indices was higher than hematological indices in piglets at the same age. It is known that the aminotransferase activity is low in suckling piglets and increased in piglets aged 2 month. The maximum activity is observed in piglets at the age of 4 month (Gudilin I.I., Petukhov V.L., Dementyeva T.A., 2000). Therefore, we will study the aminotransferase activity in Large White piglets of different age.

Theatre NCS5.1

Increasing the utilisation of forage protein in ruminant diets

R.J. Dewhurst, A.H. Kingston-Smith, R.J. Merry and F.R. Minchin, Institute of Grassland and Environmental Research, Plas Gogerddan, Aberystwyth SY23 3EB UK*

This paper reviews potential to improve the level and properties of forage protein. There is renewed interest in this area because of the need for cheap, traceable, plant-derived proteins for ruminants. Legumes typically contain 1.5 to 2 times more protein than grasses and have the further benefit of producing this without requiring N fertiliser. The efficiency of conversion of forage protein into milk and meat protein is often low (15-25%). Opportunities to increase N utilisation include strategies to reduce protein degradation in the rumen, increase rumen microbial protein synthesis, and improve energy supply to the animal tissues. Plant defence mechanisms provide some natural protein protection mechanisms: tannins (e.g. in *Lotus* spp.) and the quinones produced by the action of polyphenol oxidase (e.g. in red clover), bind to proteins and provide some protection from rumen degradation. There is scope to select grasses on the basis of activities of the proteolytic enzymes associated with plant cell death which is induced when grass is ingested and living cells enter the rumen. Providing additional energy for rumen micro-organisms, either through breeding grasses with a high sugar content, or feeding complementary low-protein forages (such as maize silage) is a further way to increase N-use efficiency. Finally, forages with high intakes, such as white clover, increase N-use efficiency without apparent improvements in rumen efficiency- confirming the importance of energy/protein balance in animal tissues.

Theatre NCS5.1

Alternative protein sources for pig feeding in Europe

B. Sève, INRA UMRVP 35590 Saint-Gilles, France

Ban on meat meals and recent decisions at EU or WTO level posed new challenges to feed industry regarding protein provision in diets for single-stomach animals. Among plant protein sources, sunflower and rapeseed oil meals could be proposed as first substitutes for soybean meal. Development of double zero cultivars of rapeseed allowed incorporation of rapeseed meal (canola meal) in diets of sows and growing-finishing pigs. Work on grain legumes first focussed on antinutritional factors (ANF), and new cultivars of peas, faba beans or lupine with low ANF contents were developed. Pea production in Europe peaked in 1993 but has been declining since then and current plant genetics research aim at improving crop safety in terms of yield and resistance to disease. Development of new concepts such as net energy and digestible amino acids, and systematic measurements on alternative oil meals or grain legumes have been powerful means for developing their incorporation in diets, together with industrial amino acids, on basis of least-cost formulation. More recent findings will lead to taking into account the impact of endogenous protein losses on amino acid availability. In addition, new data are currently obtained allowing appreciation of the risk for gut health of using alternative protein sources. Furthermore, more attention is currently paid to the impact of current feed manufacturing processes (grinding, pelleting and heating) on the protein, energy and health values of cereals and grain legumes.

Theatre NCS5.2

Improving the nitrogen value of proteagineous and oleagineous grains for ruminants

C. Poncet[1], M. Doreau[1], D. Rémond[2], [1]URH, [2]UNMP - INRA - Theix - 63122 St-Genès-Champanelle, France

Proteagineous and oleagineous grains can contribute with forage legumes to increase the protein autonomy for farmers. These grains exhibit high contents in energy and proteins. However these proteins are rapidly degraded in the rumen and the fraction actually supplied to the animal is low.
A variety of methods are available to decrease protein degradability without altering intestinal digestibility; their applications in Europe are scarce, because of the omnipotence of soybean meal.
The nitrogen value of grains can be improved through conventional breeding. Until now, breeding efforts aimed at improving the feed value for monogastric animals. Now they are also directed towards ruminant feeding, selecting very different potentially effective genetic features.
Grains can be fed whole or coarsely cracked. Production trials showed the efficiency of this simple way of protein protection. This solution has to be optimised and its impact on nitrogen value needs to be quantified.
Numerous heat treatments have been proposed. Few of them are used in Europe, except extrusion and heat treatments included in the process of oil extraction. Oil extraction methods can be easily adapted to grain processing.
Chemical treatment with formaldehyde is successfully used, but the future of this chemical agent remains clouded because of questions on its health hazards. Conversely, protein protection by adding natural tannins is gaining in interest, but the process is not yet settled.

Theatre NCS5.3

Dry matter intake,milk composition and production in dairy cows fed whole and processed cottonseed

A.R. Foroughi[1], R. Valizadeh[2], A.A. Naserian[2] and M. Danesh Mesgaran[2] , [1]Education centre of khorasan Jihad-Agriculture, Animal Science Department,Mashhad, [2]Ferdowsi University of Mashhad , Animal Science Department,Iran

The objective of this study was to evaluate the effect of grinding and heating of whole cottonseed (WCS) on milk composition and production of Holestein lactating cows during early lactation. Multiparious cows (n=8) averaging 84.50±10.34 days in milk and 36.10±4.46 milk yield (MY) were used in a 4×4 Latin square design. Cows were divided into four groups, receiving one of the following treatments: 1) WCS; 2) Ground cottonseed (GCS); 3) GCS heated in 140°C and steeped for 2.5 minute (GHCS2); or 4) GCS heated in 140°C and steeped for 20 minute (GHCS2). Diets were similar CP, NDF and NEL. The percentage of whole or processed was fixed at 14%. The mean DMI was significantly ($P<0.01$) affected by diets and in treatments of 1,2,3 and 4 were 25.97, 27.24, 27.63, and 27.63 (kg/d), respectively. MY was significantly ($P<0.01$) affected by the diets and was greatest for HGCS2 (34.13 kg/d) and the lowest for WCS (31.13kg/d). Supplementation with HGCS2 resulted in increased milk fat percentage ($P<0.05$) and milk fat yield ($P<0.01$). Milk protein percent was progressively increased, averaging 3.21%, 3.02%, 3.42% and 3.42% for 1,2,3 and 4 treatments, respectively. Milk protein yield tended ($P<0.05$) to be higher for cows fed CHCS1 and GHCS2 than for cows fed WCS and GCS.

Theatre NCS5.4

Performance, carcass traits and blood constituents of growing lambs fed rations containing bread wastes

A.A.Gabr, A.E. Abd El-Khalek,and M.S. El-Haisha, Animal Production Department, Faculty of Agriculture, Mansoura University. 35516, Egypt

Ten lambs 3-4 months old (1/2Romanov×1/2 Rahmani), and 21.2 Kg live weight (LBW) were randomly divided into 2 equal groups to study the effect of partial replacement of concentrate feed mixture-protein (CFM) by bread wastes protein (BW) at the rates 0.0 (R1), 12.5 (R2), 25 (R3) and 37.5 (R4)% along with *ad lib* feeding on berseem hay on their productive performance. Increasing the replacement of CFM only by a level of 37.5% BW significantly increased digestibility coefficients of DM, EE and NFE. Digestibility coefficients of CP was tented to reduce and that of CF tended to insignificantly increase in the rations R2, R3 and R4 compared with R1.The highest TDN, DE and ME concentrations were recorded with ration R4, while those of R2, R3 did not significantly differ than R1. However, DCP values were tented to be decreased as the level of BW was increased. Lambs fed on R4 showed higher average daily gain (170 vs. 159 g/h) than R1. The differences in all blood parameters and carcass characteristics between both groups were not significant. There was a pronounced increase in cost of feeding on R1 to produce kg gain as compared to those fed R4. Double economic efficiency was recorded as a result of feeding lambs on BW ration compared with CFM.

Theatre NCS5.5

Productive capacity of grass-clover sward as a source of substrate for microbial protein synthesis in the rumen of sheep

J. Verbic, D. Babnik, J. Verbic and M. Resnik, Agricultural Institute of Slovenia, Hacquetova 17, 1000 Ljubljana, Slovenia*

Herbage dry matter (DM) yield, microbial protein (MP) synthesis in the rumen, protein degradability in the rumen and digestibility of protein and organic matter were continuously measured during the first growth of grass-clover sward. MP synthesis in the rumen was measured by means of urinary purine derivative excretion and protein degradabilitiy by means of the nylon bag technique. Between 21 April and 8 June the concentration of crude protein (CP) in herbage decreased from 185 to 80 g kg^{-1} DM. Protein degradability in the rumen decreased from 0.72 to 0.61, true protein digestibility from 0.96 to 0.89, organic matter digestibility from 0.83 to 0.56 and MP synthesis from 136 to 92 g kg^{-1} DM intake. The estimated concentration of digestible undegradable protein (DUDP) in herbage decreased from 45 to 23 g kg^{-1} DM and the concentration of metabolizable protein from 130 to 82 g kg^{-1} DM. DM, CP, DUDP, MP and metabolizable protein yields increased during the experiment from 1983 to 9974, from 359 to 740, from 82 to 213, from 261 to 879 and from 250 to 780 kg ha^{-1} respectively. MP contributed from 0.62 to 0.73 of metabolizable protein. The results indicate that the potential of grass-clover sward for the production of microbial protein in the rumen is similar to its potential for direct protein production on the meadow.

Theatre NCS5.6

Safety measurements in utilizing poultry westes in sheep feeding

M. Khorshed[1], M. El-Ashry[1], H. Saleh[2] and H. El-Fouly[2], [1]Animal Production, Faculty of Agriculture, Ain Shams University, Cairo, Egypt, [2]Nuclear Research Center, Atomic Energy Authority, Cairo, Egypt*

The main objectives of this work are: (1) to study the safety aspects when using poultry waste in ruminant diets, (2) to evaluate the nutritive values of rations containing 0 and 14% poultry litter (BL, DM basis). The results indicated that, there is no significant differences between poultry litter (BL) and caged layer dropping (CLD) for the percentage of DM, CP, true proteins, NPN, As and Cd, but the differences for OM, ash, pb and Cu were significant ($P<0.05$). There is significant effect for progesterone activity, the concentrations were 117.2 and 192.6 ng/g for BL and CLD, respectively. Mean total bacterial count was higher ($P<0.05$) in CLD than BL. Also, the results of metabolism trial indicated that, the differences between the two rations (0 and 14% BL) were not significant ($P>0.05$) for digestibility of CP, CF, EE, NFE and ash. The starch value (SV) and total digestible nutrients (TDN) were higher ($P<0.05$) in control diet than BL diet. However, the difference between the two diets in the digestible protein (DP) was not significant ($P>0.05$). The treatment group (BL) was higher ($P<0.05$) in rumen liquor TVFA, NH_3-N and NPN. But no differences ($P>0.05$) was detected between treatments in rumen pH, total-N and true protein. The blood analysis indicated that, all experimental animals were healthy.

Theatre NCS5.7

Effects of extruded linseed substitution to linseed oil and/or soybean meal, on milk yield and fatty acid (FA) composition in goats receiving a high-forage diet

Y. Chilliard[1], J. Rouel[1], L. Leloutre[2], E. Bruneteau[3], P. Capitan[1], A. Lauret[4] and A. Ferlay[1], [1]INRA-Theix, 63122, France, [2]INZO-Château-Thierry, [3]INRA-Lusignan,[4]ITPLC-Surgères*

Thirty-six multiparous mid-lactation goats received alfalfa hay *ad libitum* (70% of total DMI) and daily either .61kg corn grain (CG)+.27kg soybean meal (SM)+.14kg beet pulp (BP)(dietC, control), or .18kgCG+.44kgSM+.06kgBP+.13kg linseed oil (LO, at 4.3% of diet DM) or .19kgCG+.13kgSM+.05kgBP+.51kg of an extruded mixture (70/30) of linseed and wheat (ELW), for 5 weeks. Diet ELW, compared to C, increased ($P<0.05$) milk fat (+7.3 g/kg), protein (+1.7 g/kg) and lactose (+2.5 g/kg) content and milk fat yield (+31 g/d), decreased milk fat % of 10:0 to 16:0 (36 *vs* 59%), odd-numbered (ON) FA, 18:2n-6, increased 4:0, branched-chain (BC) FA, 18:0, *trans10*- and *trans11*-18:1, *cis*9,*trans*13-18:2, *trans*11,*cis*15-18:2, *cis*9,*trans*11-18:2 (2.1 vs .3%) and 18:3n-3 (2.7 vs .8%), and decreased the normalized delta-9 desaturation ratios for 14:0, 16:0, 17:0, 18:0 and *trans*11-18:1. Diet ELW, compared to isolipidic isoproteic LO, increased ($P<0.05$) milk protein content (+1.4g/kg), decreased milk fat % of 16:0, *trans*11-18:1 and *cis*9,*trans*11-18:2 (2.1 *vs* 3.1%), increased 18:0, 18:2n-6, *cis*9,*trans*13-18:2 and 18:3n-3 (2.7 *vs* 1.7%) and decreased the normalized delta-9 desaturation ratios for 17:0, 18:0 and *trans*11-18:1. In conclusion, diet ELW, compared to C, improved dairy performances, decreased 10:0-16:0 and increased *trans* and n-3 FA. Diet ELW, compared to LO, increased less *cis*9,*trans*11-18:2 but more 18:3n-3, and decreased more 16:0 (*This work was funded by EU BIOCLA Project QLK1-2002-02362*).

Grain legumes, rapeseed meal and oil seeds for weaned piglets and growing/finishing pigs

E. Royer[1], J. Chauvel[2], V. Courboulay[2], R. Granier[3] and J. Albar[1], Institut Technique du Porc, Pôle Techniques d'Elevage, [1]34 boulevard de la Gare, 31500 Toulouse, France, [2]La Motte au Vicomte B.P. 3, 35651 Le Rheu, France, [3]Les Cabrières, 12200 Villefranche de Rouergue, France*

The maximum inclusion rates of grains legumes, rapeseed meal and oil seeds in wheat-soybean meal basal diets have been studied at the ITP experimental unit in Villefranche de Rouergue (France) in several trials using piglets or pigs.
Rapeseed meal and peas can be used in post weaning phase 2 diets at rates of 15 and 35 % respectively, and in growing-finishing feeds at rates of 18 % and 40 % respectively.
Their association allows to limit the soybean meal to 2 % and 0 % in growing and finishing diets containing 32 and 35 % of peas, 15 and 18 % of rapeseed meal. Maximum rates of 10 % of white (*Lupinus albus*) or blue (*Lupinus angustifolius*) sweet lupins in phase 2 diets appear careful. The inclusions in phase 2 diets of 7 % of full-fat rapeseeds or sunflower seeds, or 15 % processed whole soybeans or 3 % rapeseed oil give similar results. The use in fattening of 8 % of oleic acid-rich sunflower seeds has no negative effect on carcass fat quality whereas 4 % of ordinary sunflower seeds will have an effect.

Poster NCS5.9

Evaluation of Duckweed as a protein supplement for ruminants

R.Tahmasbi, Department of Animal Science, University Of New England, Armidale, NSW, 2350 Australia

The family of duckweeds *(Lemnaceae)* are the smallest flowering plants. These plants grow floating in still or slow-moving fresh water around the globe, except in the coldest regions. Duckweeds have the ability to remove relatively large quantities of soluble nutrients from the water .An alternative use of duckweed as a protein rich feed source for animals. Rates of growth and production of duckweed, in the field, are influenced by factors such as nutrients, sunlight, wind, temperature, and plant density. The growth of duckweed can be very rapid and, under optimal conditions, the quantity of duckweed biomass can double in less than 2 days. Under experimental conditions, the production rate of duckweed DM has been estimated about 180 tonnes/ha per year, although the actual yields of DM in the field more usually range from 10 to 20 tonnes/ha per year. The growth rate of duckweed is higher than that of any other higher plants and this confers a high potential for production of DM and protein. Nutrients available in the water are removed by duckweed through all surfaces of the fronds. The plant can tolerate and grow on a wide range of nutrient concentrations. Duckweeds grow better with ammonia-ammonium than nitrate as the main source of N .In cultivated populations, N can be supplied by urea or animal manures. The pH range for optimum growth is between 6.5-7.5.

Poster NCS5.10

Protein degradability of pea and faba bean in the rumen of sheep

J. Verbic[1], D. Babnik[1] and T. Znidarsic, [1]Agricultural Institute of Slovenia, Hacquetova 17, 1000 Ljubljana, Slovenia*

Protein degradability of untreated and heat treated samples of pea, high tannine faba bean variety (HTFB) and low tannine faba bean variety (LTFB) were examined. Heat treatment was carried out for 40 min at 125 °C in a laboratory oven. Protein degradability was determined by means of nylon bag technique using two sheep. Samples were incubated in the rumen for 3, 6, 12, 18, 24 and 36 h. Effective protein degradabilities (EPD) in the rumen were calculated on the basis of degradation characteristics and theoretical outflow rate 0.05 h^{-1}. Concentrations of crude protein in pea, HTFB and LTFB were 218, 295 and 286 g kg^{-1} dry matter respectively. Pea sample was characterised by lower EPD (649 g kg^{-1}) than HTFB (700 g kg^{-1}) and LTFB (687 g kg^{-1}). EPD of heat treated pea (651 g kg^{-1}), HTFB (692 g kg^{-1}) and LTFB (700 g kg^{-1}) were not significantly different from untreated samples (P>0.1). The results indicate that neither mild heat treatment nor faba bean variety affect protein degradability in the rumen.

Poster NCS5.11

Effect of texture, maturity and ensiling of maize grain on ruminal protein degradability

D. Babnik and J. Verbic, Agricultural Institute of Slovenia, Hacquetova 17, 1000 Ljubljana, Slovenia*

The effect of grain texture, maturity and ensiling on protein degradability in the rumen was studied using nylon bag technique. Ears of four hybrids, two dent and two flint types, were harvested at three maturity stages, characterized by the whole plant dry matter concentrations 28.7, 31.3 and 35.7% respectively. Kernels which were removed from frozen ears were ground and either ensiled in laboratory silos for 85 days or stored frozen at -20 °C. Samples were weighed into nylon bags and incubated in the rumen of two cows for 3, 6, 12, 24 and 48 hours. Protein degradability was higher in fresh samples of dent type grain than in samples of flint grain (77.3 *vs.* 68.8%, $P < 0.001$). With advanced maturity the protein degradability of fresh grain decreased from 74.8 to 71.7% ($P < 0.05$). A consistent decrease was found only in dent maize. In comparison with fresh samples the protein degradability of ensiled samples was considerably higher (87.3 *vs.* 73.0%, $P < 0.001$). The increase in protein degradability during the ensiling process was higher in flint type grain (from 68.8 to 84.4%) than in dent type (from 77.3 to 89.4%). The results indicated that the supply of undegradable protein to ruminants is considerably affected by texture, ensiling and maturity of maize at harvest.

Poster NCS5.12

Lupin seed as a substitute to soybean meal in broiler chicken feeding: incorporation level and enzyme preparation effects on performance, digestibility and meat composition

E. Froidmont[1], Y. Beckers[2], F. Dehareng[3], A. Théwis[2] and N. Bartiaux-Thill[1], [1]Agricultural Research Centre, Animal Production and Nutrition Department, Rue de Liroux 8, 5030 Gembloux, Belgium, [2]Agricultural University, Passage des Déportés 2, 5030 Gembloux, Belgium, [3]Agricultural Research Centre, Quality of Agricultural Products Department, chaussée de Namur 24, 5030 Gembloux*

During 22 days, 324 chicks (56.3 ± 4.8 g LW) received 3 iso-energy and iso-first limiting amino acids (Met, Lys, Thr and Trp) diets composed of 35% of soybean meal (diet 1), 30% of lupin seeds (Lupinus *albus*, Arès, diet 2) or 58% of lupin seeds (diet 3), with or without an enzyme preparation (pectinase and hemicellulase), in a randomised blocks design. All diets contained β-glucanase, cellulase and xylanase activities. ADG, FCR, N retention, apparent nutrient digestibility and AME were significantly depressed with diet 3. Proportionally to LW, gizzard weight and ileum lengths were increased with diet 3. Chyme viscosity in the small intestine increased with diet 3 compared to diet 1. No effect of enzyme preparation was observed on performances or nutrient digestibility. Fat and protein contents in chicken legs did not differ among diets but $C_{16:0}$ and $C_{16:1}$ contents were lower for diets 2 and 3, which reflects the fatty acid pattern of lupin seeds compared to soya. It was concluded that the enzyme preparation was not efficient to improve lupin utilization and that lupin seed could not replace all soybean meal in broiler chicken feeding. Other enzymes (protease, galactanase) or enzymes more specific to lupin pectines should be investigated in the future.

Poster NCS5.13

Extracted shredded meal of Oenothera biennis as an alternative protein source in pig nutrition

J. Bojcuková, F. Krátky, J. Seifert and S. Kucharová, Research Institute of Animal Production, Prague, workplace 51741 Kostelec nad Orlicí, Czech Republic

Extracted shredded meal of Oenothera Biennis (EOBM) is a waste from the production of the physiologic effective oil for cosmetic or pharmaceutical industry. EOBM has relatively high content of crude protein and amino acids, therefore it is possible to utilize as an alternative protein source in animal nutrition. The effect of the addition of EOBM into complete feed mixtures for finishing pigs on digestibility of nutrients and body weight gain was investigated. There were 30 pigs divided into 3 groups in the experiment. The addition in amount of 3% and 6% to basic mixture was compared with control group. The addition of 3% did not evidently influence the coefficient of digestibility of crude protein and metabolized energy. The significant decreasing of digestible crude protein and metabolized energy occurred at the addition 6%. At the evaluation of weight gain of pigs there was no statistically significant difference among the groups. Then the addition of 3% of EOBM did not influence parameters of fattening, hence it is possible recommend EOBM in this amount as an alternative protein source for finishing pigs.

 Poster NCS5.14

Electrophoretic analysis of un- and xylose treated rapeseed meal proteins in the rumen

A.A. Sadeghi, A. Nikkhah, P. Shawrang and M.M. Shahre-babak, Dep. of Animal Science, Faculty of Agriculture, Science & Research Branch, Islamic Azad University, Hesarak, Ponak, Tehran, Iran*

This study was carried out to determine the effects of xylose treatment (2 gr/100g DM) of rapeseed meal (RSM) on protein degradation characteristics and fractionation in the rumen. Major proteins in untreated and xylose treated RSM and the residues of the same feeds, incubated for 0, 2, 4, 6, 8, 12, 24 and 48h in the rumen of three Holstein steers were extracted and fractionated by SDS-PAGE. From the 14 % slab gel analysis, RSM proteins are composed of napin and cruciferin accounting for 38 and 51 percent of the total meal protein, respectively. The molecular weight of 10.3 and 8.2 KDa for two subunits of napin and 31.2, 26.8, 21.1 and 20.5 KDa for four subunits of cruciferin were estimated. Napin sub-units of untreated and xylose treated RSM were degraded completely within 2 and 4h incubation, respectively. About 31 and 52 percent of the untreated and xylose treated RSM proteins incubated in the rumen were not degraded after 12 h incubation, respectively. SDS-PAGE indicated that two and four subunits of cruciferin when untreated and xylose treated RSM is fed to ruminants, make an appreciable contribution to metabolizable protein, respectively. From *in situ* trial, ruminal protein degradation characteristics of untreated and xylose treated RSM were significantly different ($p<0.05$). In conclusion, the results of this study indicated that xylose treatment can reduce RSM protein degradation in the rumen effectively.

Theatre N6.1

Growth performance and body composition of dromedary camels raised on rations having varying levels of energy: protein ratio

A.Y. El-Badawi[1] and M.H.M. Yacout[2], [1]Dept. of Animal Production, National Research Center, P.O.Box 12311, Dokki, Giza, Egypt, [2]Dept.of By-products Utilization, Animal Research Institute, P.O.Box 16401, Dokki, Giza, Egypt*

Twelve male Sudanese camel calves (*Camelus dromedarius*) aged 20-24 mths. old and weighing 367.9 ± 38.1 Kg were randomly blocked by weight in a 3 x 4 factorial designed feeding experiment. The rations were manipulated to have R_1 11:1, R_2 9:1 and R_3 7:1 TDN: DCP ratio and were fed on individual basis for a feeding period that lasted 20 weeks. The mean daily DM intake per 100Kg body weight was 1.77, 1.73 and 1.76 and ADG was 0.749, 0.765 and 0.802 Kg for camels in groups R_1, R_2 and R_3, respectively. Dressing percentage relative to EBW was not significantly influenced by the feeding regime and ranged between 69 to 69.5%. Edible offals (heart, liver, defatted kidneys and spleen) comprised nearly 3% of EBW for all slaughtered camels. Chemical composition of chilled boneless-carcasses was nearly similar among slaughtered camels and ranged from 59.9-61.4% for water, 23.1-23.5 % for protein, 10.7-13.0% for ether extract and 0.89-0.94% for ash. The hump fat consisted 7.0-7.3% of the whole boneless-carcass or about 70% of the total knife separable fat for all animals. It is feasible to note that, camels could be a promising source of high quality meat from rations containing lower protein (10%CP) when they fed under intensive feeding systems.

Theatre N6.2

Substituting bread by-product for barley grain in fattening diets for Baladi kids

S. Haddad[1] and K.I. Ereifej, Department of Animal Production, Jordan University of Science and Technology, P.O. Box 3030, Irbid 22110, Jordan

The effects of substituting bread by-product (BBP) for barley grain in fattening diets for kids on growth performance, and nutrient digestibility were studied. Twenty-eight Baladi kids (body weight = 17.1 ± 1.0 kg) were assigned randomly to 4 finishing diets for 70 days. The control (CON) diet contained 20, 60, 11, 7, and 2% alfalfa hay, barley grain, soybean meal, corn grain, and mineral and vitamin mix, respectively. Bread by-product substituted barley grain by 10, 20, and 30% of the diet DM in the LBBP, MBBP, and HBBP diets, respectively. Dry matter intakes for the CON, LBBP and MBBP diets were similar (avg. = 592 g / day), however, kids fed the HBBP diet had a lower (P<0.05) DM intake (451 g / day). Dietary treatments did not affect (P>0.05) average daily gain for kids fed the CON, LBBP and MBBP diets (avg. = 150 g / day). Feed to gain ratio was greater for the CON, LBBP and MBBP diets (avg. 3.9) compared with the HBBP diet (5.0). No significant (P >0.05) effect of the dietary treatment was observed for DM, OM and NDF digestibility. Substituting BBP for barley grain up to 20 % of the diet DM did not affect nutrient intake, growth performance and nutrient digestibility of kids and resulted in a decrease in feed cost.

Theatre N6.4

Intake, ruminal fermentation, *in situ* kinetics and digestibility on goats fed a fibrous diet supplemented with slow release 5% urea

M.A Galina, Facultad de Estudios Superiores Cuautitlan, UNAM, Cuautitlan, México

Four cannulated goats (55 ± 2 kg BW) were used to study the effects of (0% T1; 10% T2, 20% T3 and 30% T4) with slow release 5% urea as an alternative nitrogen supplement (ANS), on a basic of 80:20% corn (*Zea mays*) alfalfa (*Medicago sativa)* diet. Apparent digestibility was measured by total faeces collection. The nylon bag technique was used and disappearance data was fitted into an exponential model $P = a + b\ (1\text{-}exp^{-ct})$. *In situ* DM disappearance differed (P<0.05) between T4 (30% ANS) and control diets at 9, 12, 24, 36, 48 and 72 hr of incubation. The ruminal NH_3 concentration in T4 was higher (P<0.05) compared with T1 at 8, 10 and 12 hr (P<0.05) . DMI ranged from 1,114; 1,172 and 1,221 g/d for 10, 20 or 30% ANS while 0% ANS was only 0.950 g/d. Apparent digestibility of DM, OM, NDF, hemicellulose and cellulose did not differ between the four treatments, however, digestibility of nitrogen was significantly higher (P<0.05) in T4 (83.5%) as compared to T1 (78.4%). The extent of digestion and indigestible fraction of T4 (30% ANS) was different (P<0.05) from all the other treatments; lowering the mean retention time (19.4 h). Results showed that high fiber forages can be used efficiently by ruminants.

Theatre N6.5

Comparison of equine faecal with colonic inocula for use in *in vitro* gas production

V.K. Morris, R.J. Gosling and J. Hill, Faculty of Applied Science, Writtle College, Essex, UK, CM1 3RR.*

Research into functionality of microbial communities in the hindgut of the horse requires samples representative of non-surgically modified healthy animals. Cannulation has been shown to increase digesta passage rate, is increasingly not acceptable on ethical grounds and requires a high level of management. Samples of colonic and faecal material were collected from four slaughter horses immediately after exsanguination, stored under anaerobic conditions, diluted and used as inoculum. Oats and medium quality meadow hay (1 gram substrate per incubation flask; on a 30:70 MJ DE ratio) was used in the gas production experiment to mimic typical feeding standards for horses under medium exercise. No significant differences in gas production (volume) and lag phase were observed between the two inocula sources over 72 hours of incubation. Small but consistent differences in rate of gas production during the first 4 to 16 hours of each experiment were observed with slightly elevated rate of production in colonic inocula samples. This may represent higher populations of starch-utilising microbial communities in colonic inocula compared with faecal inocula. The results imply equine faecal inocula is representative of colonic inocula for *in vitro* gas production and would allow investigation of samples without invasive procedures that may effect host health and therefore stability of microbial communities.

Theatre N6.6

Spelt, an ancient cereal of interest for modern horse feeding

A. Delobel[1], J.L. Hornick[1], L. Istasse[1], I. Dufrasne[2], [1]Nutrition, [2]Experimental Station, Faculty of Veterinary Medicine, University of Liège, Sart Tilman, B4000 Liège, Belgium

Samples of cereals -spelt, winter barley, oat, maize- harvested during the 2003 summer season, were collected for chemical composition analysis. On the whole, the chemical composition of spelt was in the normal range of values found for others cereals.

In spelt, the concentrations in ash, crude protein, NDF, ADF, hemicellulose and P, were among the highest values while the ether extract, Ca and Na contents were among the lowest. Although spelt is a cereal with a large husk, the NDF content was in the same range as in barley and oat (34.9, 31.0 and 31.0% DM). Surprisingly, spelt showed a content in ADF lower than in oat (16.3 vs 18.1% in DM) and a content in hemicellulose lower than in barley (18.6 vs 21.1% DM).

In the concentrate diet, the protein content was rather low (12.1%DM) and the fibre content intermediate (28.1, 11.2 and 16.9 %DM) for NDF, ADF and hemicellulose respectively. Energy and protein allowances and requirements were calculated and compared from feeding data obtained on a group of 5 dressage horses and another group of 5 leisure horses. For both types of horses, the energy allowances were lower than the requirements. By contrast, protein was provided by an excess of 12% in dressage and an excess of 15 % in leisure.

Models for growth simulation and feed utilization by broilers

R. Burlacu, The University of Agriculture Sciences and Veterinary Medicine - Bucharest, 59 Marasti street, sec. 1, Bucharest, Romania

Broiler raising enjoys particular attention at the national level both in the large farms and in peasant households. Therefore, making efficient use of the vegetal resources by providing for a scientific feeding of broilers is one of the priority objectives of animal production in Romania.

This paper proposes a new method to use broiler forages. This is done with a mathematical model of energy and protein metabolism simulation, which yields a better accuracy in assessing the productive potential of the forages because it provides additionally:

- correction concerning the biological value of the protein, which shows the amounts of energy, urine, deamination, variables according to the level of essential amino acids;
- correction of the microbial fermentescible matter and sugar according to forage content of crude fiber and mono- and disaccharides;
- dissociation of the net energy in protein energy and lipid energy that may be produced b the analysed forage.

 Theatre MCL1.1

Dairy cows health and production performances according to the concentrate level

V. Brocard[1], N. Bareille[2], V. Jégou[3] and Ph. Roussel[1], [1]Institut de l'Elevage, 149 rue de Bercy 75595 Paris Cedex12, France, [2]Veterinary School & INRA, BP 40706, 44307 Nantes cedex 3, France, [3]EDE et CA de Bretagne, Rue du Chalutier-sans-pitié 22195 Plérin Cedex, France

This presentation aims to summarize a 9-year experiment on the consequences on farm economical efficiency and dairy cows performances (milk, milk solids, body condition, reproduction) and health of a large concentrate decrease. Three 3-year trials took place from 1992 to 2001 in the experimental farm of Trévarez (Brittany, western France) in order to compare different concentrate levels (1600 vs 650; 1350 vs 650; 1100 vs 650 vs 300 kg per cow per year). Data from 998 lactations were considered altogether.

In average, the decrease in concentrate led to a lower milk production by 0.9 kg/kg of saved concentrate, a slight decrease in the protein content, and an increase in the fat content. The lower concentrate level had no effect on body condition and reproduction performances. It led to -0.7 health intervention/cow/year per t of saved concentrate and a lower culling rate. Statistical analyses were carried out in order to precisely explain fertility and diseases frequencies according to general factors such as production at peak, protein/fat ratio in milk, body condition score at calving (or maximal body condition loss), fertility and health problems during the precedent lactation, calving season, forage diet and parity.

Theatre MCL1.2

Assessing herd health of dairy cattle in different housing systems using systematic clinical examinations

I.C. Klaas, P.T. Thomsen and M. Bonde, Dep. of Animal Health and Welfare, Danish Institute of Agricultural Science, P.O. Box 50, DK-8830 Tjele, Denmark*

Herd health in dairy farms is often assessed using farm records of veterinary treatments. The detection of clinical diseases as well as the threshold when to initiate veterinary treatment varies among farmers. Systematic clinical examinations reveal herd health status independently from farm records. The objective of this paper is to discuss the results from applying clinical protocols in different housing systems with focus on their ability to identify and quantify farm specific health risks. The protocols included assessment of lameness, udder health, hock lesions, body condition, and overall health status. A sample of 40-50 cows can be examined within 1-1.5 hours. Practical problems like missing ear tags, a bull in the herd or higher flight distance in grazing cows or cows housed in deep litter systems have to be considered. Farmers accepted the method when the herd was undisturbed under examination. Especially in AMS-farms cows should have unlimited access to the robotic milking unit. Clinical examinations were especially useful to detect chronic diseases such as chronic mastitis, lameness, and diseases related to loss of body condition. Implementation of systematic clinical examinations in herd health management can be a valuable tool to identify and quantify specific health problems and to monitor the effect of changes in management on the herd's health.

 Theatre MCL1.3

Investigation of claw health of dairy cows in Switzerland

J.C. Bielfeldt[1], K.-H. Tölle[1], R. Badertscher[2] and J. Krieter[1], [1]Institute of Animal Breeding and Husbandry, Christian-Albrechts-University, 24098 Kiel, Germany, [2]Swiss Federal Research Station for Agricultural Economics and Engineering, 8356 Tänikon, Switzerland*

Claw health of dairy cows was investigated in an observational study in different housing systems in Switzerland. Twenty-five professional hoof trimmers examined lameness (LN) and claw disorders on 4,621 cows in 290 farms within routine hoof trimming. 82 farms had tie-stall barns without exercise (T1) and 166 had tie-stall barns with exercise (T2), another 42 farms kept their animals in loose housing systems with exercise (L2). Observation period lasted from September 2001 until June 2002. Single claw disorders were joined together to four different diagnosis-complexes: Sole disorders (SD), white line disorders (WD), heel erosions (HE), and disorders of skin and interdigital space (ID). Environmental and management factors were documented in a questionnaire for analysing possible risk factors on claw health. Data from three breeding associations were available, including animal information and performance parameters. Prevalences were 15.7 % (SD), 13.6 % (HE), 10.0 % (LN), 6.1 % (WD), and 5.3 % (ID). LN and SD showed highest prevalence (13.2 %; 16.4 %) and highest odds ratio (OR = 1.89; 1.33) in T1. WD were more often detected in L2, accounting for 9.4 % (OR = 1.0). HE was identified most in T2 (17.1 %, OR = 4.72) and T1 (13.2 %, OR = 4.45). ID were most frequently found in T2 (7.5 %, OR = 1.55).

Theatre MCL1.4

Outdoor exercise reduces the incidence of hock lesions in dairy cows kept in tie stalls

B. Wechsler, N. Keil, T. Wiederkehr and K. Friedli, Swiss Federal Veterinary Office, Centre for proper housing of ruminants and pigs, FAT, 8356 Ettenhausen, Switzerland

Hook lesions are a major health problem in dairy cows kept in tie stalls. This study aimed at investigating the influence of the frequency (number of days per month) and duration (hours per month) of outdoor exercise on the incidence of such lesions. Over a one-year period, cows on 66 Swiss dairy farms with tie stalls were examined every two months (six times in total) for the occurrence and severity of hock lesions (hairless patches, scabs and open wounds, swellings). In addition, general farm characteristics (cow breed, milk yield, length of the stalls, type and amount of bedding material) were recorded. Data were analysed using a linear mixed-effects model allowing for the nested random effect of the farm.

The incidence of scabs and wounds was negatively and positively associated with the duration ($P<0.0001$) and frequency ($P<0.001$), respectively, of outdoor exercise and was also significantly affected by time of the year ($P<0.0001$), the length of the stalls ($P=0.01$) and the type of bedding provided on the stalls ($P=0.05$).

It is concluded that the incidence of hock lesions in dairy cows housed in tie stalls can be reduced by outdoor exercise. The results also indicate that sufficient duration of each exercise period seems to be important, whereas increasing the frequency alone is not beneficial.

Theatre MCL1.5

Influence of fattening systems on *Eimeria* infections and weight gain in sheep lambs

M. Gauly[1], J. Reeg[2] and G. Erhardt[2], Institute of Animal Breeding and Genetics, University of Göttingen[1], Giessen[2], Albrecht-Thaer-Weg 3, 37075 Göttingen, Germany*

233 lambs were used to access the effect of production systems on *Eimeria* infections and weight gain. All lambs were kept with their dams in a barn on straw and fed with hay and concentrate ad libitum until they were 40 days old. Then they were divided in the following systems: 1. and 2. no weaning until slaughtering, 3. and 4. weaning at an age of 40 days. Weaned lambs were kept as singles (3.) or in groups (4.). Lambs were fed with hay and a standard pellet diet ad libitum. Animals in group 1 were given no concentrate. Male lambs were slaughtered with a weight of 42 and females with 39 kg.

Multiple *Eimeria* infections were common with up to nine *Eimeria* species in one animal. No clinical signs of coccidiosis were observed in any of the groups. Significant differences were found between the groups for weight gain. The mean fattening period was significantly shorter in groups 2, 3 and 4 (88 to 104 days) compared with the extensive managed (265 days) one (group 1). The output of *Eimeria* oocysts was significantly different under the various conditions of husbandry. It was higher in the extensively managed group (1). Estimated correlations between bodyweights at different ages and oocyst counts were significantly negative (-0.11 to -0.25).

Theatre MCL1.6

The influence of pen design and grouping strategy on behaviour, stress, immune competence, health and production

K.H. Jensen[1], M.E. Busch[2], B.M. Damgaard[1], E. Jørgensen[1], J. Nielsen[3], V.A. Moustsen[2], B.K. Pedersen[4], M. Studnitz[1] and E.M. Vestergaard[1], [1]Danish Institute of Agricultural Sciences, P.O.Box 50, DK-8830 Tjele, Denmark, [2]The National Commitee for Pig Production, Axeltorv 3, DK-1609 Copenhagen, Denmark, [3]Danish Institute for Food and Veterinary Research, Lindholm, DK-4771 Kalvehave, Denmark, [4]Egebjerg Maskinfabrik, Egebjerg Hovedgade 27, DK-4500 Nykøbing Sjælland, Denmark*

Pen design and grouping strategy were studied in groups of 60 pigs. Pens were either clearly zonated or divided in few, less defined zones. Grouping was done by mixing either intact or divided litters. Pigs were weaned into the experimental pens at 4 weeks of age. Leukocyte counts, behaviour, clinical observations, and floor conditions were registered at week 0, 4, 9 and 14 after weaning. Responses to novelty were measured at week 4 and 14. Response to Porcine Parvovirus vaccine was measured at week 12 and 14. Data were subjected to analysis of variance using the mixed model methods. The results showed that zonation reduced stress during the first 4 weeks after weaning. However, by increasing pig size, zonation resulted in unrest, decreased hygiene, increased snout to faeces contacts, activation of the unspecific immune response, depressed specific immune response, and decreased general condition without affecting medical treatments, mortality and productivity. Grouping of intact litters in contrast to divided litters improved the social behaviour, reduced stress, and increased daily weight gain, percent of lean meat and general condition, whereas feed conversion, medical treatments, mortality and immune competence was not affected.

Aggressive behaviour of dry sows fed by means of an electronic feeding system without feeding stalls

R. Weber[1] and B. Wechsler[2], [1]Swiss Federal Research Station for Agricutural Economics and Engineering, Taenikon, 8356 Ettenhausen, Switzerland, [2]Swiss Federal Veterinary Office, Taenikon, 8356 Ettenhausen, Switzerland*

The study investigated the behaviour of dry sows fed restrictedly by means of an electonic feeding system where the sow is not locked up in a feeding stall during feed intake.
For the study, three feeding systems were installed for a dynamic group of a maximum of 36 sows. On six days in intervals, the frequency of "aggressive displacements of a feeding sow" and "displacements not involving aggressive interactions" was recorded. For each visit, it was differentiated whether food was distributed to the sow or not.
On average, the sows visited the feeding system 55.6 times per day with and 99.9 times with-out feed intake. Per sow and day a mean of 16.6 incidences of "aggressive displacements of a feeding sow" and of 9.3 incidences of "displacements not involving aggressive interactions" were recorded. Older sows (parity number three or more) accounted for 89 % of all such displacements at the feeding system.
The results show that the restricted feeding of dry sows by means of an electonic feeding system without locking the sows during feeding may result in animal welfare problems. Especially low ranking sows are displaced very often and have to visit the feeding system very frequently to consume their daily ration.

Theatre MCL1.8

Qualitative and quantitative risk factors concerning the introduction and spread of Salmonella in different pig housing systems

C. Meyer[1], E. große Beilage[2] and J. Krieter[1], [1]Institut of Animal Breeding and Husbandry, Christian-Albrechts-University, 24098 Kiel, Germany, [2] Bakum Field Station for Epidemiology, School of Veterinary Medicine, Hannover, Germany*

Subject of this study was the estimation of Salmonella seroprevalence in different pig housing systems as well as the detection of specific risk factors and analyzing them in consideration of their qualitative and quantitative importance. All together 4343 blood samples were taken from 78 conventional, 17 ecological and 13 outdoor farms and serologically examined by SALMOTYPE® meat juice ELISA (Cut-off at 40%).
1498 sows from farrowing and farrow-to-finish farms showed a seroprevalence of 17.1%. 12.3% of sows from 39 conventional farms, 7.6% of sows from nine ecological farms and 35.1% of sows from 13 free range farms were serological positive. In fattening farms altogether 301 (11.4%) of 2642 samples showed positive findings. 13% of animals from 43 conventional and 1.9% of animals from 12 ecological farms reacted positively to ELISA. Risk analyses showed that the integration of gilts without preceding quarantine, pasture housing systems, the presence of a hygienic sluice, non-use of acids, leaving ill animals with the herd and not cleaning the feed storage increased the Salmonella seroprevalence in conventional farrowing farms. Risk factors for fattening pigs were the regular antibiotic treatment in the beginning of the fattening period, partially slatted floors, feeding of granulate, fouled feeding facility as well as a location in regions with high pig density.

Poster MCL1.9

The effect of body condition on milk production

E. Báder[1], I. Györkös[2], J. Szili[3], A. Muzsek[1], P. Báder[1], Z. Gergácz[1] and A. Kovács[1], [1]University of West Hungary, Faculty of Agriculture, Vár 2, 9200 Mosonmagyaróvár, Hungary, [2]Research Institute for Animal Breeding and Nutrition, Gesztenyés str. 1, 2053 Herceghalom, Hungary, [3]Aranymező Plc, 7671 Bicsérd, Hungary*

The influence of the first lactation yield on the body condition at the end of the first lactation was studied. The effect of body condition just before calving on milk yield in the second lactation was also scrutinized. The experiments were carried out at the dairy farm of the Bicsérd Aranymező Plc. 401 purebred and crossbred Holstein-Friesian cows were involved in the trial. The judgement of condition scores was based on a 5-grade scale.
Body condition becomes poor following by a very high (9275 kg) first lactation yield. However, this poor condition does not prevent cows to increase their milk yield by 300 kg in the second lactation. The first lactation yield of cows with condition scores between 2.6 and 3.4 was between 8925 and 8877 kg. The second lactation yield of these cows exceeds their first lactation yield by 300-700 kg. The highest yields (9121 kg for the first and 9853 kg for the second lactation) can be found in the group of cows with ideal body condition (3.5 scores). Poor milk production (8158 kg) results in excess-condition (4.1-5.0 scores) at the end of lactation. Despite of it, milk yield increases in the next lactation significantly (9281 kg), but it falls away from the average. In conclusion, it can be highlighted that there is a remarkable relationship between body condition and milk production.

Poster MCL1.10

Social hierarchy in unweaned buffalo calves

F. Grasso[1], F. Napolitano[2], G. De Rosa[1], A. Braghieri[2], G. Migliori[1] and A. Bordi[1], [1]DISCIZIA, Università di Napoli "Federico II", Portici (Napoli), Italy, [2]DISPA, Università della Basilicata, Potenza, Italy*

28 buffalo female calves were equally allocated to 4 groups (2.6 indoor m^2 + 2.0 outdoor m^2, 2.6 indoor m^2, 1.5 indoor m^2, 1.0 indoor m^2 per calf) to examine the effects of space allowance on dominance hierarchy. 4 weekly competition tests were performed for each group. Rubber teats were reduced from 4 to 2 per pen, when acidified milk substitute was prepared and offered to the animals. For 15 min all occurrences of competitive interactions were recorded. For each group, a dominance score was calculated based on the agonistic interactions recorded during the four competition tests. The relative dominance of each member of a dyad was determined by the ratio of wins to total encounters. Dominance values were indexed for each animal on the basis of the fighting index described by Clutton-Brock *et al.* (1979; *Anim. Behav.*, 27:211). Landau's index of linearity for groups A, B, C and D was 0.36, 0.75, 0.30 and 0.62, respectively, indicating that hierarchies were not strongly linear. A significant correlation between social rank and daily weight gain was found ($\tau = 0.319$; $P<0.01$). Conversely, no association between hierarchy and humoral, cellular and cortisol responses was observed. In conclusion, dominant animals have a priority access to the food, thus to avoid an increment of aggression, feeding places should be increased.

Effect of age and sex on plasmatic levels of cortisol and some haematic parameters in sheep subjected to shearing

V. Carcangiu[1], G.M. Vacca[1], A. Parmeggiani[2], M.C. Mura[1]; M.L. Dettori[1]; P.P. Bini[1], [1]Dipartimento di Biologia Animale, via Vienna 2, 07100 Sassari, Italia, [2]Dipartimento di Morfofisologia Veterinaria e Produzioni Animali, via Tolara di Sopra 50, 40064 Ozzano dell'Emilia (Bo), Italia*

Many breeding techniques produce stress in the animals and shearing is surely one of these. With the aim of investigating how much shearing stress could be affected by age or sex, cortisol and some haematic parameters were studied in Sarda sheep breed. For this project n.10 one year old ewes (Group A), n.10 three years old ewes (Group B) and n.8 three years old males (Group C), were used. From each sheep 4 blood samples were collected: the first, the day before and the remaining throughout shearing (isolation from the group, in the middle and at the end of shearing). Cortisol level was evaluated by RIA, glucose, Mg, Na and K by colorimetric assay. The data were submitted to ANOVA. Cortisol and glucose haematic levels showed increasing values (P<0,01) during shearing in all the groups. Mg and Na haematic concentrations displayed increasing values (P<0,01 and P<0,05 respectively) in the second blood withdrawal in A and B groups. The males showed the lowest cortisol and glucose levels. Result obtained indicate that the reaction to shearing stress is affected both by age and sex of the animals.

Poster MCL1.12

Effect of transport on animal welfare in commercial rabbits in Spain

G.A. María[*], M. Villarroel, G. Liste, M. López,T. Buil, S. Garcia and G. Chacón, Faculty of Veterinary Medicine, University of Zaragoza, Zaragoza, Spain

We tested whether transport time (1 or 7 hours), position on the lorry (upper, middle, lower decks) and the season (summer or winter) affects plasmatic stress variables in commercial rabbits in Aragón (NE Spain). Each journey was repeated three times and 196 rabbits were sampled in total. We used an unarticulated lorry with spring suspension, natural ventilation and a hydraulic loading ramp. The density per cage was 360 cm^2 per animal. The stress indicators analysed were corticosterone, glucose, lactate and creatine kinase. Transport was an important stressor for the animals (corticosterone levels above 70 ng/ml) although the differences were not significant between journeys. Position on the lorry also affected stress variables (p<0.01). Rabbits on the bottom level had higher levels of corticosterone (82.99±4.98 ng/ml) than the middle (66.84±5.50 ng/ml) or upper levels (58.44±5.57 ng/ml). In general, transport was a medium stressor, independently of transport time. Nonetheless, the adaptive mechanism of the animals was able to minimise the risk of a loss of welfare of the animals transported. Transport did not affect meat pH at 24 hours *post mortem*. The effect of transport does not appear to affect the final price of the meat that is sent to market.

 Theatre M4.1

Milk yield of cows and Bovine Viral Diarrhoea Virus (BVDV) infection in 7,252 dairy herds in Bretagne (western France)

F. Beaudeau[1], C. Fourichon[1], A. Robert[1], A. Joly[2] and H. Seegers[1], [1]Unit of Animal Health Management, Veterinary School-INRA, BP 40706, F44307 Nantes cedex 03, [2]UBGDS, BP110, F56000 Vannes*

No previous study has investigated the possible difference in milk yield associated with BVDV-infection in French commercial dairy herds at a large scale. This study aimed at quantifying the reduction in test-day milk yield (TDMY) of cows according to BVDV-infection status of their herd. Five BVDV-infection statuses were defined, based on levels of BVDV-antibodies measured in bulk tank milk twice four months apart (BVDV-status definition-period): presumed (1) not-infected for a long time (NI), (2) not recently-infected (NRI), (3) past-infected recently-recovered (PIRR), (4) past-but-still-infected (PSI), (5) recently-infected (RI). A total of 982,745 test-days in 7,252 herds located in Bretagne were considered. The effect on TDMY of the BVDV-status, adjusted for herd (random), lactation number and days in milk was assessed using a mixed linear model. The BVDV-status was significantly associated with TDMY. Considering test-days within the BVDV-status definition-period, the reduction in milk yield was 0.41 ($P<0.001$), 0.58 ($P<0.001$) and 0.21 (NS) kg/day for cows in PIRR, PSI and RI herds respectively, compared with cows in NRI herds. A carry-over effect (at least 1 year) of BVDV-infection on TDMY was also evidenced in PIRR, PSI and RI herds, possibly in relation to an increased incidence of other health disorders (e.g. intramammary infections) due to the BVDV-immunodepressive effect.

Theatre M4.2

Efficiency of strategies to control BVDV spread within a dairy herd: Assessment by simulation

A.-F. Viet, C. Fourichon and H. Seegers, Unit of Animal Health Management, Veterinary school-INRA, BP40706, 44307 Nantes Cedex 03, France*

At the farm level, strategies available to farmers to control infection by the bovine viral-diarrhoea virus (BVDV) include vaccination, combined test-and-cull programs (elimination of persistently infected (PI) animals) with biosecurity actions (prevention of virus introduction and transmission). The objective was to assess *ex-ante* the expected efficiency of strategies to control BVDV spread within a dairy herd without vaccination. A stochastic simulation model was developed. It represented horizontal and vertical virus transmission in a population with controlled size, structured into subgroups. Virus transmission from herd's neighbourhood and purchase of PI animals were avoided. A single virus introduction by an immune heifer carrying a PI foetus was simulated. Four scenarios were studied: (1) no action, (2) transmission prevention between subgroups, (3) test-and-cull program and (4) combination of (2) and (3). Simulation experiments were done for a typical western-France dairy herd (38 cows, 34 youngstock). The duration (clearance was obtained when no shedding or PI-carrier animal was present anymore) and the extent of virus spread (number of new infections, of PI animals and of immune animals) were compared. Without test-and-cull program, duration and extent of spread differed according to the scenarios: in 50% of the replications, for scenarios (1) and (2), respectively, clearance occurred before 261 and 35 days and less than 19 and 3 animals were infected.

Theatre M4.3

The effect of lameness on feed intake, feeding behaviour, liveweight change, milk yield, milk let down, milking duration and reproductive performance of Holstein Friesian dairy cattle

J. Margerison, J. Hollis, A. Snell, G. Stephens and B. Winkler, University of Plymouth at Seale Hayne, Newton Abbot, Devon, UK*

This study assessed the effect of lameness on yield, milking duration, feed intake, body condition and live weight. Over 3 years 160 multiparous Holstein Friesian dairy cows were selected at random and paired retrospectively according to calving date, milk yield, body condition and live weight. Cows were allocated at 110 days postpartum into two groups, 80 lame (L) and 80 non lame (NL) according to locomotion score (<3 = Non lame, 4 and above = Lame). Data was found to be normally distributed and analysed using ANOVA General linear model and a stochastic model was made from the data. Lame cows had a significantly greater (P<0.001) locomotion score compared with non-lame cows, NL 2.01, L4.47 (sem 0.445), significantly (P<0.05) lower body condition score (scale 1 to 5), NL 3.24, L 2.87 (sem 0.064), a significantly (P<0.05) lower mean milk yield, NL 42.3, L 38.7 (sem 0.60) and significantly (P<0.05) lower milking duration (mins), NL 7.03, L 6.06, (sem 0.224) compared with non-lame cows. There was no significant difference between live weight per se, NL 654, L 646, (sem 5.5), live weight change was significantly different (P<0.05) lame cows lost weight compared with non lame cows which gained weight. Lame cows had lower dry matter intakes (L 23.01, NL 25.5 kg/h/d DM), had fewer meal bouts (L 2.2, NL 4.1/d).

Theatre M4.4

Mechanical properties of the solear hoof horn of heifers before and during the first lactation: a prediction of lameness susceptibility

Betina Winkler, Jean K. Margerison and Charles Brennan, University of Plymouth, Seale Hayne, Newton Abbot, Devon, UK

Mechanical tests were completed on samples of sole hoof horn taken from 20 heifers at 2 months before parturition (p1) and 100 days postpartum (p2). Simultaneously, all hoof claws were assessed for the level of haemorrhage or lesions score (LS) in sole horn. Heifers were kept at pasture prepartum and housed loose in a straw bedded yard postpartum. Hoof samples were collected from all claws and analysed for elastic modulus (ELM) and puncture resistance (PR) using a texture analyser, each measurement was replicated five times on the same area (5) of each claw. Data was analysed by ANOVA - GLM using period and claw as fixed effects. PR force of the sole horn was significantly greater in front claws (FC) compared to hind claws (HC) (P<0.05) (p1 - FC 8.2, HC 7.4N, p2- FC 11.1 HC 10.3N). The PR force and ELM significantly increased postpartum p2 compared with prepartum p1 (P<0.01) (p1 07.8, p2 10.7N and p1 86, p2 118.0N/mm^2), while the LS of the claw horn increased between periods (p<0.001). No significant difference in sole bruising and LS were found between FC and HC in the prepartum period, however LS was significantly greater HC compared with FC (p<0.001) (HC 223.7, FC 149.3). Prepartum, ELM and PR force were not correlated with lesion score either pre or postpartum. However, postpartum ELM and PR force were significantly negatively correlated (p<0.01) to the increase in lesion score between periods (R=0.65).

Theatre M4.5

Impact of climatic conditions on dairy cows kept in open buildings

B. Wechsler[1], M. Zähner[1], M. Keck[1], R. Hauser[1], W. Langhans[2] and L. Schrader[2], [1]Swiss Federal Veterinary Office, Centre for proper housing of ruminants and pigs, FAT, 8356 Ettenhausen, Switzerland, [2]Swiss Federal Institute of Technology (ETH) Zurich, Institute of Animal Sciences, Physiology and Animal Husbandry, 8092 Zurich, Switzerland

The aim of this study was to assess whether dairy cows kept in open buildings are able to cope with the range of climatic conditions which may occur in such buildings in central Europe. On each of four commercial farms, ten lactating cows were observed over a total of five weeks in winter, spring and summer. For each cow, both behavioural and physiological parameters were recorded simultaneously. In addition, air temperature (range -13.8 to 28.7°C) and relative air humidity (range 25.8 to 98.8%) were measured continuously, and a temperature humidity index (THI) was calculated.

THI had significant effects (General Linear Model, $p < 0.0001$) on skin and body surface temperature during both night and day. With regard to rectal temperature, duration of lying and cortisol concentration in the milk, there were significant effects ($p < 0.0001$) of THI for the day but not for the night time. Heart rate and frequency of lying did not covary significantly with THI. For most physiological and behavioural parameters, there were also significant differences between farms ($p < 0.05$) and significant interactions between THI and farm ($p < 0.05$). In conclusion, the results indicate that within the measured range of climatic conditions the cows were hardly exposed to severe cold or heat stress.

Theatre M4.6

A cross-sectional survey about influenza in pigs in France

F. Madec[1], E. Eveno[1], L. Mieli[2] and J.Cl. Manuguerra[3], [1]AFSSA Ploufragan BP53, Ploufragan, France, [2]LDA22 BP54, Ploufragan, France, [3]Pasteur Institute, Paris, France

Influenza and influenza-like syndromes are common health disorders affecting pigs especially in densely populated pig areas. But Influenza is an issue that needs consideration far beyond the sole pig sector especially because of its interspecific and zoonotic components.

An epidemiological survey was carried out in a sample of 15 farms in France where acute respiratory problems were detected. On the farm the compartment where the pigs were at the critical and typical stage of disease expression was considered. Nasal swabs (for virus isolation) and blood samples (for specific antibody detection) were taken from 12 sick pigs. The pigs were identified and bled again 3 weeks later to look at seroconversions. The results showed the high prevalence of influenza A viruses (45 isolates). They were all of A/H1 subtype. Clear seroconversions occurred against A/H1N2 (n = 7 farms) and A/H1N1 (n = 3 farms). No reaction was found against A/H3N2 and other subtypes. Some confusing results were found when comparing virology and serology. In addition to influenza, seroconversions against other pneumotropic viruses did occur: PRRSV (n = 1), PRCV (n = 2), PCV2 (n = 1). All the farms were pseudorabies-free. The paper draws attention on the need to use appropriate laboratory tools and methods when assessing influenza infections. It also emphasizes the usefulness of epidemiosurveillance.

Theatre M4.7

Salmonella DT 104 in Denmark - a case study of the interplay between risk management, uncertain knowledge and risk communication

K.K. Jensen, Danish Centre for Bioethics & Risk Assessment (CeBRA), University of Copenhagen, Department of Education. Philosophy and Rhetoric, Njalsgade 80, DK-2300 Copenhagen S., Denmark

By deciding in 1997, as the only industry in Europe, on a radical strategy of eradication, the Danish Bacon and Meat Council (DBCM) signalled to both the authorities and consumers that it perceived *Salmonella* DT 104 to be far more dangerous than other *Salmonellas*. A TV-documentary by the end of 1997 led to severe criticism of the Danish Veterinary and Food Administration (DVFA). The authorities responded to this during 1998 by setting up the requirement of zero tolerance in food to protect consumers. Later that year, the DBMC started getting second thoughts about the strategy. By then, DT 104 did not appear to be all that dangerous. But now it was very difficult for the DVFA to change policy. A change of policy would seem to suggest that the DVFA now wanted to accept a greater risk for the consumers, which might cause the DVFA to be perceived as untrustworthy by consumers.
Why did the DBMC end up this situation? And what can be learned from this course of events? The paper will answer these questions by presenting the results of a rational reconstruction of the DBMC's decision problem regarding DT 104, based on available case study material. The reconstruction will demonstrate how interaction with other stakeholders and risk communication is inherent in the value base of a risk management decision.

Poster M4.8

Genetic and environmental effects on claw disorders in large-scale dairy farms

S. König, R. Sharifi, D. Landmann, M. Eise and H. Simianer, Univ. of Göttingen, Institute of Animal Breeding and Genetics, Albrecht-Thaer-Weg 3, 37075 Göttingen, Germany*

Feet and leg disorders are a source of considerable concern for dairy producers because lameness causes direct and indirect economic losses. Much of the variability in feet and leg health is associated with environmental effects, but several studies have revealed genetic impact in such traits.
Data comprised test day records and claw disorders recorded in 2003 of rear legs from 7790 Holstein cows in 9 large-scale dairy farms from one region within Eastern Germany. Diseases were divided in 3 categories (dermatitis, sole ulceration, horn erosion) and analysed separately. According to incidence of disorder cows were assigned a score of 0 or 1. Estimation of variance components for categorical traits was done using univariate and bivariate animal models applying REML and threshold methodology. Models included fixed effects of parity and herd visit. Heritabilities were 0.15 for dermatitis and sole ulceration and 0.19 for horn erosion. Genetic correlations between groups of diseases ranged from 0.33 to 0.56. There is a positive genetic correlation between claw disorders and the production level in milk yield in the first third of lactation and between claw disorders and SCS, resp. Analysis of fixed effects on claw disorders was carried out using the Glimmix Macro (SAS) and revealed significant differences between husbandry and feeding systems and parities of cows.

Poster M4.9

Analysing serial data for mastitis detection by means of local regression

D. Cavero, K.-H. Tölle and J. Krieter, [1]Institut of Animal Breeding and Husbandry, Christian-Albrechts-University, Olshausenstraße 40, 24098 Kiel, Germany*

The aim of this study was to detect incidence of mastitis in an automatic milking system using serial information. Data from 11,677 milkings (20 cows with at least one mastitis case) of the research dairy herd Karkendamm of the University of Kiel were available. A local regression method was applied using the SAS-procedure LOESS. Weighted least squares are estimated to fit linear or quadratic functions of the predictors at the centres of neighbourhoods. The method was modified to regard only previous information related to the prediction point.
The incidence of mastitis was defined both on therapies carried out and on weekly somatic cell count measurements. The time series of electric conductivity of quarter milk were analysed to find deviations as a sign for mastitis. The goodness of alerts varied dependent on the used confidence limits. A confidence interval (CI) of 95% led to a sensitivity of nearly 100%, however the specifity was only 31% and thus the error rate was high (74%). By application of KDE (kernel density estimation)-method (CI = 95%) to residuals estimated by means of LOESS-procedure specifity increased to 84% and sensitivity was reduced to 73%.

Poster M4.10

Effect of the early part of the first lactation to the udder health of Holstein cows in their later age

J. Bacher, Zs. Kompolti and J. Vagi, Szent Istvan University, Institute of Environmental Management, Dept. of Applied Ethology, Hungary, Gödöllö, Páter K. u.1. H-2103, Hungary

On the basis of investigations on six farms with more than 800 cows was proved that, the early part of the first lactation has a significant effect on the udder health condition in the latter milking periods.
In this study the udder health condition was grouped according to the somatic cell count (SCC) as follows: under 76 thousand SCC; between 76,1-200 thousand SCC and 200,1-400 thousand SCC; over 400,1 thousand.
The results proved that, the difference between SCS in the first test days, also prevailed as a tendency during the later test days. This means that, groups with the most advantageous udder health condition kept their condition in the later stages of lactation, and groups with the worst udder health condition remained in their bad condition.
Cows grouped according to the udder health condition in the first test month showed significant difference on the base of SCS. The differences in the daily milk production were significant till 5-7 month of the lactation.
It was found that, the earlier developed sub-clinic and clinic mastitis occur in the early part of the first lactation than these illnesses can occur in the latter milking periods as well.

Poster M4.11

Investigation of mastitis susceptibility values in Hungarian Holstein populations by using Statistic Process Control (SPC)

J. Vagi, Szent István University, Institute of Environmental Management Dept. of Applied Ethology, Working Group of Informatics and Methodology for Animal Breeding; H-2103 Gödöllö, Páter K. u.1., Hungary

The udder health condition of first lactation holstein cows was controlled in four farms with more than thousand cows by using methods of Statistic Process Control (SPC). The mastitis susceptibility values were investigated on the common basis of udder health monitoring mean values of the Somatic Cell Score (SCS), and Statistic Process Control (SPC).
The cow populations can be divided into five groups on the mastitis susceptibility values. Using the SPC methods, the five values of the mastitis susceptibility are the following:

1. Nearly always excellent health condition
2. Well judged and good udder health tendency, good condition
3. Uncertain udder health condition
4. Well judged udder health tendency, but only acceptable condition
5. Nearly always bad udder health condition

The effect of changes in environmental condition is demonstrated by the ratio-changes between groups. The preliminary results show that the applied new method can be useful for developing less-mastitis. This way we can establish herds with better milk quality, and make more economic production.

Poster M4.12

The use of homeopathic preparations for the treatment of mastitis of cows

E.M. Boldyreva and I.A. Porfiriev, Russian University of People's Friendship, 26 Bakinskikh comissarov, 7-6-28, 119571, Moscow, Russia

Mastitis is one of the most spread diseases of dairy cows. This disease leads to the decrease in production and milk quality and causes considerable damage to livestock farming. Preparations used for treatment are not always effective and harmless for animals and for people consuming milk.
In the research conducted different methods of treatment of cows with serous and catarrhal mastitis were studied. Mastitis was caused mainly by improper milking machine work and had traumatic etiology.
Two experiments were conducted. In control groups of both experiments antibacterial preparations were administered. In experimental groups different homeopathic remedies with antibacterial preparations of control groups were used. An average duration of illness in experimental groups of the 1st and 2nd experiment was 2.4 and 1.9 days less than that in control ones correspondingly. In control groups the period of illness was prolonged and 27.3-31.7% of cows recovered only after the 10th day of illness (up to 3-4 weeks) while in experimental groups all the animals had been cured to the 10th day. The treatment in experimental groups was more economical than in control groups.
Homeopathic preparations are harmless, do not cause side effects. Milk after the treatment with homeopathic remedies may be used without restrictions that makes their administration preferable.

Poster M4.14

The effect of climatic factors on the milk production of Holstein-Friesian cows

I. Györkös1, E. Báder[2], Ms. K. Kovács[1] and A. Kovács[2], [1]Research Institute for Animal Breeding and Nutrition, Gesztenyés str. 1, 2053 Herceghalom, Hungary, [2]University of West Hungary, Faculty of Agriculture, Vár 2, 9200 Mosonmagyaróvár, Hungary*

Authors studied the effect of temperature, atmospheric pressure and the number of sunny hours on the milk yield, fat and protein production in the second and third lactation of 120 Holstein-Friesian cows. Experimental data in milk production and climate were analysed between 1995 and 1998 and were processed per seasons. The influence of climate was compared to the estimated milk production by multivariate regression analysis. The highest milk loss was found in summer, mainly in periods above 25°C. The average lactation loss per cow exceeds the 522 kg milk due to the high temperature. No significant milk fat loss could be found in cows with lactation yield lower than 6000 kg. Climatic factors hardly influenced the milk protein production. Sunshine had a direct positive effect on the milk yield of cows in spring, autumn and winter.

Poster M4.15

Environmental effects on production and fertility traits in Northern Thai dairy cattle

N. Chongkasikit, S. König and H.-J. Langholz, University of Göttingen, Institute of Animal Breeding and Genetics, Albrecht-Thaer-Weg 3, 37075 Göttingen, Germany*

Milk production in Northern Thailand has been growing as an important agricultural sector, but it still faces numerous difficulties in the area of environmental effects and feeding management. Environmental component of variance can have a great variety of causes and its nature depends very much on the trait and the organism studied. The aim of the study was to identify significant environmental effects on production and fertility traits, because adjusting records for known causes of variation is essential in making selection accurate.

The data were collected by regular farm visits over a period of 2 years and consist of lactation performances and somatic cell score, fertility traits and body measurements from 2764 Holstein upgrade cows in 252 farms in Northern Thailand. With a body weight of 415 kg and a production level of 3668 kg milk (305 d) Thai Holsteins only reach approximately 70 % of performances of Holstein cows in temperate zones. Body development as well as body condition score show a certain effect on calving interval and services per conception ($p<0.05$) but not on milk performances. Calving in rainy seasons and high quality in roughage sources improve fertility significantly. Despite seasonal effects on milk yield are not very pronounced, there is a strong interaction ($p<0.01$) between years and calving seasons.

Poster M4.16

Supplementation of ω3 fatty acids and immune competence of Italian Friesian cows

C. Pacelli[1], F. Napolitano[1], G. De Rosa[2], A. Braghieri[1], A. Girolami[1], F. Surianello[1] and F. Grasso[2], [1]DISPA, Università della Basilicata, Potenza, Italy, [2]DISCIZIA, Università di Napoli "Federico II", Portici (Napoli), Italy*

Twenty-one lactating Italian Friesian cows were equally allocated to 3 treatments in order to evaluate the effect of dietary ω3 fatty acids on humoral and cellular immune responses. For 4 weeks group C received no supplementation, group D was offered 136 g of DHA gold (Omegatech, Inc. Boulder, Colorado) per day and group E was supplemented with 136 g of DHA gold + 2.000 U.I. of vitamin E per day. The phytohemagglutinin skin test was performed at the end of the supplementation period and 8 weeks later assessing the skinfold thickness 24 h after intradermal injection of 500 μg of mitogen. Five mg of keyhole limpet hemocyanin were subcutaneously injected at the end of the supplementation period and 2 weeks later. Antibody titer was tested at weekly intervals for eight weeks by ELISA. *In vivo* cell-mediated immune response was higher in groups D and E compared to group C ($P<0.05$). A significant interaction time x group was found ($P<0.05$) for humoral immune response. Both supplemented groups showed evidence of increased antibody response 2, 4, 6, 7 and 8 weeks after antigen administration. We conclude that both treatments (DHA and DHA + vitamin E) were able to increase cellular and immune responses of lactating cows.

Poster M4.17

Dietary inclusion of natural antioxidants in periparturient dairy cows: effects on oxidative status

D. Tedesco[1], S. Galletti[1], S. Rossetti[1], M. Tameni[1], S. Steidler[1] and P. Morazzoni[2], [1]Department of Veterinary Sciences and Technologies for Food Safety, Via Celoria 10, 20133 Milano, Italy, [2]Indena S.p.A., Viale Ortles 12, 20139 Milano, Italy*

In late pregnancy and early lactation, dairy cows can experience oxidative stress, which may contribute to periparturient disorders, such as retained placenta and mastitis.
Silymarin, a hepatoprotective and antioxidant natural substance, has shown a positive effect on productivity and health in transition cows. Lycopene is a caroteinoid with high oxygen radicals scavanging activity. Twenty cows selected according to parity, previous production and BCS, were divided into two groups. From 7 d before calving to 14 d after calving, 10 cows received 50 g/d of a mixture of silymarin + lycopene (Indena S.p.A.) by oral drench. The BCS was evaluated at -7, 0, 7, 14, 21d from calving. Milk production was recorded and samples collected on 7, 14, 21 DIM. Blood samples were collected at -7, 0, 14 d from calving. Oxidative status was assessed by determining thiobarbituric acid reactive substances (TBARS) in serum.
Treatment increased milk production. No difference was found in BCS between groups. In both groups TBARS increased with time, suggesting a stress status in periparturient cows. In control animals the TBARS value was higher ($P<0.05$) compared to treated animals in all days considered. This result suggest that treatment with natural antioxidants enhance oxidative status of periparturient cows.

Poster M4.18

Possibilities of prevention of hypochromic anaemia in pregnant dairy cows

J. Bíres[1], L. Hetényi[2], M. Húska[1], J. Buleca Jr.[1], M. Hiscáková[1], [1]University of Veterinary Medicine, Komenského 73, 041 81 Kosice, Slovakia, [2]Research Institute of Animal Production, Hlohovská 2, 949 92 Nitra, Slovakia

In the work influence of injection copper supplementation on concurrence of hypochromic anaemia in dairy cows 30 days before parturitions was aimed. In the beginning and on the 20th day of experiment, experimental animal were treated with Copperdep inj. at the dose of 1 mg Cu/kg b.w.. This preparation was tested before its commercial production. Increase of serum copper levels was recorded on 4th day after administration and significant differences between experimental and control group persisted till 15th day after calving ($p<0.05$, $p<0.01$, resp.). In the experimental group. increase in serum Fe concentration was observed from 7th day after the treatment. Administration of the preparation to pregnant cows suffering from hypocupraemia, sideropenia and hypochromic anaemia resulted in improvement of serum levels of Cu, as well as Fe, and values of RBC, haemoglobine and PCV. Moreover, in the experimental cows gradual restitution of hair and mucosal membranes were observed from 20th day after the treatment.

Poster M4.19

Amino acid profile of colostrum of charolais cows in the first week after calving

R. Zándoki1, J. Csapó2, Zs. Csapó-Kiss2, I. Tábori1 and J. Tözsér1, 1Szent István University, Cattle and Sheep Breeding Department, 2100 Gödöllö, Páter K. u. 1. Hungary, 2University of Kaposvár, Faculty of Animal Science, Institute of Chemistry, Department of Biochemistry and Food Chemistry. Guba S. u. 40, 7400 Kaposvár, Hungary

Authors determined amino acid profile of colostrum and milk protein of Charolais cows (n=15) in a Hungarian herd. Samples were taken by hand milking right after calving, 24, 48, 72 hours and 1 week after calving. Increasing tendencies ($p<0.05$) were obtained during the whole week for glutamate (0h: 16.98%; 1week: 23.16%); until the 72th hour for methionine (0h: 1.47%; 72h: 2.74%) and phenylalanine (0h: 3.89%; 72h: 4.48%); in the first 48 hour for isoleucine (0h: 3,10%; 48h: 4,26%), and in the first 24 hours for lysine (0h: 6.80%; 24h: 7.50%). Threonine- (0h: 6.53%; 72h: 3.94%), cysteine- (0h: 2.28%; 72h: 0.80%), and serine- (0h: 8.00%; 72h: 4.83%) content decreased ($p<0.05$) till the 72th hour after calving; and asparaginate (0h: 8.41%; 48h: 7.36%) and arginine (0h: 4.42%; 48h: 3.40%) lessened in the first 48 hours. Protein in first-milked colostrum of cows younger (Y) than 5 years (n=5) contained more histidine (Y:2.64g/100g, E:2.33g/100g; $p<0.005$), leucine (Y:8.46g/100g, E:8.01g/100g; $p<0.05$), lysine (Y:7.04g/100g, E:6.69g/100g; $p<0.05$), methionine (Y:1.66g/100g, E:1.37g/100g; $p<0.01$) and phenylalanine (Y:4.08g/100g, E:3.79g/100g; $p<0.05$), but less threonine (Y:5.88g/100g, E:6.67g/100g; $p<0.05$) and serine (Y:7.58g/100g, E:8.19g/100g; $p<0.05$) than that of elder ones (E (elder) >5years, n=10).

Microelement profile in colostrum of charolais cows in a Hungarian herd in the first week after calving

R. Zándoki[1], J. Csapó[2], Zs. Csapó-Kiss[2], I. Tábori[1], B. Zándoki[2] and J. Tözsér[1], [1]Szent István University, Cattle and Sheep Breeding Department, 2100 Gödöllö, Páter K. u. 1. Hungary, [2]University of Kaposvár, Faculty of Animal Science, Institute of Chemistry, Guba S. u. 40, 7400 Kaposvár, Hungary

Authors determined microelement profile in colostrum of charolais cows (n=15) in a hungarian herd. Colostrum samples were taken by hand milking right after calving (0h), 24, 48, 72 hours and 1 week after calving. Microelements were determined by AAS. Manganese content of colostrum did not change significantly during the first week after calving (0h:0.01mg/l, 1week:0.016 mg/l, p>0.05). Copper increased by the 24th hour after calving (0h:0.086mg/l, 24h:0,158mg/l, p<0,001), then it decreased continuously until the 72nd hour (24h-48h:0,158mg/l-0,132 mg/l, p<0,06; 48h-72h:0,132mg/l-0,094mg/l, p<0,01). There was no significant difference between the copper content of colostrum milked 72 hours and 1 week after calving. In the 48th and 72nd hours, copper content in colostrum of younger cows (Y<5 years, n=5) was larger than those of elder (E>5years, n=10) ones (48h: Y=0.18mg/l, E=0.11mg/l, p<0.001; 72h: Y=0.14mg/l, E=0.70mg/l, p<0.001). Concerning of zinc, a strong decreasing was observed by the 24th hour (0h:17.26mg/l, 24h:5.12mg/l, p<0.001), and a slighter change between the 72nd hour and 7th day (72h:5.06mg/l, 1week:3.95mg/l, p<0.06). Iron decreased in the first 48 hours after calving, then it stayed constant during the second part of the first week (0h-24h:0.86mg/l-0.60mg/l, p<0.001; 24-48h:0.60mg/l-0.41mg/l, p<0.005).

Poster M4.21

Efficiency of the control of chloramfenicol residues in milk powder in Ukraine

Y. Kosenko, D. Yanovych, D. Zasadna, A. Kostyuk and G. Teslyar, State Scientific-Research Control Institute of Veterinary Preparations and Fodder Additives, Donetska Str., 11, 79019, Lviv, Ukraine*

Ukraine is carrying out harmonization of the national legislation according to the European and world norms. By the order of State department of veterinary medicine in 2002, there was forbidden the use of series of veterinary medicine products, including chloramfenicol, which was forbidden in Europe as well. The long previous application of preparations, containing chloramfenicol and presence them in the market of medical preparations demand the amplified control for keeping of interdiction. For prevention of exports the products of an animal origin, which content residues of forbidden preparations, all of them passes the obligatory control on the contents of chloramfenicol. In State Scientific-Research Control Institute of Veterinary Preparations and Fodder Additives, which is national reference-center of the control veterinary drug residues, during 2002-2004 there were carried out researches of large quantity of milk powder series. The method ELISA was used for screening . Due to updating of a methods of sample preparation, the sensitivity of method was 0,15 ppb. From all the products, which had passed the control, in 28,9 % was detected chloramfenicol in concentration above 0,3 ppb. The analysis of the data received during 2002-2004 shows the decrease of amount of positive assays, testifies efficiency of control action.

Poster M4.24

Proteins from bovine tissues and biological fluids: defining a reference electrophoresis map for liver, kidney, muscle, plasma and red blood cells

C. D'Ambrosio, S. Arena, L. Ledda, G. Zehender, L. Ferrara and A. Scaloni, I.S.P.A.A.M., National Research Council, via Argine 1085, 80147 Naples, Italy

A number of high resolution 2-DE reference maps for bovine tissues and biological fluids have been determined for animals in basal state. Differences in protein profiles were evident depending on the tissue/fluid analysed. Among the 1863 distinct protein features detected in samples of liver, kidney, muscle, plasma and red blood cells, 509 species were identified and associated to 209 different genes. Difficulties in the identification were related to the poorly characterized *Bos taurus* genome and were solved by a combined MALDI-MS and LC-ESI-MS-MS approach. The nature and the differences in protein expression profiles reflected the role of each protein in organ biochemistry and physiology. The experimental output allowed establishing a 2-DE database accessible through the World Wide Web network at the URL address (http://www.iabbam.na.cnr.it/Biochem). These reference maps may serve as a tool in future veterinary medical studies aimed to the evaluation of changes in protein repertoire for altered animal physiological conditions and infectious diseases, to the definition of molecular markers for novel diagnostic kits and vaccines, as well as to the characterization of protein modifications in bovine materials following the technological processes used in food industry.

Poster M4.25

Pathomorphological pattern of the liver in Pomeranian lambs and crossbreeds of Pomeranian ewes with Ile de France rams

J. Szarek[1], P. Brewka[2], H. Brzostowski[2], A. Andrzejewska[3], Z. Tański[2], J. Lipinska[1], M.Z. Felsmann[1] and B. Babinska[1], [1]Faculty of Veterinary Medicine, [2]Faculty of Animal Bioengineering, University of Warmia and Mazury in Olsztyn, 10-717 Olsztyn, Poland, [3]Medical Academy, 15-269 Bialystok, Poland*

The production of good quality lamb meat can be obtained, among the others, by various crossbreeding. The aim of this study was to evaluate the effect of the breeds used for crossbreeding on the pathomorphological pattern of the liver in their crossbreeds.

The investigations were carried out on 100-day old young rams of meat and wool sheep line origin (Pomorska - P) and young rams at the same age (n = 8) that are crossbreeds of P breed lamb and meat breed lamb - Ile de France (IF). After macroscopic evaluation, the liver was sampled for microscopical and ultrastructural examination.

According to the observed microscopic and ultrastructural lesions (particularly retrogressive changes and circulation disturbances) the most pathological alterations of the liver were observed in young rams of F_1 P x IF. They were statistically less significant in purebred lambs of P breed. The same dependence was also noted regarding the degree of intensity of morphological changes. This study has shown that the differences in breed growth-rate influence the pathomorphological pattern of the liver in lambs. It is known that crossbred lambs have higher productivity in comparison to the purebreds. This could be the main reason why the liver of the crossbreds is more susceptible to morphological lesions than the liver of the purebreds.

 Poster M4.26

The effect of sheep crossbreeding on the morphological and histochemical lesions of the liver in Pomeranian lambs and crossbreeds of Pomeranian ewes with Texel rams

J. Szarek[1], H. Brzostowski[2], P. Brewka[2], Z. Tanski[2], J. Fabczak[1], A. Andrzejewska[3], J. Lipinska[1] and M. Blum[1], [1]Faculty of Veterinary Medicine, [2]Faculty of Animal Bioengineering, University of Warmia and Mazury in Olsztyn, 10-717 Olsztyn, Poland, [3]Medical Academy, 15-269 Bialystok, Poland*

Intensification of the production and improvement in the quality of lamb meat is obtained by various crossbreeding. It may lead to the morphological lesions in their off-springs due to the metabolic differences in these animals.
Investigations were carried out on 100-day-old rams (n = 8), Pomorska (P) meat-wool breed, Texel (T) meat breed lamb and their crossbreeds (P X T). After the macroscopic evaluation the liver was sampled for histopathological (hematoxylin and eosin staining, Sudan III according to Lillie Ashburn method) and ultrastructural (TEM, Opton 60 kV) examination.
Microscopic examination of the liver as well as ultrastructural analyses of this organ revealed retrogressive lesions, circulation disturbances, inflammations and progressive changes. They were noted frequently, rarely and sporadically regarding the breed. P X T rams were affected with lesions in more than 50% in comparison with T lambs. The highest level of glycogen was noted in the liver of P X T rams. Its lower level was observed in animals of P race. These studies suggest that the differences in breed growth-rate have effect on the morphological lesions of the liver of the lambs examined.

Poster M4.27

Prevalence of nematodes in sheep and goats at LPRI Bahadurnagar

M. Afzal, A. Ahmad and F. Ahmad, Livestock Production Research Institute, Bahadurnagar, Okara Punjab, Pakistan

During July, 2001-June, 2002 f aecal samples from 386 sheep and 276 goats were collected. Prevalence of gastro intestinal parasites in sheep and goats maintained at Livestock Production Research Institute and area around Bahadurnagar (Okara) was investigated. Various species of nematodes were recorded. Haemonchus contortus (sheep 83.165% and goat 85.14%) followed by Trichostrongylus spp. (sheep 73.05% and goat 69.56%) and Nematodirus spathiger (sheep 44,56% and goat 37.31%) had the highest prevalence. Albendazole @ 5 mg/kg. Levamisole @ 7.5 mg/kg was 100% effective against internal parasites. Ivermectin @ 1ml/50 kg injected subcutaneously was equally effective against gastro intestinal parasites with additional benefit against ectoparasites.

Poster M4.28

Physiological indicators and the quality of the obtainable production by goats

E. Birgele, D.Keidane and A.Mugurevics, Latvian University of Agriculture, Faculty of Veterinary Medicine, Preclinical Institute, K.Helmana 8, Jelgava, LV-3004, Latvia

The aim of the research was to investigate the dynamics of functional gastric condition, the blood biochemical indicators and the quality of the obtainable production for goats in control and experimentally infected with strongylus of digestive organs. In this series of the research two groups of animals were made up:

1) five 2 to 4 month old kids; 2) five 9 month to 2 years old goats.

Chronic fistulas were operated into the rumen and abomasum of all goats.

In the morning and after feeding intra-ruminal and intra-abomasal reiterated and lasting (8 to 12 hours) pH-metry was carried out. Grown-up goats were infected with artificially grown larvae of strongylus in order to investigate the influence of widely-spread in Latvia parasitocenosis on physiological indicators of the goats organism and the quality of production. Dynamics of the blood morphological and biochemical indicators was also investigated for all animals involved into the experiment. Some part of the research was carried out on the Jelgava region goat farm "Līcīši". Fat content was fixed in milk and the amount of cholesterol, urine acid, the number of somatic cells and the amount of 17 aminoacids was also analysed. 2 from 10 goats were not infected, The results testify that indicators of the milk quality changed for the goats infected with the by strongylus according to the invasion degree. The amount and ratios of some separate/definite aminoacids change; the amount of cholesterol changes too.

Poster M4.29

Effect of heat stress on protein requirements for maintenance of goats

M.H. El-Shafie[1], G. Ashour[2],F. Abou Ammou[1] and T.M. El-Bedawy[2], [1]Animal Production Research Institute, Cairo, Egypt, [2]Department of Animal Production, College of Agriculture, Cairo University, Giza, Egypt

The study was conducted to determine protein requirements for maintenance (RCP)of goats under 25°C and 35°C environmental temperature. Metabolic fecal nitrogen (MFN) and endogenous urinary nitrogen (EUN) were determined from fasting metabolism trails in which feces and urine were collected from eight male goats in climatic chambers for 14 days. The EUN and MFN values were used for RCP determination.

Protein requirements for maintenance of goats were also determined under the two environmental temperatures from nitrogen balance data applying statistical, factorial and graphical methods. Goats were fed four ration differed in dietary protein concentrations (9.38, 11.81, 13.22 and 14.74 %).

Results indicated that there was difference in protein requirements for maintenance (RCP) among the used methods and RCP calculated by any method was lower for goats under 35°C than those under 25° C.

The RCP of heat stressed goats was about 77% of that of goats under 25°C, averaging 4.84 g/Kg $W^{0.075}$ for goats under 25°C and 3.71 g/Kg $W^{0.075}$ for those under 35°C.

Theatre ML5.1

"Happy pigs are dirty" - conflicting perspectives on animal welfare

P. Sandøe, Centre for Bioethics and Risk Assessment, Royal Veterinary and Agricultural University, Royal Veterinary and Agricultural University, Groennegaardsvej 8, DK-1870 Frederiksberg C, Denmark

There is a growing awareness by all stakeholders that there is need for more focus on the well-being of farm animals. These animals have in several respects been badly hit by the intensification of farm animal production. They typically get less space per individual than they did previously and many live in barren environments that do not allow them to exercise their normal range of behaviour, while genetic selection has been accompanied by increased problems with production-related diseases.

However, underneath the apparent agreement there may be very different perceptions of what it is for farm animals to be offered a decent life. Thus the quotation in the title of the paper is from an interview with Danish citizens who were being shown pictures of what is supposed to be animal friendly pig production.

In the presentation an attempt will be made to characterize and discuss the different points of view found in the debate about how animals are treated in farm animal production. These differences in point of view both concern ideas of what is a good animal life and the ethical framework within which the balancing of concerns for animals and human interests is to be discussed.

Theatre ML5.2

We and they: animal welfare in the era of advanced agricultural biotechnology

Assya K. Pascalev, 2039 Selva Marina Drive, Atlantic Beach, FL 32233, USA

The paper discusses central moral issues involved in the treatment of animals in agriculture and introduces the major ethical concepts and principles that pertain to animal bioethics. It explores critically the concept of animal rights, animal suffering, animal welfare, and the moral values behind such movements as vegetarianism and animal liberation. Special attention is given to the issue of animal welfare in light of the latest advances in biotechnology such as cloning, genetic engineering and xenotransplantation. Some of the questions to be addressed are: What are the main ethical challenges that animal agriculture faces today? Is it moral to genetically engineer farm animals and can the need for greater productivity justify the genetic modification of such animals? Should we change the natural capacities of animals e.g., to reduce their ability to feel pain and increase their resistance to disease? What is the moral status of animals with human genes or genes from other animal species? What is involved in respecting animals? The paper also considers relevant aspects of the currently existing practices and policies concerning animal welfare in the United States of America such as the Animal Welfare Act and the regulations of the Food and Drug Administration, and subjects them to a critical analysis.

Theatre ML5.3

Ethics and animal welfare considerations in extensive pig production

M.A. Aparicio and J.D. Vargas, Veterinary School, University of Extremadura, University Campus, s/n. 10071 Cáceres, Spain*

This paper is focused to reflect about ethics and animal welfare aspects required in the extensive pig production. The extensive pig production in Spain is traditionally characterized by: the use of the Iberian pig, an autochthonous breed perfectly integrated into the environment in which they have developed; a long duration of the productive cycle for about 23-24 months; a high level of animal welfare level, specially in the fattening process with freedom of movement and feeding base on natural sources: acorns and grass, and a equilibrated "dehesa" agroforestry system where this activity has been developed.
Nowadays the introduction of more intensificated methods is debt to the increasing to demand. Thus, important changes are being made, such as: the shortening of the productive cycle (10-12 months); freeing from the territorial base; changes during the fattening period, fattening with mixed feed and less animal freedom. All these facts may implicate a loss of the animal welfare condition. These circumstances lead us to question it from an ethical point of view.

Theatre ML5.4

A proper intensive dairy farming can be optimal for welfare and milk yield

E. Trevisi, M.Bionaz, F. Piccioli-Cappelli and G. Bertoni. Istituto di Zootecnica, Piacenza, Italy

The intensive breeding of dairy cows is often considered at risk of low welfare for the animals. Namely, the high yield has been suggested to be, per se, a cause of well-being reduction. On the contrary we have in many farms demonstrated that high milk yielding cows are not necessarily in a bad welfare condition, while many environmental and management problems can cause a distress situation and a reduced milk yield. To confirm these results, in an average yielding dairy farm, the welfare was assessed according to an Integrated Diagnostic System which considers health status, milk yield and quality, feeding strategy, nutrition status, blood profiles etc. The same check-up has been repeated a year later, after some attempts to correct the main mistakes that were previously observed: dry and lactating cow diets, n° of cubicles, hygiene conditions, preparation and milking procedures etc.. In one year only part of the previous mistakes have been fully corrected; in spite of the animal response was definitively improved suggesting a better welfare situation. Namely better teat and body condition scores, an increase of calve number (fertility), the almost complete disappearance of legs and feet lesions. At the same time, the milk yield (23,4 vs 21,1 kg/d) and its quality were also improved (fat content 3,60 vs 3,40%) or remained almost unchanged (protein and SCN contents).
Again it is confirmed that better breeding technologies can optimize the animal welfare and raise milk yield in the intensive systems also.

Theatre ML5.5

Livestock farmers and technicians representations of animals and animal welfare

A.C. Dockès[1], F. Kling-Eveillard[1] and C Wisner-Bourgeois[2], [1]Institut de l'Elevage 149 rue de Bercy - 75595 Paris Cedex 12, France, [2]Chaire de Sociologie, INA-PG 16 rue Claude Bernard 75231 Paris cedex 5, France

Animal welfare constitutes an increasing social demand, and is an ethical preoccupation for farmers. We studied the representations of livestock, pork and poultry farmers and of technicians about their profession, about animals and animal welfare, which have not yet been studied until now. Breeders and technicians share some ideas: animals is the main aspect of the definition of the breeder profession; their relationship to animals is professional, observing is an essential activity for a breeder. But we also pointed out the diversity of the representations of breeding, of animals, and of practices. It concerns affection to animals, ethic vision of the profession, consideration of the animal needs, definition of animal welfare. This diversity relies on the personal history of the farmers or the technicians and on the type of animals bred. Beyond the philosophical distinction between the animal described as a machine and the animal as a sensitive being, we think that the relations between breeders or technicians and animals can be organised in three dimensions: "the animal as a machine" which is seen through it's productive functions; the "communicating animal", in relation with the human beings and the "affective animal", able to develop a real attachment relationship with man.

Theatre ML5.6

The relativity of ethical issues in animal agriculture related to different cultures and production conditions

R.L. Doerfler and K.J. Peters, Humboldt University Berlin, Faculty of Agriculture and Horticulture, Department of Animal Breeding and Aquaculture in the Tropics and Subtropics, Philippstr. 13, 10115 Berlin, Germany*

At present ethical issues are increasingly on the agenda not only in relation to novel technologies, but also in relation to animal welfare in livestock agriculture; being mainly concerned with the human obligation towards animals. In many western countries animal welfare legislation has been revised recently and there are on-going activities to incorporate these standards in international trade agreements. However, the adoption of ethical norms by countries with vastly differing socio-cultural background, living conditions, and livestock production systems adapted to the prevailing ecological circumstances remains highly controversial: Ethics and animal welfare considerations are related to issues of ethical relativism. This paper analyzes arguments for the relativism paradigm linked with the human-animal interaction in selected tropical livestock production systems. While in pastoral systems the human-animal relation can be termed affective and highly interdependent, smallholder livestock production systems are characterized by a strong multi-functional mutual dependency between humans and animals. In these systems only ethical concerns related to the prevention of cruelty to animals are appropriate. Only if the human animal relationship turns to a solely utilitarian, as it is the case in market driven production systems, a shared responsibility for the mismanagement of human relations to animals is demanded and the application of western views on animal welfare legislation is justified.

 Theatre ML5.7

Public perception of farm animal welfare in Spain

G.A. María, Faculty of Veterinary Medicine, University of Zaragoza, Zaragoza, Spain

The social claim in favour of animal welfare has produced changes in the European legislation controlling livestock industry. Modifications in production systems will be necessary for the compliance of the new requirements affecting production costs. The question is to determine whether people in countries such Spain will accept the increment in the price in order to improve animal welfare. The objective of this study was to assess the human attitude and perception of farm animal welfare in Spain. People living in the urban area of the University of Zaragoza was surveyed (n=3978). The questionnaire comprises 3 sections with a total of 12 questions. The first section refers to the general attitude on animal welfare. The second asked to give a note about their perception about the treatments of the animals in the farm. The third part asked about their agreement to pay more for a product to improve welfare and their actual consumption of welfare friendly products. Descriptive statistics were calculated and the fixed effects of age, sex and occupation were analyzed. A high proportion of the people agree to pay more for a product, if this greater price would guarantee a better welfare. There was a trend indicating a positive response in women, young and students. There is a disagreement between the expression of availability to pay more and the declaration of the actual consumption of welfare friendly products.

Theatre ML5.8

Culture, values and ethics of animal scientists

John Hodges, Lofererfeld 16, A-5730 Mittersill, Austria

Culture is defined as the shared worldview of a sub-set of humanity: race, nation, or professional group. In practice Culture means "The way we do things around here". Values are the objectives that matter most to a person or to a cultural group to which priority of interest is consistently given in decisions allocating time, energy, resources, wealth and education. Ethics defines the moral component of each decision reflecting self-interest and/or concern about the well-being of other individuals or groups in society. Thus, in any sub-set of humanity, including professional animal scientists, Culture, Values and Ethics are closely linked.

The normative cultural assumptions and commonly-held values of animal scientists guide group and individual decisions on the research and application of scientific knowledge. Strong links between animal scientists and business interests mean that the culture and values of commerce also inform and steer decisions by animal scientists. The food chain is increasingly watched by society as a whole, by governments and by special interest groups to determine the extent to which our behaviour is ethical or serving only our special interest group.

The changing culture and values of societies in Europe and North America and Developing Countries are examined and compared with those of animal scientists. It is proposed that more radical changes in the culture and values of animal scientists are needed to match the assumptions of all societies for their food supply.

 Theatre ML5.9

The welfare implications of animal breeding and breeding technologies in commercial agriculture

Judy A MacArthur Clark, Chairwoman, UK Farm Animal Welfare Council (FAWC), London, UK

The commercial applications in agriculture of new breeding technologies, as well as conventional breeding strategies, have the potential to influence animal welfare in both positive and negative ways. For example, the sexing of cattle semen might be used to reduce the number of unwanted male dairy calves provided that the technique had not been shown to produce adverse effects. On the other hand, inappropriate use of some breeding technologies may create new problems, or exacerbate welfare problems that may already have arisen within conventional livestock breeding. It is the impact of any breeding technology or strategy that is important to welfare, whether it is the quality of life of the offspring that is compromised, or whether it is the application of the technology itself that causes pain, distress or lasting harm to the subject animal.
The aim of this paper is to provide clear and practical advice on the establishment of an appropriate framework within which developments in animal breeding and breeding technologies, and the outcome of such processes, may be considered, monitored and, where necessary, regulated. It will build on recommendations recently made by the FAWC to UK Government, and will consider the wider public interest in these technologies.

Theatre ML5.10

Code of good practice for European farm animal breeding and reproduction

A.M. Neeteson[1], D. Flock[2], S.A. Korsvoll[3], J. Merks[4] and H. Stålhammar[5], Farm Animal Industrial Platform (European Forum of Farm Animal Breeders), Benedendorpsweg98, 6862WL Oosterbeek, Netherlands, [2]Lohmann Tierzucht, Am Seedeich 9-11, 27454 Cuxhaven, Germany, [3]AquaGen, P.O.Box203, 7200 Kyrksæterörа, Norway, [4]IPG, Institute for Pig Genetics, P.O.Box43, 6641SZ Beuningen, The Netherlands, [5]SvenskAvel, Önsro, 53294 Skara, Sweden

The aim of CODE-EFABAR is to develop a verifiable Code of Good Practice for Farm Animal Breeding and Reproduction, carried and implemented by European farm animal breeding companies, critically guided by ethics, and communicative to society. CODE-EFABAR is a follow up of SEFABAR (QLG7-2000-01368). Breeding organizations have developed the technical details, Centre for Bioethics and RiskAssessment ethical and society elements, European Federation of Biotechnology the communication strategy, and Société Générale de Surveillance advises on verifiability and certification requirements. The draft Code will be discussed at a workshop with representatives of NGOs and European breeding organisations (October 2004, Barcelona), after which the Code will be finalised.
In June 2005, a training workshop will be given on the implementation of the Code of Good Practice in breeding organizations: on the web site only or also as a verifiable Code.
The Code of Good Practice will be available from the FAIP(EFFAB) web site, and disseminated to breeding and reproduction organisations and, among others, policy makers and society organisations (e.g. environmental, welfare and consumer organisations). In the future, FAIP will take the responsibility for the maintenance of the Code of Good Practice.

Theatre ML5.11

The role of ethics in assessing acceptable animal welfare standards in agriculture

C.M. Mejdell, Norwegian Council on Animal Ethics, Arnemoveien 6, 2500 Tynset, Norway

Veterinary and agricultural sciences can tell us how to prevent and cure diseases, how to effectively breed, feed and keep animals for production purposes. The last decades, ethologists have provided information on the behavioural needs of livestock. The optimal choice, from the animal point of view, is however usually not an option, because of practical or economical factors. Most husbandry practices set limits for the life of animals, in terms of restricted room or high stocking densities. The efficient production may lead to an increased risk of stress, disease and suffering. Science alone does not provide the answer to where to draw the line between acceptable and unacceptable welfare levels, i.e. what is an ethical way of keeping farm animals. The answer to this depends on values and attitudes towards animals in the society. These attitudes are formed in a complex way by history, religion, culture and tradition. Attitudes are slowly changing over years, influenced by new scientific knowledge, but also by urbanisation and life standard of people. "How should animals be treated in our society", is a question that should be discussed by the public, since the answer can be said to reflect the humanity of the society. In Norway and many other countries the authorities have appointed independent, advisory bodies on animal welfare and ethics to discuss these issues. The composition of such committees should be as broad as possible and lay people should always be included.

Theatre ML5.12

Building a new governmental authority based on public demands for improved animal welfare

C. Berg and M. Hammarström, Swedish Animal Welfare Agency, PO Box 80, SE-532 21 Skara, Sweden

Sweden has a long history of detailed and progressive legislation related to animal welfare for laboratory, farm and companion animals. Historically these issues have been the responsibility of the Veterinary Administration and later the Swedish Board of Agriculture (SBA). As a certain proportion of the public opinion felt that the animal welfare related issues were not given proper attention at the SBA, a political decision was recently made to separate animal housing, management and welfare from the SBA and create an independent Animal Welfare Agency. This Agency, which was formally launched on January 1st 2004, also includes a section for research animals. The government has commissioned the Agency to improve animal welfare by developing, enforcing and evaluating legislation. The agency should consider scientific evidence when writing new legislation, while most of the purely ethical considerations in relation to legislation will be made at the political level. However, the Agency will incorporate an external Animal Welfare Council, which will discuss ethical aspects in relation to existing or proposed legislature. Also, the new Agency must deal with a diversity of public expectations. Animal rights groups have high expectations regarding new and stricter legislation, for example related to fur animals, while some farmers fear that production aspects may be completely lost in discussions about improving welfare standards for farm animals.

The expanding role of animal welfare within EU legislation and beyond

R. Horgan and A. Gavinelli, European Commission, Directorate General Health and Consumer Protection , Unit E2 Animal Health and Welfare, Zootechnics Rue Froissart 101, B-1040 Brussels, Belgium*

Animal welfare is being accorded an increasingly important role in today's civil society. Within the EU this has been enshrined within the specific "Protocol on the Protection and Welfare of Animals" of the Treaty of Amsterdam, obliging Member States and the EU Institutions to pay full regard to the welfare of animals when formulating and implementing Community legislation. There is a growing body of EU legislation on this issue, founded on sound scientific principles (including the role of EFSA in providing independent scientific advice), taking account of public concerns, stakeholder input and possible socio-economic implications. Recent CAP reforms also testify to animal welfare's growing stature in policy-making, with the introduction of the principle of cross-compliance regarding eligibility for direct payments and additional financial incentives for producers to achieve higher standards. Animal welfare is being increasingly perceived as an integral element of overall food quality, having important knock-on effects on animal health and food safety. On a worldwide level the OIE (World Organisation for Animal Health) has embarked on an initiative to develop global animal welfare guidelines and standards mandated by 166 member countries and the issue remains on the agenda for future WTO negotiations. Consumers demand higher standards of animal protection and it is incumbent upon policy-makers and legislators to respond accordingly.

Theatre Ph2.1

The role of immune defences in udder health

A. Zecconi and R. Piccinini, University of Milan, Dept.Animal Pathology, Hygiene and Health, Via Celoria 10, 20133 Milano Italy

The udder immune defences play a pivotal role in preventing intramammary infection and in reducing the impact of bacterial toxins and metabolites. Immunological components of the mammary gland have been traditionally classified as anatomical, humoral and cellular factors. This classification is suitable for an analytical description of immune defences factor by factor, however it is less useful when the focus is to describe the influence of different internal and external factors on the immune defences. The review will cover some of the most important factors involved in mammary gland immune defences. The dynamics of immune defences during the periparturient period in different farm conditions will be described. The paper will also supply an update on the available opportunities to modulate immune defences by vaccination or by biological response modifiers.

Theatre Ph2.2

Genomic identification of bovine beta-defensin genes and their expression in mammary gland tissue

S. Roosen, C. Looft and E. Kalm, Christian-Albrechts-University, Faculty of Agricultural and Nutritional Science, Institute of Animal Breeding and Husbandry, Olshausenstr. 40, D-24098 Kiel, Germany*

β-defensins are cationic peptides of 38–42 amino acids and are part of the innate immune system of vertebrates, insects and plants. They are multifunctional peptides, having an antimicrobial activity against bacteria, viruses and funghi. Because mastitis is influenced by genetic components, the relevance of bovine β-defensin genes concerning udder health and mastitis resistance has to be analysed. We isolated total RNA from tissue samples of the mammary gland of nine cows with different clinical findings and amplified complete cDNA using defensin-consensus-primers. For genomic characterization we isolated 20 BACs from two bovine BAC libraries, RZPD number 750 and 754, with defensin-consensus-primers based on published β-defensin sequences. By means of subcloning of PCR-products from BACs generated with consensus-primers and primers derived from a bovine EST which shows similarity to human β-defensin-103, seven novel β-defensin-genes and four putative pseudogenes have been identified. One gene and two pseudogenes represent a new subgroup of bovine defensin-genes, because they are orthologous to *hBD103*. Furthermore, the assembly of two contigs using BAC-endsequencing, results of a PCR-strategy, subcloning, determination of insert-length and digestion by restriction endonucleases is presented. In conclusion, differential expression of defensin-genes strongly supports the hypothesis that they are involved in local host defense during udder infections and therefore further functional analysis is required.

Theatre Ph2.3

Haptoglobin gene expression in bovine and human leukocytes in blood and bovine somatic cells in milk

M.A. Thielen, M. Mielenz, S. Hiss and H. Sauerwein, Institute for Physiology, Biochemistry and Animal Hygiene, University of Bonn, Katzenburgweg 7 - 9, 53115 Bonn, Germany*

Local expression of haptoglobin (Hp) in tissue samples collected from different sites of the bovine mammary gland has recently been established (Hiss et al., 2003). The aim of this study was to investigate whether bovine leukocytes are capable of synthesising Hp and might thus contribute to the Hp mRNA transcripts found in the mammary gland through migration into this organ. For this purpose, RT-PCR was carried out with RNA extracted from mononuclear and polymorphonuclear cells in blood and from somatic cells in milk of dairy cows. In comparison, human leukocytes were examined. Both subpopulations of bovine blood cells and the bovine somatic cells in milk exhibited Hp mRNA transcripts in contrast to human leukocytes in which no Hp mRNA transcripts were detectable by our method. In conclusion, blood derived leukocytes express Hp and, hence, might at least be partly responsible for the Hp mRNA transcripts found in the bovine mammary gland. It remains to be clarified whether other cells within the mammary gland are also capable of Hp synthesis.

Theatre Ph2.4

Incidence of pathogens involved in clinical cases of mastitis and the effectiveness of differing antibiotics in specific mastitis pathogens

K. Clemens[1], Y. Hunt[2], J.K. Margerison[2], P. Northway[2] and R. Shepherd[3], [1]Milklink, Okehampton Business Park, Okehampton, Devon EX20 1UB, UK, [2]School of Biological Sciences, University of Plymouth, Seale Hayne, Devon. TQ12 6NQ, UK, [3]Agrifood Centre, University of Plymouth, Seale Hayne, Devon. TQ12 6NQ, UK

This study monitored bulk milk somatic cell count (SCC) levels in milk supplied to a co-operative over a 12 month period and select a sub-sample of producers according to SCC, analyse clinical incidence of mastitis, establish pathogens involved and response to antibiotics. 2150 bulk milk SCC levels were monitored and a sub-sample of 10 dairy farms were selected according to SCC and matched according to herd size, housing type, feeding levels and milk yield. 410 milk samples were collected from cows exhibiting clinical symptoms of mastitis. Organism(s) responsible were detected by spreading 0.2ml of milk on selective culture media (Baird Parker medium for *Staphylococcus aureus,* Edwards Medium for *Streptococci agalactiae, dysgalactiae and Uberis,* Macconkey Agar for *E Coli*) and incubating (37 C for 24/48 hrs). Antibiotic sensitivity was completed using 'Multidiscs'. *S. uberis and S. Aureus* were found most frequently and incidence of *S. uberis and S. aureus* were similar over 12 m. Synulox was found the most effective antibiotic compared with Furazolidone and Neomcyin. Furazolidone was found more effective than Neomcyin.

Milk flow rate at the beginning of milking as an indicator for the subsequent milk let-down stages

N. Livshin[1], E. Maltz[1], M. Tinsky[2] and E. Aizinbud[1], [1]Agriculture Research Organization, Volcani Center, PO Box 6, Bet-Dagan 50250, Israel, [2]S.A.E. Afikim, Kb Afikim 15148, Israel*

Proper milking is prerequisite for proper cow's performance. This work's objective was to relate the initial milk flow rate in first 15s of milking (FR15s) with the selected subsequent milk let-down parameters: peak flow rate, percent of milk at first 2 min, and mean flow rate of milking. The analysis was performed for 7140 milkings during 5 consecutive days in 14X2 herring bone parlor, where an average of 476 Holstein cows were milked thrice daily using Afiflo system (S.A.E. Afikim). The milkings were divided, according to FR15s, into 3921 low initial flow (≤1 kg/min, LIF) and 3219 higher initial flow (>1 kg/min, HIF) groups. The analysis showed that let-down values, including FR15s, had strong positive correlation with milking yield. Hence, each of selected milk let-down parameters was computed for LIF vs. HIF milkings of the same yield interval. In different milk yield intervals, the peak flow, milk 2 min percent, and mean flow rate in HIF were higher than in LIF: 37.2-55.2% (mean 51.0%), 29.5-56.5% (mean 33.0%), and 11.4-45.0% (mean 29.2%), relatively. These differences between LIF and HIF milkings were significant (P<0.001) in all yield intervals. It is suggested that FR15s can serve as an important indication for efficiency of the whole milk let-down process.

Poster Ph2.6

The milk flow curve of bovine in the prevention decrease of the somatic cell counts of milk

P. Mijic[1], I. Knezevic[1], M. Domacinovic[1], A. Ivankovic[2], M. Baban[1]. [1]Faculty of Agriculture Osijek, P.O.B. 719, 31000, [2]Faculty of Agriculture, 10000 Zagreb, Croatia*

Milk flow curve studies and its application within the selection of cattle are expanding towards the higher number of European countries. The researches made in Croatia have shown the existence of the relationship between the milk flow curve and the number of the somatic cell. At a short sustained incline (dIFR < 0,40 min) and the high quotient between the peaks flow rate and the sustained decline (QPD > 1,20) there was the lowest number of the somatic cell counts in milk (LSCC = 3,47, respectively 3,30). This kind of relationship is an acceptable one and the reason is that with its increase the peak flow rate is prolonged and the sustained decline is shortet. With the correlation coefficients we have the possibility to improve milk ability through breeding where we could decrease the somatic cell counts in milk through the prolongation of the peak flow rate (r = -0,25), the reduce of the sustained incline and the sustained decline (r = 0,05).

Through the breeding selection we could make a selection of bull fathers and mothers which would have the most desirable shape of the milk flow curve and the duration of the individual milking phase depending on the lowest number of somatic cells in milk.

Poster Ph2.7

Mammary glad health asociated with machine milking

M. Rákos, L. Stádník and F. Louda, Czech University of Agriculture in Prague, Department of cattle breeding and dairy management, Kamycká 129, 165 21 Praha 6, The Czech Republic

Objective of this attempt was to establish an influence of machine milking on the teat end condition and somatic cells count. The observation was carried out on sixty cows at three farms. Each of these three groups involved 20 cows. Two groups consisted of Holstein cows and the last one from Czech spotted cows. Using ultrosonograph with 7,5 MHz probe, we assessed lenght of teat canal, teat wall thickness and area of teat end. Sonografic frame was taken before milking, immediately and three hours after milking. It was performed twice during laktation period first time between the 40th and 70th day and second time between the 200th and 230th day. We recorded actual yield and milking time in the test day. The data were analysed by procedure GLM. Diferences between values of measured charakteristics obtained before and immediately after milking were increasing in accordance with increasing time of milking. We did not find any significant diference between breeds and between values obtained first time and second time during lactacion period. Variance of somatic cells count was high. We estimated diferencies between values obtained before and imediately after milking and those obtained imediately and three hours after milking. Than we found out coralation coeficients between the above diferencies and somatic cells count.

Poster Ph2.8

Pathogenic microorganisms elevating somatic cells count in cow's udder secretions

A. Jemeljanovs, I.H. Konosonoka and J. Bluzmanis, Latvia University of Agriculture, Research Centre "Sigra", 1 Instituta Street, Sigulda LV-2150, Latvia*

Udder secretion samples were taken from cows four to five months after calving with SCC over 300 000 cells ml^{-1} to determine mastitis pathogens. In total, 394 samples were investigated. Gram-positive microorganisms from the genera *Staphylococcus (S.), Aerococcus, Micrococcus, Streptococcus* and *Corynebacterium* were isolated in 93.9 % of cases. Gram-negative microorganisms from the family *Enterobacteriaceae* were isolated in 6.1 % of cases. Using the BBL Crystal Identification System for gram-positive microorganisms the coagulase negative staphylococci were identified. Acquired results showed that 54.8 % of all isolated staphylococci were *S. aureus*, 41.1 % - coagulase negative staphylococci (int.al. *S. haemolyticus, S. kloosii, S. simulans, S. xylosus, S. capitis, S. saprophyticus, S. epidermidis, S. vitulus)* , but 4.1 % - coagulase positive *S. intermedius. S. haemolyticus* (61.2 %) were the most isolated coagulase negative staphylococci. The corresponding average SCC were 1 676 200 and 1 467 600 cells ml^{-1} for *S. aureus* and coagulase negative staphylococci, respectively. Coagulase negative staphylococci similar to major pathogens *S. aureus* promote an increase in SCC statistically significantly (α = 0.05). Acquired results testified that coagulase negative staphylococci can cause inflammatory reaction of cow's organism and their diagnosis is important.The polyvalent staphylococcal vaccine is elaborated in the Research Centre "Sigra" to control bovine mastitis.

Poster Ph2.9

Changes of plasma components concentrations in dairy cows artificially infected by *Escherichia coli*

Y. Yamada[1], H.Okada[2], N.Katoh[3], M. Sakaguchi[1] and H. Kadokawa[1], [1]National Agricultural Research Center for Hokkaido Region, Hitsujigaoka-1, Toyohira-ku, Sapporo-city, 062-8555 Japan, [2]Rakuno-Gakuen University, Faculty of Veterinary Medicine, Ebetu-city, 065-8501, [3]National Institute of Animal Health, Hokkaido branch, Sapporo-city, 062-0045*

The objective of this study was to clarify the changes of plasma components in dairy cows artificially infected by Escherichia coli. Three multiparous Holstein cows were used. Five ml of 10^4~10^5CFU/ml of *Escheria coli* (E. coli) were injected in one mammary cistern of each cow. Somatic cell counts in milk and plasma components concentrations were measured. They were slaughtered on 7 days after injection. Each cow showed the elevation in body temperature, increase in heart rate and respiration rate and the decrease in appetite after the injection. The somatic cell count in milk increased gradually until 8hr. and increased rapidly 10hr. after the injection. They reached to 5,000,000~22,000,000/ml at maximum level. Almost all of the plasma components showed the changes after injection. It appeared that the patterns of the changes of glutamic-oxaloacetic transaminase and alkaline phosphatase were corresponding with the pattern of the changes of the somatic cell count in milk. One cow had a comparatively low level in total cholesterol when E. coli was injected, and it showed relatively severe symptoms after injection compared to other two cows.

Poster Ph2.10

Effect of Echinapur on functional proteins' content and hygienic standards of goat milk

E. Bernatowicz[1]*, B. Reklewska[1], R.R. Pinto[2], K. Zdziarski[1], [1]Warsaw Agricultural University, Animal Science Faculty, Cattle breeding Department, 02-786 Warsaw. Ciszewskiego 8, [2]Universidade de Trás-os-Montes e Alto Duoro, Vila Real, Portugal*

Abstract: The aim of the study was to estimate the effect of immune system stimulation on somatic cell (SCC), bacteria counts (CFU) and content of functional whey proteins ie.: α-lactalbumin (α-LA), β-lactoglobulin (β-LG) and lactoferrin (Lf) in goat milk. Goats (n=10) with variable milk hygienic standards received twice daily a dose of Echinapur containing 300mg *Echinacea purpurea* extract for two weeks. Immediately after termination of EP treatment a significant decline in SCC and CFU with the pre-treatment level was observed. These changes progressed further for the next two weeks and a month after the last dose of EP milk SCC was over 9 times lower than the initial level. The difference in CFU was not as spectacular, but it was also significantly ($p<0.01$) lower than at the beginning of treatment. The content of functional whey proteins was also significantly affected by EP. Compared with the initial levels, the milk Lf, α-LA and β-LG content increased in response to EP treatment by 84.8, 48.8 and 23.2 %, respectively. It was concluded that EP is effective in the improving milk hygienic standards in goats. Apart from this, EP treatment may serve as a tool for producing milk with the increased content of important functional proteins.

Supported by State Committee for Scientific Research grant no 6 PO 6Z 052 21.

Theatre PhN3.1

Energy intake, energy expenditure and energy balance: different views on causes and effects of changes in animal body reserves

B.J. Tolkamp and I. Kyriazakis, Animal Nutrition and Health Department, Scottish Agricultural College, West Mains Road, Edinburgh EH9 3JG, UK*

Dynamic models of long-term changes in body reserves of farm animals are generally based on descriptions of daily energy intake and energy expenditure. One view is that animals aim at long-term fitness goals and try to achieve these by adapting short-term intake. Some data sets obtained with different animal species are difficult to interpret from this perspective. An alternative view is that animals regulate energy intake on the basis of actual state and food quality. From this perspective, short-term feeding behaviour that has an adaptive value under natural conditions (e.g. a variable food supply) can lead to behaviour that is detrimental for fitness when animals are exposed to environments they are not adapted to (e.g. continuous good food supply). A feedback model was developed based on the above assumption that actual state (i.e. fatness) of the animal affects both energy expenditure and energy intake and, therefore, long-term reserves. The model was parameterised with data obtained from sheep subject to various nutritional treatments. The model was tested against data of sheep subject to other nutritional treatments that resulted in different combinations of gain and loss of weight and condition. The model gives a reasonable description of the observations, leads to additional testable predictions and provides a framework for modelling body reserves.

Theatre PhN3.2

Leptin plasmatic levels during lactation in Sarda goat breed

V. Carcangiu[1], G.M. Vacca[1], A. Parmeggiani[2], M.C. Mura[1] and P.P. Bini[1], [1]Dipartimento di Biologia Animale, via Vienna 2, 07100 Sassari, Italia, [2]Dipartimento di Morfofisologia Veterinaria e Produzioni Animali, via Tolara di Sopra 50, 40064 Ozzano Emilia (Bo), Italia*

Of the different hormones regulating lactation, leptin also could be involved in modulating energetic metabolism of the animal. In order to assess which role this hormone plays in lactation, leptin haematic level and its correlation with GH, NEFA, BCS and milk production were determined in Sarda goat breed. Blood samples were obtained from each goat fortnightly, from day 35 to day 180 of lactation, milk production and BCS were recorded. Leptin and GH were analysed by RIA, NEFA by colorimetric assay. The data were submitted to correlation and variance analysis. Leptin haematic levels showed increasing values ($P<0.01$) during observations, while GH and NEFA showed decreasing values ($P<0.01$). BCS values resulted higher ($P<0.01$) at the end of observations, while production started lowering from blood withdrawal n. 11. Leptin showed a negative correlation with GH, NEFA and milk production, and a positive correlation with BCS. The pattern of leptin haematic concentration and its correlation with the parameters considered indicate that leptin is involved in regulating metabolism during lactation.

 Theatre PhN3.3

Estimation of mature protein mass and relative growth rate of 920 growing pigs *ad libitum* from measurements of live weight and backfat thickness

S. Schiavon. Bailoni, L. Gallo, F. Tagliapietra and C. Ceolin, M. Noventa, Dipartimento di Scienze Zootecniche, Viale dell'Università 16, 35020 Legnaro (PD), Italy

As preliminary step to predict feed intake, a farm dataset was used to estimate mature protein mass (Pm) and relative growth rate (B) of 920 ad libitum fed male growing pigs. The information used were: body weight (BW), measured at 71±4 (t_1), 126±5 (t_2) and 184±5 (t_3) days of age and backfat thickness (P2), measured at age t_2 and t_3. Body lipid mass was estimated as L=(9.16+0.69*P2)*BW/100. Assumptions: at birth (t_0) BW=1.5 and L =BW*0.04; at age t_1 L=BW*0.95- (1.2*(0.152*BW)+5.193*(0.152*BW)$^{0.855}$. Fat free empty BW was computed as FFEBW=BW*0.95-L. Individual estimates of Pm and B were obtained by solving the equation FFEBW= 1.2*P+5.193*P$^{0.855}$; where P= Pm*exp(-exp(-B*(t_i- t^a), t_i=age and t^a=age at the inflexion point. Mature lipid mass (Lm) and relative growth rate of lipid (B_l) were estimated by resolving L= Lm*exp(-exp(-B_l*(t_i-t^a). The resulting Pm and B mean values were 43.4±5.9 kg and 0.0117±0.0011 d^{-1}, respectively. Rsd and R^2 of the 920 curves averaged 1.4±1.3 kg and 0.997+0.003, respectively. Lm and B_l averaged 78.5±14 kg and 0.0099±0.0002 d^{-1}, respectively. Rsd and R^2 values were 1.7±1.1 kg and 0.991±0.011, respectively. The estimated ratio Lm/Pm was 1.82±0.28. The results are in agreement to those described in literature for improved pig genotypes.

Theatre PhN3.4

Heat stress influence on feeding behaviour of the dairy cows fed a TMR

E. Maltz, N. Livshin and I. Brukental, Agriculture Research Organization, Volcani Center, PO Box 6, Bet-Dagan 50250, Israel*

Under heat load, cows decrease feed consumption. The objective was to study diurnal feeding activity (FA) in its' relation to management routine, in order to evaluate possibilities to stimulate feeding in summer. FA was monitored automatically in August and November (max33^0, and 24^{0}C respectively) for 30 multiparous Israeli Holstein cows (DIM>90), housed in a roofed pen, fed TMR distributed once daily and milked thrice daily. In both seasons, most of FA was triggered by routine management actions: about 70% of feed was consumed within approximately one hour after feed distribution and milkings. In August there was an additional FA peak after heat relieve that continued up to the post milking (21:30 h) feeding, resulting in high level of feed intake in 1700-2300 (32.5% from diurnal intake vs. 20% in November). This was followed by low FA in the night hours - 1.6 meals per cow vs. 2.5 in November between evening and morning milkings. It appears, that FA during night in August was suppressed due to combination of intensive FA after heat relieve immediately followed by the feeding after evening milking. Reduced feed intake in the night (3.1 kg/cow less than in November) was only partly compensated by FA after morning milking (2 kg/cow more in August than in November). It is suggested that changes in management routines' timing may stimulate feeding activity and encourage summer food consumption.

Theatre PhN3.5

Animal and feeding factors affecting voluntary feed intake of reproductive rabbit does

J.J. Pascual, Unit of Animal Nutrition, Department of Animal Science, Universidad Politécnica de Valencia, PO Box 22012, Valencia 46071, Spain

This paper try to develop a framework describing the main animal (breeding, age, physiological state, body condition, health status) and feeding (energy level and source, fibre level, and rearing management) factors affecting the voluntary feed intake of reproductive rabbit does under farm conditions. The effect of some of these factors (breed, parity order, dietary fat and fibre levels) on feeding and productive level of rabbit does have been frequently studied. However, there are not works enough about the possible effect of other relevant factors, as continuous genetic selection for litter size or the individual body condition and reserves mobilisation ability, in the performance of does on current reproductive cycle and in the success of future cycles. Negative effects of an excessive fat mobilisation during the first reproductive cycles on health and reproductive life of rabbit does is presented. Relationships between body fat mobilisation, reproduction success and longevity of animals are also described. On the other hand, genetic selection focussed on prolificacy, insemination with semen from males selected by high growth rate, and the increasing reproductive rates have clearly changed female demands, requirements and balances in recent years. Feeding programmes developed to improve energy intake of females by increasing voluntary feed intake or the energy concentration of diets are analysed, with an especial remark to their long-term effects on performance, body condition and health status.

Theatre PhN3.6

Investigations on water intake of dairy cows

U. Meyer, M. Everinghoff and G. Flachowsky, Federal Agricultural Research Centre, Institute of Animal Nutrition, Bundesallee 50, 38116 Braunschweig Germany*

The aim of the study was to determine relationships between water intake and influencing factors such as average ambient temperature, humidity, milk production, dry matter intake, dry matter content of the ration, body weight, roughage portion of the ration, lactation rank, lactation day, Na- and K-intake. In two trials the individual daily water intake of 60 German Holstein cows, kept in a thermally non isolated loose housing system, has been logged under varied conditions. The cows were fed two rations different in roughage type and roughage proportion which were calculated to meet the energy and protein requirements according to the German recommendations (GfE 2001). Feed intake was recorded by an automatic feeding system. The cows had free access to water. The individual water intake of cows was logged electronically by measuring the mass of the water vat before and after drinking.

The subjection of data (n=12821) to multiple regression analysis leads to the following equation: Water intake (kg/day) = - 26.12 + 1.516 * average ambient temperature (°C) + 1.299 * milk production (kg/day) + 0.058 * body weight (kg) + 0.406 * Na-intake (g/day). The r^2-coefficient was found to be 0.60.

It is supposed that the presented equation implies the most important factors for estimating the water consumption of dairy cows under housing and feeding conditions existing in Central Europe.

Theatre PhN3.7

Effects of ambient temperature, genotype and nitrogen supplementation on intake and digestibility of a low quality grass hay fed to mature ewes

A.L.G. Lourenço and A.A. Dias-da-Silva, CECAV-UTAD Department of Animal Production, Apartado 1013, 5000-911 Vila Real, Portugal*

The effects of ambient temperature (25±3.3 °C, HT *vs* 11±0.3 °C, LT), genotype and nitrogen (N) supplementation on dry matter intake (DMI), organic matter digestibility (OMD) and live weight (LW) variation were studied with 48 non pregnant female adult ewes; half of the animals belonged to a local Churra breed (CH), and half to Ile-de-France breed (IF). The ewes were individually fed *ad libitum* with chopped (4 cm) meadow hay (H) supplemented or not with soybean meal (S) (85 H:15 S, DM basis) over a period of nine weeks followed by a five week period of diet restriction (11 g hay DM kg^{-1} LW). Temperature and N supplementation did not affect hay DMI when ewes were fed *ad libitum* ($P > 0.10$). However, HT and N supplementation reduced weight loss ($P < 0.05$ and $P < 0.001$, respectively). CH breed exhibited higher hay DMI per kg LW ($P < 0.01$), and lower weight loss ($P < 0.05$) than IF breed. N supplementation increased OMD ($P < 0.001$), irrespective of feeding level. For both levels of feed intake OMD ($P < 0.001$) was higher in IF breed (from 1.95 to 4.75 percentual points). The data suggest that there are inherent differences between breeds in ingestive and digestive abilities irrespective of ambient temperature.

Theatre PhN3.8

Intake and weight gain effect of using alfalfa hay in Romanian Black Spotted young bulls feeding

C. Bunghiuz, C.V.Calin, S.Grigoras, M.Gras and D. Georgescu, Research and Development Institute for Bovine Balotesti, Ilfov 077015, Romania*

We will presenting information obtain from an experiment who use Romanian Black Spotted young bulls with average corporal weight 160 kg at start. On these young bulls, we establish intake value of alfalfa hay obtained from bottom stage alfalfa on the I-III harvest. Obtain information have high variations: 68-140 g DM/kg $G^{0.75}$ this inform we offer occasion to make optimal rations who include maximal quantities from alfalfa hay (4.8-6 kg) and minimal quantities of concentrates (1-1.3 kg corn meal). Daily gain weight obtains from 250 kg level it was 1150 g on all experimental period.

Poster PhN3.9

Effect of protein/lysine ratio on the performance and meat quality of fattening-finishing Duroc and Large White pigs

M. Habeanu[1], V. Hebean[1], I. Moldovan[1], M. Neagu[1], [1]Institute of Biology and Nutrition Animale-Balotesti*

A total of 48 Large White and Duroc fattening-finishing pigs weighing in average 71kg respectively 70 kg, were assigned to two groups per breed and were given dietary treatments with different levels of protein and lysine: 15.15% crude protein (CP) and 0.704% lysine (L) for the control group (C) and 17.10% CP and 0.901% L for the experimental group (E) from each breed. The data of the 2×2 factorial design experiment were analysed with Anova. There were no significant differences (P>0.05) in the average daily gain (ADG) of the groups or in the interaction of the two factors. Large White pigs displayed a higher ADG (0.983 kg in group C and 0.931 kg in group E) compared to Duroc pigs (0.933 and 0.977 kg, respectively). The significant differences (P<0.05) observed in the meat/fat ratio were determined by the treatment and not by the breed. Meat/fat ratio was 48.78 and 51.4 in the control and experimental groups of the Duroc pigs, while it was 45.77 and 49.9, respectively in the Large White pigs. Intramuscular fat was higher in Duroc pigs (29.135 g/100 g dry matter (DM) in group C and 23.165 g/100 g DM in group E) compared to Large White pigs (24.10 g/100 g DM and 20.01 g/100 g DM in groups C and E, respectively).

Poster PhN3.10

Diet selection, intake and digestibility by two breeds of sheep offered two hays in the long form

A.A. Dias-da-Silva, M.J.M. Gomes, A.L.G Lourenço and A. Nogueira, CECAV-UTAD, Department of Animal Production, Apartado 1013, 5000-911, Portugal*

Two experiments with female sheep of a native Churra (CH) and the Ile-de-France (IF) breed (initial LW 31 and 48 kg, respectively), having a similar degree of maturity (70%), were performed. Animals (10 per hay and breed) were individually fed alfalfa or oat hay (AH and OH; 19.5 and 8.7 % CP, 45.1 and 74.2 % NDF, respectively, on a DM basis) as the sole feed in the long form during 14 (alfalfa) and 9 (oat) weeks. A minimum of 30 % refusals was allowed.

There was no difference between breeds in diet selection of both hays as assessed by CP and NDF content of the refusals (P> 0.05). DM intake of AH (g/kg LW) was similar between breeds and 35 to 65 % higher than observed with OH. Breed did not affect DM digestibility of AH. Intake of OH was significantly higher and digestibility significantly lower in CH breed (P< 0.05).

The data suggest that most of the difference observed in digestibility of OH between breeds (8 points) may be explained by an inherent higher turnover rate of feed particles in the rumen of CH breed since no difference was observed with AH, a feed having more then twice of the DM as cell contents compared to OH.

Poster PhN3.11

Investigations on water intake of fattening bulls

U. Meyer, W. Stahl and G. Flachowsky, Federal Agricultural Research Centre, Institute of Animal Nutrition, Bundesallee 50, 38116 Braunschweig Germany*

The experiment was carried out to investigate relations between the water intake impacting factors such as average ambient temperature, humidity, average daily life weight gain, body weight, dry matter intake, dry matter content of the roughage, roughage portion of the ration, Na- and K-intake. 62 German Holstein bulls with an initial mean body weight of 193kg were housed on slatted floor in a thermally non regulated stable in groups of 7 or 8 bulls. The experimental ration consisted of corn silage and concentrate mixtures with wheat, dried beet pulp, minerals, soybean oil and soybean meal or peas. The bulls had free access to water at any time. Feed intake and daily live weight gain as well as daily water intake were registered individually. The experiment was finished with a life weight of approximately 550kg. The live weight gain was 1267g/d. Applying a multiple regression analysis to the data results in the following equation: Water intake (kg/day) = - 3.85 + 0.507 * average ambient temperature (°C) + 1.494 * dry matter intake (kg/day) - 0.141 * roughage fraction (%) + 0.248 * dry matter content of the roughage (%) + 0.014 * body weight (kg). It could be concluded, that the presented equation considers the most significant factors for predicting the water consumption of fattening bulls under housing and feeding conditions existing in Central Europe.

Poster PhN3.12

Influence of high fat diet on food intake and body fat in growth selected mice

M. Langhammer[1], M. Derno[1], U. Renne[1], G. A. Brockmann[2] and C.C. Metges[1], [1]Research Institute for the Biology of Farm Animals, Dummerstorf, [2]Institute of Farm Animal Science, Humboldt University, Berlin, Germany*

In comparison to farm animals, we analysed how inbred mice (DU6i) descending from an over 100 generation high body weight selected line regulate food intake (FI) and body fat content (BF%) in response to a high fat diet (HF).

After weaning 8 males and 8 females of line DU6i and their unselected control (DUKsi) were fed a control (5% fat, C) or a high fat diet (20% fat, HF) for 12 weeks. Body weight (BW), daily gain (DG) and FI were monitored between day 35 and 42. BF% was determined by dual energy X-ray absorptiometry (Lunar PIXImus densitometer).

Fed HF DU6i showed higher BW35 compared with C (males 53.4 vs. 46.9, females 46.8 vs. 40.3 g) while DG of DU6i was not influenced by sex or diet in that period. FI was not different between DU6i and DUKsi but lower in HF by ~30% in DU6i and ~24% in DUKsi (g/g $BW^{0.75}$; $P<0.05$). In both lines the voluntary energy intake was significantly decreased when HF diet was consumed. But in contrast, BF% increased with HF feeding by ~115% in DU6i and only ~40% in DUKsi males. This indicates that growth selection in DU6i altered the lipid metabolism but not the regulation of food intake.

Poster PhN3.13

Grass silage amino acids content in daily ration of dairy cows

A. Jemeljanovs[1], M. Beca[1], J. Miculis[1] and J .Spruzs[2], [1]Latvia University of Agriculture, Research centre "Sigra", 1 Instituta Street, Sigulda LV-2150, Latvia, [2]Latvia University of Agriculture, Department of Animal Science, 2 Liela Street, Jelgava LV-3001, Latvia*

To evaluate silage as a basic feed investment in high productivity dairy cows total daily rations amino acids amount, the feeding trial with Latvian Brown Breed cows (n=10) was carried out. Two years investigations results showed that wilted timothy-red clover silage made in early vegetation stage ensure 70.2% of daily ration protein, but total (AA) and essential (EAA) amino acids investment in total amino acids amount of ration was comparatively high correspondingly 66.1 and 68.2% by feeding out in average 29 kg silage. Methionine, histidine and lysine amount in daily ration was ensured by silage in 72.0%, 67.9% and 75.3% extent.
EAA of timothy-red clover silage made in late stage ensured only 44.0% of total AA determined in daily ration but methionine and histidine - 49.0% and 41.5% of AA amount correspondingly. During the trial silage AA level decrease was observed in total AA amount but especially for phenylalanine, cystine, histidine, glutamic acid and thyrosine. Comparatively high content of silage alanine (in average 80.7%) of ration total AA amount was ascertained. Therefore wilted timothy-red clover silage made in budding stage must be considered as feedstuff of high level EAA content. Timothy-red clover silage made in fullflower stage caused significant ($p<0.05$) amino acids quantitative level changes in daily ration of dairy cows.

Poster PhN3.14

The effects of biological inoculant and combination of biological and chemical additive on microbial composition of grass silage during storage and feeding

A. Jemeljanovs, D. Kravale, B. Osmane and I. Ramane, Latvia University of Agriculture, Research Centre "Sigra", 1 Instituta Street, Sigulda LV-2150, Latvia*

We evaluated the effect of the inoculant SIL-ALL^{4x4} (Alltech Biotechnology Centre, USA) on count of microorganisms in grass silages. The lactic acid count in grass silages was significantly effected by grass species, conservation method ($p<0.01$); but butyric acid count was significantly effected by grass species and stage of maturity, conservation method ($p<0.01$). Mould fungi CFU count was significantly effected by grass species, stage of maturity, conservation method ($p<0.01$); yeast-like fungi CFU count was significantly effected by stage of maturity ($p<0.05$) and conservation method ($p<0.01$).
The aerobic stability of silage was very important. Aerobic deterioration generated losses and reduced hygenic quality during the feed-out. We evaluated the effect of the combination of *Lactobacillus plantarum, Enterococcus faecium, Pediococcus acidilactici, Lactococcus lactis*, enzymes and sodium benzoate - Lactisil 200 NB (manufactured by Medipharm, Sweden) on aerobic stability of timothy/red clover silage and performance of dairy cows. After 5 days exposure to air, the untreated silages contained higher counts of moulds (7.5 log CFU g^{-1} FM) than treated silages (5.2 log CFU g^{-1} FM). After 6 days exposure pH-values of untreated silages increased from 4.01 to 4.92.The silage with additive of Lactisil 200 NB and untreated silage were fed out to dairy cows. The milk yield was significantly higher (by 1.2 kg day^{-1}) for cows fed out silage with the additive ($p<0.05$).

Poster PhN3.15

Factors influencing voluntary feed intake of growing cattle

J.Várhegyi and I.Várhegyi, Research Institute for Animal Breeding and Nutrition, 2053 Herceghalom, Hungary*

Data from 60 experimental groups of growing-finishing bulls were used to evaluate the factors affecting feed intake. Mean liveweight (W), energy density (NEm and NEg MJ/kg DM) crude protein (CP) % and daily gain (ADG) varied between 249 and 549 kg; 6,44 and 8,50 NEm; 4,06 and 5,73 NEg; 9,7 and 15,6 %; 897 and 1549 g/day, respectively. Dry matter intake (DMI) expressed as % of liveweight was between 1,89 and 2,81. Multiple regression analysis was used to evaluate the data. Voluntary feed intake was negatively related to weight, increasing energy density and crude protein increased feed intake. There was a positive relationship between intake and daily gain Feed intake can be predicted from W, W^2, NEm, NEg (MJ/kg DM), CP % and ADG (R^2=0,75 $P<0,01$). Protein balance in the rumen influences feed intake. Feed consumption increases as protein balance turn from negative to positive value.

Trials conducted with growing heifers showed that cell wall (NDF) content was negatively related to feed intake. Comparison of grass and lucerne hay showed that feed intake is higher from lucerne than from grass of similar digestibility, which might be due to the slower NDF degradation of grass than lucerne (3,7 vs 6,5 %/h).

Poster PhN3.16

Influence of dietary barley silage level on young Brown cattle performance

Smaranda Pop, M. Nicolae, C. Dragomir and D. Voicu, Institute of Biology and Animal Nutrition, Balotesti, Romania, 077015*

The effect of feeding barley silage at three levels, next to adequate compound feeds, on Brown steer performance (feed intake, gain, feed conversion ratio) was assessed for 80 days in an experiment on three groups of 9 steers each (average initial weight 360 kg). Group 1 had unrestricted access to both forages, group 2 had unrestricted access to silage and received a limited amount of compound feed, while group 3 received a limited amount of silage an had free access to compound feeds. The structure of corn + soybean meal-based compound feeds was adapted to the nutrient supply of the barley silage and to its dietary level. The highest feed intake was observed in the group with unrestricted access to compound feeds, while the lowest feed intake was o served in the group with free access to barley silage, which resulted in a higher weight gain for group 3 (1237.5 g/day) compared to 1134.7 and 1042.2 g/day in groups 1 and 2, respectively (not significant differences, $p<0.05$). Group 2 displayed the best feed conversion ratio.

Theatre PhL6.1

Consequences of genetic selection for high milk production for seasonal pasture based systems

P. Dillon, F. Buckley, R.D. Evans, D.P. Berry and B. Horan, Teagasc, Dairy Production Research Centre, Moorepark, Fermoy, Co. Cork, Ireland

During the past 20 years the percentage of North American Holstein-Friesian (NAHF) genes in the Irish dairy cow population has increased from 8% in 1990 to 63% in 2001. The use of the NAHF genetics was increased to incorporate traits of high productivity and broaden the genetic base of the British Friesian in Ireland. Analysis of Irish dairy farms participating in Dairy Management Information Systems have shown that calving rate to first service has reduced from 55% in 1990 to 44% in 2001, while at the same time average lactation number declined from 4.3 to 3.5. Over this period milk production per cow has increased 5033 kg in 1990 to 5755 kg in 2001. Results from controlled experiments comparing genetic groups across a range of feeding systems have shown that dairy cattle selected solely for higher milk production have higher milk production, higher liveweight, lower condition score, poorer fertility and lower survival than cows of lower genetic potential for milk production. The lower reproductive performance is associated with lower energy balance in early lactation, greater partitioning of additional nutrients toward milk production and inability to achieve desired grass DM intake from pasture. The results suggest that reproductive and health traits should be included in the national breeding objectives. The results also demonstrate the importance of progeny testing future sires within the environment in which they are to be used.

Theatre PhL6.2

Long term effect of energy supplement to dairy cows fed roughage ad libitum

T. Kristensen, Danish Institute of Agricultural Sciences, Department of Agroecology, Research Centre Foulum, P.O. Box 50, DK-8830, Denmark

This paper gives the results from an experiment during three years, with the objective to study the effect of different level of concentrates given to cows with ad lib access to roughage. Concentrate, H (6 kg), M (4 kg) and L (2 kg), was fed in feeding parlours 6 times per day. Roughage was fed ad libitum to the herd twice a day in the period with silage feeding. In the summer period the herd was at pasture. The tree feeding groups were housed and managed in one group. A cow was kept in the same group during the whole experiment or until replacement. Heifers were randomly allocated to one of the three groups. Annual production per cow was H 7667 kg, M 7376 kg and L 7070 kg milk and H 32 kg, M 36 kg and L 27 kg live weight gain. Calving interval was H 382 days, M 395 days and L 399 days, but percent of cows that calved again was lowest in H 73 against M 78 and L 75. The cows did not adapt to the level of supplement as effect on milk production and live weight was independent on the duration of the treatment. The heifers with the highest genetic merit had a significant larger (4,8 kg) response (group H-L) to concentrates than low merit heifers (0,7 kg).

Theatre PhL6.3

Environmental effects on dairy cow fertility measures derived from progesterone profiles

K.-J. Petersson[1], H. Gustafsson[2], E. Strandberg[1] and B. Berglund[1], [1]Dept. of Animal Breeding and Genetics, Swedish Univ. Agric. Sci., PO Box 7023, SE-750 07 Uppsala, Sweden, [2]Swedish Dairy Assoc., PO Box 7039, SE-750 07 Uppsala, Sweden*

Traditional measurements of fertility in dairy cows are highly influenced by environmental factors and the farmer's management decisions, and have therefore low heritabilities. Endocrine measurements such as progesterone in milk estimate the physiological capacity of the cow better and could be utilized in genetic evaluation for fertility. We have studied environmental effects on fertility measures early in lactation, such as interval from calving to first luteal activity, proportion of samples with indication of luteal activity during the first 60 days after calving, and interval to first ovulatory oestrus, and their association to traditional measurements of fertility such as pregnancy to first insemination, number of insemination per service period and interval from first to last insemination. Data were collected from the university experimental herd during 1987-2002 and included 1106 *post partum* periods from 516 cows of two different breeds (Swedish Holstein and Swedish Red and White). Individual progesterone samples were taken twice a week. In total, the data contained 30,415 progesterone samples. First parity, winter season, tie-stall, high milk yield, mastitis, and lameness all had negative effects on the early fertility measurements. The early measurements had higher repeatabilities compared with late ones. A few associations between early and late measurements were detected.

Theatre PhL6.4

Effect of once daily milking of dairy cows on milk production and nutritional status

B. Rémond[1] and D. Pomiès[2], [1]Ecole Nationale d'Ingénieurs des Travaux Agricoles, 63370 Lempdes, France, [2]Institut National de la Recherche Agronomique, URH, 63122 Saint Genès-Champanelle, France*

Once daily milking (ODM) could be used as a tool to limit the yield of milk or to reduce the constraint of milking. This review summarizes the ODM effects through a set of ten recent trials.
ODM decreases milk yield all the more as lactation stage is earlier (40% to 20%). If twice daily milking is resumed, milk yield remains all the more lower than that of control cows as the duration of ODM was longer (10% for 10 weeks of ODM). Milk yield decreases (%) differ between breeds (lower for Montbéliarde than for Holstein) and cows (15% to 40%), and are repeated from one period of ODM to the following one. Importance of loss (kg/d) is not related to the level of milk yield.
ODM increases milk contents in fat and protein. Somatic cell count generally increases (< 100,000 cells/mL). Free fatty acid content decreases as well as lipoprotein lipase activity. Plasmin and plasminogen derived activity increases.
ODM does not change food intake during the first weeks of its implementation. Hence, nutritional status is improved, as appreciated by changes in body weight, body condition, energy balance, and blood profile. Cows get used to this rythm of milking in 2-3 days and do not seem to be bothered by it.

Poster PhL6.5

A survey of descriptive and reproductive characteristics of the Libyan camel herds

S. Hermas, University of Al- Fateh Faculty of Agriculture, Dept. of Animal Prodution P.O.BOX 82909 Tripoli, Libya

This study was conducted using a large stratified random sample of 83 herds in a large region sought west of Tripoli. Data were collected using questionnaires and by taking observations and measurements. Traits studied included: reproductive performance assessed by % of females serviced (PFS), % conception (PC), calving rates (CR) ,years of sire use(SU), age of sires (AS) ,% of abortion (AB), and sex ratio (SR). Body measurements were hump, chest and thigh circumferences (HC,CC,TC), body width (BW) front and rear heights (FH,RH), and nick length (NL). Mature and yearling weights (BWM,YW) were estimated using formulae (Bove1949, Hermas1988). Data collected were arranged and analysed. The average herd size was 67.6±3.3. The reproductive traits were 51.8 ±1.7 ,51 ± 1.9, 35.5 ±1.9, 42.6 ±2.5, 24.8 ±1.4, 9±0.9, 47.5 ± 1.6, 52 ± 1.5, 3.8 ± 0.4, 9.2 ± 1.3 for PFS, PC, CF, IF, AB, SU, and AS respectively. The mean of BWM was 376.5 ± 3.6. The effect of color was significant on BWM ($P<0.05$). The means were 417 ± 9.8, 393.4 ± 8.6 , 382 ± 23, 370 ± 19 and 369.6 ± 17 for hamra, safra, shakra, sawda, and zargha respectively. The estimated YW was 186.9 ± 3.8 kg .Reproductive performance was law. PFS and PC were only 50% and CR was 35%. 15% loss after conception 9% abortions and 6%as fetal absorption and early mortality . The aged sires and extended sire use caused inbreeding and low reproduction. There was a large variation in all traits which make selection possible if these traits are fairly heritable, but the mean hindrance to this process is the low reproductive performance.

Poster PhL6.6

Assessment of cold tolerance in lambs exposed to low temperature and rain simulation

G. Malá, I. Knízková, P.Kunc, V.Mátlová, J.Knízek and E.Nemcová, Research Institute of Animal Production, 104 01 Praha 10-Uhríneves, Czech Republic

Cold tolerance in lambs of breeds considered for low-input grazing systems (RM - Romney, ML - Merinolandschaf, S - Sumavska, SFxML - Suffolk x ML crossbreds, SF x S - Suffolk x Sumavska crossbreds) was assessed by thermographic method. During series of experiments in a climatic chamber with controlled air temperature (4.1 ± 1.8 °C) three days old lambs were exposed to 30 min. rain simulation. Body surface (BST) and rectal (RT) temperature were recorded before, right after and 60 min after the exposition. The highest BST before the rain exposure (22.8 °C, $P<0.001$) and 60 min after (21.3 °C, $P>0.01$) showed SFxML lambs, the lowest BST both before (16.9 °C, $P<0.001$) and 60 min after (16.3 °C, $P>0.01$) the exposure showed S lambs. The extreme decline in BST right after the rain exposure was recorded in SFxML lambs (10.3 °C), the lowest (5.1°C) in S lambs.
The increase in RT just after the rain exposure recorded in all breeds (0.3 - 1.0 °C) however was not significant. In all breeds after 60 min the RT recovered to the initial level. Similar breed differences indicate also observed values of simultaneously carried blood biochemical analyses. Based on these results, cold tolerance of S lambs appears to be the best advantage in low-input (zero housed) sheep grazing systems.
The project supported by NAZV QD1236.

The effects of microbial phytase and acids mixture on broiler performance fed with three corn hybrids

D. Grbesa[1], L. Bacar-Huskic[2], G.Kis[1] and M. Kaps[1], [1]University of Zagreb, Faculty of Agriculture, Animal Nutrition Department, Svetosimunska 25, 10000 Zagreb, Croatia, [2]Veterina, Svetonedjeljska 2, 10436 Rakov Potok, Croatia

The effects of Acimiks® F, Veterina ltd aromatic mixture of microbial phytase Natuphos 5000 and organic acids (Neubacid A Dry, Neuber) fed with diets of three corn hybrids on broiler performance and toe ash content was evaluated. Data were analyzed as a 2 x 3 x 2 factorial design with two levels of Acimiks F (0 and 1% Acimiks F), three levels of corn endosperm types (flint, semi-flint and dent), and two levels of nutrients (with ≥100% or 90% of the metabolizable energy, crude protein, amino acids, calcium, and 70% of phosphorus broiler requirements). The experiment was conducted in a 40-day trial using 648 sexed one-day old Cobb 500 chickens distributed randomly in 54 pens (12 chickens per pen and 6 pens per treatment). Supplementing phytase and organic acids from 0 or 500 FTU/kg and 0,3 % organic acids significantly ($P<0,05$) increased body weight from 1816 to 1881 g/chick, feed intake from 3461 to 3548 g/chick, and toe ash from 34,26 to 35,98. Results idicate that addition 1% Acimiks®F can reduce the variability of broilers performance fed various corn hybrids.

Theatre C3.1

Quality Management Systems (QMS) as a basis for improvement of milk quality in extension services

L. Döring[1] and H.H. Swalve[2], [1]MRA of Saxony-Anhalt, Angerstraße 6, D-06118 Halle, Germany; [2]Institute of Animal Breeding, University Halle, A.-Kuckhoff-Str.35, D-06108 Halle, Germany*

The implementation of a quality management system on farms is a basic prerequisite for a safe, sustainable and transparent production and also needed for an optimization of processes on the farms. Since 2000 the Milk Recording Agency (MRA) of Saxony-Anhalt is offering support to farms for a basic quality management system (QMS) specifically designed for farms with different production lines. The certified quality management system comprises the four components animal health, animal welfare, ecological aspects and documentation. For the implementation and operation of the QMS, farmers are supported by production guidelines using clearly defined control points, manuals, registers and forms.
The implementation of the QMS on farms in general is only successful with the support of the extension service of the MRA. The extension service includes the analysis of production processes, suggestions for improvement and supplementation of the documentation. Based on the QMS, also complex herd management extension services are offered which are on a new level of extension services and enhance the success of these measurements. This is explained using examples.
At present, 178 dairy farms with 76,602 cows in Saxony-Anhalt use the basic quality management guidelines.

Theatre C3.2

Integrating production and financial data to improve farm profitability in consultancy

D.E. Little, DVM, DairyNet2000 Inc, 151 Airport Avenue, Brookings, SD 57006, USA

Successful implementation of on-farm consulting services is dependent on the perceived value of these services by the farm owner or manager. Historically, product and service sales were built solely on relationships. Later, sales success shifted to vendors that could produce the lowest price. While relationships and price remain as important contributors to the successful sale, the ability to produce value for the operation is paramount in securing a long-term advisory role with the farm. Confidentiality and client profitability are also key outcomes of the client-consultant relationship. Program value is based on the ability of the consultant to integrate knowledge and data to produce information of high value that is specific for the client's operation. The RADAR management system, in conjunction with, accurate financial records is an effective way to integrate financial and production data to provide meaningful information for management decisions. This approach includes the repeated assessment of measurable factors that ultimately result in the historical data commonly used to benchmark managers with their peers. The financial impact of management changes is best understood through the adaptation of the DuPont model of financial analysis, with emphasis on changes in profitability for the operation.

Theatre C3.3

Knowledge transfer in Slovak cattle production during the transformation period

S. Mihina[1], B. Mitchell[2] and V. Brestensky[1], [1]Research Institute for Animal Production, Nitra, Slovak Republic, [2]West Yorkshire, UK

In Slovakia, as in other Central and Eastern European countries, major changes have occurred in all areas of life. The conditions for cattle breeding have changed significantly, mainly in relation to product markets. Domestic consumption has decreased, and the milk and meat markets have been liberalised. Pressure on product quality has increased. Simultaneously, however, the opportunities for technological and biological modernisation have improved. A system of extension services was required which would be widely available. A flexible Extension Services Network was therefore introduced in which all could engage - independent advisers as well as advisory organisations. Advisers, as well as users, have become familiar with new ways of communication with the aim of utilising the most recent knowledge in practice as effectively as possible.
The paper will describe examples of production system approaches to the modernisation of cattle farms in Slovakia. Also the role of the Animal Production Institute in this process of knowledge transfer will be discussed. Some findings of the EU-project, in which the dissemination of research into practice under the conditions of the enlarged European Union were studied, will be presented.

Theatre C3.4

Use of longevity data for breeding and management of sustainable dairy cattle in the Netherlands

C. van der Linde and G. de Jong, NRS, P.O. Box 454, 6800 AL Arnhem, The Netherlands*

Longevity becomes increasingly important in breeding and management of dairy cattle. In the Netherlands longevity is defined as the time between first calving and last milk recording date of the cow. The average longevity of culled dairy cows in the Netherlands per year of culling varied from 1083 to 1204 days in the past 15 years. No unfavourable systematic trend in longevity can be observed over the past 15 years and results from the last few years even suggest a positive trend. The average lifetime milk production of culled dairy cows increased by 38 percent from 19,593 to 27,109 kilograms from 1988 to 2003.
The average longevity of culled Dutch dairy cows is yearly published in herdbook year reports. A new management information product will be introduced this year in the Netherlands to provide farmers more insight in longevity of their cattle.
Data on longevity is not only used for management purposes but also for improvement of longevity by breeding. In the Netherlands, breeding values for longevity have been computed and published since August 1999 and have also been included in the total merit index. Longevity is weighted 26 percent in the Dutch total merit index, production 58 percent and other health traits 16 percent.

Theatre C3.5

Labour efficiency and multi-functionality on Irish dairy farms

B. O'Brien[1]*, K. O' Donovan[1,2] and D. Gleeson[1], [1]Teagasc, Moorepark Research Centre, Fermoy, Co. Cork, Ireland; [2]Department of Agribusiness, Extension and Rural Development, University College Dublin, Ireland

Current inflexibility of dairy systems in terms of labour requirement means that operators cannot easily adopt a multi-functional approach, which may assist in maintaining family farm income. This study investigated the labour invested on dairy farms and the feasibility of reducing that labour to provide opportunity for alternative enterprises.
Ninety-four dairy farms participated in the study. Proportionally 0.32, 0.28, 0.21 and 0.19 of farms were within milk quota ranges 135 x10^3 to 250 x10^3 litres (Group 1), >250 x10^3 to 320 x10^3 litres (Group 2), >320 x10^3 to 500 x10^3 litres (Group 3) and >500 x10^3 to 1,500 x10^3 litres (Group 4), respectively. Participant farmers recorded labour input on consecutive 3 or 5-day periods once per month.
The average dairy labour input per day for farms in milk quota groups 1, 2, 3 and 4 over the 12-month period was 7.0 h, 7.9 h, 9.6 h and 13.3 h, respectively. A daily time saving of 3.0 h and 2.2. h at the milking process and calf care, respectively, was observed on the efficient farms within quota group 1. The data indicated the possibility of reducing dairy labour input on these farms to 3.8 h per day or by 65 %. Good infrastructure and practices on farms should facilitate labour efficiency and feasibility of multi-functionality.

Poster C3.6

Online-available milk-recording data for efficient support of farm management

B. Logar, P. Podgorsek, J. Jeretina, B. Ivanovic and T. Perpar, Agricultural Institute of Slovenia, Animal Science Department, Hacquetova 17, 1001 Ljubljana, Slovenia*

For the purpose of milk recording and herdbook keeping in Slovenia, a central database has been established at the Agricultural institute of Slovenia. Following the recording scheme, the farmers receive reports of productive and reproductive status of their herds and individual cows in a few days. When requested, the reports include urea and somatic cell count info. Milk composition provides valuable information on the management practice of dairy herd. On the basis of fat:protein ratio, animals with digestive disorders and excessive body reserve mobilization, can be detected. Information about urea concentration and its relation to milk protein concentration enables optimization of ruminal nitrogen balance and metabolizable protein supply. Based on somatic cell count and lactose concentration, farmers are warned about eventual udder infections. Since 2003 the milk-recording data are available online as well. Special application for farmer use was developed. After authorization, the farmers can browse trough tabular and graphical presentations of whole herd or individual animal data. Data on current and previous lactations are available. Reports are in pdf or txt format. Data can be sorted, filtered, and exported to other programs for further analysis. The system offers an efficient support to farm management.

Poster C3.7

Influence of work routine elements of milking on milking parlour performance

B. O'Brien, K. O'Donovan and D. Gleeson, Teagasc, Moorepark Research Centre, Fermoy, Co. Cork, Ireland*

In future, herd-size expansion is likely to be limited by the labour input required for the milking process. This study measured work routine times (WRTs) for different work routine elements of milking and established the factors affecting milking parlour performance.
The time taken to carry out each element of the milking routine was recorded for 50 rows of cows (5 rows x 10 milkings) over a five-day period. Measurements were conducted in a 14-unit parallel, mid-level parlour, with swing-over arms, automatic feeding and sequential bailing. Measurements were taken with one operator, during the May/June period (2002) when average milk yield/cow was 27.4 litres/day. A psion personal organiser was used for data recording.
The maximum predicted number of cows milked/operator-hour was 79 and 88 with a routine incorporating a complete teat preparation technique (washing, pre-milking, drying of teats) in the absence and presence of automatic cluster removers (ACRs), respectively. Predicted milking performance was increased to 92, 100, 118, 122 and 150 cows/operator-hour when the pre-milking, drying, washing and drying, washing and pre-milking and drying, elements were eliminated, respectively, in the presence of ACRs. Predicted optimum unit number for these routines was 16, 17, 21 and 25, respectively. The ability to accurately predict parlour performance associated with changes in milking parlour size/design and milking routine will be necessary for future herd-size expansion.

Poster C3.8

Transfer of knowledge to practice in Slovenia

M. Klopcic[1], J. Osterc[1], M. Cepon[1] and B. Ravnik[2], [1]University of Ljubljana, Biotechnical Faculty, Zootechnical Department, Groblje 3, 1230 Domzale, Slovenia, [2]Chamber of Agriculture and Forestry, Miklosiceva 4, 1000 Ljubljana, Slovenia*

The extension service and experts from the area of milk recording and animal selection as well as researchers and experts from agricultural faculties have contributed the most to the transfer of knowledge to practice for the last decades. Most of the counseling is connected to technology followed by ecology. New ecological criteria require different knowledge and diverse approach to agricultural production. Milk quotas and negotiated rights to premiums have contributed to changes in agricultural production, which requires help of all expert services in agriculture and above all of extension service. In knowledge transfer group forms like lectures, courses, circles, seminars and workshops have proved to be the most effective methods of education of agricultural population. Such instructions are especially provided during winter. A lot of individual counseling is provided in office and in field-work. In 2003 extension service performed 4,750 group forms of education and above 200,000 of individual consultations and made above 1,400 investment, technological and adjustment programs. Agricultural experts provide regular information for farmers using mass media on local and national level. Last year the extension service published above 2,000 expert articles and 27,000 notices. Expert recommendations are accessible on telephone responder. Furthermore, farmers are instructed by organized exhibitions, demonstrations of technology and skills, sample plants, organization of expert excursions and various tasting, appraisals and promotional presentations.

Theatre C4.1

Environmental and age effects on the semen quality of Austrian Simmental bulls

B. Fuerst-Waltl, H. Schwarzenbacher, Ch. Perner, J. Sölkner, University of Natural Resources and Applied Life Sciences Vienna, Division Livestock Sciences, Gregor Mendel-Str. 33, 1180 Vienna, Austria*

In Austria's dual purpose Simmental population more than 90% of the breeding stock is artificially inseminated. Knowledge of factors affecting sperm production and semen quality supports AI centres in increasing productivity and profitability. Hence semen data from two Austrian AI centres collected in the years 2000 and 2001 were evaluated. In total, 3627 and 3661 ejaculates from 147 and 127 AI bulls, respectively, were analysed regarding various effects on ejaculate volume, sperm concentration, proportion of viable spermatozoa in the ejaculate, total spermatozoa/ejaculate and motility. Effects accounted for were temperature on day of semen collection, in the course of germinal epithelium cycle (average temperature of days 1-13 before collection), and during spermatogenesis (average temperature of days 14-67 before collection), length of day, humidity, day of week, assistant leading the bull, assistant collecting the semen, number of collection on collection day, days of rest before day of collection and age of bull.
Age of bull and number of collection were significant for most traits and both AI centres. Temperature, either on day of semen collection or during germinal epithelia cycle or spermatogenesis had an important effect on semen production and sperm quality while length of day and humidity had distinctly less impact. Results for other effects on semen production were inconsistent between AI centres.

Theatre C4.2

Effect of media and LH hormone on *in vitro* maturation of Egyptian buffallo oocytes

H.M. El Ashmaoui[1], E.EL.Nahass[1], A. Barkawi[2], M.S. Hassanane[1], S.A Ibrahim[2] and I.A.H. Barakat[1],* [1]Cell Biology Department, National Research Center, [2]Department of Animal Production, Faculty of Agriculture, Cairo University. Egypt

This study was carried out to evaluate the maturation buffalo oocytes *in vitro*. In the first experiment, oocyte (N = 3551) were cultured for 22-24 h. in TCM-199 OR TCM-199+LH, CR1aa + LH, and mSOF or mSOF+LH. Results showed that all media with LH hormone were better than media without LH hormone; maturation rate was low (40.43±0.72) in mSOF without LH hormone and high (95.49±0.41 and 91.74±0.75) in TCM-199 and CR1aa with LH hormone, respectively. Moreover, when compared TCM-199+LH with CR1aa+LH, 1282 oocytes were cultured. The maturation rate was assessed by evaluation of degree of cumulus cells expansion and meiotic development. The results indicated that maturation rate, telophase and metaphase II stage were higher for oocytes natured in TCM-199 +LH (94.78±0.48, 8.8±0.36, and 16.4±0.45, respectively) than oocytes matured in CR1aa+LH (91.72±0.40, 7.2±0.33, and 14.1±0.35, respectively). In the second experiment, 1117 oocytes were matured in vitro in TCM 199 +LH for investigation the effect of oocytes quality and culture system. Excellent oocyte and group culture system were significantly higher in maturation rate (90.64±2.19) and metaphase II stage (2.7±0.3 and 2.2±0.2, respectively) In conclusion, use of TCM-199 +LH, produced excellent oocytes and the group culture system for maturation of oocytes increased oocytes matured *in vitro*.

Theatre C4.3

Routine quality monitoring of embryo transplantation in dairy cattle based on NR56

R.M.G. Roelofs and E.P.C. Koenen, NRS, P.O.Box 454, 6800 AL Arnhem, The Netherlands*

The aim of this study was to develop a monitoring procedure that routinely provides accurate and early information on the success of embryo transplantation. Data included observations on 54,622 transplantations from June 1997 until June 2001. Non-return status at 56 days after transplantation (NR56) was defined as a measure of success. In the full model NR56 was determined using data on inseminations, transplantations, culling and calving. The alternative excluded calving data, which allowed an earlier availability of NR56 data. Rank correlations between estimated effects of ET-technicians (n = 43) were used to evaluate the differences between the two alternatives for calculating NR56 and between the effect of using a linear or binary model.

Unadjusted NR56 was 59.4% for the full model and 69.6% for the alternative. The statistical model included 10 fixed and 4 random effects. Third and higher parity recipients had a 7.1% lower NR56 compared to first parity recipients. NR56 of the highest quality recipients was 12.1% higher than of the lowest quality recipients. The average of the 10% worst ET-technicians is -6.8% whereas the average of the best 10% is 6.7%.

Rank correlations were 0.86 between the two alternatives and 1.00 between the binary and linear model. In conclusion, a routine evaluation of ET-technicians can adequately be based on a combination of insemination, transplantation and culling data using a linear model.

Theatre C4.4

Factors affecting *in vitro* oocyte maturation and early embryonic development in Egyptian buffaloes

Omaima M. Kandil and A.S.S. Abdoon, Dept. of Animal Reproduction and Artificial Insemination, Tahrir st., Dokki., 12622 Giza, Egypt*

The present work was designed to improve in-vitro embryo production in Egyptian buffaloes through examining the effect of: 1- Different cultured media (TCM-199, TCM-199 + GH, CR1aa or CR1aa + GH.) on cumulus expansion, nuclear maturation and embryo development. 2- Addition of PHE to fertilization TALP during IVF. 3- Co-culture by using BOEC or granuloza cell. 4- Addition of IGF-1 to IVC media .

Results revealed that CR1aa medium with or without GH significantly increased ($P<0.01$) buffalo oocyte maturation and embryo development to a greater extent than TCM-199 with or without GH. Addition of PHE to IVF medium significantly improved ($P<0.01$) cleavage rate, transferable embryo % when compared with control. BOES was significantly ($P<0.05$) higher than granuloza cell and being significantly ($P<0.01$) higher than control. Moreover, addition of IGF-I to IVC medium significantly ($P<0.01$) increased cleavage rate & transferable embryo % compared with non treated control.

In conclusion, we can improve in vitro embryo production in buffaloes by using CR1aa medium with addition of GH in maturation media, addition of PHE to fertilization medium, co-culture with BOEC or supplementation with IGF-I in IVC medium.

Theatre C4.5

Measurement of Pregnancy-Associated Glycoproteins (bPAG) in cow's milk

R. Metelo and F. Moreira da Silva, University of the Azores, 9700 Angra do Heroísmo, Portugal

The pregnancy-associated glycoproteins (PAGs) are placental antigens that were initially characterized as pregnancy markers in the maternal circulation of bovine species. The use of milk to determine levels of hormones has been investigated in many studies as the milk sampling avoid stressful effects of venipuncture, does not require special experience and the milk collection is easier than the blood.

The aim of the present study was to validate a new bovine pregnancy method evaluating the bPAG in milk. Blood and milk samples were collected weekly from a group of 60 Holstein-Friesian cows for one year.

Milk profiles where obtained after evaluation by the method described by Zoli *et al* (1992). The drop of bPAG concentration was faster and more accentuated in the milk, than in the blood. A good correlation was obtained between the results obtained from the blood and the milk ($r=89.2\%$; $P<0.0001$) and it was possible to determine by the statistical method described by Snedecor and Cochran (1967) when the animals where pregnant.

This study clearly indicates that milk can be used for pregnancy diagnosis in cows, giving a very helpful tool to farmers, especially in countries e.g. New Zealand and Portugal - Azores, where the animals spend whole year in the field. Milk can be use as a later diagnosis of gestation, a non invasive and non stressed method to follow easily a gestation.

Theatre C4.6

Fertility in U.S. Holsteins by states

I. Misztal, S. Oseni, S. Tsuruta and R. Rekaya, University of Georgia, Athens, GA 30605, USA

This study looked at differences in management and genetics of fertility of U.S. Holsteins in several states. Traits included days open (DO) limited at 250 d, and pregnancy rate defined as PR=1/[(DO-X] /HI+1], where HI is heat interval and X is an approximate voluntary waiting period (VWP). X was set to 50, 80 and 120d, with PR upper bound set at 1.0. The model included the effects of herd-year-month of calving, age at calving and the animal effects. The use of PR emphasises records with smaller DO. Higher heritability at longer X was indication of small genetic variation for records with DO < X. Heritabilities for DO and PR were 3-6%. For WI and NY, which have colder climates, the highest heritability was for DO, with heritability for PR at X=80 slightly smaller. For FL, which experiences heat stress all year, the highest heritability was for DO, with heritability for PR at X=120 slightly smaller. For GA and AZ, where heat stress is seasonal, the highest heritability was for PR with X=50. Larger VWP in colder states could be associated with higher profitability for longer lactations. For states with seasonal heat stress, low VWP during colder seasons maximizes the chance of pregnancy before the hot season. For FL, larger VFW maximizes the chance of pregnancy under heat stress. High DO may occur for different reasons, including management and heat stress.

Theatre C4.7

Relationship of muscular scores and carcass grades with killing-out rate, meat proportion and carcass value of bulls

M.J. Drennan and B. Murphy, Teagasc, Grange Research Centre, Dunsany, Co. Meath, Ireland*

The objective was to examine the relationship between live animal muscular scores and carcass conformation and fat scores with killing-out rate, carcass meat proportion and carcass value (€/kg based on value of cuts). Four experiments involving a total of 176 continental breed cross suckled bulls slaughtered at 15/16 months were used. Records included muscular scores taken pre-slaughter, carcass grades and meat proportion (following dissection into meat, fat and bone of either a half carcass (Experiment 2), the entire carcass (Experiment 3) or an 8 rib pistola (Experiments 1 and 4)). In all four experiments significant correlations were obtained between muscular scores and killing-out rate (r = 0.68 to 0.73), meat proportion (r = 0.62 to 0.69) and carcass value (r = 0.47 to 0.73). Corresponding correlations with conformation score were 0.59 to 0.68 for killing-out, 0.49 to 0.72 for meat proportion and 0.38 to 0.71 for carcass value. Regression analysis showed that a 1 unit improvement in conformation (scale 1 to 5) increased carcass value by 4, 14, 11 and 7c in Experiments 1, 2, 3 and 4 respectively. Corresponding affects of a 1 unit increase in fatness (scale 1 to 5) were -9, -9, -14 and -1c. In conclusion, good relationships were obtained between both muscular scores and carcass grades with killing-out rate, meat yield and carcass value.

Theatre C4.8

Changes in carcass and meat composition during the finishing of Belgian Blue double-muscled cull cows

J.F. Cabaraux, I. Dufrasne, L. Istasse and J.L. Hornick, Nutrition Unit, Veterinary Faculty, Liege University, Belgium

One of the characteristics in the finishing of cull females is a propension to deposit more fat than protein. This is a rather expensive process in terms of energy. Production of excessively fat animals did not fully agree with the consumers claims and the objectives of the meat market. On a total of 78 Belgium Blue double-muscled cull females, 15 were slaughtered at the beginning of the trial and 63 were fattened on a maize silage based diet. The initial average live weight was 571.2±12.8 kg. The fattening gain was 91.1±3.3 kg for a period of 91.4±2.1 days. The warm carcass weight was 405.9±7.8 kg and the killing-out proportion, 63.3±0.3%. The carcasses were characterised by a high proportion of muscles (67.6±0.4%) and rather low proportions of adipose tissue (18.2±0.5%). The meat protein and fat contents were 88.5±0.3 and 6.6±0.3% DM. During finishing the main changes at carcass level were a 2%unit decrease in muscle proportions and a 4.6%unit increase in fat proportions. In such conditions, muscles and fat accretions were respectively 36.0 and 28.2 kg in the carcass. Similarly, the protein content was reduced by 2.6%unit and fat content was increased by 2.9%unit. In terms of colour, L* and a* increased (37.9 vs 37.3% and 19.9 vs 19.3%) with finishing, the meat being less dark but more red.

Comparison of rearing systems at low environmental impact: quality meat of beef cattle in central Italy

M. Iacurto, A. Gaddini, S. Gigli, G.M. Cerbini, S. Ballico and A. Di Giacomo, Animal Science Research Institute - Via Salaria, 31, 00016 Monterotondo (Roma), Italy*

We analyzed the physical quality of 150 meat samples of Chianina bulls reared in different conditions: slatted floor (SF, 50 samples), feed-lot (FL, 60 samples) and pasture (PA, 40 samples). The samples belonged to 5 different muscles (30 samples each): *Longissimus dorsi* (LD), *Gluteobiceps* (GB), *Semimembranosus* (SM), *Semitendinosus* (ST) and *Caput longum Tricipitis Brachialis* (CloTB).
The Warner-Bratzler shear force wasn't different between groups on raw, while on cooked was higher in PA (11.15 kg vs 10.02 kg); drip losses were lower in FD (1,34% vs 1.94%); PA had less cooking losses (27.00% vs 29.94%). SF had higher Lightness (42.91 vs 41.24), no difference was found in Chroma, while Hue was different between the systems (39.52 vs 35.91).
The comparison between muscles showed a ST significantly tougher (20.63 kg vs 11.89 on raw, 12.65 vs 8.18 on cooked).
ST showed higher drip losses (2.79% vs 1.51%) and higher cooking losses (31.42% vs 25.98).
Lightness was higher in ST (46.00 vs 40.12) while Chroma was lower in LD (25.84 vs30.21) and Hue was significantly higher in ST (38.45 vs 33.08).
The meat from PA animals was, as expected, tougher, darker and showed less cooking losses; between muscles ST was tougher, with an higher lightness and lesser water losses, result also expected.

Theatre C4.10

Comparative analysis of beef production from young versus older steers originating from Portuguese milk herds

Sara M. Figueiredo, João Santos e Silva, Nuno Salvador, José A.Maia and Carmen S. Roberto, Estação Zootécnica Nacional, Sistemas e Técnicas de Produção Animal, Fonte Boa, 2005-048, Portugal*

The aim of the present work was to compare "young steers" vs "steers" production in 42 male animals of *Holstein-Friesian* breed, being 21 destined to produce young steer beef (aged 236 days; slaughtering at 324 Kg BW) and the other 21 to steer beef (aged 538 days; slaughtering at 634 Kg BW). The animals were distributed into 3 testing herds, they were fed commercial food plus hay. The steers were also fed maize silage. Growth (ADG), feed intake and food conversion ratio (FCR) were assessed during the fattening period. After slaughtering, carcass killing-out percentage, conformation and fat were evaluated. They were cut into pieces, weighed and sold in the Portuguese market. For economical comparison the Gross Margins Method was used.
Young steers had significantly higher growth (P<0, 05) than steers (1,524 *vs* 1,172 Kg *per* day, respectively) and better FCR (4,7 *vs* 7,8), for any of the herds considered .
Young steers carcasses showed better quality (conformation and fat) in, major killing-out percentage of first quality pieces and less bone and fat, suggesting that slaughter weight traditionally used in steers should be reduced.
Calculated Gross Margins were higher in young steers, even when the higher financial support attributed to steer producers is considered. Conclusion: "young steers" production represents an alternative for Portuguese dairy milking herds.

 Theatre C4.11

Economic weights for beef traits in the Slovakian Simmental population

D. Peskovicová[1], E. Krupa[1], J. Dano[1], J. Kica[1], M. Wolfová[2] and L. Hetényi[1], [1]Research Institute for Animal Production, Hlohovská 2, 949 92, Nitra, Slovak Republic;[2]Research Institute for Animal Production, PO Box 1, 104 01, Praha - Uhríneves, Czech Republic*

Economic weights of the integrated beef cattle production system in Slovakian Simmental were calculated for the following growth traits (birth weight, weight at 120 days, weaning weight, yearling weight) and carcass traits (dressing percentage, fleshiness and fat covering). The economic weights were calculated for three different marketing strategies: A/ commercial herds producing breeding heifers and breeding bulls for own replacement with surplus male and female calves fattened in intensive feedlot; B/ breeding herds producing breeding heifers and breeding bulls for own replacement with every surplus male and female calves exported for feedlot after weaning; C/ breeding herds producing breeding heifers and breeding bulls for own replacement with selling surplus breeding heifers to other herds and fattening of surplus male calves. A bio-economic simulation model of an integrated beef cattle enterprise was used to simulate economic efficiency for the mentioned marketing strategies. The relative economic importance of weaning weight comparing to weight at 120 days was the same, while the economic importance of yearling weight was lower in system with export of animals after weaning. The production and economic parameters of the system are discussed more in detail.

Poster C4.12

Semen quality of CVM carriers and CVM free Estonian Holstein young bulls

Padrik and T. Bulitko, Animal Breeders' Association of Estonia, 79005 Keava, Raplamaa, Estonian Agricultural University, Kreutzwaldi 64, 51014 Tartu, Estonia

The aim of the research was to determine the quality of semen of CVM (Complex Vertebral Malformation) carriers and CVM free bulls. A total of 123 ejaculates from 5 CVM carriers (Group 1) and 248 ejaculates from 22 CVM free bulls (Group 2) was studied. Semen concentration was determined with a photo-electrical colormeter (SDM-5, Germany). One hundred spermatozoa (stained with SPERMAC™) in each preparation were examined under the phase contract microscope (x 1000). Different morphological abnormalities (detached heads, abnormal heads, abnormal necks, proximal and distal cytoplasmic drpolets, abnomal midpieces and abnormal tails) were registered in each preparation as a percentage of the total number of counted spermatozoa. Sperm motility of post-thawing semen was determined with a computer assisted motility analyser (CMA, Computer Assisted Cell Motion Analyser, Germany). The following figures were determined: percentage of total motile and progressive spermatozoa, velocity average path-VAP (μm/sec), velocity curve line-VCL (μm/sec), velocity straight line-VSL (μm/sec), linearity-LIN (VSL/VCL), beat cross frequency-BCF, amplitude of lateral head displacement ALH (microms). The research showed that the 1st group had ejaculates with bigger capacity ($P<0.0001$) but the sperm concentration was higher in ejaculates of the 2nd group($P<0.0001$). The semen of the 1st group contained relevantly more spermatozoa with abnormal heads, detached heads and abnomal midpieces sperme ($P<0.0001$). The 2nd group had higher % of total and progressive sperm motility ($P<0.05$) and sperm motility speed (VCL, VAP and VSL) ($P<0.0001$).

Poster C4.13

Evaluation of reproductive performance in selected Holstein breeding farms

A.Jezková, M.Parilová and M.Rothová, Czech University of Agriculture in Prague, Department of Cattle Breeding and Dairying, Kamycka, 165 21, Czech Republic

The reproduction indices reported in the Czech Republic cannot be evaluated positively.The percentage level of repeated insemination indicates a low level of heat detection, or low interest and motivation by the staff of cow raising operations concerned with this situation. The faulty management of the reproduction of herds (faulty nutrition, unfavourable conditions in the cow shed, imperfect detection of heat, inadequate hygiene at parturition) causes the unsatisfactory state of the results of cow reproduction in CR. The work assessed the effect of selected factors (number of lactation, year and season of mating, year and season of insemination, sex of calves, bull name-insemination dose, insemination labour code) on indices of reproduction (age at first mating, age at first calving, length of calving interval, insemination index, insemination interval, service period, interinsemination interval, gestation lenght, and milk production factors - milk yield, index of persistence) in selected herd (192 cows)of Holstein cattle (H).Subsecquent average results of selected H cows were: milk yield = 7 614 kg, index of persistence=89,8; insemination index=2.62; intercalving interval=419,6 days, service period=142,7 days, insemination interval=66,8 days, age at first mating =532 days (17,8 months), age at first calving=868 days (28,9 months).The results were evaluated statistically using a linear model in SAS software, with a correlation analysis for evaluating dependencies between the selected factors.

Poster C4.14

The effect of different calving intervals on production and some reproductive traits in high-yielding cows in second consecutive standard and extended lactation

N. Strzalkowska, J. Krzyzewski, Z. Reklewski and E. Dymnicki, Institute of Genetics and Animal Breeding Polish Academy of Sciences, Jastrzebiec, 05-552 Wólka Kosowska, Poland*

The aim of this study was to examine the cumulative effect of calving interval length in high-yielding lactating cows on milk yield, its chemical composition and some reproductive traits in second standard and extended lactation. The experiment was carried out on 90 HF cows (in 2 equal groups). The cows in the 1st group were inseminated during the first oestrus which occur after 42, and in the 2nd - after 120 days since the calving. The cows were kept in a loose barn with an outside run and fed according to TMR system. Individually milk samples were taken once a month for estimation of fat, protein and lactose.

The average length of period from calving to conception in the 1st and 2nd group was (respectively): 106 and 149 days. There were no differences in the insemination indexes between groups. The cows from the 2nd group produced about 297 kg milk more per lactation. There were no differences in chemical composition of milk. The percentage of mastitis was lower in the 2nd group. Our results showed, that cows, which produce about 9000 kg of milk per 305-d lactation, should be inseminated about 110 days after calving.

Poster C4.15

Estimates of genetic and phenotypic parameters for days to calving in Nelore cattle

S. Forni and L.G. Albuquerque, Universidade Estadual Paulista, Faculdade de Ciências Agrárias e Veterinárias de Jaboticabal, Departamento de Zootecnia, Via de Acesso Prof. Paulo Donato Castellane s/n , Jaboticabal, São Paulo, Brasil*

A set of data from 53,181 Nelore animals was used to estimate genetic and phenotypic parameters for days to calving (DC) and days to first calving (DFC) and their correlations with scrotal circumference (SC), age at first calving (AFC) and weight adjusted to 550 days of age (W550). Covariance components were estimated using restricted maximum likelihood method fitting bivariate animal models. For DC a repeatability animal model was applied. The fixed effects considered for DC and DFC were contemporary group (CG), month of last calving and age at joining season as covariable (linear and quadratic effects). Contemporary groups were composed by herd, year, season and handling group at joining; sex of calf and mating type (multiple sires, single sire or artificial insemination). Heritability estimates were 0.06 for DC and 0.13 for DFC. Genetic correlation estimate between DC and SC was low and negative (-0.10), between DC and AFC was high and positive (0.76) and between DC and W550 was almost null (0,07). Similar results were found for genetic correlation estimates between DFC and AFC (0.93) and DFC and W550 (-0.02). The favourable genetic associations with AFC indicate that selection for DC or DFC could promote desirable correlated responses in that trait.

Poster C4.16

Relationships among pelvic measurements, weight of calves and course of parturition in Charolais cows

D. Bures, V. Teslík, L. Barton, D. Rehák, R. Zahrádková and M. Krejcová. Research Institute of Animal Production, Prátelství 815, 104 00 Prague - Uhríneves, Czech Republic*

Pelvic measurements of 164 Charolais cows from 4 herds were obtained in 2003. The taken measurements were external (pelvic length, width at hips, thurls and pins) and internal (pelvic height and width). The pelvic area was calculated as the product of internal pelvic height and width. The course of parturition evaluated according to the degree of required assistance was analysed using a linear model with herd, parity, pelvic area and birth weight of calves included as fixed effects. The course of parturition was significantly affected by birth weight of calf ($P<0.001$) and pelvic area ($P<0.05$). The cows that delivered still-born calves had smaller pelvic area than the cows with calves born alive. A significant interaction ($P<0.001$) was found between pelvic area and birth weight of calves. The highest frequency of difficult calvings was observed in the cows with a below-average ($x<\bar{x}-1/2$ SD) pelvic area and an above-average ($x>\bar{x}+1/2$ SD) birth weight of calves. This group of cows also exhibited the highest rate of still-born calves. Correlations ($P<0.001$) were determined between internal pelvic measurements, pelvic area, external pelvic measurements and age. The study was supported by the project MZE 0002701403.

Poster C4.17

Possibilities for improving superovulatory results with embryo donor cows

N. Pacala, I. Bencsik, N. Corin, D. Dronca, J. Stanculet and V. Caraba, Banat's University of Agricultural Science Timisoara, Faculty of Animal Science and Biotechnologies, Calea Aradului 119, 300645 Timisoara, Romania*

On 6 donors we have induced superovulation with FSH; gestagene hormones and PgF2α. We used 3 mg Norgestomet implants (Syncro Mate B - SMB) and Folltropin-V. Hormonal administrations have started with SMB subcutaneous implants in the ear and SMB injection (day 0). Starting on the 5th day from the implant administration, we have given 8 doses of 50 mg Folltropin-V, every 12 hours (in total 400 mg NIH-FSH-P_1). Together with the last 2 doses of FSH were administrated 25 mg Lutalyse (PgF2α). The Norgestomet implant was taken of in the morning of the 9th day. Donors were inseminated twice in the 11th day. After 18 days of administrating the implants we have recovered in media 18.5 embryos, out of which 10.5 were transferable. The majority of transferable embryos (88.6%) were in stage 6 (blastocysts). From donors that were superovulated using only FSH we recovered in media 12 embryos, from which only 6 were transferable. The advantage of the superovulatory process in which are used FSH and gestagene hormones concern in application on cycling cows without knowing the stage of estral cycle.

Poster C4.18

Effect of oxytocin and FSH injection on reproductive function of post-partum dairy cows

N.M. Rechetnikova, T.A. Moroz, A.M. Malinovsky, All-Russian Research Institute of Animal Breeding, Box Lesnye Polyany, Pushkin District, Moscow, 141212, Russia

The goal of this study was to investigate the effect of oxytocin and FSH injection on reproductive function of Holstein cows at post-partum period. Cows group 1 (n=45) were injected i.m. with 50 I.U. oxytocin on day 3,5 and 7 after parturition. Cows group 2 (n=28) were single injected i.m. with 100 I.U. FSH (FSG:LG >1000) on the day 15 after parturition. Control group (n=40) received no injection of hormones.

The treatment with oxytocin caused shortening of interval from calving to 1st ovulation by 8 days and shortened time of uterine involution by 6 days compared in control ($p<0,05$). In group 1 conception rate was increased by 10% in comparison with the control and the average interval from calving to first service was shorter by 10 days in comparison with the control. The treatment with FSH caused shortening of interval from calving to 1st ovulation by 10 days and shortened time of uterine involution by 8 days compared in control ($p<0.05$). In group 2 conception rate was increased by 13% in comparison with control and the average interval from calving to first service was shorter by 12 days vs. control.

These results indicated that the administration of oxytocin and FSH in dairy cows at post-partum period may improve their reproductive performance.

Poster C4.19

Fertility improvement of dairy cows with estrus synchronization protocols using PGF2α, GnRH and estrogen in conjunction with progesterone or HCG

H. Karami, M. Moeini, A. Ahmadisefat. M. Mamoe and A. Moghadam, Department of Animal Science, Razi University, Kermanshah, Iran

Fourty lactating dairy cows within 30 to 40 days postpartum with the same milk yield, number of parity and body weight was assigned to four groups. Cows in 1st group injected intramuscularly with progesterone (100mg) +1 mg estradiol benzoate, after one week injected with 500µg cloprostenol sodium, then after 30-32 hours injected with 500 ug Gonadorelin and 18-20 hours later inseminated at fixed time. In 2nd group after synchronized injection of progestrone and estradiol ,and then 500µg cloprostenol sodium cows were injected with 1mg estradiol benzoate 30 h. later and then after 18-20 hours inseminated . Cows in 3rd group received 500 µg cloprostenol sodium intramuscularly and 12 hours later injected with 1 mg estradiol + 250 Iu hCG, and then after 36 h. were inseminated. Cows in 4th group (control) received two injection of 500µg cloprostenol sodium and then inseminated. The result has shown that only significant difference observed in pregnancy rate between control and third group ($p< 0.05$). There is a synergism between hCG and estradiol benzoate in inducing ovulation and aproved estrus synchronization. It can be concluded that the combination of hCG and estradiol benzoate injection 12 hours after cloprostenol sodium was effective on pregnancy rate in dairy cows compared to other synchronization protocols.

Poster C4.21

Effect of media and presence of Corpus luteum on *in vitro* maturation of buffalo *(Bubalus bubalis)* oocytes

Marwa S. Faheem, A.H. Barkawi, G. Ashour and Y. Hafez, Cairo University, Faculty of Agriculture, Animal Production Department, Giza, Egypt*

This work is aiming at studying the effect of culture media and presence of Corpus Luteum (CL) on in vitro maturation of buffalo oocytes. Two experiments were conducted to achieve the study goals. Experiment 1 was to test the effect of media (TCM-199 + epidermal growth factor- EGF, TCM-199 + FSH,LH,E2 hormone and synthetic oviductal fluid-SOF medium) on maturation rate. Each medium was supplemented with 10% FCS and 50µg/ml gentamycin. Experiment 2 was to test the effect of CL presence on maturation rate. Oocytes were collected separately from two types of ovaries (with or without CL) and cultured for 24 h at 38.5°C in 5% CO_2 in air in the same media used in experiment 1. Some oocytes were artificially denuded and their diameters (external, internal diameters and thickness of zona pellucida) were measured in 0 time and 24 h after culturing for maturation. Average (LSM) and percentage were used to analyse data. Results showed that maturation rate of TCM-199 with EGF medium (96.0%) and TCM-199 with hormones (93.9%) were higher ($P<0.05$) than SOF medium (87.6%). Presence of CL had a positive effect ($P<0.05$, 90.8 vs. 85.8% in absence of CL) on maturation rate. External-oocyte diameter and previtelline space were significantly ($P<0.05$) larger after maturation period.

Poster C4.22

Ovarian follicles and corpora lutea evolution dynamics in embryo donor cows

I. Bencsik, N. Pacala, N. Corin, A. Bencsik and J. Stanculet, Banat's University of Agricultural Science Timisoara, Faculty of Animal Sciences and Biotechnologies, Calea Aradului 119, 300645 Timisoara, Romania*

The study on ovarian follicles and corpora lutea was done on 28 cows with transrectal examination and using an ecograph (SONOVET 600) with a 7.5 MHz sound. Superovulatory treatment started on the 9th - 12th day of the estrus cycle in presence of small (2-5 mm) and medium (6-9 mm) follicles. For superovulation we used FSH and PMSG. Ovarian follicles and corpora lutea evolutions study was done twice a day starting with first day of hormonal treatment (day 1) during 18-20 days. We have measured using the ecograph ovarian diameters, follicular cavities diameters and corpora lutea dimensions. Follicular cavities diameter started to grow after 3rd day of treatment, in the 4th day it is 7-11 mm, and in the 6th day, insemination day, is 11-13 mm. After ovulation in 7th and 8th day the diameter decreases very much. Corpora lutea are starting to form in the 10th day. In the 11th day we have observed the appearance of small follicles (1-2 mm). The corpora lutea did not change their dimensions until the prostaglandin PgF2α administration (day 18-20). The ovarian follicles evolution is the same in both superovulatory treatments. The only difference between these two methods is that when we used PMSG for superovulation we have observed 17.6% more cystic follicles.

Poster C4.23

Population-genetic changes in cattle generations, born in conditions of chronic low doze ionizing irradiation (Chernobyl's zone)

V.I. Glazko, Institute of Agriecology and Biotechnology UAAS, Metrologicheskaja st.12, Kiev, 03143, Ukraine

Research of dynamics of population structure and cytogenetic anomalies in 3 cattle generations of experimental herd (in farm near Chernobyl's NPP, 200 Ci/km^2) were carried out. The changes of genetic structure in cattle generations were analyzed by family analysis of allele's transfer in structural genes and ISSR-PCR markers. Increases of mutant animals were not detected, but reversal of genetic structure in cattle generations from an initial Holstein breed to ones, typical for more primitive breed (for example, Grey Ukrainian) was revealed. Fertility of cows in the first generation, born in the experimental farm, was reduced in comparison with the parent generation (on average, from 0,9 up to 0,2 calves per cow per year). The sterile cows were absent in parent's population. In F1, between 36 cows 21 ones (58%) were sterile; only 15 cows of F1 born animals of F2 generation (0,73±0,06 calves per cow per year). Four cows of F2 in summary born 10 calves (F3), in average, 0,94±0,06 calf per year per cow. It allowed to suppose, that the fertility of F2 cows could increased in comparison with F1 cows. The data obtained can result from selection pressure in F1 generation, related with new conditions of cattle reproduction (the increase level of ionizing irradiation), which lead to elimination of some genotypes and changes of genetic structure in cattle generations.

Poster C4.24

Seasonal climatic changes affect fertility of Holstein Friesian cows on large-scale dairy farms in Malawi

M.G.G. Chagunda[1] and C.B.A. Wollny[2], [1]Department of Animal Health and Welfare, Danish Institute of Agricultural Sciences, Foulum, Denmark, [2]Institute of Animal Breeding and Genetics, Georg-August University of Göttingen, Germany*

Challenges limiting reproductive efficiency of the potentially high producing Holstein Friesian cows performing in tropical conditions include the environmental effects, which play their role either directly through the extreme effects of temperature and humidity or indirectly through feed and management. Affected may be the onset of normal oestrus cycles, detection of oestrus, and embryonic survival. Under harsh environmental conditions, generally, most mammals develop important strategies to modulate the timing of conception to take advantage of resource abundance hence depress fertility or produce fewer offspring. The current analysis investigated factors affecting fertility of Holstein Friesian cows performing on large-scale farms in Malawi. Period of intense reproductive activity calculated retrospectively from date of calving for 2362 cows between 1986 and 1997 was used as proxy for fertility. Results indicated seasonality in fertility with the highest ($p<0.001$) estimates (41.61%) occurring in the cold-dry season compared to 31.29% for hot-wet season and 27.10% for hot-dry season. Further, fertility was affected ($p<0.001$) by cow, parity, and year-effects indicating individual cow's coping mechanisms. The hot-wet season, characterised by high humidity, high temperatures and low wind speed, seem to exert some stress on the Holstein Friesian cows. Sound management strategies should be developed to optimize the seasonal fertility rhythms for reproductive efficiency.

Poster C4.25

The use of world genetic resources to increase the production of high-quality beef

V.I. Sharkaev. All-Russian Institute of Animal Breeding, Lesnye Polyany, Pushkin district, 141212 Moscow Region, Russia

The genetic resources of beef cattle Charollais and Limousin were used to increase the production of high-quality beef. For that cows of local breed Sychovskaya and Black-and-White cows were inseminated by sperm of Charollais and Limousin sires imported from France. Contemporary type of Charollais and Limousin animals is characterized by easiness of calvings. The weight of bulls-crossbreds F_1 Sychovskaya x Charollais at the age 18 mo.was higher by 49,9 kg ($P<0,001$) and that of bulls-crossbreds F_1 Black-and-White x Charollais was higher by 42 kg ($P<0,001$) in comparison with the weights of their purebred analogs - 451,8 and 427,3 kg respectively. The crossing with Limousin sires improved the growth performance of crossbreds. The morphological composition of carcass was better in ocrossbreds compared with purebred bulls and yield of high-quality meat was higher in crossbreds too (18,0-20,2 vs. 17,1-17,3). Meat of crossbreds was characterized by the higher qualitative traits for chemical composition, calorie content, physico-chemical traits and ratio full-valuable to unfull-valuable proteins.

Poster C4.26

Use of endangered dual purpose cattle for a beef production scheme

M.D. Henning, Chr. Ehling and P. Köhler, Institute for Animal Breeding Mariensee, Federal Agricultural Research Centre (FAL), Höltystraße 10, D-31535 Neustadt, Germany*

Genetic diversity in livestock has been reduced dramatically as a consequence of a high degree of specialisation in animal production. The number of the old type German Blackpied cattle (DSB) in Germany as a representative of a dual purpose breed went down to 2235 Herdbook cows after replacement by Holsteins for a more efficient milk production. At the same time beef breeds mainly from France, Italy, the U.K. and Ireland were spread all over Germany, they have been proven as useful sires in numerous cross breeding trials.

So far DSB gene reserve cows in Mariensee were kept for milk production, beef from a cross breeding scheme with Limousin sperm is meant as an alternative product for covering the costs of maintenance. All cows in the programme should produce one female offspring for remount. Then they are available for carrying out Lim x DSB calves. Production goal is beef from cattle (heifers and steers) with a maximum age of 20 month. The first trial with 20 heads is finished. As average daily gain from birth to weaning at the age of 4.5 month 1225 g/d were achieved. Pasture period was influenced by an extremely dry summer in northern Germany 2003 which resulted in 520 g/d weight gain. Two different feeding regimes indoors were applied to either 10 animals, one group was fed corn silage and concentrates the other grass silage and hay, only. Carcass and meat quality were determined, a calculation for economic value will follow.

Theatre C4.27

Implications of the early weaning of artificially reared Brown Swiss Calves for health and growth performance

A.A.ElShahat, Animal Prod. Dept. National Res. Centre, Dokki (12622), Cairo, Egypt

Early weaning is essentially the process whereby the transition from the pre-ruminant to the ruminant state is accelerated. Provided good nutritional and management practices are followed early-weaning of individually-penned calves at an average age of 6 weeks is consistent with adequate performance for most systems of production.

Early weaning, however, is affected by several factors such as: providing a suitable starter, restricting milk intake to encourge the intake of solid food, having an effective roughage source and providing a reliable water supply.

However, the critical issues which arise are: (1) Composition and processing of starter, (2) Quality level and duration of milk feeding, (3) providing sufficient roughage to ensure normal digestive function, (4) Improving animal health and decreasing mortality rate, and (5) criteria for weaning and weaning strategy. These aspects of this open scientific field will be discussed on the basis of animal nutrition and cattle production.

Poster C4.28

Feeding hay to the preruminant calf: effects on growth and economics

L.T. Cziszter, G. Stanciu, S. Acatincai. Banat University of Agricultural Sciences, Calea Aradului 119, 300645 Timisoara, Romania*

The aim was to assess the growth and economics of calves fed hay later in their life. Research were carried out on 25 Romanian Black and White 2-wks old calves. Control group (C, n=12) received the alfalfa hay *ad lib* from the beginning and experimental group (E, n=13) received it only from 8th week of age. Calves were housed outdoors on deep litter. Water and a concentrate mixture was given *ad lib* to all calves from the first day. Calves were bucket-fed a total of 69.5 kg milk replacer, and weaned at 70 days of age. Body weight, heart girth and feed consumption were recorded individually, every second week. Mann-Whitney U test was used for the comparison between groups. Concentrates intake was double ($p<.001$) at 8 wks of age in E compared to C, and 1.5x higher ($p<.01$) at weaning (898.1 and 2198.1 vs. 402.9 and 1499.6 g/d, respectively). At 10 wks of age experimental calves consumed 15% more hay than control calves (771.4 vs. 668.6 g/d, $p<.05$). Body weight showed a tendency ($p=.08$) to be 6% higher in E than in C at 10 wks of age. ADG from 8 to 10 wks of age was 21% higher in E than in C (940 vs. 777 g, $p<.05$). Production costs were 0.15 EUR/kg ADG (8%) lower in E than in C.

Poster C4.29

Feeding level and duration of finishing for bulls

M.G. Keane and R.J. Fallon, Teagasc, Grange Research Centre, Dunsany, Co. Meath, Ireland*

The decoupling of the special beef premium has increased interest in bull beef production in Ireland. The effects of concentrate level with grass silage, and length of finishing period, on performance of Holstein-Friesian bulls were examined. Fifty-four Autumn born calves were reared indoors for 130 days and turned out to pasture for a 203-day grazing season. They were then assigned to a 3 (feeding levels) x 2 (finishing periods) factorial experiment. The 3 feeding levels were grass silage plus 3 (L) or 6 (M) kg/day concentrates, or concentrates *ad libitum* (H). The two finishing periods were 179 and 272 days. Daily liveweight gains from start to 179 days, 179 to 272 days, and from start to 272 days were 908, 1164 and 1395 (s.e. 28.2)g, 961, 1022 and 1200 (s.e.32.5)g, and 941, 1141 and 1358 (s.e. 25.1)g, for L, M, and H, respectively. Corresponding daily carcass gains to slaughter were 548, 741 and 895 (s.e. 14.5)g. Up to 179 days, efficiency of metabolisable energy utilization increased from L to M but did not change further from M to H. It is concluded that only the H and M feeding levels for 272 days and the H feeding level for 179 days produced commercially acceptable carcasses. To produce a 320 kg carcass required dry matter intakes of 1.25t concentrates plus 1.0t silage for M, and 1.80t concentrates plus 0.20t silage for H.

Poster C4.30

Examination of fattening and slaughter results of Hungarian Simmental bulls

P.J. Polgár[1], I. Füller[2], B. Húth[3], Z. Lengyel[1] and J. Stefler[3], [1]University of Veszprém, Georgikon Faculty of Agriculture, H-8360 Keszthely, Deák F. str. 16, Hungary. [2]Federation of Hungarian Simmental Breeders, H-7150 Bonyhád, Zrínyi str.3., Hungary, [3]University of Kaposvár, H-7400 Kaposvár, Guba Sándor str. 40., Hungary

The examination in the interest of determine meat breeding value was started due to The Federation of Hungarian Simmental Breeders. In every generation 12-15 male offsprings of 6-12 sires are fattened from the age of 120day to slaughtering. During the fattening period the feeding of animals is based on ad libitum maize silage and concentrate on a ration of 100kg live weight / 1 kg concentrate. At the end of fattening the slaughter is done in the slaughter house of Zalahús Sc. The data of slaughter and carcass weights, and ph values (1 and 24 hours after the cutting) are collected. Carcasses are judged on the base of EUROP. The database is examined by SPSS 9.0.
In the project the data of 244 slaughtered male offsprings has already been calculated. Daily gain (refered on the age of slaughter) and daily gain under the fattening period are found to be 1139 g/day and 1177 g/day, respectively. The overal mean value of killing out percentage is 58.62%, the EUROP quality is R and fat score is 2.56. According to these findings the gain was but killing percentage, and EUROP quality was not significantly influenced by sire.

Poster C4.31

Daily gain, feed conversion and carcass composition of Limousin and Charolais heifers

L. Barton, V. Teslík, V. Kudrna, M. Krejcová, R. Zahrádková and D. Bures, Research Institute of Animal Production, Prátelství 815, 104 00 Prague - Uhríneves, Czech Republic*

Live-weight gain, feed conversion and carcass composition were compared in 46 purebred Limousin (LI) and Charolais (CH) heifers given two isocaloric and isonitrogenous diets differing in content and source of dietary fat. At the start of the trial, the animals were blocked on weight within breed to form a 2 (breed) x 2 (diet) factorial experimental design. They were individually fed via electronically controlled feeding troughs. Mixed diets consisted of maize silage, hay, straw and concentrates (wheat and soybean meal). Diet 1 was, in addition, supplemented with extruded linseed. Mean live-weight gains for LI and CH were 0.790 and 1.012 kg/day, respectively ($P<0.001$). Although CH heifers consumed on average 0.621 kg/day DM of feed more than LI ($P<0.01$), they gained more efficiently ($P<0.001$). Total internal fats, fat trimmed off in the dissection and fat thickness over MLLT were lower for CH ($P<0.05$ or less). LI tended to produce more carcass lean meat ($P=0.0514$) and had significantly lower proportion of bones in carcass ($P<0.001$). No diet effect on the observed traits was displayed. It is concluded that CH heifers were superior to LI for live-weight gain and feed conversion and produced less internal and carcass fat while LI heifers averaged higher meat to bone ratio. The study was supported by the project MZE0002701403.

Poster C4.32

In vivo prediction of intramuscular fat content in Slovak Simmental and Holstein bulls

N.E. Blanco Roa[1], A.M. Pérez Artuch[2], P. Polák[1], J. Huba[1], J.A. Mendizabal[2] and D. Peskovicová[1], [1]Research Institute of Animal Production Nitra, Slovakia, [2]Departamento de Producción Agraria, Universidad Pública de Navarra, 31006 Pamplona, Spain*

The aim of the present work was to evaluate the use of ultrasound and image analyses as methods for *in vivo* prediction of intramuscular fat content in beef meat. Thirty 15-month-old calves belonging to Holstein (n =18) and Simmental (n=12) breeds were used. An ecography over the 13th rib was taken on each animal before slaugther and grey value was calculated by image analysis for the *longissimus dorsi* (LD) muscle area in each ecography. Finally, after slaughter LD fat content was calculated by chemical analysis.
Obtained results showed that intramuscular fat percentages in LD were 2.40 and 2.17 g.100g^{-1} for Simmental and Holstein breeds respectively ($P>0.05$) and the grey values obtained in ecographies 51.8 and 60.6 (scale 0-255) ($P>0.05$). Values of $r=0.81$ ($P<0.001$) and $r=0.68$ ($P<0.05$) were found when correlating *in vivo* measured grey values of ecographies and intramuscular fat percentage determined *post-mortem* in Simmental and Holstein bulls respectively. In conclussion, these results showed the suitability of ecography and image analysis used together as a technique for *in vivo* prediction of the amount of intramuscular fat in beef meat.

Poster C4.33

Comparison of muscle and fat thicknesses obtained by ultrasound in four different breeds of cattle in Slovakia

P. Polák, E.N. Blanco Roa, E. Krupa, J. Huba amd M. Oravcová, Research Institute for Animal Production, Hlohovská 2, 949 92 Nitra, Slovak Republic*

The aim of the work was to compare muscle and fat thicknesses obtained by ultrasound method from 51 fattened bulls at the age of 450 days. Bulls of 4 different breeds: Pinzgau (12), Pinzgau with proportion of Red Holstein breed to 25 % (13), Swiss Brown (12) and Black and White Holstein (14) were fattened in the same conditions. Ultrasound measures were taken at 5 positions: on the shoulder blade, on the back behind the shoulder blade, on the last thoracis vertebrae, on the last lumbal vertebra and on os ischia. Means and standard deviations were calculated for live weight and age at ultrasound measurement, and for muscle and fat thickness. Scheffe´s test was used to test differences between mean values. The highest live weight was found out at Pinzgau cattle (481.17 kg). Animals of Holstein breed showed significantly lowest thickness of muscle and tended to have the highest layer of subcutanous fat. On the contrary, bulls of Pinzgau breed had the highest thickness of muscles and moderate thickness of fat. It could be concluded that thickness of muscle was depend on breed specialization to milk production whereas thickness of fat was independ.

Poster C4.34

Relationship of muscular scores and carcass conformation with the proportions of higher value cuts in the carcass

M.J. Drennan and B. Murphy, Teagasc, Grange Research Centre, Dunsany, Co. Meath, Ireland*

The objectives were to examine the relationships between (1) live animal muscular scores and carcass conformation score with the proportion of some high value cuts in the carcass and (2) high value cuts with carcass meat proportion and carcass value. Following fat removal, the fillet, striploin and cube roll only account for 8% of carcass weight but over 30% of carcass value. A total of 176 continental breed cross suckled bulls in 4 experiments slaughtered at 15/16 months of age were used. Records collected included muscular scores pre-slaughter, carcass grades, and meat proportion (following dissection into meat, fat and bone of either the half carcass (Experiment 2), the entire carcass (Experiment 3) or an 8 rib pistola (Experiments 1 and 4). Over the 4 experiments correlations for the proportion of fillet in the carcass varied from 0.19 to 0.34 with muscular score, from 0.10 to 0.33 with conformation score, 0.32 to 0.49 with carcass meat proportion and 0.44 to 0.66 with carcass value. Corresponding correlations for the proportion of striploin were 0.08 to 0.43, -0.14 to 0.29, 0.24 to 0.50 and 0.39 to 0.68. In conclusion, the correlations between these two high value cuts were generally positive but low with muscular and conformation scores but were highly significant with meat proportion and carcass value.

Poster C4.35

Accuracy of the Video-Image-Analysis-System (VBS 2000) to assess the composition of cattle carcasses

D. Brinkmann[1], M. Soennichsen[2], E. Tholen[1] and W. Branscheid[2], [1]Universtity of Bonn, Institute of Animal Breeding Science, Endenicher Allee 15, 53115 Bonn, Germany, [2]Federal Research Centre for Nutrition and Food, Institute for Meat Production and Market Research, E.-C.-Baumann-Str. 20, 95326 Kulmbach Germany Federal

Automatic video-image-analysis-systems (VIA) are a promising and suitable technique to predict the true composition of cattle carcasses. The accuracy of the VBS 2000 (E+V-Technology, Oranienburg) is verified in a cooperation project between the German Central Marketing Agency (Bonn) and the slaughter-company Westfleisch eG (Münster). Based on digital cameras, the VBS 2000 uses stripe-projection to digitalize the left side of the carcass. Surface and volume measurements are used to estimate the true carcass composition. In a first trial the accuracy of predicting the conformation and fat-cover of young bulls was examined. The underlying classification setup covers fifteen both conformation (EUROP) and fat-cover classes. On total, 550 validated carcasses were categorised by a reference panel of experienced BAFF-classifiers and the VBS 2000 using the preset estimation functions. Satisfactory correlation (r=.9) was achieved for the conformation-classes between VIA-system and the subjective reference. However, the correlation between the fat classes did not exceed .7. In further projects regression formulas for estimating the conformation- and fat-cover classes used by the VBS 2000 will be validated for cattle- and calve-carcasses. In addition, the VBS 2000 System should be complemented by techniques to measure veal colour, which is an important value determining factor of calve-carcasses.

Poster C4.36

Chemical and technological quality of meat (m. long. dorsi) in the population of slaughter bulls in Slovakia

K. Zaujec, J. Mojto and K. Novotná, Research Institute of Animal Production Nitra, Hlohovská 2, 949 92 Slovakia

The differences in quality of carcass (meatiness, fattiness) among different breeds of cattle are known and they are used when animals are chosen for particular breeding, climatic and production conditions. The differences in meat quality among breeds are not so unambiguous because they can be affected by a whole range of intravital and pre-mortal factors. Beef production in Slovakia is provided by three main breeds. In a comparative experiment with adult slaughter bulls of Slovak Spotted (SS), Slovak Pinzgau (SP) and Holstein breed (H) were evaluated the differences in some chemical, physico-technological and sensorial parameters of meat quality. The animals were kept under identical conditions before slaughter and they came from tying fattening stalls. The quality of m. long. dorsi was evaluated 48 hrs post mortem, and the shear force on 7th day. Statistically significant differences ($P < 0.5$) were detected in the content of total water with the highest value in SS breed (75.30 $g.100g^{-1}$), further in the content of intramucular fat (SS - 1.94; SP - 2.47; H - 2.41 $g.100g^{-1}$) and in meat colour (SS - 8.67; SP - 9.49; H - 9.35 % remission). The differences in further parameters were statistically non-significant. The shear force (Warner - Bratzler) was in SS - 3.93; SP - 3.96; H - 3.92 kg, and the value of pH_{48} in SS was 5.78; SP - 5.72; H - 5.77. The results showed that the Holstein breed has not only a fatter carcass but also higher content of intramuacular fat and more intensive marbling of meat.

Poster C4.37

The effect of gender on the amino acid composition and biological value of beef

*G. Holló[1], *E. Szücs[2], J. Csapó[1], G. Pohn[1], É. Visi[1], J. Tözsér[2], I. Holló[1] and I. Repa[1], [1]University of Kaposvár, Kaposvár, H-7400, [2]St. István University, Gödöllö, H-2103, Hungary*

The influence of gender on amino acid and biological value (BV) of beef was compared using 12 male and 15 female Holstein-Friesian fattening cattle. Animals were fed with maize-silage, grass-hay and mixed-grain. Average slaughter weights for male and female animals were 463+25 kg and 458+23 kg, respectively. After chilling for 24 h samples were taken from right half carcasses, between 11-13th ribs. The amino acid profile of samples was made by LABOR MIM amino acid analyser, protein was hydrolysed in 6 mol hydrochloric acid for 24 h at 110°C. 17 amino acids were identified and quantified (Arg, His, Ile, Leu, Lys, Met, Tyr, Val, Ala, Asp, Glu, Gly, Pro, Ser, Tre, Cys, Phe). The BV of the protein was calculated by the 2:1 potato-egg mixture reference. Statistical analysis (SPSS 10.0) reveal significant differences between genders for the total amount of essential amino acids ($P<0.05$), methionine ($P<0.01$), isoleucine ($P<0.001$), leucine ($P<0.01$), lysine ($P<0.01$), histidine ($P<0.05$) and arginine ($P<0.05$) but there was no significant effect for non-essential amino acids, and for the BV of proteins in meat. Gender may be a factor that significantly influences the amino acid profile in beef, as the quantity of essential amino acids turned out is significantly larger in females than in males.

Poster C4.38

Analysis of length of productive life in Holstein cattle in Czech Republic

E. Páchová[1,2], L. Dedková[1] and I. Majzlík[2], [1]Research Institute of Animal Production in Prague, Prátelství 815, CZ 104 01 Prague, Czech Republic, [2]Czech University of Agriculture in Prague - CUA, Kamycká 129, CZ 165 21 Prague, Czech Republic

An analysis of productive life in Holstein cattle in Czech Republic was performed using programs package Survival Kit, based on Cox regression. Data included 290 834 cows from the official milk recording database. The length of productive life (LPL) was observed between years of first calving 1976 and 2000. The mean LPL in this data set was 1 222 days. Records on cows that were still alive in the end of study were treated as censored (7.9 %). The risk of being culled was defined as product of a baseline hazard function and a function of explanatory variables. The model included two time-dependent effects, lactation number and stage of lactation and three time-independent effects, age at first calving, herd and milk production in the first lactation. The highest risk of being culled was found for the cows on the first lactation and cows in the second part of lactation (from 150 to 300 days of lactation). Generally, the hazard decreased with parity. Age at first calving showed significant impact on LPL. The cows older at first calving had higher risk of being culled. There are differences between LPL in herds in this study. Management of herds is specific and it plays a role in culling police. Milk production yielded very important impact on LPL. The highest risk of being culled existed for cows with the low milk production.

Poster C4.39

Influence of dry period feeding on body condition, and in earlier milk production on Romanian Black Spotted Cattle

C.V. Calin, S. Grigoras, C. Bunghiuz, D. Georgescu and M. Gras, Research and Development Institute for Bovine Balotesti, Ilfov 077015, Romania*

Results are presented from an experiment organised on 3 experimental lots formed from cows in dry period with 3 levels of body condition using the Body Condition Score (BCS) system. BCS system uses a scale between 1 (very thin) and 5 (very fat). All 3 lots with correspondent body condition score was: 2, 75 BCS; 3 BCS; 3, 25 BCS.

Rations were derived using information on the body weight of animals, maintenance requirements of that and body weight gain necessary to achieve the the target body condition score at calving. For detection of metabolic deviations, we make tests periodically. All 3 categories of body condition were achieved at the end of the dry period. From energy deposits in the cow body that accumulated over two months of dry period, productive levels below were noted:

- for BCS 2, 75 average milk productions were 37 l;
- for BCS 3 average milk productions were 37, 3 l;
- for BCS 3, 25 average milk productions were 35 l.

Poster C4.40

Selection of the Russian Red Steppe breed for high fat and protein in milk

T. Knyazeva, E. Bogomolova and S. Shnaider All-Russian Institute of Animal Breeding, Lesnye Polyany, Pushkin district, 141212 Moscow Region, Russia

The breeding of the protein type animals in the Red Steppe breed is carried out in 11 herds in Altai Krai and Omsk region. About 20,000 cows and calves of the different genotypes of ameliorated breeds (the Angler, Red Danish) resulted from breeding 'in itself' has been obtained. The animals of this population meet the type's requirements.
The minimal requirements for the useful traits of the fat and protein type animals are as follows: milk yield for 305 lactation days - 4,000-5,000 kg, protein content - 3,35-3,50%, fat portion - 4,0-4,2%, cows body weight - 550-550 kg, intensity of letting down of milk - 1,7 kg / min.
The efficiency of the interbreed type's breeding are based on the directional rearing of the young animals, valuable nutrition of the adult animals and perfection of the selection system of the sires: genetic testing of the bovine origin and their daughters, energy control of their growth and development, protein content control in milk, the kappa-casein genetic testing of bulls-sires and dams, breed value estimates by BLUP.

Poster C4.41

Methodology problems of improvement the milking ability in Hungarian Simmental breeding stocks

B. Huth[1], E. Szücs[2] and I. Hollo[1], [1]University of Kaposvar, Faculty of Animal Science, H-7401 Kaposvar, P.O. Box 16, Hungary, [2]St. Istvan University, H-2100 Gödöllö, Pater K. u. 1., Hungary*

As a result of the selection to increase the milk yielding capacity, the specific milk production of all dairy and dual-purpose breeds has definitely been increased in the last decades. However, in parallel with it, physiological load of the udders also has increased. The udder's health condition is one of the essential parameters that determine the genetically fixed milk yielding capacity and the realization of the quality milk production. Mastitis is a multi-factorial disease that causes important economical damages. The factors causing and making susceptible to this sickness include the improper milking ability, too. From udder's health aspects disproportion of the udder-quarters (blind milking), very slow and very fast milk flow all are unfavourable.
Report is based on results obtained for milking ability on breeding bull producing farms.
It is found that the investigated cow population shows significant difference and heterogenity according to milking ability. Average milk flow is by 0,20 kg/min below the breeding goal. Only at 30% of cows could be experienced 2,00-3,00 kg/min milk flow, which is accepted at Simmental populations.
Between the average and maximum milk flow there is a close (r=0,84) correlation, which is statistically significant. This means that that according to milking ability determination and utilisation of average milking ability is satisfactory in improvement.

Poster C4.42

The effect of the paternal origin and other conditions of milking capacity of cows

B. Sitkowska and S. Mroczkowski, University of Technology and Agriculture, Genetics and Animal Breeding, Zootechnical Department, Mazowiecka 28, 85-084 Bydgoszcz, Poland

The research was carried out based on breeding documentation which covered 15 528 milk yield trials for cows. The cows were milked over the period 1997-2002.

The numerical data were verified statistically with variance analysis of the GLM procedure, incorporating the effect of herd, successive lactation, calving season, share of HF in the cow genotype and paternal factor. The effect of these basic factors on all the traits of their female offspring was highly significant.

With the method of the least squares the following effects were analysed: paternal origin on the traits of milk capacity of daughters; including the country of origin of the father, giving birth by mother or acceptor, the share of HF in the paternal genotype and the dates the father was born. Analysis of milk performance showed that the progeny of bulls from Netherlands, Canada and USA were better than progeny of Polish and German ones. The effects of most of the factors studied on milk performance, its chemical composition and the content of somatic cells were highly significant.

Poster C4.43

Wither height index in Holstein-Friesian heifers from born birth to 1 year old of age

E. Báder[1], I. Györkös[2], J. Bartyik[3], P. Báder[1], M. Porvai[3] and A. Kovács[1], [1]University of West Hungary, Faculty of Agriculture, Vár, 2, 9200 Mosonmagyaróvár, Hungary, [2]Research Institute for Animal Breeding and Nutrition, Gesztenyés str. 1, 2053 Herceghalom, Hungary, [3]Enyingi Agrár Company, Farm Kiscséripuszta, 8130 Enying, Hungary*

The wither height index besides condition scores is a good method for controlling the development and growing vigour of heifers. Moreover wither height index indicates precisely minus, optimal and plus condition.

The changes in wither height index in Holstein Friesian heifers from born birth to 1 year old of age is was studied. Experiments were carried out in Kiscséripuszta, the cattle farm of Enying Agrár Plc. 13 measurements were made until 1 year old of age (in every 30 days). The total number of measurements was 5984. Live weight and wither height were measured. The wither height index is the rate of live weight and wither height. This index is 0.48 at born birth (weight: 36 kg and wither height: 75 cm) and 0.61 at 1 month of age. Live weight of heifers at 3 months of age (86 kg) exceeds the wither height (81 cm), so the index is 1.06. Live weight at 15 months of age is anderthalb times more than wither height (index: 1.56) and the index will increase up to 2.01 at 7 months of age (weight: 208 kg, wither height: 105 cm). Wither height index will further increase, it is 2.56 at 11 months of age and 2.81 in 12 months of age (weight: 328 kg, wither height: 117 cm). Wither height indices of the examined stock are equal with the American references, so the condition of heifers is optimal.

Poster C4.44

Ultrasound evaluation of the mammary gland tissue structure in preparturient heifers vs. performance of first calvers

J.A. Strzetelski[1], K. Bilik[1], B. Niwinska[1], G. Skrzynski[2] and E. Luczynska[3], [1]Department of Animal Nutrition, National Research Institute of Animal Production 32-083 Balice n. Kraków, [2]Department of Cattle Breeding, Agricultural University, 30-059 Kraków, Al. Mickiewicza 24/28, [3]Oncology Centre, Department of Imaging Diagnostics, 31-115 Kraków, Garncarska 11, Poland*

To determine the development of secretory tissue of the mammary gland and to avoid the use of invasive methods, tissue structure of the mammary gland was examined in vivo with Aloka SSD-500 ultrasound device (Aloka Co., Ltd., Japan). The ultrasound images were binarized and brightness level was determined on a grey scale of 0 to 255. The image brightness scale of 128 (lower threshold) to 255 (upper threshold) was applied. Pixels with brightness values of 0 to 127 were assigned to the adipose tissue, and pixels with brightness values of 128 to 255 were assigned to the secretory tissue of udder quarter. Then, percentage of bright image area was calculated in a predetermined image area. The results were recorded with regard to the area of the predetermined ultrasound image portion, the grey area and the percentage of grey area, and was output in the results file. Highly significant correlations were found between the percentage of secretory tissue in the mammary gland and the milk yield of first-calving cows.

Poster C4.45

Effect of time of day, milk yield, and lactation phase on milking time and milk flow rate in Holstein-Friesian dairy cows

E. Szücs[1], Cs. Ábrahám, G. Holló[2], I. Holló[2] and Tran Anh Tuan[1], [1]Szent István University, Gödöllö, H-2103, [2]University of Kaposvár, Kaposvár, H-7400, Hungary*

The impact of time of day, milk yield, and lactation phase on milking time and milk flow rate was investigated in Holstein-Friesian cows in this study. Means and SE for lactation milk yield, butterfat and milk protein content were 10365±7,2 kg. 3.74±0.002 %, and 3.40±0.009 %, respectively. Cows were milked three times a day. Milking started at 0500 AM, 1300 and 1900 PM. Data of 42833 milkings were recorded electronically. Means and SE for daily milk yield, milking time and milk flow rate were 34.91±0.04 kg, 18.57±0.02 min and 1.91±0.002 kg/min, respectively. Significant differences in milk flow rate across daily milk yield categories (<20; 21-30; 31-40; 41-50; and 51< kg) as well as phases of lactation (<100; 101-200; 201-300; and 301<) were established (P<0.001) for each of the three milkings. Coefficients of correlation between (1) yield and milk flow rate, (2) milking time and milk flow rate, or (3) milk yield and milking time ranged from 0.52 - 0.72, -0.47 - -0.32, and 0.50 - 0.39, respectively. (P<0.001). Between milking time and milk yield, as well as milk flow rate and milk yield curvilinear relationship with moderate quadratic effect has been established the extent of which varied depending on the order of milkings of the day.

Poster C4.46

Effects of an automatic milking system on feed intake as well as on milk yield and composition of high lactating cows

M. Spolders, U. Meyer and G. Flachowsky, Federal Agricultural Research Centre, Institute of Animal Nutrition, Bundesallee 50, 38116 Braunschweig Germany*

In four experiments with 60 animals each the influence of an automatic milking system (AMS) on milk yield and composition as well as on animal behaviour was studied in comparison to a conventional milking system.
The type of milking process did not significantly affect the amount of FCM. There was only a tendency towards an improved milk yield in the AMS. But rating for milk composition milk fat content was predominantly affected by the applied milking system. The cows milked automatically showed a low-grade decrease of milk fat.
The quantity of roughage and concentrate intake was not affected by the milking system. But differences of intake during the course of the day were detected harmonized with the distribution of milking processes over the day. Rest behaviour was influenced by the applied milking system: fewer cows rested in the lying cubicles, when they were milked automatically due to the fact that the movement cycles such as the way to the milking unit and to the feeding trough were spread evenly over the day. In addition to this the AMS proved to be an extra course of disturbance, which was evident that the 3 lying cubicles next to the milking unit were less occupied by the cows.

Poster C4.47

The effect of genotype, feeding system and milk yield on the activity of some glycosidases in whole cow's milk and somatic cells separated from the milk

A. Józwik, A. Sliwa-Józwik, J. Krzyzewski, N. Strzalkowska, E. Bagnicka and A. Kolataj Institute of Genetics and Animal Breeding Polish Academy of Science, Jastrzebiec, 05552 Wólka Kosowska, Poland

The aim of this study was to estimate the influence of genotype (races), feeding system and milk yield on the glycosidase activity in the whole cow's milk and in somatic cells, which were separated from the whole milk. It is known these enzymes take part in the control of homeostatic changes in cell. Changes on the activity these enzymes are coefficients of homeostatic disturbances and of sickness - processes.
The experiment was carried out on 40 cows, divided into two groups (20 cows in each): 1 - Polish Red (native race) - the average milk yield about 4200 kg per lactation 2 - HF - the average milk yield about 8500 kg per lactation. In the whole milk and in somatic cells, basing on substrates from SIGMA-ALDRICH Co., the activities of mannosidase (E.C. 3.2.1.24), α-glucosidase (E.C. 3.2.1.20), β-glucosidase (E.C. 3.2.1.21), β-galactosidase (E.C. 3.2.1.23) were estimated. The activities were determined according to the method Barrett and Heath (1972). The activity of the enzymes was expressed as nMol/mg of protein/hour. The activity of examined glycosidases was significantly higher in HF cows whole milk and somatic cell in comparison to Polish Red cows.

Poster C4.48

The effect of the season of the year on the yield, composition and quality of milk of Jersey cows

I. Antkowiak and J. Pytlewski, Chair of Cattle Breeding, Agricultural University 60-625 Poznan, Wojska Polskiego 71A, Poland

The investigations were conducted in the years 2000 - 2002 in a herd of Jersey cows. In the experiment the calendar year was divided into 4 quarters - i.e. seasons of the year: winter (December, January, February), spring (March, April, May), summer (June, July, August), autumn (September, October, November). In the study the effect of the season of the year was analyzed on the daily milk yield, daily milk yield in terms of FCM, the percentages of butterfat and protein, the protein to butterfat ratio, the somatic cell count and the urea content in milk. The results were subjected to statistical analysis using the SAS® software package. The significant and highly significant statistical differences were found between the seasons of the year for all the analyzed traits (except the SCC). The highest daily milk yield (20.86 kg), the yields of butterfat (0.93 kg) and protein (0.65 kg) were observed for cows in the spring season. Autumn turned out to be the least advantageous for the above mentioned traits of milking performance. The highest butterfat percentage (5.91) was found for milk obtained in the winter, and the most advantageous protein content was observed in the autumn (4.12%). Milk collected in the summer season was characterized by the highest SCC (12.52) and the lowest value of this parameter was observed in the milk obtained in the winter (12.16).

Poster C4.49

Prediction of daily milk protein and urea content using alternating (AT) recording scheme

M. Klopcic, G. Gorjanc, S. Malovrh, M. Kovac and J. Osterc, University of Ljubljana, Biotechnical Faculty, Zootechnical Department, Groblje 3, 1230 Domzale, Slovenia

The introduction of AT4 method in milk recording instead of the present A4 method was studied. During trial period milk composition in daily (D), evening (PM) and morning (AM) milk samples was studied. Correlations of protein and urea content between PM, AM and D milking were estimated on the base of PM and/or AM milking. Throughout the trial period average protein and urea content were both higher in the evening milking, regardless milking interval. Phenotypic correlation for protein and urea content was significant and ranged between 0.95 (PM - AM) and 0.99 (D - PM) for protein, and between 0.82 (PM - AM) and 0.91 (D - PM) for urea content. Determination coefficient (R^2) and residual variance were considered in the model selection. For protein content, R^2 ranged from 97.47% to 97.65% for PM, and from 95.48% to 95.74% for AM milking, depending on the model. For urea content R^2 ranged from 82.50% to 83.07% for PM, and from 80.82% to 81.61% for AM milking. The AM protein and urea content proved to be better for prediction of daily content. Differences among models were small. Chosen model for protein content included AM or PM content and milking interval, while milking interval had no significant effect for urea content. Bias in urea and protein content was the highest at extreme values and it was considerably higher for urea content compared to proteins. Complex models with several explanatory variables slightly reduce bias at extreme values therefore simpler models are sufficient for practical usage.

Poster C4.50

Oscillation of somatic cell count in milk recorded cows in years 1995-2003

M. Klopcic[1], M. Klinkon[2] and J. Osterc[1], [1]University of Ljubljana, Biotechnical Faculty, Zootechnical Department, Groblje 3, 1230 Domzale, Slovenia; [2]University of Ljubljana, Veterinary Faculty, Gerbiceva 60, 1000 Ljubljana, Slovenia

Oscillation of somatic cell count (SCC) in milk recorded cows was studied in Slovenia for the period 1995 to 2003. Measurements of 1,662,615 test days were taken predominantly of Simmental (41.6 %), followed by Black and White (34.6 %) and Brown breed (23.8 %). The analysis showed substantial seasonal differences in milk yield and in milk composition. Breeders are faced with problems of low milk quality and frequent mastitis in summer months. It has been concluded that the average SCC was the highest in Black and White cows (477 x 10^3/ml milk) and the lowest in Simmental cows (324 x 10^3/ml milk). Since 1996 SCC has been considered one of the criteria for milk price determination, therefore it has been reduced and health status of udder improved in recent years. Both are the result of better herd management and wider knowledge that breeders achieved. The average SCC dropped from 431 thousand in 1996 to 397 thousand cells/ml milk in 2002. Lower SCC was noticed also on family farms compared to the former large scale farming enterprises. Monthly recording analysis proved that 24.0 % cows had more than 400 thousand SCC/ml milk, and 64.6 % cows up to 250 thousand SCC/ml milk at recording. During the trial period milk yield/milking day increased and fat, as well as protein content improved.

Poster C4.51

Some biological and farming qualities of Lithuanian native cattle and measures for conservation of their genetic resources

V. Juskiene, R. Sveistiene and R. Juska. Institute of Animal Science of Lithuanian Veterinary Academy, R. Zebenkos 12, LT-82317, Baisogala, Radviliskis distr., Lithuania

In order to preserve the gene pool of Lithuanian native cattle, two preservation herds of relic ash-grey and white-backed cattle have been formed for pure-breeding and evaluation of the biological and farming qualities of these cattle.

Milk recording of the cows indicated that Lithuanian native ash-grey and white-backed cattle are no less productive than wide-spread Lithuanian Black-and-White and Lithuanian Red cattle. Body conformation measurements indicated that Lithuanian ash-grey cattle are larger in size and higher than white-backed ones: their withers and back height were, respectively, by 4.1 ($P < 0.01$) and 5.6% ($P < 0.025$) and hip width 3.5% ($P < 0.05$) higher than those of white-backed cows. The indices of compactness, long-leggedness and chest were also higher by 0.6-5.1%.

The breeding plan based on the circular mating scheme has been developed to avoid inbreeding consequences and to consolidate separate lines.

A large number of semen doses of the native bulls are stored at the breeding enterprises. However, studies indicated that some of the bulls do not produce typical progeny and, therefore, should be culled.

 Poster C4.52

Breed composition of cattle in West Herzegovina canton

S. Ivankovic, A. Zelenika and Z. Knezovic, Faculty og Africulture, Biskupa Cule 10, 88000 Mostar, Bosnia and Herzegovina*

Breed composition of native animals in whole BIH has changed during the war. After-war renewal is based on import of productive breeds, that with management and marker demand leading to decreasing population of autochtonous breeds. 201 country estates that possess 390 milch-cows whose milk is buying off in the villages in townships Posusje and Siroki Brijeg is analysed. Busa intact is represented with 1,8%, bastards in type of busa and gatacki breed with 28,4%, grey breed with 5,9%, brown with 11,0%, black-white and HF with 18,7%, simental breed with 15,6% and indefinite types with 18,4% in total number of milch-cows. The average number of milch-cows in analysed country estates is 1,94 in West Herzegovina Canton, in township Siroki Brijeg it is 1,48 and in township Posusje 2,22. 45,1% country estates have only one cow, and 37,8% of them have two milch-cows in total number of analysed estates. The goal of this research has been to evaluate situation in the West Herzegovina Canton. Measuring of height of crest ($\bar{X}$= 106,4 *cm*) and lenght of trunk ($\bar{X}$ = 119,2 *cm*) is made for busa. As conclusion the necessity intrudes itself whole engagement of all performers of developement in this Canton and Bosnia and Herzegovina to try enable the producers of milk and also necessity for urgent preservation of imperilled genom autochtonous busa from daying away in this area.

Poster C4.53

Genetic distances among Burlina, Holstein Friesian and Brown Swiss cattle breeds

C. Targhetta, M. De Marchi, C. Dalvit, P. Carnier, F. Gottardo, I. Andrighetto and M. Cassandro, University of Padova, Department of Animal Science, Viale dell'Università, 16, Agripolis, 35020, Legnaro, Padova, Italy*

Aim of this study was to assess genetic distances among Burlina and two world-wide breeds, such as Holstein Friesian an Brown Swiss. Burlina is a small size cattle breed with a black and white coat, reared on pastures of the Veneto region hills. About 273 dairy cows, reared in 20 herds of 6 provinces, are under national milk recording system. Milk yield of Burlina is used to produce local cheese. Phenotypic performances of Burlina dairy cows are (AIA, 2001): 4,083 kg of 305-lactation milk yield, 3.56% of fat, 3.21% of protein, 62 months of age at calving, 1.5 inseminations for calving and 389 days of calving interval. AFLP fingerprinting technique was applied to 50 animals per each breed considered in this study. Genomic DNA was extracted from whole blood and AFLP analysis was carried out according to the protocol described by Barcaccia et al. (1999). Burlina showed an higher genetically homogeneity respect to Holstein and Brown Swiss and a similar genetic distances with the two world-wide breeds. In addition, a factorial analysis allowed to distinguish clearly Burlina from the other two breeds suggesting that AFLP fingerprinting technique could be an efficient tool for characterisation of local populations and also to define a genetic traceability system for products of cattle breeds.

Theatre C6.1

High-fat rations and lipid peroxidation in ruminants; consequences on animal health and quality of products

D. Bauchart, D. Durand, V. Scislowski, Y. Chilliard and D. Gruffat, Herbivore Research Unit, INRA Clermont-Ferrand/Theix, 63122 Saint-Genès-Champanelle, France

Ruminant products (milk, meat) provide 20 to 35% of lipids and 40% of saturated fatty acids (SFA) consumed by humans. They are criticized by the medical profession because they are rich in atherogenic SFA and poor in polyunsaturated FA (PUFA) that prevent atherothrombosis. Therefore, experiments in ruminants aimed at favouring incorporation of dietary PUFA into muscle or milk fat. However, the high sensitivity of PUFA to peroxidation can be detrimental for the health and performances of animals but also for the nutritional (risk of toxicity) and sensorial (undesirable tastes) qualities of their products. Compared to corn silage, rye-grass, very rich in 18:3n-3, did not favour PUFA peroxidation in muscles, probably due to the efficient protection of their antioxidants (AO) (vitamin E, β carotene, ..). On the other hand, addition of oils or oil seeds (rich in PUFA) in diets reduced resistance of PUFA to peroxidation in muscles since AO in feedstuffs (dry forages, concentrates) are generally low. A similar tendency was observed in milk of cows given oil supplements but less marked since mammary gland takes up circulated AO more efficiently. Nutritional quality of FA in ruminant products can be improved by lipid supplements both in fattening and lactating animals. This strategy must integrate level of AO in diets and their protective activities towards lipoperoxidation. This guaranteed for better health of animals and better stability of lipids and their PUFA in products during their storage and processing.

Theatre C6.2

Milk quality and automatic milking: effects of free fatty acids

B.A. Slaghuis, J.A.M Verstappen-Boerekamp and C.H. Bos, Applied Research of the Animal Sciences Group of Wageningen UR, PO Box 2176, 8203 AD Lelystad, The Netherlands*

With the introduction of automatic milking (AM) systems, increased levels of free fatty acids (FFA) in milk were observed, which might result in off-flavours in milk and dairy products.

The aim of this study was to investigate the factors contributing to elevated FFA levels: influence of the milking frequency, technical parameters of the milking system, and finally, other farm management aspects.

Milking frequency was studied in a Latin square design with milking intervals of 4, 8 and 12 hours and showed increased FFA-levels for the shorter intervals. Milk fat and protein contents of milk were equal.

Technical factors were studied in a laboratory study using the milking machine components of an AM-system and a conventional milking system. The AM milking machine components caused an extra increase of 0,09 mmol/100 gram fat compared with conventional milking. These differences were found using susceptible milk. Susceptible milk is defined as milk with a high initial FFA level, caused by e.g. short milking interval or end of lactation of the cow.

Although most FFA problems were solved by adjusting milking frequencies and solving technical problems, in some situations high FFA levels remained. Other farm management aspects, like feeding regime, breeding and animal health are still subject of ongoing research and will be incorporated in the final paper.

Theatre C6.3

Protein binding properties of a polymorphic AP-1 element in the 5'-flanking region of the bovine α_{S1}-Casein gene are closely correlated with α_{S1}-Casein yield and milk composition

A.W. Kuss, T. Peischl, J. Gogol, H. Bartenschlager and H. Geldermann, Department of Animal Breeding and Biotechnology, University of Hohenheim, D-70593 Stuttgart, Germany*

In this study bovine milk proteins coded by specific loci were to be quantified and investigated in view of their individual association with DNA-variants in the 5'-flanking region, focussing on polymorphic protein binding sites. Blood and milk samples from cows of the German Holstein Friesian and Simmental breeds were analysed. Isoelectric focussing was used for allelic identification of polymorphic milk proteins, and alkaline Urea-PAGE in combination with densitometry was applied for quantification of milk protein fractions (α-Lactalbumin, β-Lactoglobulin, α_{S1}-, β- and λ-Casein) in a number of milk samples per cow. DNA variants were analysed in the 5'-flanking region of the α_{S1}-Casein gene by PCR-RFLP. Binding capacities of a variable AP-1 binding site were investigated by EMSA. The findings reveal differences in protein binding capacities between the specific DNA variants which show associations with the relative proportion as well as the amount of α_{S1}-Casein. The results represent an example for a potentially specific influence of variable regulatory elements on gene expression and gene product quantities, therefore being of relevance for further breeding on milk protein yield and composition.

Theatre C6.4

Raw milk for Protected Denomination of Origin (PDO) and long ripening cheeses: how to establish his suitability

G. Bertoni, L. Calamari, M.G. Maianti and B. Battistotti, Istituto di Zootecnica, Piacenza, Italy

Protected Denomination of Origin (PDO) cheeses have distinctive sensorial characteristics. When long ripening, these cheeses can be made from raw milk, on condition that possessing specific features, and that is processed through the "art" of the cheesemaker. The specific milk characteristics that ensure a high success rate for PDO cheeses are high protein content and good renneting properties, appropriate fat content with appropriate fatty acid composition and the presence of chemical flavours originating from local feeds. Moreover, an appropriate microflora is also of major importance. All these aspects would be precisely assessed in the milk. Unfortunately not always and not for all the aspects this is possible. However, quite recent DNA studies have allowed the definition of species and breeds; furthermore some organic volatile compounds of specific plant origin can be useful to trace the management and feeding systems. With concern to the milk properties to evaluate its suitability for cheese-making, some new indicators (enzymes, bacteria DNA ecc.) can be useful for a better evaluation of creaming activity, bacteriological traits and fermentation suitability of milk. Moreover, considering the interest of mastitis on the renneting property of milk, some new indicators, i.e. biosensors that can be applied on line during milking, could be of great interest; still for renneting, the Fourier transformed infrared analysis is very promising for casein and titratable acidity.

Muscle metabolism in relation to genotypic and environmental influences on consumer defined quality of red meat

D. Pethick[1], M. Cake[1], G. Gardner[2], G. Harper[3], J.F. Hocquette[4] and J. Thompson[2], Cooperative Research Centre for Cattle and Beef Quality : [1]Murdoch University, Perth, 6150, Western Australia, [2]University of New England, Armidale, 2351, New South Wales, Australia, [3]CSIRO, Livestock Industries, St Lucia, 4067, Queensland, Australia, [4]INRA, Herbivore Research Unit, Theix, France

The red meat Industry within Australia has recently undertaken large consumer based analyses of on farm and post farm gate factors influencing beef and sheep meat quality. This paper will initially describe the principle factors influencing consumer defined eating quality of meat derived from *Bos taurus* cattle including ossification score, intramuscular fat, muscle temperature at pH6, carcass hanging method, ageing time, muscle group, ultimate pH and cooking method . The paper will then discuss the influence of genetic selection and nutrition on the metabolism of muscle with particular reference to influences on bone maturity, intramuscular fat, glycogen metabolism and pH decline post slaughter. The paper will show that the metabolic activity of skeletal muscle is influenced by genotypic and environmental factors that powerfully influence the ability of the beef industry to meet optimal slaughter management and also to produce product which is appreciated by the consumer.

Theatre C6.6

C18:3n-3 content in beef meat as influenced by breed, diet and muscle location

C. Cuvelier, O. Dotreppe, J.F. Cabaraux, I. Dufrasne, L. Istasse and J.L. Hornick, Nutrition Unit, Veterinary Faculty, Liege University, Belgium

C18:3n-3 is an essential fatty acid present in animal products such as meat or milk. Beef meat could be considered as a source of C18:3n-3 for the consumer. The aims of the present work was to compare the C18:3n-3 content in different beef meat samples. Three groups of young growing fattening bulls (6 Belgian Blue double-muscled (BB), 6 Limousin (L) and 6 Aberdeen Angus (AA)) were offered a concentrate diet based either on barley or on sugar beet pulp. At slaughter, samples from the *Longissimus thoracis* (LT), from the *Rectus abdominis* (RA) and from the *Semi-Tendineus* (ST) were obtained for fatty acids analysis.

The fat content of meat was significantly affected by the breed (2.19, 4.50 and 6.26% DM for BB, L and AA respectively; $P<0.001$) and by the muscle location (6.45, 4.10 and 2.40% DM for LT, RA and ST respectively; $P<0.001$). The C18:3n-3 contents in the fresh meat followed the similar ranking in terms of breed (9.14; 12.14 and 15.94 mg/100g; $P<0.001$) and in terms of muscle location (16.3, 10.50 and 10.42 mg/100g; $P<0.001$). There were no significant dietary effects on both fat and C18:3n-3 contents, $P<0.326$ and 0.064). From the present results, it was concluded that, due to fat content changes, both breed and muscle location influenced the C18:3n-3 contents.

Theatre C6.7

Proteomics, an approach towards understanding the biology of meat quality

E. Bendixen[1], R. Taylor[2], K. Hollung[3], K. Ivar Hildrum[3], B. Picard[2] and J. Bouley[2], [1]Danish Institute of Agricultural Sciences, 8830-Tjele, Denmark, [2]INRA, Theix, 63122 Saint-Genès-Champanelle, France, [3]Matforsk, 1430-Ås, Norway

Sequencing the complete genome of many species within the last few years has allowed proteomics, like all other approaches to functional genomics, to rapidly expand. Proteomic technologies such as 2D gels, image analysis and mass spectrometry are presently developed with unexpected speed, and these technologies promise a quantum leap in many applications of life sciences.

Within the field of animal production the current focus is to use these technologies to describe the function of individual genes, and how heredity and environment act together to control cellular functions and thereby the physiological traits that are relevant for production of farm animals.

Understanding the biology behind the complex traits of meat quality still remains a major challenge in cattle breeding and production. A number of current projects combine genomic, transcriptomic and proteomic technologies with the aim to link the genotype and phenotype behind meat quality, and thereby develop molecular markers for meat quality traits.

Theatre C6.8

Assessment of the impact of production systems on sensorial quality of charolais heifers meat

M.P. Oury[1], D. Micol[2], H. Labouré[3], M. Roux[1], R. Dumont[1], [1]ENESAD, BP87999, 21079 Dijon Cedex, France, [2]INRA Clermont-Ferrand Theix, URH, 63122 St-Genès-Champanelle, France, [3]UMR ENESAD-INRA, BP86510, 21065 Dijon Cedex, France*

Meat industry and some consumers criticize rather often beef meat since sensorial quality and tenderness are variable and difficult to predict. The aims of the present research were, firstly, to assess the variability of meat quality, secondly, to relate this variations to management practices, finishing status of the animals and genetic of animals.

Muscles m. *rectus abdominis* of 41 charolais heifers were sampled and tested by a sensory panel. For each heifer, the production system was studied at farm level, and carcass data were studied.

Two different tenderness meat levels could be separated. The most tender level (I) have a higher carcass weight, more muscle and less bone in the 6th rib.

Heifers from level I have a quicker growth rate before weaning than those from level II. It seems that the genetic type of sire is significantly related with meat tenderness, in favour of beef type against breeding type. Finishing management, slaughter and weaning age didn't affect tenderness. Since differences between management practices were observed on meat tenderness and carcass characteristics in charolais finished heifers, one has to advise farmers to focus on sire genetic type, heifers growth and slaughter weight in order to produce meat with a higher tenderness.

Poster C6.9

Effects of dietary herb supplements for cows on feed conversion, milk yield and technological quality of milk

J. Kraszewski[1], J.A. Strzetelski[2] and B. Niwinska[2], [1]Department of Technology and Ecology of Animal Production, National Research Institute of Animal Production, 32-083 Balice, Poland, [2]Department of Animal Nutrition, National Research Institute of Animal Production, 32-083 Balice, Poland

Thirty Black-and-White cows were studied for the effect of dietary herb supplementation on the yield, composition, technological usefulness and nutritive value of milk. Cows yielding about 22 kg milk/day were fed a partial mixed ration (PMR). Diets were fed without herbs or with a herb supplement of 0.10 and 0.20 kg/day. The herb mixture contained: peppermint, wild pansy, chamomile, nettle, common yarrow and thyme. The herb supplement was shown to increase total protein, casein and fat in milk ($P \leq 0.01$). Analysis of technological usefulness of milk for cheese making revealed that the milk of cows receiving 0.20 kg herbs/day showed higher thermal resistance and better results in the fermentation-renneting test. It was shown that 70% of milk from the cows of that group was suitable for high-quality cheese production. The milk fat of cows receiving the higher herb supplement had the highest contents of CLA, $C_{18:2}$ and $C_{18:3}$ ($P \leq 0.01$) of all the groups. A daily supplement of 0.20 kg herbs to the diet had a favourable effect on the milk characteristics analysed.

Poster C6.10

Milk quality and automatic milking: effects of system and teat cleaning

B.A. Slaghuis, J.A.M Verstappen, R.T. Ferwerda, C.H. Bos and H. J. Schuiling, Applied Research, Animal Sciences Group, Wageningen UR, PO Box 2176, 8203 AD Lelystad, The Netherlands*

With introduction of automatic milking (AM) systems, some increases in total bacterial count (TBC) and free fatty acids were found. Contamination of milk originates from four main sources: inside udder, outside udder and interior of milking equipment and bulk tank.

The aim of this study was to investigate the efficacy of teat cleaning devices and to study the effect of two or three times per day system cleaning on milk quality.

The effect of teat cleaning was studied by artificial contamination of teats and by sampling teats before and after the teat cleaning procedure. AM systems have special teat cleaning devices. Per brand of AM-System, two or three farms were sampled.

The effect of system cleaning was studied on 13 farms by performing two or three system cleanings per day. From bulk tank milk quality was determined.

All systems showed effects. Differences were found for the effect of teat cleaning between brands of AM systems. The level of housing hygiene influenced the level of teat contamination.

Three systems cleanings per day resulted in a significant lower TBC (10.000 vs. 13.000), number of coliforms, thermodurics and psychrotrophs compared with two system cleanings per day. However the levels found for TBC were far within the penalty levels.

Poster C6.11

Breed effect on milk coagulation and cheese quality parameters of dairy cattle

M. Cassandro[1], M. Povinelli[1], D. Marcomin[1], L. Gallo[1], P. Carnier[1], R. Dal Zotto[2], C. Valorz[2] and G. Bittante[1], [1]University of Padova, Department of Animal Science, Viale dell'Università, 16, Agripolis, 35020, Legnaro, Padova, Italy, [2]Superbrown consortium of Bolzano-Bozen and Trento, Italy*

This study aimed to estimate the effect on milk coagulation and cheese quality traits of Holstein Friesian (HF) and Brown Swiss (BS) cattle reared in the Trento province. Data from 157 dairy cows (BS: 85 and HF: 72), herded in 9 mixed-breed dairy farms, were used for the estimation of the breed effect on the following milk coagulation traits: casein yield (CY), casein content (CC), acidity (SH50), calcium (Ca) and phosphorus (P). The milk produced by a random sample of 44 dairy cows (BS: 23 and HF: 21), herded in 3 mixed-breed dairy farms, was processed to "*Casolet*" cheese, a local 60 d aged product of Trentino region, using either milk by BS or HF or mixed milk by BS+HF. Cheese quality traits considered were: acidity (pH), yellow index for cheese paste (b*)and tenderness (Te). Results on milk coagulation traits did not show any statistical difference between BS and HF for CY, whereas for CC, SH50, Ca and P, BS performed better than HF (P<0,05). Concerning cheese quality traits, the breed effect was negligible for pH and Te, whereas cheese obtained by BS milk showed higher b* with respect to those obtained by HF and BS+HF milk (P<0,05).

Poster C6.12

Milk SCC and PMN as indicators of milk processability and subsequent cheese quality

B. O'Brien[1], B. Gallagher[1, 2], P. Joyce[2], W.J. Meaney[1] and A. Kelly[3], [1]Teagasc, Moorepark Research Centre, Fermoy, Co. Cork, Ireland; [2]Zoology Department, University College Dublin, Ireland; [3]Food Science and Technology Department, University College, Cork, Ireland*

This study investigated the effect of SCC and PMN (neutrophils) on the processability of milks from individual udder quarters.

Four milks of SCC $6x10^3$/ml and $920x10^3$/ml and PMN $<1x10^3$/ml and $700x10^3$/ml were selected. The low and high SCC and PMN milks were mixed in various proportions to give four artificial mix milks of SCC and of PMN $200x10^3$/ml and $400x10^3$/ml approximately. All milks were analysed for gross composition, renneting properties (RCT, K20, A60) and N-fractions. Miniature cheeses were manufactured and analysed for composition after 30 d and for proteolysis after 30 d and 90 d of ripening.

Fat, protein and lactose contents of the milks generally decreased with increasing SCC and PMN and there was no effect on nitrogen fractions in the milks or on total solids, moisture, NaCl or pH in the miniature cheeses. Elevated SCC and PMN had a negative effect on rennet coagulation of milk. The urea-PAGE electophoretogram indicated that the patterns of proteolysis differed quantitatively and qualitatively with increasing milk SCC and PMN during cheese ripening. High SCC and PMN milk resulted in reduced levels of residual intact α_{s1}- and β-caseins, indicating cell-associated proteinase activity.

In conclusion, high SCC and PMN milks had inferior composition and processing characteristics.

Poster C6.13

CLA content and n-3/n-6 ratio in dairy milk as affected by farm size and management

L. Bailoni, G. Prevedello, S. Schiavon, R. Mantovani and G. Bittante, Department of Animal Science, University of Padova, Agripolis, Viale dell'Università 16 - 35020 Legnaro (PD), Italy

Conjugated linoleic acid (CLA) and fatty acids from the n-3 series have been related to several beneficial effects on human and animal health. The objective of this study was to evaluate fatty acid profile and, particularly, CLA content and n-3/n-6 ratio in milk samples collected in 250 dairy farms located in Veneto region (northeastern Italy). The farms belonged to 21 classes according to 7 milk production levels (PL: <20, 20-40, 40-60, 60-100, 100-150, 150-300, and >300 t/year) and to 3 hygienic conditions of milk (HC: <20%, 20-50%, and >50% of milk test day records with SCC higher than 400.000/ml and/or with total bacterial count higher than 100.000/ml). Milk fatty acids were analysed by gas chromatographic method as methyl ester derivatives. Mean and standard deviation for CLA and n-3/n-6 ratio were: 0.467±0.123% on the total fatty acids and 0.336±0.156, respectively. Both parameters were affected significantly ($P<0.001$) by milk LP. CLA content and n-3/n-6 ratio decreased linearly ($P<0.001$) from the lowest LP (on average 0.552% on the total fatty acids and 0.591, respectively) to the highest LP farms (on average 0.415 % on the total fatty acids and 0.194, respectively). No significant effects of HC and PLxHC interaction were found.

Poster C6.14

Fatty acid composition and CLA content of milk fat from Italian Buffalo

P. Secchiari[1], M. Mele[1], A. Serra[1], M. Del Viva, L. Amante[2] and Zicarelli[2], [1]DAGA, Sezione Scienze Zootecniche, Università di Pisa, via del Borghetto, 80 Pisa. Italy, [2]Dipartimento di Scienze Zootecniche e Ispezione degli Alimenti, Università di Napoli, via Delpino, 1 Napoli, Italy

Aim of the present work was to characterize fatty acid composition and CLA content of milk from buffalo fed hay or green forage. Milk samples from 88 lactating buffalo belonging to seven herd located in South Italy were collected. Two samplings have been performed: the first when animals fed a ration included green forage and the second when hay was the only kind of forage in the ration. Milk samples were analysed for fatty acid composition. Results showed that when buffalo fed green forage the milk fat content of short chain fatty acids, medium chain fatty acids and saturated fatty acids significantly decreased, while that of long chain fatty acids, monounsaturated fatty acids and polyunsaturated fatty acids increased. The conjugated linoleic acid average content (CLA) of milk was enhanced by the inclusion of green forage in the diet (0.76 vs 0.42), in a similar way to what reported for dairy cattle. Also in buffalo species a wide individual variation of milk CLA content may be observed, when animals were submitted to a similar dietary regimen. The relationship between rumenic acid (RA, *cis*9, *trans*11 CLA) and vaccenic acid (VA, *trans*11, 18:1) resulted comparable to that reported for dairy cattle ($RA = 0.12 + 0.23 (VA+RA)$; $p< 0.01$; $R^2 = 0.89$).

Poster C6.15

How pasture-based systems may influence beef characteristics

I. Cassar-Malek, C. Jurie, I. Barnola, D. Dozias, D. Micol and J.F. Hocquette, INRA, Herbivore Research Unit, Theix, France

Extensive beef production systems at pasture are promoted to improve animal welfare and beef quality. However, the respective effects of animal mobility and grass feeding are unknown. Thus, this study aimed to compare the influence of the nature of the diet and physical activity on muscle characteristics. Four groups of 6 steers were fed either cut grass or maize-silage indoors with or without 1 hour of walking per day and one group of 6 steers was fed at pasture. Activities of glycolytic (Lactate dehydrogenase [LDH], phosphofructokinase) and oxidative (Isocitrate dehydrogenase [ICDH], citrate synthase [CS], hydroxyacyl-CoA dehydrogenase [HAD]) muscle enzymes were assessed in *Rectus abdominis* (RA) and *Semitendinosus* (ST) muscles. A transcriptomic analysis was performed to compare gene expression profile in RA and ST muscles between the pasture and maize-based diet without any mobility groups. Activities of oxidative enzymes were higher and LDH activity lower ($P<0.05$) in muscles from grass-fed animals, especially in RA for ICDH ($P<0.01$) and in ST for HAD ($P<0.01$). CS and HAD activities were higher in muscles from steers with high mobility, especially HAD in ST ($P<0.001$). Preliminary results suggest that some genes may be differentially expressed in pasture- *vs* maize-fed RA muscle. Among the latter genes, Selenoprotein W could be considered as a new indicator of grass feeding. In conclusion, muscle-specific changes and enzyme-specific adaptations were observed in response to changes in diet or physical activity.

Poster C6.16

Serum metabolites levels and meat water holding capacity in double muscled young bulls

C. Lazzaroni and A. Brugiapaglia, Department of Animal Science, University of Turin, Via Leonardo da Vinci 44, 10095 Grugliasco, Italy*

Water holding capacity (WHC) in one of the major aspects influencing the commercial evaluation of meat, as it has a great influence on its shelf life and eating qualities. Between factors influencing the meat WHC there is the animal physiology, and in this trial the influence of the level of some haematochimic parameters was studied. Two homogeneous groups of double muscled Piemontese (P, 24 subjects) and Belgian Blue (B, 24 subjects) young bulls, reared in the same environmental conditions until the same age and fattening degree, were slaughtered at the same weight (555.73±47.95 *vs.* 557.68±27.91 kg, in P and B). The week before the starting of slaughters a blood sample was drawn from each animal to determine the levels of Na, K, glucose, cholesterol, NEFA, urea, total protein, creatinine, ALT, AST, LDH and CK. After slaughters and 7 days of ageing in the same condition, from the right side of each animals was taken a sample of *Longissimus thoracis et lumborum* m. on which the WHC was determinate as water boiling losses (WBL). Positive and significant correlation were found between WBL and Na ($r=0.382$, $P<0.01$), K ($r=0.360$, $P<0.05$), glucose ($r=0.360$, $P<0.05$), cholesterol ($r=0.361$, $P<0.05$), urea ($r=0.459$, $P<0.01$), total protein ($r=0.415$, $P<0.01$), creatinine ($r=0.403$, $P<0.01$), ALT ($r=0.412$, $P<0.01$), so it could be interesting improve the knowledge of such relationships to be able to foresee the WHC of meat.

Poster C6.17

An association of leptin polymorphisms with carcass traits in cattle

J. Oprzadek, E. Dymnicki, U. Charytonik and L. Zwierzchowski, Institute of Genetics and Animal Breeding PAS. Jastrzebiec 05-552 Wólka Kosowska, Poland*

Leptin (LEP) is the hormone product of the obese gene that acts on central and peripheral tissues to modulate appetite and energy metabolism. The aim of this study was to characterize the genetic variations at LEP locus and to evaluate their allelic effects on carcass traits in Polish Friesian cattle. Genotypes of leptin (LEP) were analysed using the PCR-RFLP technique. Crude DNA was prepared from blood samples according to Kasai (1994). Primer sequences and PCR conditions used for LEP/*Hph*I was according to those described by Haegeman et. al. (2000) and for LEP/*Kpn*2I described by Buchanan et al.(2002). The digestion fragments were identified on 2% agarose gel. The allele frequencies at the loci studied were 0.60/0.30 for LEP C/T variants and 0.77/0.23 for LEP A/B variants. The weight of valuable cuts in carcass was highest in TT genotype bulls - 50.4 kg as compared to 49,2 in CC and 49,4 in CT animals. The genotype was significantly associated with higher fat content in carcass. In this study the effect of LEP/ *Kpn*2I genotype was observed only for dressing percentage. AA homozygotes had a highest dressing percentage as compared to AB and BB animals. In conclusion, the results presented here confirm the value of LEP as a marker for carcass traits in cattle.

Poster C6.18

Adipocyte fatty acid-binding protein expression and mitochondrial activity as indicators of intramuscular fat content in young bulls

I. Barnola[1], J.F. Hocquette[1], C. Jurie[1], I. Cassar-Malek[1], J.F. Cabaraux[2], C. Cuvelier[2] L. Istasse[2] and I. Dufrasne[3], [1]INRA, Herbivore Research Unit, Theix, France, [2]Nutrition Unit and [3]Experimental Station, Veterinary Faculty, Liège University, Belgium

Intramuscular fat (IMF) deposition influences many quality attributes of beef meat, especially flavour. Meat produced from young bulls contains low amounts of fat. IMF can be regulated by fat oxidation within mitochondria. Fat is mainly stored within intramuscular adipocytes. Adipocyte fatty acid binding protein (A-FABP) is considered to be a marker of intramuscular adipocytes This work aims to assess the contribution of markers of intramuscular adipocytes and mitochondria to variability in IMF. Belgian Blue (n=12) and Aberdeen Angus (n=12) young bulls were slaughtered at 18-20 months of age. Samples of *Longissimus thoracis* (LT) muscle were taken at slaughter. Quantification of the mRNA coding for A-FABP (in units per µg RNA) was performed in quadruplicate using Light Cycler technology relative to a standard curve. Cytochrome-*c* oxidase (COX) activity was determined as a mitochondrial marker. IMF content, A-FABP mRNA content and COX activity were 3.9, 7.9 and 2.0 fold higher in LT from Angus than from Belgian Blue young bulls. A-FABP mRNA content and COX activity explained 46 and 41% respectively of variability in IMF content ; together, they explained 72% of this variability. In conclusion, A-FABP expression and COX activity may be indicators of the ability of bulls to deposit intramuscular fat.

Poster C6.19

The fatty acid profiles of meat from calves fed two varieties of linseed

M.B. Zymon and J.A. Strzetelski, Research Institute of Animal Production, Krakow, Poland*

The objective was to determine the effects of two linseed cultivars with different FA proportions on the dietetic - health related quality of meat.

The studies comprised 18 bull calves (3 groups of 6 animals) of the Black-and-White breed (67.8% HF) aged from 7. to 90. days. The *ad libitum* fed concentrates contained: ground cereals (81-60%), soybean meal (15-8%), minerals (4%) and additionally 11% linseed var. Opal (group LO) or 10.5% linseed var. Linola (LL). The diet for group C contained no supplemental feeds. Calves were given milk (from 7. to 56. days of age) and fed liquid and constant feeds individually. At 42. and 90. days, 3 calves from each group were slaughtered and samples of *MLD* were taken. Content of SFA was higher in group C, and there were also lower UFA/SFA and DFA ratios and higher n-6/n-3 ratio in the meat fat. Content of the acids $C_{18:0}$, $C_{18:2\ n\text{-}6}$, $C_{18:3\ n\text{-}3}$, and CLA as well as $C_{20:4\ n\text{-}6}$, EPA and DHA was greater in experimental groups (esp. LL). During the period of milk feeding, meat fat contained more SFA than UFA, while a reverse ratio was found after weaning. Fat of meat obtained from calves at 42. day of age was characterized by lower UFA/SFA ratio than fat of meat obtained from calves at 90. day of age.

Poster C6.20

Understanding the effect of age and gender on the pattern of fat deposition in cattle

A.Pugh[1], B. McIntyre[2] and D. Pethick[1], Cooperative Research Centre for Cattle and Beef Quality: [1]Murdoch University, Perth, 6150, Western Australia, Australia, [2]The Department of Agriculture, South Perth, 6151, Western Australia, Australia

This experiment investigated changes in fat deposition, including intramuscular fat (imf), and total rib fat (subcutaneous, and intermuscular), with respect to the growth of Angus heifers and steers as they reach maturity. This was a serial slaughter experiment where 90 Angus steers and 75 heifers were grown in a feedlot and randomly allocated to a slaughter weight in a 200kg to 400kg carcass weight range. Measurements taken on each of the cattle at slaughter were total rib fat from a 6 rib fat dissection (4th/ 5th to $10^{th}/11^{th}$ ribs), imf content of the *m. longissimus thoracis et lumborum* (LTL), and Meat Standards Australia carcass grading data. Analysis of the results using general linear models showed that at any given fatness level, steers had a significantly higher level of intramuscular fat ($p<0.0001$) when compared to heifers, though at the same carcass weight there was no significant difference ($P >0.05$). There was a significantly linear increase ($P <0.0001$) in imf of the LTL as proportion of total dissectible fat. It can be concluded that steers are more efficient than heifers with respect to accumulation of imf, as they have a higher level of imf at the same fatness level.

Dietary tea catechins and lycopene in rabbits: effects on meat lipid oxidation

D. Tedesco[1], S. Rossetti[1], S. Galletti[1] and P. Morazzoni[2], [1]Department of Veterinary Sciences and Technologies for Food Safety, Via Celoria 10, 20133 Milano, Italy, [2]Indena S.p.A., Viale Ortles 12, 20139 Milano, Italy*

Twenty White New Zealand male rabbits aging 45 days and weighing on average 1.360±0.04 kg were equally divided into two groups. A control group was fed a basal commercial diet. The other group was fed the basal diet with 600mg/kg diet of a mixture of green tea extract (Green Select^{TM}) and lycopene (Indena S.p.A., Milan, Italy). Feed intake and body weight were monitored weekly. After slaughtering and after 7 days of storage in commercial conditions, muscle brightness and colour indices (lightness, redness, yellowness), pH, and thiobarbituric acid-reactive substances (TBARS) were evaluated on the *Longissimus dorsi*.
Feed intake, body weight and feed conversion ratio were not negatively affected by treatment. No differences between groups were found in meat pH and color indices both at slaughtering time and after 7 days of storage. TBARS were similar in the two groups at slaughtering time. After 7 days of storage TBARS were higher ($P<0.05$) in control group with respect to treated group, indicating an improvement in lipid oxidation status of meat from treated animals.
We conclude that treatment with green tea extract and lycopene have positive effects on oxidative status of rabbit meat and do not negatively affect animal performances.

SHEEP AND GOAT PRODUCTION [S]

Theatre S1.1

Sustainability of sheep and goat production in North European countries - from the Arctic to the Alps

Ólafur R. Dyrmundsson, The Farmers Association of Iceland, Bœndahöllin, 107 Reykjavík, Iceland

Sheep, and to a lesser extent goats, have contributed substantially to the grassland - based agricultural production in North European countries for centuries. Most of these countries are now members of the European Union and both EU policy and global trade negotiations with the World Trade Organisation are bound to influence in many ways the future development of the present production systems. This paper will review briefly the present situation with special reference to the sheep meat sector. Sustainability of production will be analysed and discussed in terms of several criteria such as economics and farm income, resource utilisation, environmental impact, landscape conservation, marketing of local value-added products and the maintenance of the rural population. Attention will be paid to the influence of increasing world trade and competition, as a result of WTO negotiations, and changes in the direction of support through EU policy reform (CAP), namely the decoupling of subsidies from production to sustainable development ("green box"). Finally, the economic problems experienced by sheep and goat farmers in countries ranging from the Arctic to the Alps will be discussed in the face of these challenges.

Theatre S1.2

Economics and profitability of sheep and goat production under new support regimes and market conditions in Central and Eastern Europe

R. Niznikowski, Warsaw Agricultural University, Ciszewskiego street 8, PL 02-786 Warsaw, Poland

Due to changes in the management system after the year 1989, headage of sheep and goats in Central and Eastern Europe underwent various changes depending on the country. Some countries were affected by considerable regress resulting in huge headage reductions, which caused changes in the significance of production as well as its support regimes

At that time sheep breeding underwent considerable changes. An especially severe regress in this branch of production could be noticed in Bulgaria and Poland. The best conditions for it were found in Slovenia. Moreover, stabilisation of the size of headage took place in the Czech Republic and Slovakia where, similarly to Slovenia, it was possible to organise a good support regime. In the remaining countries such as Hungary and Romania, headage underwent regress. Nevertheless, it was not as significant as in Poland. Fat lamb production is the leading direction of production after production of milk and wool.

Goat breeding, mainly of milk breeds, dominates in Romania and Bulgaria despite a fall in the level of e.g. milk production. Also, in Poland it is developing relatively fast after years of regress. A rather stabilised situation in this matter is noticed in the Balkan and Baltic states as well as in the Czech Republic and Slovakia. The leading direction of production is obtaining milk for consumption and processing.

Theatre S1.3

Mediterranean sheep and goat: an uncertain future

M. de Rancourt[1], N. Fois[2], M.P. Lavin[3], E. Tchakérian[4] and F. Vallerand[5], [1]ESA Purpan, Toulouse, France, [2]IZCS, Sardinia, Italy, [3]CSIC, Leon, Spain, [4]Institut de l'Elevage, Montpellier, France, [5]Thessalie University, Greece*

European sheep and goat production, in the Mediterranean areas, are important economic, environmental and sociological issues.

Our article aims at comparing the situations of the main small ruminant systems in South European regions (Spain, France, Sardinia and Greece), and their possible future evolutions.

On average, the income of milk systems tends to be higher than for meat systems thanks to a good market price. Moreover, the subsidies' dependence is higher for meat systems and for extensive systems.

However, milk systems seem more sensitive to the market situation, (US market for Italy, some foreign companies for Greece).

Sheep and goat productions being frequently the only possible production in less favoured areas are often fundamental to maintain social activities and to keep the vegetation out of fire danger. However, in Greece and Spain some intensive systems are settling in more fertile plain areas.

The success paths are rather uncertain considering the new CAP reform, but nearly all these systems seem to be dependant on the important European subsidies.

Nevertheless, we identified different success paths according to the systems and regions.

Such an uncertain future could be better anticipated if sheep and goat actors worked more collectively in Europe.

Theatre S1.4

Economics and profitability of sheep and goat production in Turkey under new support regimes and market conditions

O. Gürsoy, University of Çukurova-Facolty of Agriculture, Department of Animal Sciences, 01130 Adana, Turkey

Turkey has been one of the major sheep and goat producers of Europe and West Asia and North Africa (WANA) in the 20th century. Sheep and goats are no longer the major meat and milk supplying species and has been in a declining trend since early 80's. There are many factors vitally operating on this situation among which population growth, low genetic potential of the indigenous breeds, inappropriate breeding strategies, decrease in the pastures and rangelands, high rate of economic development, intensification of agriculture and livestock production, better education and demand for high status jobs, support regimes favoring poultry and dairy production, unfavorable market conditions for sheep and goat production.

Small ruminant production is practiced extensively and traditionally dominated by low input/low output system. Hence it is bound to diminish due to its stagnant and resistant to change nature. Unfortunately there are no serious efforts and support regimes under the liberal market conditions.

Theatre S1.5

Overview of the economic and social importance of the livestock sector in Cyprus

C. Papachristoforou and M. Markou Agricultural Research Institute, P.O. Box 22016, Lefkosia 1516, Cyprus*

The value of livestock production in Cyprus exceeds the amount of 220 million EURO and accounts for about 40% of the value of total agricultural production. The country is self-sufficient in milk and milk products, eggs, pig and poultry meat while production covers the demand for beef by 70-75% and for sheep and goat meat by 90%. Over the last 40 years, the production of meat increased more than 10-fold, of milk 6-fold and of eggs doubled. These achievements were the result of the gradual transformation from low to high input production systems in an effort to improve productivity to satisfy the increasing demand, to reduce production risks associated with frequent droughts, to decrease pressure on the environment from overgrazing, and to lower production costs. To day, in dairy cattle, pigs and poultry, the production is based on a small number of high input and medium to large size commercial farms using employed labour force, while in sheep and goats, farms are smaller and rely on family labour. In cattle and sheep, one predominant breed in each species is utilized for production, while in goats, two breeds and their crosses are used. Pig and poultry farms rely on imported breeds and hybrids. The per capita consumption of livestock products is among the highest in Europe leaving little room for further increases. The present trends relate to quality aspects of livestock products, introduction of new technology, improved production management and production methods friendly to the environment.

Theatre S1.6

Common agricultural policy reform and its effects on sheep and goat market and rare breeds conservation

G. Canali[1], Econogene Consortium[2], [1]Università Cattolica, Istituto di Economia Agro-alimentare, via Emilia Parmense, 84, 29100 Piacenza, [2]http://sirs.epfl.ch/projets/econogene/

In June 2003 the EU approved a very important reform of the CAP that will strongly affect the entire European agriculture.

This paper analyses, mainly in qualitative terms, the major issues related to the reform and its various effects on these sectors, also taking into account its possible direct and indirect effects in terms accelerating or decelerating the loss of biodiversity due to the extinction of already rare breeds. In fact, the Single Farm Payment, decoupled from any production, will include from 50% to 100% of the present amount of aid paid to sheep and goat breeders, according to decisions to be taken by each Member State; these decisions could deeply affect breeders' decision about continuing to keep sheep and goats or not. Another issue is the one related to the possibility of "regionalisation" of the SFPs, i.e. the possibility to distribute the same SFP per hectare to all farmers of a given homogeneous region, instead of applying the new scheme with reference to historical payments received by each farmer.

On the other hand new regulations will also allow Member States to introduce supplementary premium in case of transhumance, for preserving rare breeds, for promoting high quality food products. These new possibilities could balance possible negative effects of previous measures.

The activity of traditional sheep and goat breeders in Europe: an econometric analysis

M. Bertaglia[1,2,], J. Roosen[1] and the ECONOGENE Consortium[3], [1]Institute of Food Economics, University of Kiel, Olshausestrasse 40, 24098 Kiel, Germany, [2]Unit of Rural Economics, University of Louvain, Croix du Sud 2/15, 1348 Louvain-la-Neuve, Belgium, [3]see http://lasig.epfl.ch/projets/econogene/partners.html*

Within the framework of the ECONOGENE Project, a questionnaire was addressed to 604 farms breeding traditional sheep and goat breeds in 10 EU-25 countries. The questionnaire permitted to gather data on the socio-economic characteristics of the farm, as well as on marketing of traditional products, revenues and on characteristics of the breed(s) raised and the type of management. We have built an ordered logit model to explain the trend in the stock of animals of the breed raised on the farm on the basis of such explaining variables as the age of the farmer, the presence of a successor, the existence of particular marketing strategies for the most relevant production, etc. After discussing the particularities of this model, we present the variables found to affect production decisions in a significative way. These findings are discussed to provide some guidelines to policies for the conservation of animal genetic resources in marginal rural areas in Europe.

Acknowledgement: This work has been supported by the European Commission (Econogene contract QLK5-CT-2001-02461). The content of the publication does not represent the views of the Commission or its services.

Theatre S1.8

Effect of new support regime on sheep production profitability in the Czech Republic

M.Milerski[1], V.Mátlová[1], J.Kuchtík[2] and V. Mares[3], [1]Research Institute of Animal Production, Prátelství 815, 10401 Prague -Uhríneves, Czech Republic, [2]Mendel University of Agriculture and Forestry, Zemedelská 1, 61300, Brno, Czech Republic. [3]Sheep and Goats Breeders Association in CR, Palackého 1-3,61242, Brno, Czech Republic

Since the year 1990 huge transformations of sheep production systems in the Czech Republic have occurred. Wool sheep breeds (Merino), which represented 63% of total number of sheep kept in CR in 1990, have been replaced by breeds with meat or combined utilization. The number of sheep decreased from 430.000 in 1990 to 84.000 in 2000. In 1998 the direct subsidies paid by government per ewe kept in less favourable areas (LFA) were introduced. In years 2000-2002 the individual payment per ewe was 2000 Kc (Czech crowns - app. 60€) in mountain areas and 1500 Kc in sub-mountain areas. As the response to that arrangement the number of sheep increased up to 103.000 in 2003. Support regime for the year 2004 is different. Number of supported ewes is quoted. The national quota for the CR is 66.733 ewes espite the fact that the number of ewes over 1 year of age registered 31.12.2003 in the central database was 96.000. Due this fact the farmers in sub-mountain areas dispose of individual limits equal to 80% of the registered ewes and farmers outside the LFA only 60% of registered ewes. The individual payment per one individual limit (ewe) in 2004 will be 700 Kc. Approximate ratio between income for products sold and direct subsidies per ewe in mountain regions was 1:1 in 2002, while under new support regime is supposed to be 1:0.35.

Poster S1.9

Characteristics of traditional Sheep & Goat breeding in marginal European rural areas

S. Jones, M. Bertaglia, C. Ligda, A. Georgoudis , J. Roosen, G. Canali and, R. Scarpa, "ECONOGENE Consortium", 1 The Sheep Trust, Biology Area-8, University of York, P.O. Box 373, York, YO10 5YW, England*

This paper presents a description of the characteristics of farms raising traditional sheep and goat breeds in marginal European rural areas. A questionnaire was submitted and data collected from 604 farms in ten different EU-25 countries as part of the ECONOGENE project http://lasig.epfl.ch/projets/econogene/. This project, funded by the European Union within the Quality of Life V framework programme, combines molecular analysis of biodiversity, socio-economics and geostatistics to address the conservation of sheep and goat genetic resources and rural development in marginal rural areas in Europe.

The questionnaire consisted of four different parts. The first one addressed socio-economic characteristics such as family structure and revenues, outside-farm activities, and the role of subsidies in total revenues. The second and third ones included data on sheep and goat breeding, as well as the structure of revenues from different agricultural activities on the farm, and marketing strategies. The fourth section was dedicated to the particular breed(s) raised on the farm and comprised data on husbandry and relevant breed characteristics. Data is analysed so that aspects of raising traditional breeds are described. The current situations is discussed and future possibilities / trends for these farms are illustrated in the context of the Common Agricultural Policy and the role of subsidies.

Poster S1.10

Profitability of Hungarian goat farming

T. Németh[1], L. Branduse[2], M. Ábrahám[2] and S. Kukovics[2], [1]Hungarian Goat Breeders' Association, Gesztenyés u. 1. H-2053 Herceghalom, Hungary, [2]Research Institute for Animal Breeding and Nutrition, Gesztenyés u. 1. H-2053 Herceghalom, Hungary*

The goat industry is the smallest among the so-called big domestic animals sectors in Hungary. At present the estimated size is about 60-70 thousand does, which are kept by a little bit more than 7,000 holders. There are 8 different breeds bred in the country including 3 imported ones (Saanen, Alpine, Boer).

In order to clean the economic situation of the goat sector a survey was carried out covering 92 farms (belonging to 8 size categories) representing more than 15% of the national herd. The income and the cost factors were evaluated along with the performance level of the animals, as well as the lands exploited by the farms.

The average milk and meat production was relatively low (270 litres and 150% kidding rate).

The 75-80% of the total input originated from milk, and less than 20% came from meat. The biggest cost factors were feedstuffs (45%), and the labour (31%). Goat farming was proved to be profitable in the smallest category and in the ones with more than 100 heads.

Profitability of goat farming was recalculated based on the expected changes in the subsidizing system. Because of the limited size of lands used by goat farmers, the profitability of farming will be improved by only a few %.

Theatre S3.1

Pattern and manipulation of ovarian follicle development in sheep and goats

A.C.O. Evans , Department of Animal Science, Faculty of Agriculture and The Centre for Integrative Biology, Conway Institute of Biomolecular and Biomedical Research, University College Dublin, Belfield, Dublin 4, Ireland

Improving our understanding of the pattern of follicle development and its manipulation by exogenous hormones in sheep and goats will improve our ability to manipulate their fertility. Repeated ultrasonographic observations of the growth of individual follicles have shown that follicles develop in a wave-like pattern during oestrous cycles, with two or three follicular waves per cycle being most common in sheep and three or four follicular waves per cycle being most common in goats. The largest follicles that are present at luteolysis continue development to ovulation and these usually derive from the last follicular wave but in some cases also develop from the penultimate follicular wave.

Follicle development is often stimulated by using eCG to enhance the recruitment of small follicles, increase ovulation rates and increase the synchrony of oestrus after oestrous synchronisation treatments. Superovulation treatments also use gonadotrophin stimulation, the success of which seems to be related to the number of small follicles at the start of treatment. Long-term (12 - 14 days) progestagen treatments used for oestrus synchronisation can promote prolonged growth of ovulatory follicles in sheep but has not been studied in goats. The ovulation of aged follicles in cattle has a detrimental effect on fertility, but this relationship is less clear and seems to be less critical in sheep.

Theatre S3.2

Control of sexual activity in goats using photoperiod and male effect

J.A. Delgadillo, Centro de Investigación en Reproducción Caprina, Universidad Antonio Narro, Periférico y Carretera a Santa Fe, A.P. 940, Torreón, Coahuila, Mexico

Reproductive seasonality observed in all breeds of goats originating from temperate latitudes and in some breeds from subtropical latitudes, is mainly controlled by changes in photoperiod. Short days stimulate sexual activity, while long days inhibit it. This knowledge has allowed the development of photoperiodic treatments to control sexual activity in goats. In Alpine bucks, seasonality can be prevented by alternations between 1 month of long days and 1 month of short days. In Mexican male goats, sexual behavior can be stimulated during the non-breeding season using only 2.5 months of long days. In females, sexual activity can also be induced during seasonal anestrus using a sequence of long and short days. Another possibility to stimulate the sexual activity of females is the male effect. Does in seasonal anestrus can respond to the male effect if an intense sexual activity is previously induced in males by a photoperiodc treatment. Indeed, local male goats from subtropical Mexico treated only with artificial long days induce estrous behavior in about 90 % of the females, while control males cause this response in only 35 % of them. These photoperiodic treatments combined to the use of the male effect can improve the out-of-season estrous induction in goats.

Economical and technical consequences on French breeding schemes of a possible ban of hormones for sheep and goat reproduction

A. Piacere[1], G. Brice[1], G. Perret[1], JM. Astruc[1], G. Lagriffoul[1], B. Leboeuf[2], F. Barillet[3], L. Bodin[3], [1]Institut de l'Elevage, BP 18, 31321 Castanet-Tolosan France, [2]INRA-SEIA 86480 Rouillé France, [3]INRA-SAGA, BP 27, 31327 Castanet-Tolosan France*

In France, artificial insemination has been used since the 1970's when breeding schemes were set up for a few goats or sheep breeds. It's a fact that two key-steps of selection are easier to control when using insemination, i.e. planned matings and progeny testing of males. At present, several breeds of sheep and goats use at least 20000 inseminations for their programme of genetic improvement by selection or crossing.
The insemination procedure includes an hormonal treatment to synchronize the oestrus of females, but for a few years such treatments have been questioned and other methods have been tried. The "male effect" allows a relative synchronisation of oestrus ; but this technique implies that the breeder looks after the animals and handles them more, and that the inseminator comes more often for not planned on interventions. In large flocks, an increase in work time and costs, and the risk of a lower fertility, are limiting factors for the development of this technique. Should the ban of hormones in Europe become effective, it would deeply question our breeding schemes. The possible changes in organisation are discussed, as well as their consequences on the efficiency of selection.

Theatre S3.4

A comparison between royal jelly and PMSG effects on reproductive parameters of Awassi ewes

M.Q. Husein R.T. Kridli, S.G. Haddad and S.S. Al-Khatib, Department of Animal Production, Jordan University of Science and Technology. Box 3030, Irbid 22110, Jordan*

Two experiments were conducted to compare the effects of royal jelly (RJ) and PMSG on reproductive responses of Awassi ewes. In experiment 1, 42 ewes were allocated to three (RJ, PMSG and control) groups of 14 ewes each. Ewes were treated with CIDR-G for 12 days. Ewes in the first group received 12 equal doses of 400 mg/ewe/day concurrent with CIDR-G treatment. At the time of CIDR-G removal (0 h, 0 day), ewes in the second group were injected with 600 IU PMSG and no further treatment was given to control ewes. Ewes were exposed to four rams from 0 h for 5 days. Expression of estrus was similar among ewes of the three groups. Intervals to estrus were shorter ($P<0.01$) and pregnancy and lambing rates were greater ($P<0.05$) in RJ- and PMSG-treated than control ewes. Experiment 2 evaluated the effects of administering increasing doses of RJ. Thirty-seven 12-day-FGA-treated ewes were assigned to 5 groups to be treated with either 0, 250, 500, 750 mg RJ/day, or 600 IU PMSG. Overall reproductive responses to the RJ and PMSG treatments were similar. The 500 mg RJ produced greater responses than other RJ-treated groups. It was concluded that the administration of RJ in conjunction with exogenous progesterone in ewes can be as effective as gonadotropin administration. In both experiments the RJ or PMSG treatment produced high levels of estrus expression, shorter intervals to estrus and improved pregnancy and lambing rates.

Theatre S3.5

Reproductive and endocrine characteristics of delayed pubertal ewe lambs after melatonin and L-Tyrosine administration

K.A. El-Battawy, National Research Center, Animal Reproduction and A.I. Dept., Cairo, Egypt

This investigation was carried out to study the impact of melatonin and L-Tyrosine administration on the onset of cyclicity in delayed pubertal ewe lambs.
Fourteen delayed pubertal ewe lambs (age >16 month) were used in this study after being assigned randomly into three groups. First group (melatonin treated group, n=5), each lamb was administered 3 mg melatonin orally at 1600 from 1st July to 15th September while the second group (L-Tyrosine treated group, n=5), each lamb was administered L-Tyrosine at the level of 100 mg/kg b.w as a single oral dose. A third group (n=4) served as control. Lambs were exposed to mature ,fertile rams daily and blood samples were collected twice weekly.
The progesterone concentrations (P_4 evaluations) were significantly higher ($p<0.05$) in treated groups than control one. Ovarian activity, assessed by P_4 evaluations , showed that all animals in the first group came in estrus from them four got pregnant (80%) while three lambs from the second group came in heat from them two became pregnant (66.6%). On the other hand, non of the control lambs showed estrus.
In conclusion, this study confirms that the oral administration of melatonin and L-Tyrosine succeeded to a great extent to induced cyclicity in delayed pubertal ewe lambs and improved their reproduction.

Theatre S3.6

A study on using *ferula communis* (Chakshir) for oestrus synchronization in Shami (Damascus) Goats under east-mediterranean condition of Turkey

M. Keskin, O. Biçer, S. Gül and E. Can, MKU Faculty of Agriculture, 31034, Antakya, Hatay, Turkey*

This study was carried out at the Research and Training Farm of Mustafa Kemal University in Hatay province of Turkey. The experimental Shami goats (20 heads) was allocated into 2 groups as control (C) and treatment (T). Animals in both group were inserted with Progesterone-Sponges (method of Chrono-gest) for a period of 17 days. C group animals were fed a diet containing 16% crude protein and 2500 kcal/ME in dry matter. The other group (T) was given the same diet with addition of 10% *Ferula communis* (Chakshir) which is known with oestrogenic effect, from 10th day of sponge application to the end of treatment. Oestrus determination was done by mean of teasing buck. The teaser was introduced to the experimental goat group three times in a day after withdrawing of sponges. During the teasing period it was detected that *Ferula communis* caused earlier oestrus surge in T group than that in C group (27 h vs 41 h). But, the time interval between first and last animal in heat was rather large in T group. Thus, the interval observed 25 h and 21 h for T and C group, respectively. On the other hand, twinning rate was calculated as 60% and 30% for the same group order. As a result it was determined that *Ferula communis* has got an important effect on the oestrus synchronization ($P<0.001$) while no effect on litter size ($P>0.05$).

Theatre S3.7

Estimation of genetic parameters for litter size after natural and hormone-induced oestrus in sheep

M. Baelden[1], L. Tiphine[2], J.P. Poivey[1], J. Bouix[1], C. Robert-Granié[1] and L. Bodin[1], [1]Station d'amélioration génétique des animaux, Institut National de la Recherche Agronomique, BP 27, 31326 Castanet-Tolosan, France, [2]Département génétique, Service Sélection, Institut de l'Elevage, 149 rue de Bercy, 75595 Paris Cedex 12, France*

Genetic parameters for litter size after natural and hormone-induced oestrus have been estimated for Ovin Ile de France, Blanc du Massif Central and Mouton Vendéen-sheep, using REML methods in animal and sire BLUP models on litter size and normal scores. Whatever the samples and the model used, the results are very stable. For the three breeds, the value of the heritability of natural litter size (h^2_{ON}=0.10) is higher than the heritability of induced litter size (h^2_{OI}=0.06). The genetic correlation between the two types of performances is closed to 0.75 : natural and induced litter size can be considered as two traits controlled by a great number of common genes. This genetic correlation is higher than the former estimation (0.40). Nevertheless, the genetic correlation between litter size on first lambing after induced oestrus and the other litter size are not the same for the three breeds. In the future, this would be worth further investigations.

Poster S3.8

Goat spermatozoa survival rate and freezability in different extenders with/without melatonin supplementation

K.A. El-Battawy and W.S. El-Nattat, Dept. Animal Reproduction and Artificial Insemination, National Research Center,Egypt

The aim of this study was to compare preservation ability of five extenders-sodium-citrate-egg-yolk (SCY), tris-citrate-glucose-glycerol-yolk (TGGY), Tris-citrate-fructose-glycerol-yolk (TFGY), Cornell University (CU-16) and 14 extenders.by means of sperm motility (SM) %, alive sperm (AS) %, sperm-abnormalities (SA) % and acid phosphatase levels (AcP) of goat extended semen stored at 5°C for seven days. Also, the effects of melatonin (at the doses of 0.0, 10.0, 15.0 and 20.0 mg/100 x 106 sperm) on SM%, AS%, SA%, AcPlevels and post-thawing motility of goat extended semen stored at 5°C for seven days were studied.
The findings of the current investigation showed that CU-16 and 14 extenders ascertained high significant difference ($P<0.0001$), in the previous parameters than Tris extenders for goat semen preservation. Melatonin particularly the high concentrations (15.0 and 20.0 mg) improved SM% ($P<0.0001$), reduced dead sperm % ($P<0.0001$) and AcP levels ($P<0.0001$) as well as post-thawing motility. In conclusion, the use of CU-16 and 14 extenders was better than. Tris extenders for goat semen preservation. Moreover the inclusion of melatonin (particularly 15.0 and 20.0 mg) improved the extended goat semen quality and its freezability.

 Poster S3.9

GnRH responsiveness and LH secretion during the hormonal treatment for estrous synchronazation and the subsequent estrous cycle in ewes

J. Menegatos, S. Chadio, E. Xylouri, D. Kalogiannis and T. Lainas, Agricultural University of Athens, Department of Anatomy and Physiology of Farm Animals, Iera Odos 75, Athens, 11855. Greece*

Six ewes were treated during anestrus with MAP sponges for 14 days, followed by PMSG injection. Blood samples were collected at 3 hrs before, on day 0 (sponge insertion) and on days 10, 14, 16, 25 and 35 after sponge insertion every 20 min for 3 hrs for LH measurement. One ewe was injected with 50μg GnRH on days 1, 10, 14, 16, 21, 25 , 29 and 35 and blood samples were collected at -10, 0, 20,40,60,90,135,180 and 270 min after injection for the evaluation of pituitary responsiveness. Mean LH concentrations on days 10 and 14 were lower than those observed before and immediately after sponge insertion ($P<0.001$), while mean LH concentration on day 16 was higher than values on day 25 ($P<0.05$) and 35 ($P<0.001$). LH response after GnRH challenge on the day of sponge insertion was greater than the responses on days 10 ($P<0.05$) and 14 ($P<0.01$). During the subsequent estrus cycle LH response on day 16 (first day, follicular phase) was significantly ($P<0.05$) higher than responses on days 21, 25 and 29 (luteal phase). In addition the response on day 35 (follicular phase of the second induced cycle) differed significantly ($P< 0.01$) from the responses on days 21, 25 and 29 (luteal phase).

Theatre S4.1

Analysis of test day data in dairy sheep with random regression model

A. Komprej, G. Gorjanc, S. Malovrh, D. Kompan and M. Kovac, University of Ljubljana, Biotechnical Faculty, Zootechnical Department, Groblje 3, SI-1230 Domzale, Slovenia*

Data on daily milk yield (DMY), daily fat (DFC) and protein content (DPC) from local breeds in Slovenia (Bovec - B, Improved Bovec - IB and Istrian Pramenka - IP) were analysed with random regression model. Altogether 39125 records of 3133 ewes from 35 flocks were used. Analyses were done for all breeds together due to lower number of records in IB (6506) and IP (6004) breed. Days in milk was modelled with modified Ali-Schaeffer's lactation curve, while quadratic and linear regressions were used for parity and number of liveborn lambs, respectively. Additionally fixed effect of breed was included in joint analysis of all breeds. Common herd environment on test month, permanent environment and additive genetic effect were included in random part and modelled with Legendre polynomials for days in milk. Legendre polynomials from linear to cubic power were fitted. The eigenvalues of covariance functions showed that individual genetic curve of ewes explained up to 17.59 %, 24.66 % and 17.73 % of genetic variability for DMY, DFC and DPC, respectively. Heritability for DMY increased from 0.22 at the beginning of lactation to 0.32 around day 140 and decreased thereafter to 0.16.

Development of a test day model for milk sheep and goats under unfavourable structural conditions in Germany

B. Zumbach[1], K.J. Peters[1], R. Emmerling[2] and J. Sölkner[3], [1]Humboldt University Berlin, Dep. Animal Breeding in the Tropics and Subtropics, Philippstr.13, Haus 9, 10115 Berlin, Germany, [2]Bavarian State Research Center for Agriculture, Institute of Animal Breeding, Prof.-Dürrwaechter-Platz 1, 85586 Poing-Grub, [3]BOKU Vienna, Institute of Animal Sciences, Gregor-Mendel-Str.33, 1180 Vienna, Austria*

Milk sheep and goat breeding in Germany is characterised by limited population sizes which are distributed among 12 to 15 breeding associations. Selection is still based on phenotypic performance. Population structure is characterised by mainly small flocks and the use of bucks within herd. However, due to the active exchange of breeding animals there exist genetic links between herds and breeding associations.
A BLUP test day model is developed based on a two-step procedure proposed by Mäntysaari (1999). In the first step genetic and residual parameters for different lactation stages are estimated by multiple trait REML runs. These parameters are combined by an approach for iterative summing of expanded matrices (Mäntysaari, 1999). In the second step covariance functions are derived. The test day evaluation comprises about 20 000 and 200 000 test day records (milk yield, protein and fat content, somatic cell score) for milk sheep and goats, respectively, collected from 1990 to 2003.

Theatre S4.3

A rationale to introduce more traits in the Latxa breeding program

A. Legarra and E. Ugarte, NEIKER, Apdo 46, 01080 Vitoria-Gasteiz, Spain*

The Latxa dairy sheep breeding program has been selecting for milk yield during the last 20 years. Nowadays there is a continuous improvement and it is arrived time to undertake the inclusion of another traits in the breeding objective. Two groups of traits are considered. Those related to cheese yield (fat and protein contents) and those related to udders (udder shape scoring traits and somatic cell score). The economic weights for all of them are unknown and their recording is very expensive.
An experimental recording involving all these traits in some flocks has been in place since 2000. Estimated genetic correlations of milk yield with fat and protein contents, udder depth and teat placement are undesirable, while those of milk yield with fat and protein yields, udder attachment and somatic cell score are desirable or null. Therefore, a worsening would be expected for some traits. However, population genetic trends and EBVs of the AI rams show that in practice there is an improvement in all udder traits, which it is thought to come from "on farm" phenotypic selection of the prospective AI rams in regard of his mother's udder shape. As for fat and protein contents, trends and EBVs indicate a worsening.
The breeding objective should include fat and protein contents as in other breeds, while the inclusion of udder shape and somatic cell score is not so urgent.

Theatre S4.4

Body tissue development in two genetic lines of lambs during growth analysed by CT

T. Kvame, U.T. Brenøe and O. Vangen, Department of Animal Science, Agricultural University of Norway, P.O.Box 5025, N-1432 Ås-NLH, Norway

The value of slaughter lamb depends on total carcass weight and quantity and distribution of fat and muscle in different lamb cuts. For genetic improvement in carcass traits, there is a clear need for a better understanding of growth pattern of different body tissues.
Computer tomography (CT) is an accurate method to quantify body tissue in live animals at a given time. In this study, quantity and distribution of body tissues were analysed by CT in male lambs of a meat line (ML) and a control line (CL) of Norwegian White Sheep. Lambs were CT scanned in mid-June (CT1), in the end of July (CT2) and at weaning in the beginning of September (CT3). An average of 18, 22 and 23 CT images were recorded per animal at CT1, CT2 and CT3, respectively. Tissue weights were predicted by CT in each animal at each scanning.
Carcass tissue made up a larger proportion of total CT weight from CT2 to CT3 than from CT1 to CT2. CL lambs had more internal fat and non-carcass weight than ML at CT3. A larger weight of lean was observed of ML lambs for all scan events. The lean/bone ratio was greater for ML than CL at all scannings.

Theatre S4.5

The effect of the prion protein genotype on fertility traits, birth weights and early growth rate in Merinoland- and Rhönsheep

H. Brandt, G. Luehken, S. Lipsky and G. Erhardt, University of Giessen, Institute of Animal Breeding and Genetics, Ludwigstr. 21 B, 35390 Giessen*

The effect of the prion protein (PrP) genotype on number of lambs born, born dead, weaning losses and the birth weight from a total of 1133 lambings was investigated on ewes of Merinoland- (273) and Rhönsheep (71) of the Resaerch station Oberer Hardthof of the University of Giessen. Additionally on lambs of the breeding season 2002/2003 (179 Merinoland- and 75 Rhönsheep) the effect of PrP-genotype on birth weight, weight at 10 weeks and the daily gain was studied. The Merinolandsheep show a low frequency of the desired ARR-allele of 9.3 % for ewes and 22.9 % for lambs, while the Rhönsheep have a much higher frequency (68.3 % for ewes and 72 % for lambs). For the analysis three genotype classes (animals with 0, 1 or 2 ARR-alleles) were built to analyse a possible correlated selection response when selecting for homozygous ARR/ARR animals only, according to EC instructions. The breeds were analysed separately. No significant effect of the genotype class was found in both breeds neither for fertility traits nor for birth weights and the early growth traits.

Theatre S4.6

Awassi lamb fattening production systems in Syria

B. Hartwell and L. Iniguez, ICARDA (The International Center for Agricultural Research in Dry Areas), P.O. Box 5466, Aleppo, Syria*

Fattening of Awassi lambs, a fat-tailed sheep breed indigenous to Syria and Middle East, provides an income source for resource-poor farmers in Syria. A three-phase research methodology is being followed with a participatory research approach to assess opportunities for diversification and income generation for smaller resource poor sheep producers in Syria. The initial phase encompassed a Rapid Rural Appraisal study of existing fattening systems assisted by GIS techniques. This appraisal involved seven provinces of Syria. In selected sheep production areas a randomly selected number of farmers (n = 241) were interviewed regarding aspects such as feeding, management, labour input and marketing. Results identified that feeding costs (33%), marketing (32%) and disease (29%) were the main production constraints faced by farmers. Five types of production systems were described. Phase two comprised of controlled on-station feeding trials of lambs (< 3 months) at ICARDA's research facilities. These on-going trials include comparisons of farmers' feeding strategies with low cost diets using urea treated wheat straw and industrial by-products such as molasses. In collaboration with two selected farmer communities in Northern Syria, phase three plans the testing of low cost diet options with farmers, evaluating in a participatory manner the results and identifying/documenting the most appropriate and viable solutions. The final phase will be conducted in spring 2004.

Theatre S4.7

Analysis of serum lipoproteins fractions in the blood of sheep and moufflons

J. Buleca Jr.[1], L. Hetényi[2], P. Reichel[1], E. Pilipcinec[1], A. Bugarsky[1], L. Bajan[1], M. Húska[1], [1]University of Veterinary Medicine, Komenského 73, 041 81 Kosice, Slovakia, [2]Research Institute of Animal Production, Hlohovská 2, 949 92 Nitra, Slovakia

Lack of utility and unassumingness of breeding of asian steppe sheep resulted in the domestication of local wild sheep-moufflonin middle Europe. At the present time only a few varietes of moufflon live in the wild in the mountains of Sardinia and Corsica, (some in parks). Long term influences of domestication and selection caused changes of the physiological parameters and morphological properties, namely in the metabolism and the interior of the animals. In work presented here, the presence of four main lipoprotein classes in the blood serum of moufflons were detected. Every class differs by its physical characteristics: size, density and electrophoretical mobility. In densitometrical evaluation of the electrophoreograms the representation of individual fractions were compared. The major ratio of beta-lipoprotein fraction (LDL=47.16 %) was found. Numericaly similar values in representation of alpha-lipoproteins (HDL=25.63%) and the highest molecules of chylomicrons (24.21 %) were detected. Minor representation was found in pre-beta-lipoprotein fraction (VLDL=2.99 %). Analysis of densitometrical examinations show differences and variability in values of serum lipoprotein fractions in the blood of wild sheep of european type and those from domesticated merino sheep of the steppe type.

Poster S4.8

Selected parameters of Pomeranian breed lambs and crossbreeds of Pomeranian ewes with meat breed rams

H. Brzostowski, J. Sowinska, Z. Tanski, K. Majewska and J. Szarek, University of Warmia and Mazury, 10-718 Olsztyn, ul. Oczapowskiego 5, Poland*

Some parameters of the meat quality were studied in fifty days old lambs of Pomeranian Sheep breed and cross-breeds F_1 of Pomeranian ewes with Berrichon du Cher or Charolaise rams. Chemical content, physical characteristic, cholesterol and collagen content participation of endo- and exogenous amino acids in the protein as well as the content of saturated and unsaturated fatty acids in the intramuscular fat were studied in samples of quadriceps muscle of the thigh (*m. quadriceps femoris*).

Crossbreeding of Pomeranian ewes with Berrichon du Cher breed rams increased dry mass and crude ash content decreased cholesterol and collagen level and improved color and water hold capacity in samples of meat obtained from crossbreeds F_1. Charolaise rams used for crossbreeding increased dry mass, protein, exogenous amino acids and crude ash content improved water hold capacity and physiological maturity of meat decreased caloric value and cholesterol and collagen level in crossbreeds F_1. The present study show that crossbreeding of the Pomeranian ewes with Berrichon du Cher and Charolaise rams improves important indexes of meat quality in 50-days old crossbreeds F_1.

Poster S4.9

Influence of livestock system on growth performance of growing "Churra Tensina-breed" lambs

M. Joy, P. Albertí, S. Tort and R. Delfa, Unidad de Tecnología en Producción Animal, CITA-Aragón, Apdo. 727, 50.080-Zaragoza, Spain*

The objective was to study the influence of livestock production system, extensive *vs.* intensive, on growth performance of lambs of Churra Tensina breed. In the extensive system, 18 lambs were reared with their ewes on pasture from 7 days old to slaughter, with no supplementary feed. In the intensive system, 18 lambs were kept indoors with free access to concentrate, while their ewes were also indoors from 17 h to 8 h, receiving 500 g of barley as concentrate and the rest of time were given access to a pasture. At 50 days old, lambs from intensive treatment were weaned. When lambs had a liveweight of 22-24 kg they were slaughtered. Once a week diet intake and lamb weights were measured. The lamb weights at birth, at 52 days old (weaning for intensive treatment) and at slaughter were not affected by treatment ($P>0.05$). However the time spent from birth to slaughter was greater in the extensive treatment (85 d), while intensive lambs only needed 74 d ($P<0.01$). The average daily gain was higher for intensive treatment with 281 g/day and 242 g/day in extensive. This higher performance of intensive treatment was the consequence of supplementation of 26 kg of barley/ewe and 39.3 kg of concentrate/lamb, while the extensive treatment did not receive any supplementation.

Poster S4.10

Influence of livestock system on carcass quality of growing "Churra Tensina-breed" lambs

M. Joy, R. Delfa, G. Ripoll and P. Albertí, Unidad de Tecnología en Producción Animal, CITA, Apdo. 727, 50.080-Zaragoza, Spain*

The objective was to study the influence of livestock production system, extensive *vs.* intensive, on carcass quality of light Churra Tensina breed lambs. In the extensive system, 18 lambs were reared with their ewes on pasture from 7 days old to slaughter, with no supplementary feed. In the intensive system, 18 lambs were kept indoor with free access to concentrate, while their ewes were also indoor from 17 h to 8 h, receiving 500 g of barley as concentrate and the rest of time had access to a pasture. At 50 days old, lambs from the intensive treatment were weaned. When lambs had a liveweight of 22-24 kg they were slaughtered and the carcass weight ranged from 8.7 to 12.4 kg. At 24 h post-mortem, carcass classification, conformation and fatness degree were carried out following the Community Scale for Classification of Carcasses of light lambs (Regulation ECC 2137/92; 461/93). The pH and color values (by spectrophotometer) of meat were recorded at the lumbar area of *Longissimus dorsi*. Color of fat was recorded from subcutaneous and perirenal fats. Results showed that carcasses from intensive treatment were better classified than the extensive mainly as a consequence of the greater fatness degree ($P<0.001$). Treatment did not affect the pH values but affected hue values, although differences could not be appreciated by consumers.

Poster S4.11

Characterization of Massese lamb: growing, carcass and sample join composition

F. Sirtori, G. Campodoni, A. Olivetti, S. Rapaccini and C. Sargentini, Università di Firenze, Dipartimento di Scienze Zootecniche, Firenze, Italy

Massese is an Tuscan dairy sheep breed. The aim of this work was to individuate the slaughtering age of lamb to optimize its productivity and exploitation of the period of maximum milk production.

Thirty-five male lambs of single parity and from primiparous ewes are submitted to the trial. They were fed exclusively milk and were slaughtered at various ages, ranging from 7 to 42d. Right side and sample join was dissected. Statistical analysis was carried out by GLM procedure of S.A.S.

Average birth weight was 5.6kg and average daily gain was 350g/d for all the growing period. At two weeks of age, the live weight was 10 kg which is the typical weight for the slaughter of milk lamb. Moreover at two weeks carcass weight was 7kg, which is the threshold separating A category (milk lamb) and B category (light lamb) of EU classification. Carcass yield (without head, thoracic organ and liver) was 54.83%.

Percentages of shoulder, neck and hind leg on side remained constant over growth period and were 17, 8.6 and 31.6 % respectively; from 7 to 42d loin percentage increased while head percentage decreased. As regard sample join composition, total fat and lean percentage increased during the period (from 4.6 to 6.3% and from 59.6 to 62.6%) while bone percentage decreased (from 31.2 to 25.3 %).

Poster S4.12

Comparison of slaughter performance of Hungarian Merino, Ile de France F1 and Suffolk F1 lambs

P. Póti, F. Pajor, E. Láczó and Cs. Ábrahám, Szent István University, Faculty of Agricultural and Environmental Sciences, Department of Cattle and Sheep Breeding, Páter Károly str. 1, 2100 Gödöllö, Hungary*

After the fattening period 30 ram lambs and 30 ewe lambs of Hungarian Merino, 10 ram and 10 ewe lambs of Ile de France F1 and Suffolk F1 genotypes each were slaughtered per genotype with a live weight of 30-34 kg. Carcasses were evaluated for dressing percentage, weight of valuable carcass cuts, percentage valuable meat, bone to meat ratio, as well as meat conformation and fat cover (S/EUROP grading).

The slaughter performance proved best for the Suffolk F1 concerned dressing percentage (ram: 51,33±1,63 %, ewe: 52,07±1,22 %), percentage valuable carcass cuts (ram: 82,12±1,93 %, ewe: 81,72±2,19 %), percentage valuable meat (ram: 62,54±1,33 %, ewe: 62,72±0,62 %), and meat conformation. Suffolk F1 proved best for fat cover. Hungarian Merinos showed less favourable results, except for meat brightness (ram: 86,67±3,26 %, ewe: 85,77±4,51 %).

In Hungarian practice, the Hungarian Merino lamb is not able to be fattened to a high live body weight (30kg), because these lambs showed better results only meat brightness. Therefore, the Hungarian Merino lambs need to be upgraded with meat sheep breeds to improve meat yield and carcass composition.

Poster S4.13

The quality of meat obtained from Pomeranian and Ile de France breed lambs and their crossbreeds, stored in modified gas atmosphere

Z. Tanski, H. Brzostowski, B. Kasprowicz and J. Szarek, University of Warmia and Mazury, 10-718 Olsztyn, ul. Oczapowskiego 5, Poland*

The fresh and stored in modified gas atmosphere (80% N_2 /20% CO_2) meat quality of 50-days-old ram lambs singles of Pomeranian (P) and Ile de France (IF) breed and their crossbreeds (PIF) were examined in the present study.

The chemical content, physicochemical and sensory properties as well as the intramuscular fatty acids composition were evaluated in samples of quadriceps tight muscle (*musculus quadriceps femoris*). The quality of fresh meat and stored in modified gas atmosphere during 10, 20 and 30 days-long period in +2°C temperature were compared.

It have been found that IF meat had higher dry mass, fat and unsaturated fatty acids content and lower pH than meat obtained from P lambs. The quality parameters of PIF meat were similar to those, found in IF samples.

Prolongation of storage period in modified gas atmosphere increased dry mass and ash content caused lightening of meat color, decreased pH and worsened intensity and desirability of taste.

Poster S4.14

Determination of carcass composition in sheep by means of MRI cross sections

U. Baulain[1], W. Brade[2], A. Schön[2] and S. von Korn[3], [1]Institute for Animal Breeding Mariensee, Federal Agricultural Research Centre (FAL), Höltystraße 10, D-31535 Neustadt, Germany, [2]Chamber of Agriculture Hannover, Johannssenstr. 10, D-30159 Hannover, Germany, [3]Nürtingen University, Neckarsteige 6-10, D-72622 Nürtingen, Germany*

In German sheep breeding carcass quality is primarily measured by means of stationary progeny testing while performance testing of rams is still on a low level. To improve the latter, in vivo methods for determination of leanness are essential. Ultrasound measures of muscle and fat depth are still not standardized between and within German breeds. Aim of this study was to ascertain morphological differences and conformation of the back as a primal cut. A total of 44 carcasses of German Black Face (BF), Meat Merino (MM) and Merino Land sheep (ML) from performance testing stations was investigated by magnetic resonance imaging (MRI). A set of parallel transverse images (slices) was acquired covering the entire back. Loin muscle areas and distances were measured in consecutive slices and the volume of the entire muscle was calculated. The largest muscle area and the highest variation were observed approx. 10 cm caudal of the 13th thoracic vertebra. As expected, loin muscle area was higher in BF and MM with 15.6 and 16.2 cm^2, respectively compared to ML with 13.9 cm^2. According to the breed differences behind the last rib BF lambs had the highest volume of the loin muscle while the minimum was found for ML.

Poster S4.15

Comparison of meat quality between five lamb hybrids

M.Milerski [1] and J.Jandásek[2], [1]Research Institute of Animal Production, Prátelství 815, 10401 Prague -Uhríneves, Czech Republic, [2]Mendel University of Agriculture and Forestry, Zemedelská 1, 61300, Brno, Czech Republic.

The pH, electrical conductivity, contents of proteins, fat, ash, hydroxyprolin, amino acids and subjective assessments of colour, odour, fibrousness, texture, juiciness, and flavour of lamb meat (*musculus longissimus lumborum et thoracis)* were investigated. Ram lambs (n=100), hybrids between Merino ewes and Oxford Down, Texel, Charollais, Suffolk and Merinolandschaf rams were slaughtered at the average age of 119.2 days. Average weight of carcases was 16.3 kg. The data were evaluated by the analysis of variance using the last square method. The average pH 1 h post mortem of meat decreased continuously from 6.57 to pH 48 p.m. 5.64 and was not affected by hybrid combination. On the contrary the electrical conductivity increased from EC 3 h p.m. 2.62 $mS.cm^{-1}$ to EC 48 h p.m. 5.52 $mS.cm^{-1}$. The highest electrical conductivity of lean meat was detected in the progeny of Oxford Down sires. The average contents of proteins, fat, ash and hydroxyprolin were 19.94%, 1.70%, 1.02% and 0.065% resp. The content of fat in meat was affected by the age of lambs. Highly significant effects of the sire breed on the content of the majority aminoacids in the lamb meat proteins were found. The meat colour of the progeny of Oxford Down sires was subjectively classified as the best. The subjective assessment of the rest of studied descriptors was not affected by the breed of sire.

Poster S4.16

The effect of live weight, genotype and sex on carcass and meat quality of lambs

A.Cividini, S. Zgur and D.Kompan, University of Ljubljana, Biotechnical Faculty, Zootechnical Department, Groblje 3, 1230 Domzale, Slovenia*

Forty lambs of two genotypes (28 improved Jezersko-Solcava with Romanov (JSR) and 12 crossbreeds JSR with Texel (JSRxT)) were used to evaluate the effect of live weight, genotype and sex on carcass and meat quality. Lambs were fed with commercial concentrate and hay *ad libitum*, weaned at around 60 days of age. JSR lambs were slaughtered at 29 or 43 kg and JSRxT lambs at 45 kg live weight. Increased slaughter weight in JSR lambs had no significant effect on dressing percentage and carcass conformation. Carcasses of heavy JSR lambs were longer, wider and fatter, had higher percentage of neck, back and rib with flank and lower percentage of shoulder and hindleg, with redder meat (higher a* value). Crossbred lambs had better dressing proportion (48.7%) than purebred lambs (46.2 %) at the same live weight at slaughter. Their carcasses were shorter and wider, had better conformation and lower fatness scores, with higher chest and lower back percentage. They also had higher percentage of muscle and lower percentage of bone in hindleg. Genotype had a significant effect on b* value, being higher in crossbreeds. Females had better dressing proportion and carcass fatness. Males had higher proportion of neck, chuck and shoulder and lower proportion of back and lighter colour of meat.

Poster S4.17

The comparision of the quality of different meat parts from crossed lambs

Zs. Várszegi, A. Jávor and M. Árnyasi, University of Debrecen, Centre of Agricultural Sciences, 4032. Debrecen, Böszörményi str. 138. Hungary*

Chemical, organoleptic and fatty acid composition analyses of three different meat parts as: leg, loin and shoulder were undertaken in lambs. The highest dry matter content was in the loin, and was significantly higher than in the case of the leg and the shoulder. The highest protein content was in the loin, while the less crude protein was in the shoulder. The crude fat content was the same in the leg and loin meat, but was higher in the shoulder meat. The shoulder had the lowest haemoglobin content, which is significantly lower than that in the leg and lion meat of the lambs. The low haemoglobin content means lighter meat, which is preferred by the consumers.

There was only a minimal difference between leg meat and shoulder samples in the connective tissue content. These results were significantly lower than in loin meat ($P<0,01$). The meat of the shoulder had the highest saturated fatty acid content, much higher than in the loin. It confirms that the loin is healthier than the meat of the shoulder.

Poster S4.18

Adipose tissue development and metabolism on lambs supplemented with retinol

A. Arana[1], M. Alzón[1], P. Eguinoa[2], J.A. Mendizábal[1], J. Ochoa[2] and A. Purroy[1], [1]ETSIA, Universidad Pública Navarra, Campus Arrosadía, 31006 Pamplona, Spain, [2]Instituto Técnico de Gestión Ganadera S.A., Serapio Huici 22, 31610 Villava, Spain*

To study the retinol effect on lamb adipose tissue development and metabolism during growth and fattening, two experiments were accomplished. In the first experience retinol was administered to a Rasa Aragonesa lamb group (221.000 vs. 42.000 UI/week in Control group) from 60 to 100 days of age. There were not observed significant differences on growth and carcass parameters, fat, size and number of adipocytes and lipogenic enzyme activities of omental, perirenal, subcutaneous and intermuscular adipose depots. In a second experience higher retinol level was administered to another lamb group of the same characteristics (1.000.000 vs. 42.000 UI/week in Control group) from birth to 100 days of age. Lamb growth parameters, fat, adipocyte size and number and lipogenic enzyme activities of omental and perirenal adipose depots were not affected. However, higher amount of subcutaneous and intermuscular fat were observed on lambs supplemented with high level of retinol. Subcutaneous adipocyte size was bigger on retinol supplemented lambs indicating a higher hipertrophy process. These results suggest that it could be necessary to administrate retinol in high quantity and from birth to slaughter in order to observe its effect on adipose tissue. New studies should be undertaken to better understand the effects of retinol on lamb adipose tissue development.

Poster S4.19

The effect of crossing ewes of Turcana breed with rams of Suffolk breed on meat production parameters

I. Calin[1], V. Bacila[1], M. Vladu[2] and Livia Vidu[1], [1]University of Agronomic Sciences and Veterinary Medicine Bucharest, Animal Science Faculty, Technology Department, 59 Marasti boulevard, sect. 1, 011464, Romania, [2]University of Craiova, Agricultural Science Faculty, Phytotechny - Animal Science Department, 15 Libertatii street, Romania*

In this paper we present the results on the effect of crossing ewes of Turcana breed with rams of the specialized meat breed Suffolk, in the first generation of halfbreds and the estimation of quantitative parameters of meat under conditions of intensive fattening. Compared with the lambs of Turcana breed, the improved halfbreds resulted a higher fattening aptitude.
The results pointed out that the young halfbreds responded well to the intensive fattening system. The fattening parameters were: 198.5 g. daily weight gain and 38.84 kg final body weight, feed consumption depending on age and weight.
These parameters show that the young halfbreds of Suffolk with Turcana have a very good fattening aptitude and this system should be practiced in Romania.

 Poster S4.20

Fattening performance and carcass characteristics of Awassi sheep and Damascus goat yearlings

Sabri Gül, Osman Biçer, Mahmut Keskin and Askın Kor, MKU Faculty of Agriculture, 31034, Antakya, Hatay-Turkey*

To compare sheep and goat fattened at the same age in terms of performance and carcass merits was the aim of this study conducted at the Research Farm of Mustafa Kemal University of Hatay-Turkey. Awassi sheep and Damascus goat yearlings were used as the animal material. They were put into individual boxes and fattened for 91 days. After fattening 4 animals from both species were slaughtered, jointed and dissected according to "EAAP Standard Method of Sheep Carcass Assessment" and "The Standard Method for Goat Carcass Jointing and Tissue Seperation", respectively. Initial and final live weights were found to be 34.7±1.7 and 59.7±1.8 kg for Awassi and 20.7±0.5 and 38.7±0.6 kg for Damascus, respectively. Daily gain and daily mean food consumption were determined as 275.5±14.6 and 2067± 51.3 g and 197.8±8.1 and 1489.3±26.0 g with the same genotype order. Slaughter weight, hot carcass weight and dressing percentage were 57.1±0.89 kg, 30.9±0.06 kg and 54.1±0.50 % for Awassi and were 35.7±0.90 kg, 18.8±070 kg and 52.7±1.69 % for Damascus, respectively. In terms of carcass composition, bone, muscle, subcutaneous fat and intermuscular fat content in percentage were 16.0±0.62, 49.3±0.58, 21.8±1.60 and 11.2±0.72 for Awassi; 18.2±0.62, 55.6±2.70, 9.9±1.93 and 15.2±1.05, for Damascus, respectively. Intermuscular fat content of goats was more than sheep carcasses ($P<0.05$) which is desired in terms of eating quality.

Poster S4.21

Sex and carcass weight effects on Serrana kids carcass and meat characteristics

S. Rodrigues[1], V. Cadavez[1], R. Delfa[2] and A. Teixeira[1], [1]Escola Superior Agrária, Instituto Politécnico de Bragança, Apt. 1172, 5301-855, Bragança, Portugal, [2]Unidad de Tecnologia en Producción Animal, Servicio de Investigación Agraria, Diputación General de Aragón, Zaragoza, Spain*

The aim of this work was to evaluate the sex and carcass weight effects on carcass joints proportion, tissues measurements and meat pH and colour of kids. Twenty seven male and 23 female Serrana kids, a Portuguese breed, were used. Kids were slaughtered after 24 h fasting. Carcasses were cooled at 4 ºC for 24 h, halved and the left side divided into eight standardised commercial joints. Colour, pH and tissues measurements were taken on the surface of muscle *longissimus* at the 12th-13th ribs level.

Female kids presented higher breast proportion ($P<0.05$), kidney, knob and channel fat ($P<0.05$) and j measurement ($P<0.01$) than males. Breast proportion increased ($P<0.01$) with carcass weight increasing. A decrease in meat luminosity was observed with carcass weight increasing. At the carcass weight ranges studied the joints proportions and meat properties do not differ significantly between sexes. Carcass and meat characteristics do not change much from 3 to 5 kg, however at 7 kg carcasses presented a higher degree of fatness.

Poster S4.22

Effects of castration on growth performance and carcass characteristics of fattening Awassi lambs

S. Haddad, M.Q. Husein and R.W. Sweidan, Department of Animal Production, Jordan University of Science and Technology, P.O. Box 3030, Irbid 22110, Jordan

The effect of castration on the growth performance and carcass characteristics of Awassi lambs was studied. Two groups of intact and castrated lambs (BW = 21.0 kg) were offered a fattening diet ad libitum for 60 days after which all lambs were sacrificed. Dry matter intake was higher ($P < 0.05$) for the castrated lambs (1226 g/d) as compared to intact lambs (1098 g/d). Feed to gain ratio for castrated lambs (4.40) was higher ($P < 0.05$) as compared to intact lambs (3.91). Cold carcass weight and dressing percentages were unaffected ($P > 0.05$) by castration and averaged 19.7 kg and 53.6%, respectively. As a percentage of cold carcass weights, intact lambs had greater ($P < 0.05$) leg weights (32.9 %) as compared to the castrated lambs (31.0 %). Castrated lambs' legs had more ($P < 0.05$) total fat (902 g) and more ($P < 0.05$) subcutaneous fat (740 g) as compared to the intact lambs (766 and 612 g, respectively). Intact lambs had greater ($P < 0.05$) percentages of lean (55.6) and bone (16.1) in their legs as compared to the castrated lambs (53.0 and 15.1, receptively).

Castration reduces feed efficiency, increases subcutaneous fat and decreases carcass leanness. Therefore, castration is not recommended for Awassi lambs under an extensive feeding system.

Poster S4.24

Growth ability of lambs of Awassi sheep breed & their crossbreeds with exotic rams of Charollais & Romanov breeds in Jordan

M. Momani Shaker[1], A.Y. Abdullah [2], I. Sáda[1] and R.T. Kridli[2], [1]Czech University of Agriculture Prague, Institute of Tropics & Subtropics, Czech Republic, [2]Jordan Univ. of Science & Technology-IRBID, Faculty of Agriculture*

The aim of the study was to evaluate the effect of crossing Awassi sheep breed with mutton Charollais and prolific Romanov sire breeds on growth ability of lambs from birth to 130 days of age in Jordan. In the years 1999 and 2002 the live weight was determined in 424 lambs (86 Awassi, 122 Awassi x Charollais F_1 crossbreds, 142 Awassi x Romanov F_1 crossbreds, 32 F_{10} crossbreds (A75Ch) and 42 F_{10} crossbreds (A75R)) at birth and subsequently every fortnight until 130 days of age. Average live weight of lambs at birth was 4.42 kg and at the age of 15, 30, 45, 60 and 130 days 8.45 kg, 11.93 kg, 15.03 kg, 18.35 kg and 26.19 kg, respectively. Genotype of lambs affected ADG, live weight of lambs at birth, 15, 30, 45, 60 and 130 days of age significantly ($P \leq 0.05$-0.001).

In generally, results obtained from this study revealed that crossbreeds F1 A x R, crossbreds F1 A x Ch and crossbreds F_{10} (A75Ch and A75R) excelled pure Awassi lambs in such daily weight gain and total weight gain.

Poster S4.25

Morphological evaluation of the testicles of young Santa Inês rams submitted to different regimes of protein supplementation and drenching

O.A Carrijo Junior, C.M. Lucci, C.M. McManus, H. Louvandini, R.D. Martins and C.F.M. Veloso, University of Brazilia, Agronomy and Veterinary Medicine Faculty, C.P. 04508, Brazilia, DF, 70910-900 Brazil*

The study investigated the effect of protein supplementation and parasite load in relation to testicular morphology, in young Santa Inês male sheep. Twenty-four lambs, were distributed in four treatments: APd (dewormed + high protein), APn (not dewormed + high protein), BPd (dewormed + low protein), BPn (not dewormed + low protein) during eight months. At the end fo the experiment histological cuts of testicles were taken and cell count analyzed by computer. The group (BPn) presented larger number of spermatogonias, but a reduced number of the other cells parameters (cells of Sertoli, spermatocytes, spermatids and spermatozoa) and presented smaller diameter and circumference of the seminiferous tubules and shorter seminiferous epithelium when compared to other treatments; there was a significant correlation among the diameter of the tubules with the height of the epithelium and circumference of the tubules. The correlations between the diameter of the lumen and the circumference of the lumen, as well as height of the epithelium and the circumference of the tubules also presented high and positive correlations (P>0.70). In conclusion, the high protein influences positively spermatogenesis, and the non-deworming associated to a poor protein diet worsens almost all reproductive parameters of young Santa Inês rams

---

Poster S4.26

Biometric appreciation of ovine animals exterior of Merino type studied at the didactic resort Banu-Maracine, Craiova

M. Cola, University of Craiova, Faculty of Agronomy, Romania

The body measurements of animals permit an objective appreciation of the process of increasing of different corporal regions dimensions which constitute the exterior of the animal. At the same time, biometry facilitates the lenow ledge of the increasing and developing corporal level of the ovine animals at certain ages, establishing thus the morphological features at different races and types of ovine animals.

The body measurements describe well-developed animals from the point of view of the corporal conformation, with widths and perimeters that show large animals capable of taking a high productive effort and a strong muscular mass.

The standard lacks underline the fact that the studied material is well-consolidated from the genetical point view. Comparatively with the data in the technical specialized literature the studied ovine animals register values which are lower than those of the Merino de Palas, especially regarding the dimensions of the widths and perimeters.

In comparison with the Transylvanian Merino, the studied material registered positive differences at all the features, excepting the depth of the thorax.

The biggest differences were registered at the withers height, trunk length, chest width and thorax perimeter.

Poster S4.27

Differences among the various domestic Tsigai sheep populations: divergences in body measurements

S. Kukovics[1], A. Molnár[1], A. Jávor[2], A. Gáspárdi[3] and Z. Dani[1], [1]Research Institute for Animal Breeding and Nutrition, Gesztenyés u.1., 2053 Herceghalom, Hungary, [2]University of Debrecen Centre of Agriculture, Böszörményi út 138., 4032 Debrecen, Hungary, [3]Szent István University, Faculty of Veterinary Sciences, István u. 2., 1053 Budapest, Hungary

The authors studied the different characteristics of the domestic Tsigai sheep populations. In the first part of the work the indigenous and the milking type of the breed was compared based on official production performance results (prolificacy, meat and milk production).
In the second part of the study the body weight and the body measurement data of 8 different flocks were compared in order to determine the differences. The individual body weight and body measurements (like height of withers, trunk length, depth of thorax, rump I and II width, length of nose, length of ear, and the biggest width of ear) of 100 ewes from each flocks were analysed in order to compare the various populations.
Microsoft Excel and SPSS for Windows 6.5 software was used in the processing of data.
Big differences were found between the two types of the breed. Based on the results of body measurements, the flocks belong to indigenous variant were differing in several traits from each others and from the milking variants.. Apart of these more transitional types were observed between the indigenous and the milking type of Tsigai sheep.populations.

Poster S4.28

The morpho-productive parameters of the Carpathine goat breed reared in small private units in the south of Romania

I. Raducuta, University of Agronomical Sciences and Veterinary Medecine, Bucharest, Str. Marasti 59, Romania

In Romania, goats are bred separately or together with sheep, and their numeric distribution in the territory varies from one area to another, depending on nutrition resources and tradition. At present, there are a few large farms (over 100 head) and many private units, where the number per herd is very small (1-5 head/unit). In these last units, the goats are kept only for family self-consumption. Almost 98% of the total numbers are owned by these units. The aim of this experiment is to study the main morpho-productive parameters of the Carpathine goat breed reared in these small units from the south of Romania (n = 50 females and 10 males). The results show that the mean body weight is 41.25 kg in the adult females and 49.40 kg in the adult males, which are smaller by about 10% than the breed standard. On the other hand, we consider that the mean milk production of 294.6 liters obtained during 217 days of lactation, is good and close to the mean of the breed standard. This milk production satisfies the breeding and development requirements of kids until weaning and also leaving about 220.9 liters of processed milk per goat. However, there is a large variability in milk production (range: 151 and 462 liters). These results recommend the urgent application of selection according to individual performance.

Poster S4.30

Genetic diversity of European goats as measured by AFLP markers

R. Negrini[1], S. Joost[2], E.Milanesi[1], J. Bernardi[1], M. Pellecchia[1], R. Caloz[2], P. Ajmone-Marsan[1] and ECONOGENE CONSORTIUM[3], [1]Università Cattolica del Sacro Cuore, Faculty of Agriculture, Institute of Zootechnics, via Emilia Parmense, 84, 29100 Piacenza, Italy, [2]Swiss Federal Institute of Technology, Laboratory of Geographical Information Systems,1015 Lausanne,Switzerland, [3]http://lasig.epfl.ch/projets/econogene/*

Preliminary data produced within the Econogene EU project are presented on the genetic diversity of 45 goat breeds from 14 European countries. Three highly polymorphic and robust AFLP primer pairs were selected and used to type over 1200 animals. A total of 102 polymorphic bands were scored across breeds. Alpine goat individuals were sampled in France, Switzerland and Italy, to investigate the power of AFLPs in assigning individuals to highly related populations. In addition, Alpine goat was double sampled in France and Switzerland, to test the effect of sampling on the output of the analyses. Data have been analysed by standard and Bayesian methods to investigate the relationship between breeds, the partition of the total diversity between individuals, sampling areas, breeds and Countries, and the extent of genetic admixture of breeds. Geographic maps of markers frequencies across Europe and clines of diversity are also investigated and presented.

Poster S4.31

Relationships between genetic distance measured by AFLP markers and ecological distance among goats breeds of Italy, Swiss, France and Spain

S. Joost[1], R. Negrini[2], E. Milanesi[2], R. Caloz[1], P. Ajmone-Marsan[2] and [3]ECONOGENE Consortium, [1]Swiss Federal Institute of Technology, Laboratory of Geographical Information Systems,1015 Lausanne,Switzerland, [2]Università Cattolica del Sacro Cuore, Istituto di Zootecnia, Via Emilia Parmense 84, 29100, Piacenza, Italy, [3]http:/lasig.epfl.ch/projets/econogene/*

In farm animals, genetic diversity has a geographic component due to the origin of the breeds. Therefore, genetic distances between breeds might be correlated to the geographic pattern. To investigate the effect of natural barriers and the spatial distribution of genetic variation, ecological distance may be more effective than the euclidean one.

In this preliminary study we have i) designed an ecological distance mainly based on a high resolution digital elevation model and ii) investigated the correlation between the ecological distance calculated between the site of sampling of 747 goats and their genetic distance calculated from binary AFLP (Amplified Fragment Length Polymorphism) molecular markers.

To accomplish this task, we focused on autochtonous and cosmopolitan goat breeds reared in Italy, France, Swiss and Spain, exploiting the information produced in the framework of the Econogene project (QLK5-CT2001-02461) that integrates molecular genetics, socio-economics and spatial analysis to deliver guidelines for a sustainable conservation of animal genetic resources in marginal rural areas.

Poster S4.33

Genetic trends of milk production traits in Slovak sheep populations

M. Oravcová[1], M. Kovac[2], D. Peskovicová[1], M. Margetín[1], J. Huba[1] and P. Polák[1], [1]Research Institute for Animal Production, Hlohovská 2, 949 92 Nitra, Slovak Republic, [2]University of Ljubljana, Biotechnical Faculty, Zootechnical Department, Groblje 3, 1230 Domzale, Slovenia*

Milk recording data were used to analyse genetic trends in Slovak sheep populations of Improved Valachian and Tsigai breeds. A three-trait test-day model was used to estimate breeding values of tested animals separately for each breed. To account for the shape of lactation curve, modelled from individual test day measurements, a fixed regression submodel was considered. Traits under study were daily milk yield, and fat and protein contents. Breeding values for daily milk yield were adjusted for standardized length of milking period (150 days). Genetic trends were computed as changes in average breeding values over the years of birth.
For both breeds, two different tendencies in genetic changes of milk yield were found. In period of 1996-1998, the average breeding values were close to zero with negligible changes over the years. In period of 1999-2002, slightly higher genetic changes were observed. Concerning phenotypic trends, no significant differences among the birth years were found out. These findings indicated that selection criterion based on dam's phenotype seems to be insufficient. More sophisticated methods need to be adopted in sheep breeding strategies in Slovakia.

Poster S4.34

Changes in some selected parameters of sheep milk in the course of lactation of crosses of Eastfriesian and Improved Valachian breeds

P. Zajicova and J. Kuchtik, Mendel University of Agriculture and Forestry Brno, MSM 4321 00001, Department of Animal Breeding, Zemedelska 1, Brno 613 00, Czech republic*

Evaluation of changes in some selected parameters of sheep milk (percentages of dry matter (DM), fat (F), protein (P), casein (C), whey protein (WP) and lactose (L)) during the lactation was carried out using samples of milk obtained from 20 ewes (crosses of Eastfriesian and Improved Valachian breeds). All ewes under study were on the first lactation. Milk was sampled in the morning on the 46th, 74th, 102th, 132th, 162th and 190th day of lactation. Analyses were performed using standard laboratory methods. The average contents of DM, F, P, C, WP and L ranged, in dependence on the day of lactation, from 16.66 to 19.98 %; 5.45 to 7.70 %; 5.24 to 6.64 %; 3.95 to 5.08 %; 1.21 to 1.57 % and 4.78 to 5.13 %, respectively. Average contents of milk components mentioned above for the whole lactation were as follows: 18.5 %; 6.44 %; 5.93 %; 4.53 %; 1.40 % and 4.99 %, respectively. The average daily milk performance for the whole lactation was 0.58 kg.

Poster S4.36

Relation between the number of psychotroph bacteria and the susceptibility of goat milk fat to lipolysis

N. Strzalkowska, J. Krzyzewski, E. Bagnicka, A. Józwik and Z. Ryniewicz, Institute of Genetics and Animal Breeding Polish Academy of Sciences, Jastrzebiec, 05-552 Wólka Kosowska, Poland*

The aim of this study was to examine the relation between the number of psychotroph bacteria and the susceptibility of goat milk fat to lipolysis. The investigations were conducted on 20 Polish White Improved dairy goats, over the whole lactation. The animals were fed according to INRA standards. The milk samples were taken from individual goats once a month and analysed for the SCC, fat, protein and lactose. Also FFA were determined directly after milking and after 72 hours of storing at 4° C. Also the total number of lipolytic bacteria was determined. The data obtained were analysed statistically using the SAS program. The highest number of bacteria was recorded during the summer season, while the lowest during spring. The storage of milk over 72 hours at a temperature of 4° C led to changes the in the number of lipolytic bacteria in milk, the most pronounced increase being observed during the spring season. After milking the level of FFA was low. However, after 72 hours of storage at 4° C, differences were observed between milk samples obtained in different seasons. The level of FFA directly after milking was significantly higher in the milk from older goats. As regards LnSCC, significant differences were observed both for individual season and for lactations.

Poster S4.37

Effect of αs_1-casein variants on yield and physicochemical properties of Sarda goat milk

G.M. Vacca[1], V. Carcangiu[1], M.L. Dettori[1], L. Chianese[2], G. Longu[1] and P.P. Bini[1]*
[1]Dipartimento di Biologia Animale, via Vienna 2, 07100 Sassari, Italy, [2]Dipartimento di Scienza degli Alimenti, Parco Gussone, 80055 Portici, Italy

In order to evaluate the effect of αs_1-casein genotype on milk yields and the main physicochemical properties of goat milk, 600 Sarda breed goats were investigated. Individual milk samples were collected in the middle of lactation and analysed to determine: fat, lactose, total protein (infra-red method), pH, titratable acidity, cryoscopic index. Phenotypical analysis of αs1-casein was performed by UTLIEF in the 2.5-5.4 pH range (CEE 690/92), vertical disc-PAGE at pH 8.6 and immunoblotting. The following phenotypes were detected: AA (9%), AB (20%), AE (0.3%), BB (63.3%), BF (2.2%), EE (0.3%, EF (3.2%), EI (0.3%), FF (0.7%) and FI (0.7%). The results were analysed by ANOVA. Statistical analysis showed significantly higher ($P<0.01$) production values, for those subjects with AB phenotypes (>1 kg/die), while AB and BB subjects displayed the highest ($P<0.01$) fat values (5.51% and 5.47% respectively), the highest protein values (4.56% and 4.45%) and the lowest pH ($P<0.05$) (6.71 and 6.70 respectively). The high protein and fat content and the high strong allele frequences indicate that the Sarda goat milk is particularly suitable for cheesemaking. Moreover, the fact that strong phenotypes can give high-yield production, points out the potentiality of Sarda goat breed, for which it is possible to develop selection strategies in order to improve quantitative aspects not disregarding quality.

Poster S4.38

Efficiency of n-3 fatty acid transfer into milk

D. Kompan, A. Oresnik, J. Salobir, K. Salobir, M. Pogacnik, University of Ljubljana, Slovenia

Four groups of goat (A, C, D, E) were used in a trial to examine theeffect of supplementation with different fatty acids (FA): alfalinolenic (ALA; group A), docosahexaenoic (DHA; group D), eicosapentaenoic (EPA; group E) and control group (group C) no supplementation. The supplementation of the diet and its effect on milk composition and fat content of goat milk fat was studied. The study included 62 goats, from 28 to 105 days after parturition. After a ten-day adaptation period, during which each milking was recorded, the animals were divided into four groups. Three groups were fed with the ALA DHA and EPA supplemented diet (20 g/day) in the five successive days, until the end of experiment. Milk yield from each milking was recorded from day 1 up to day to day 20 and after that each five day recording, two days milking was done. We concluded that the supplementation of separate FA to the diet affects the FA composition of lipids in goat milk. Depending on the type of added FA, the effects appeared differently during the supplementation and at different times after the supplementation. The passage rate of supplemented FA to milk was from 13 to 21 %. The addition of n-3 FA decreased middle chain FA (from 55% to 45 %), and monounsaturated FA (from 24 % to 22 %), but increased poliunsaturated FA (PUFA) from 4 % to 14%.

Poster S4.39

Volatile compounds of ewes' cheese as affected by different pastures

M. Povolo[1], G. Contarini[1], M. Mele[2], L. Casarosa[2], A. Serra[2] and P. Secchiari[2], [1]Istituto Sperimentale Lattiero-Caseario, via A. Lombardo, 11, 26900 LODI, Italy, [2]Dipartimento di Agronomia e Gestione dell'Agroecosistema - Sezione di Scienze Zootecniche Università di Pisa, Via del Borghetto, 80 - 56124 Pisa, Italy

The research deals with the effects of different pastures on milk and cheese flavor compounds. A herd of Sarda sheep, bred in a lowland of the Tuscany region, was fed with three different types of pasture (mixture of *Lolium perenne* and *Trifolium squarrosum*; natural pasture; *Avena sativa*) and the milk obtained was used to produce a six months old cheese and whey cheese (Ricotta). The volatile fraction of the grass, milk and cheese samples was extracted by Solid Phase Microextraction (SPME) and analyzed by GC/MS. Moreover, flavor composition of cheeses was investigated during ripening. Data were subjected to multivariate statistical analysis. The presence of particular plant species in the natural pasture significantly affected the volatile composition of the cheese. Differences in the volatile compounds development during ripening were also observed according to the type of feeding.

Poster S4.40

Leptin secretion can explain that the fecundity in sheep is more influenced by live weight variation than by the live weight itself

J.L. Bister, F. Wergifosse, C. Pirotte, P. Pirot and R. Paquay, The University of Namur, Animal Physiology, B-5000 Namur, Belgium*

An experiment conducted with 48 Ile de France adult ewes submitted to various nutritional schemes shows that the fecundity (litter size 0 to 4) is significantly related to the live weight (LW, n=48, R^2=0.19, SS) and more closely to the weight change during the 2 weeks before fertilization (Δ LW, n=48, R^2=0,25, SS).
These well-known relationships are related to the Leptin plasma level at the time of fertilization: the determination coefficient (R^2) between Leptin level and the Litter Size is 0.24 and highly significant, between Leptin and LW its value is 0.26 (n=138, SS) and significant, as well between Leptin and Δ LW (n=138, R^2=0.18, SS).
Further analyses showed that a significant relationship between LW and the plasma Leptin level exists only when the Δ LW is positive. For this statistical analysis, the data were sorted according to Δ LW in 7 groups of 12 couples: Leptin-LW from a Δ LW = -6±1kg to Δ LW = +6±1kg. The slopes of the relationships between Leptin and LW in these couples are closely and linearly related to the Δ LW (R^2=0.82, n=7) as well as their linear correlation coefficients r (R^2=0.62, n=7).

Poster S4.41

Dietary preference of grazing sheep and fallow deer for white clover and fescue in different proportion in adjacent monocultures

E. Piasentier, Elena Saccà, S.Filacorda and S.Bovolenta, Dipartimento Scienze Produzione Animale, Università di Udine, Via S.Mauro 2, 33010 Pagnacco, Italy*

The experiment aimed at evaluating the grazing selectivity in terms of both herbage intake and feeding behaviour, under vegetation conditions allowing animals to gather monospecific diets. It was performed on a pasture consisting of two plots, which comprised monocultures of *Festuca arundinacea* alongside monocultures of *Trifolium repens* either 25% fescue: 75% clover or 75% fescue: 25% clover by area. Each plot was divided in two subplots, equal in area (0.50 ha) and spatial distribution of vegetation, respectively assigned to six one-year-old animals of either ovine or *Dama dama* species. The trial was repeated at a 9-day interval by exchanging the subplot of herds grazing on the same plot. Diet composition, herbage intake and digestibility were estimated by the n-alkane method while the time spent grazing either grass or legume was evaluated from visual scan of each animal.
On both pasture types, mixed diets were composed, largely dominated by the most extended plant species. Deer selected higher proportions of clover than sheep (53 vs 37%; P<0.05) and displayed a preference for the legume. In comparison with the lighter deer, sheep showed a higher total intake (1.62 vs 1.12 kg DM/d; P<0.01), thus attaining a comparable feeding level per unit of metaboloic wight (90 vs 91 g DM/kg $LW^{0.75}$), by consuming more fescue at a higher rate of intake.

Poster S4.42

Evaluation of performance qualities of Romney Marsh sheep kept under a pasture system in the Czech Republic

L. Stolc, J. Maxa, B. Dolenská, V. Drevo, L. Nohejlová, Czech University of Agriculture, Department of Cattle Breeding and Dairying, Prague, Kamycka, 165 21, Czech Republic

After the year 1990 crucial changes in the composition of sheep breeds occurred in the Czech Republic. The whole breeds have been substitued with breeds with improved meat and milk production. As well, the number of breeds of the combined production has almost doubled (57 %) since 1990. One of these breeds introduced recently in the Czech Republic is Romney Marsh.
The objective of this article is to publish the results of performance of this breed kept on pasture for all the year in marginal areas (average altitude 700 m) of the Czech Republic.
As parameters of performance, we chose the number of new-born lambs (2,170 ±0,222) and the number of weaned lambs (1,734 ±0,217) per lambed ewe as influenced by year, ram, the age of ewe and the order during the birth. We also observed average birth weight (4,7 kg) and average weight at 100 days of age, (30,4 kg ±1,0 kg). We evaluated average daily gains (260 g ±9,9g) with regards to sex, number of lambs in litter, according to ram and age of ewe.
Our results show, that this introduced breed (Romney Marsh) can be successfully used in the severe climatic conditions of the Czech Republic.

Poster S4.43

Research activities for improving small ruminant production in State Farms of the Ministry of Agriculture in North Cyprus

Okan Güney, Çukurova University, Faculty of Agriculture, Dept. of Animal Science,01330, Adana-Turkey*

This paper deals with small ruminant production systems at Ercan, Margo and Guzelyurt State Farms in North Cyprus. In addition to this, the results of our research (Güney et al., 2002; Görgülü et al., 2003; Güney et al.2003a; Güney et al. 2003b; Güney et. al, 2004) presented to national and international meeting are summarized.
The first study was carried out to compare the fattening performance of Awassi, Chios and Awassi x Chios crossbred lambs reared at Margo State Farm. The second study was conducted to compare feeding methods in respect to milk yield and milk composition and to assess dietary preferences of lactating suckling Damascus goats receiving feed ingredients as multiple choices under confinement conditions in Northern Cyprus. The objective of the other study was to determine haemoglobin and transferrin genotypes of Chios, Awassi and Chios x Awassi crossbred ewes at the Margo State Farm in Northern Cyprus. The objective of the 4th study was to determine the types of haemoglobin and transferrin found in Damascus goats and to demonstrate their relationship with different performance traits. The purpose of the last study was to increase milk production, prolificacy and growth performance of the Damascus goat using a selection programme.

Poster S4.44

Working of sheep breeding systems associated with agriculture in Algeria: management of reproduction and dynamic of the pastoral resources

Khaled Abbas, Unité de Recherche de Sétif (INRA Algérie)

The association of sheep and cereals in Algerian semi arid zones represents a very varied agropastoral system. The yearly rain level plays a major role in the availabilityof pasture. The ovine system valorizing these availabilities adjusts by the seasonal modulation of reproduction performance. In this work, treatments of reproduction improvement (synchronization, super ovulation and desaisonnement) were tested. Body score has been used to judge the level of the animal - pastoral resources interphase and to explain the observed results. These are even partial but show that the adaptation of the studied techniques to these farming systems remained complex. The pursuit of the research will lead to the definition of decision support tools for farmers.

Poster S4.45

The influence of management on compartmental pastured at Corriedale sheep breed in Roumania

C. Dinu[1], M. Parvu[1], M.Th. Paraschivescu[2] and I. Radoi[3], [1]University Spiru Haret, Veterinary Medicine Faculty, Bucharest, Jandarmeriei 2, sector 2, [2]Institute of Bovine Breeding, Balotesti, sos. Bucuresti-Ploiesti km.21, [3]USAMV, Veterinary Medicine Faculty, Bucharest, Splaiul Independentei 210, Roumanie*

It was study, comparatively, the compartmental pastured of a flock with 200 sheep Corriedale and Tigaie. The sheep was keep up on a hill pasture, the feed based only on grass. In the place of sedentary pastured characteristic at Corriedale sheep, it was use the active pasturing in common front, the change of place was making with the dogs. As a resultat of these managerial conditions, at Corriedale sheep it established that the total pastured duration decrease with 3 hours, the ruminant periods frequency decrease with 55% (as a resultat of ruminant interruption), but their time increase with 35 minutes. The water intake was not influence by the news-pastured conditions. The bringing forth at Corrriedale sheep are phasing whole the year, but at Tigaie sheep they are grouping in the springtime. Those in why it appears a different growth rate with establish an increase of mortality. Because of permanent move that Corriedale sheep are expose they appear metabolic disturbs and adapted modifies at posterior limbs.

Extensive or easy-care management systems for sheep flocks: A contradiction in terms?

P.J. Goddard[1], C.M. Dwyer[2] and A. Waterhouse[2], [1]Macaulay Institute, Craigiebuckler, Aberdeen, AB15 8QH, UK, [2]Scottish Agricultural College, West Mains Road, Edinburgh, EH9 3JG, UK*

Human-animal interactions represent a crucial interface in animal management. Recognition is increasing of the importance of understanding how human action can affect animal behaviour and welfare and how animals perceive human behaviour. In extensive systems with little regular opportunity to habituate the sheep to the stockperson, aversive handling mitigates against the development of "positive" relationships. Human-animal interactions have an especially important impact at lambing time. In more extensive systems, the stockperson has limited contact with the lambing ewe and it is appropriate that genotypes chosen exhibit good survival traits. Ewes expressing behavioural survival traits may fare better if there is no human disturbance as this may prejudice the progress of parturition and mother-young bonding. Appropriate intervention reduces ewe and lamb mortality for genotypes that show poorer survival behaviours.
For easy care systems, very low levels of intervention have animal welfare costs until selection has been successfully achieved. The alternative is to seek systems where human intervention is carefully targeted, providing support for genotypes well adapted for their systems. Knowledge of interactions between genotype and management could direct breed selection, increasing productivity and animal welfare, whilst reducing dependence upon human intervention.
High standards of animal welfare in extensive systems require significant management inputs, with emphasis on targeting inputs rather than high regular supervision costs, as availability of labour declines.

Theatre SM6.2

Economics and animal welfare in extensively managed sheep production systems

A.W. Stott and C.M. Milne, Scottish Agricultural College, Craibstone, Aberdeen, AB21 9YA, UK

Recent reform of the CAP aims to help EU farmers become more market-oriented and competitive by de-coupling subsidy payments from production. For extensive sheep farming systems this may require a reduction in inputs such as labour, supplementary feeding or veterinary treatment. Such reductions might aim to maximise farm business profitability or minimise the costs of cross-compliance and may have consequences for animal welfare. These consequences must be considered part of the decision making process as high standards of animal welfare may attract a premium in the market place and be supported through the Rural Development Measures, minimum standards will be enforced through legislation including cross-compliance. However, identifying and quantifying the different elements of animal welfare in systems of different intensity is very difficult. To address this issue we used adaptive conjoint analysis (Green and Srinivasan (1990), Journal in Marketing. Oct. 3-19) to rank alternative management policies for UK sheep farming systems in terms of impact on animal welfare. Scores were obtained by asking respondents (sheep farmers or sheep welfare specialists) via a computer-mediated survey to make trade-offs between alternative policies and between different levels of 5 attributes (labour, housing, veterinary treatment, nutrition, gathering). The impact on costs and farm incomes was then estimated in order to examine conflict or complement between economics and animal welfare as different components in sheep systems are reduced in input.

Theatre SM6.3

Vigour and maternal behaviour of sheep in two different husbandry systems

E. Moors[1] and R. Wassmuth[2], [1]Institute of Animal Breeding and Genetics, University of Göttingen, D-37075 Göttingen, [2]Thuringian State Institute of Agriculture, D-07743 Jena, Germany*

The economy of sheep production is determinably influenced by the number of lambs being reared. One of the main factors of lamb vitality is the maternal behaviour of the ewe. The aim of the present study was to examine vigour and maternal behaviour in 110 ewes and 287 lambs in two different husbandry sytems. One group was kept outside on pasture, the other being housed in a barn. The evaluation of vigour based on measurement of birth weight, daily gain, first suckling and first standing. The evaluation of mothering ability consisted on licking behaviour (i.e. start of licking activity, licking intensity) and a maternal behaviour score. No significant differences between the two husbandry systems were observed. Maternal behaviour and vitality of lamb were positively correlated. Lambs which were intensively and early licked after birth were significantly faster in standing and sucking than the lambs which were treated less intensively. The phenotypic correlation between start of licking and first standing was 0.21 ($p \leq 0.05$), between start of licking and attempt to suck 0.31 ($p \leq 0.001$) and between licking intensity and attempt to suck the phenotypic correlation was 0.25 ($p \leq 0.01$). There was a significant influence of birth type, birth progress and faecal egg count (FEC) on the maternal behaviour. The licking intensity was significantly higher in ewes with lower FEC.

Theatre SM6.4

Ewe fear response and lamb growth

E. Moors[*] and M. Gauly, Institute of Animal Breeding and Genetics, Albrecht-Thaer-Weg 3, University of Göttingen, D-37075 Göttingen, Germany

In several studies maternal behaviour have been positively associated with lamb growth rates. In sheep maternal behaviour is characterized by the rapid establishment of individual recognition of the lamb through the use of different sensory modalities. The aim of the current study is to investigate the effects of ewe fear response to handling on growth rate of lambs ($n = 185$). Ewes behaviour was described based on two different parameters on 4-point scales. A - the flight response of the ewe to a person approach on the day of birthing (1 - flees at approach of the person, shows no interest in her lamb(s) and does not return when person leaves, 2 - as 1, but returns when person leaves, 3 - stays with lamb(s), shows aggressions against person, 4 - not influenced by the approach of a person) and B - the behaviour of the ewe if their lambs on day of birthing are taken away (1 - does not follow, 2 - does follow in distance, 3 - follow close, 4 - attacks person). No significant correlations were found between the age of the ewes and the scores. Daily weight gain of the lambs until 4 and 8 weeks of age were in tendency ($p > 0.05$) higher in animals with scores A 2 and A 3 and B 3, when compared with the others.

 Theatre SM6.5

Effect of heat stress on production in Mediterranean dairy sheep

R. Finocchiaro[1], J.B.C.H.M van Kaam[1], B. Portolano[1] and I. Misztal[2], [1]Dipartimento S.En.Fi.Mi.Zo-Sezione Produzione Animale, Viale delle Scienze, 90128 Palermo, Italy. [2]Animal and Dairy Science Department, University of Georgia, 30605 Athens, USA*

Sicilian Valle del Belice dairy sheep were investigated with the following aims: 1) to establish the relationship between production and weather conditions using information from weather stations 2) to locate the heat stress point for dairy sheep and 3) to determine a heat stress function suitable for studying genetic tolerance against heat stress. The data consisted of 59,661 test-day records belonging to 6,624 lactations in 17 flocks. The model included the fixed effects flock X year of test-day, days in milk X lactation number, THI. Results shown pertain to daily productions and weather of the preceding day. Milk and fat+protein production showed negative phenotypic correlations of -0.26 and -0.33 respectively with maximum temperature. Furthermore milk and fat+protein production had positive correlations with relative humidity of 0.26 and 0.31 respectively. Fat+protein production started to decline above a temperature-humidity index (THI) of 23. For THI≥23, phenotypic correlations of THI with milk and fat+protein production were -0.29 and -0.35 respectively. This resulted in a decrease of milk and fat+protein production of about -55.7 g (5%) and -8.6 g (6%) respectively per unit of THI increase. These results indicate that Valle del Belice sheep, although originating from a hot environment, are affected by heat stress resulting in a decrease of production.

Theatre SM6.6

Space requirements of horned and hornless goats

B. Wechsler, C. Loretz, R. Hauser and P. Rüsch, Swiss Federal Veterinary Office, Centre for proper housing of ruminants and pigs, FAT, 8356 Ettenhausen, Switzerland

The behaviour of horned and hornless goats was compared to investigate their space requirements at the feed barrier and in the lying area. Two experiments were carried out with eight groups each of ten females, four groups with and four groups without horns. In experiment 1, the number of feeding places (width 35 cm) was restricted stepwise from an initial 20 to 15 and 10. In experiment 2, the size of the lying area was stepwise reduced from 2.0 m2 to 1.5 m2 and 1.0 m2 per animal.

The distance between the animals at the feed barrier was lower in the experimental condition with only 10 feeding places ($p < 0.002$) and in groups with horned goats ($p < 0.05$). The proportion of time the animals spent feeding decreased significantly with increasing animal/feeding place ratio ($p < 0.001$) and was significantly lower in groups with horned goats ($p < 0.02$). The average distance between lying animals was not influenced neither by the presence of horns nor by the size of the lying area. The proportion of time the goats spent lying decreased significantly with decreasing space allowances ($p < 0.05$), but was not influenced by the presence of horns. It is concluded that the space requirements of horned goats at the feed barrier are higher than those of hornless goats, whereas space requirements in the lying area do not differ between horned and hornless goats.

Poster SM6.7

Impact of an artificial *Haemonchus contortus* infection on the behaviour of lambs

M. Gauly[1], S. Brunn[2] and G. Erhardt[2], Institute of Animal Breeding and Genetics[1,2], University of Göttingen[1] and Giessen [2], Albrecht-Thaer-Weg 3, D-37075 Göttingen, Germany*

Various studies have described the effects of parasites on behaviour of animals. Behaviour is in many cases the first line of defence against a variety of parasitic assaults. Behavioural changes can be the result of direct and/or indirect parasite-host interactions. No informations are so far available about the influence of *Haemonchus contortus* infections on the basic behaviour in lambs. Therefore twenty-four Merinoland male lambs were weaned at an age of 12 weeks and randomly divided into two groups. Animals of group 1 were orally infected with 5000 *Haemonchus contortus* larvae 5 weeks after weaning, while group 2 was used as an control. The animals were kept in a barn on straw and fed with hay, concentrate and water ad libitum. Total behaviour activity patterns were observed and recorded from an age of 13 to 30 weeks using the time sampling method. Animals were slaughtered at the end of the study and worms were counted from the abomasum. Different activities changed significantly with the age of the animals in both groups. But no significant difference in any of the recorded behaviour patterns were found between the groups indicating that *H. contortus* had no influence. The established worm number (12 %) was very low. Therefore higher worm burden may have an impact.

Poster SM6.8

Mating behaviour of thin and fat tailed rams on fat tailed ewes

*S.G. Deligeorgis**, J.A. Bizelis and P.E. Simitzis, Department of Animal Breeding and Husbandry, Agricultural University of Athens, Votanikos, GR 11855, Athens, Greece*

The mating behaviour of inexperienced (6) and experienced (6) Chios fat tailed and Karagouniko thin tailed rams examined by video recording for 19 days after their introduction, at the beginning of mating period (August), to groups of Chios fat tailed mature ewes and ewe lambs, in a factorial 2x2 experiment. The groups of females were consisted of 12 animals and the rams were introduced every morning from 7.00 to 10.00 hours. The following behaviour patterns of rams were analyzed in terms of number and duration: resting, non-courtship behaviour, searching, sniffing, flehmen, mating attempts and mountings. The number of females on oestrus in each experimental group was also recorded.

The breed of ram had a significant ($P<0.05$) effect upon non-courtship behaviour, searching, sniffing and mating. Age of rams showed significant effects upon searching, sniffing, flehmen and mating attempts, while age of ewes upon searching, sniffing, mating attempts and mating. Significant interactions were also found between the age of ewe and breed of ram upon searching and flehmen.

Theatre PNH1.1

A survey on Aflatoxin B_1 incidence of animal feeds in Greece

S.Vlachou[1] and P.E. Zoiopoulos[2], [1]Feedingstuffs Control Laboratory, Lykovrisi Attikis 141.23, Greece,[2] University of Ioannina, Seferi 2, Agrinio 301.00, Greece*

Three hundred and two samples of straight and compound animal feeds were taken from farms throughout Greece to assess the aflatoxin B_1 status. Seven of 183 (3.8%) raw materials and none of the 119 compound feed samples were positive for aflatoxin B_1. Five of the seven positive samples were at the level of 10ppb. Aflatoxin B_1 was present only in maize (4 samples) and cotton seed cake (3 samples), whereas none of the barley, wheat and compound feed samples contained aflatoxin B_1. Two values measured for aflatoxin B_1 were higher than 10ppb, i.e.90ppb for maize and 20ppb for cottonseed meal. The aflatoxin B_1 content in 6 of the 7 positive samples was lower than the maximum permitted limit for feed materials in Community Directive 1999/29/EC for undesirable substances in animal nutrition. However the maize batch with 90ppb is not allowed to be used in animal feeding after the issueing of Directive 2002/32/EC which ruled out the possibility by Recognized Feed Manufacturers to make use of the "dilution principle" with raw materials contaminated with aflatoxin B_1. In general the presence of aflatoxin B_1 does not appear to constitute a troublesome case for feed materials under the Greek agricultural conditions.

Theatre PNH1.2

Fumonisin contamination of feeds sampled in forty riding centres in northern Italy

R. Mantovani, I. Cerchiaro, S. Schiavon and L. Bailoni, Department of Animal Science, University of Padova. Agripolis, Viale dell'Università 16 - 35020 Legnaro (PD), Italy

Seventy-two samples of simple and mixed feeds for horses were collected from 40 riding centres located in north-east of Italy in order to evaluate the level of fumonisin contamination. Feed samples were stored at -18°C and analyses were performed in duplicate by immuno-enzymatic method (ELISA), using a commercial test kit for B_1 fumonisin quantitative detection. As expected, the levels of B_1 fumonisin were very low in oat (from <0.01 to 0.84 ppm) and barley samples (from 0 to 0.95 ppm). In mixtures of several cereals, containing variable proportions of corn, fumonisin concentration was always lower than 5.00 ppm, i.e. the maximum level recommended by U.S. Food and Drug Administration in corn and corn by-products used for equids. Fumonisin levels of mixed feeds were highly variable among feed industries and, within industry, among different commercial products. High levels of fumonisin (>10.00 ppm) were found in mixed feed formulated for horses with intense activity in one industry, only. This could be due to an occasional and undetected corn contamination. The results of this study indicate the need of a continuous and careful control of the raw materials used in the feed industry in order to reduce the risks of adverse effects on horse health.

Theatre PNH1.3

Mycotoxicological challenges to European animal production: a review

S.A. Chadd, Royal Agricultural College, Department of Agricultural Science, Cirencester, Gloucestershire GL7 6JS, England, UK.

The ingestion by a range of livestock of mould-contaminated feedstuffs and associated mycotoxicity, can represent a challenge to their achievement of optimal productivity. Other negative impact factors at the commercial farming level include economic loss due to reduced crop yields and the possible compromise to human welfare through handling contaminated crop products and food chain residual effects. Mycotoxins are toxic secondary metabolites of moulds belonging predominantly to the *Aspergillus*, *Penicillium* and *Fusarium* genera. Understanding the complexity of mycotoxicological studies remains a continual challenge, the chemistry and biological adaptations of fungal organisms being so diverse. Consequentially, the toxicological effects also vary widely and include factors such as duration of exposure to toxins ingested or inhaled, species, age, physiological status and the observable synergies between different feed mycotoxins. Clinical toxicological syndromes which characterise affected animals range from acute mortality to slow growth and reduced reproductive efficiency to impaired immunity and decreased resistance to infectious diseases. In terms of prevention and control, research indicates the value of more sophisticated additives such as organic polymers (esterifed glucomannans) such extracts having a strong affinity for some mycotoxins. In addition, the manipulation of brain neurochemistry through the involvement of nutritional strategies which include the evaluation of large neutral amino acid supplementation shows promise. In the absence of a total remedy for mycotoxin contamination scenarios and their consequences, 'best agricultural practice' remains one of the most effective solutions.

Theatre PNH1.4

Comparative aspects of *Fusarium* mycotoxicoses in swine, poultry and horses

T.K. Smith[1], S.R. Chowdhury[1], H.V.L.N. Swamy[1] and S.L. Raymond[2], [1]University of Guelph, Department of Animal and Poultry Science, Guelph, Ontario, Canada N1G 2W1, [2]Equine Guelph, Guelph, Ontario, Canada, N1G 2W1*

A series of experiments have been conducted to determine the effect of feeding blends of grains naturally-contaminated with *Fusarium* mycotoxins on behavior and metabolism of starter pigs, broiler chickens, laying hens and mature horses. Blends of contaminated wheat and corn were found to contain deoxynivalenol, 15-acetyldeoxynivalenol, zearalenone and fusaric acid. Feed intake was very significantly reduced when contaminated grains were fed to swine and horses. Much higher concentrations of mycotoxins were required, however, to reduce feed intake in poultry. It was determined that elevated brain concentrations of serotonin were seen in both swine and poultry but the anorectic effects of this neurochemical change were cancelled in poultry by a simultaneous elevation in brain catecholamine concentrations. Mycotoxin-induced metabolic changes were more obvious in poultry than in other species. This was likely due to the lack of a protective effect of anorexia. In broiler chickens there were elevations in red blood cell counts, hemoglobin concentrations and blood uric acid concentrations. Decreased biliary immunoglobulin A concentrations were also seen. An even greater elevation in blood uric acid concentrations was seen in laying hens. The only mycotoxin-induced metabolic change seen in horses was elevated serum gamma-glutamyltranserferase activities indicative of liver damage. It was concluded that species differences in sensitivity to feed-borne *Fusarium* mycotoxins are mainly due to differing degrees of appetite suppression.

Theatre PNH1.5

Oxidative stress in bovine peripheral blood mononuclear cells exposed to mycotoxins

B. Ronchi, L. Colavecchia, U. Bernabucci, N. Lacetera, and A. Nardone, University of Tuscia, Department of Animal Production, via De Lellis, 01100, Viterbo, Italy

Mycotoxins are metabolites produced by moulds in foodstuffs or feeds. One of the possible mechanism through which can cause cytotoxic effect is the induction of oxidative stress. This in vitro study was performed to verify the potential oxidative effect of four mycotoxins by use bovine peripheral blood mononuclear cells (PBMC) culture as model. Blood samples were obtained from six healthy not pregnant and not lactating Holstein cows. The PBMC were isolated by density gradient centrifugation and incubated for 2 and 7 days with different concentration of four mycotoxins (AFB_1: 0, 5 and 20µg/ml; T-2: 0, 2.5 and 10 ng/ml; DON: 0, 1 and 5µg/ml; FB_1: 0, 35 and 70µg/ml). To evaluate oxidative status concentration of reactive oxygen metabolites (ROM), intracellular thiols (SH), malondialdehyde (MDA) and gene expression of superoxide dismutase (SOD) and glutathione peroxidase (GSH-Px) were determined on PBMC pellet. All mycotoxins tested induced oxidative stress. Exposure of PBMC to AFB_1 reduced SOD mRNA and SH and increased ROM and MDA. T-2 decreased SOD and GSH-Px mRNA, and SH and increased MDA. DON reduced GSH-Px mRNA and increased MDA. FB1 decreased SOD and GSH-Px mRNA and increased MDA. Concentration of mycotoxin was more important factor than time of exposure in inducing oxidative stress. Among mycotoxins AFB_1 showed higher cytotoxic effects.

Theatre PNH1.6

Determination of fumonisin B_1 content of porcine tissues after feeding diet of high toxin concentration for the sake of risk assessment

M. Kovács[1], J. Fodor[1], K. Meyer[2], K. Mohr[2], J. Bauer[2], I. Repa[1], F. Vetési[3], P. Horn[1] and F. Kovács[1], [1]University of Kaposvár Faculty of Animal Science, Research group of the Hungarian Academy of Sciences, Kaposvár, Guba S. u. 40., Hungary, [2]Technical University of Munich, Institute of Animal Hygiene, Freising-Weihenstephan, Germany, [3]Szent István University Veterinary Faculty, Budapest, István u. 2., Hungary

The residues deriving from the uptake fumonisin B_1 (FB_1) were determined in growing pigs fed 100 mg/animal (7,5 -7,6 mg/kg bw) daily FB_1 for 5-10 days. The average total FB_1 intake in the first 5 days was 403,8 (365,8-465,8) mg, the daily toxin intake was 30,4-35,9 mg/kg b.w. Among the haematological parameters examined, elevated red blood cell count (9,3-10,8 x10^6 /µL), haemoglobin concentration (15,8-17,2 g/dL), haematocrite value (54-66 %) and decreased MCH value (14,3-16,6 pg) were observed. Among the clinical chemical parameters examined the high aspartate aminotransferase activity (116-330 U/L) revealed to hepatic injury.

Particular high levels of FB_1 could be measured in kidney (81,6-4762,4 ng/g), liver (73,6-709,6 ng/g), lung (6,4-1144,8 ng/g), spleen (28-7975,2 ng/g) and pancreas (24-464 ng/g). Muscle and fat samples showed negligible contamination (43 and 6 ng/g, respectively). Considering the highest levels in edible tissues (liver, kidney, muscle and fat) and the TDI value recommended by SCF (2000) the consumers risk through a carry-over from swine seems to be negligible.

Reduction of aflatoxin M_1 in milk with a dietary magnesium bentonite

S. Sidler[1], F. Escribano[1], R. Bordini[2], V. Freddi[2], M. Casagrandi[3] A. Tampieri[3], A. Canever[4] and F. Cinti[5], [1]Grupo Tolsa, Madrid, Spain, [2]Crippsar S. p. A., Cambiago, Italy, [3]AgriOK S.p.A., Gruppo Granarolo, Bologna, Italy, [4]Granarolo S. p. A., Bologna, Italy, [5]DiSTA, Università di Bologna, Italy*

Aflatoxin M_1 (AfM_1) was categorized by IARC as a possible human carcinogen (class 2B) and a legal maximum limit of 0.05 µg/kg AFM_1 in milk has been set.
Mycotoxin adsorbents are commonly used feed additives to avoid aflatoxicosis and carry-over of AfM_1 into the milk, but their efficacy is not well accepted and requires more in depth investigations.
A magnesium bentonite (ATOX^{TM}) has been proven effective as adsorbent of AfB_1 to reduce AfM_1 in milk. Efficacy of adsorption is related to cation exchange capacity (CEC) and surface area (SA) of the adsorbent.
Several trials were carried out to corroborate the efficacy of ATOX^{TM} bentonite to reduce AfM_1 levels in milk. Cows were fed with naturally AfB_1-contaminated mixed rations through the experiments. Feed contamination levels ranged from 2.5 to 30.0 µg/kg. Milk was monitored for AFM_1. In all cases, dietary addition of 60 to 120 gr. per cow per day of ATOX^{TM} bentonite reduced between 60 to 75% AfM_1 levels in milk and was generally enough to bring AFM_1 concentration well below the 0.05 µg/kg legal limit. Response time was always faster than 2 days.

Poster PNH1.8

Heavy metals level in the hair of cows of disappearing breeds

S.A. Patrashkov, V.L. Petukhov, O.S. Korotkevich and O.A. Zheltikova, Research Institute of Veterinary Genetics and Animal Breeding of Novosibirsk State Agrarian University, 160, Dobrolubov Str., Novosibirsk 630039, Russia

The aim of our investigation was the determination of heavy metals (HM) content in the hair of cows of disappearing breeds: yakutskaya, galloweiskaya, seraya ukrainskaya. Two of the main essential elements - Zn and Cu and two of the ecotoxical HM - Pb and Cd were chosen for analyses.
The investigations were conducted in Novosibirsky region (Russia) in 2002 - 2004. Hair samples of animals (m = 0.2 g) were studied by the stripping voltammetric analysis (SVA) method using TA-2 analyzer to determine Zn, Cu, Pb and Cd concentrations.
The levels of essential elements in the hair samples of animals of all breeds (Zn: 102.72 ± 5.59 ... 107.44 ± 3.27 mg/kg and Cu: 2.91 ± 0.24 ... 8.97 ± 0.75 mg/kg) were dozens time higher than similar indexes of ecotoxical HM (Pb: 0,83 ± 0,10 ... 1,73 ± 0,37 mg/kg and Cd: 0.05 ± 0,03 ... 0.23 ± 0,06 mg/kg).
The parts of variability of Cu and Cd concentration stipulated by the animal breed were discovered with using analyses of variance (r_w = 0.301, P < 0.001 & r_w = 0.111, P < 0.05 respectively).
Thus, animals hair can be used for the relative study of breed peculiarity and the integrated evaluation the degree of environment and feed contamination of HM.

Poster PNH1.9

Anaerobic ruminal bacterial isolates from the rumen of different ruminant animals and their response to different levels of tannic acid

A.Z.M. Salem[1] and Y.M.Gohar[2], [1]Animal Production Department, Faculty of Agriculture (El-Shatby), Alexandria University, Alexandria, Egypt, [2]Division of Microbiology, Faculty of Science, Alexandria University, Alexandria, Egypt*

This work was carried out to investigate the order of different ruminant animals in their capacity to tolerate tannic acid (TA). Rumen samples from forty-eight animals of sheep, cattle, buffalo and buffalo-calves (12 animals of each one) were collected to isolate the anaerobic ruminal bacteria. Three hundred and forty pure colonies were selected. Sixty-two representative cases (9, 20 24, and 9 for our ruminant animals, respectively) were selected from all bacterial isolates by cluster grouping based on different TA levels sensitivities. Disc diffusion assay was used to determine the susceptibility of ruminal bacterial isolates to 0.0063, 0.0125, 0.025, 0.05 and 0.1-mg TA per disc. Data were analyzed by ANOVA using the general model procedure. All ruminal bacterial isolates were affected negatively by increasing the dose of TA in the growth medium. At higher dose of TA (0.1-mg disc^{-1}), sheep appeared to have a higher (SEM 1.718, $P<0.001$) susceptibility to TA than the other animals. Higher tolerance to TA levels was observed in buffalo and buffalo-calves than other animals. Higher variability (basis on F-test values) between all ruminal bacterial population was observed in response to the different levels of TA. In conclusion, ruminal bacterial isolates of each one of ruminant animals had a different characterization in degradation of TA. Animals had the following order in their tolerance to TA; buffalo-calves> buffalo> cattle>sheep.

Poster PNH1.10

Performance and some plasma biochemical constituents of Khaki-campbell ducklings fed diets containing biogen or garlic (Allium sativum l.) as feed additives

Mohamed, Khalifah[1], A.A. Abdallal[1] and A.A. Abdelhamed[2], [1]Animal Production Research Institute, Agricultural, Research Center, Ministry of Agriculture, Cairo, Egypt. [2]Poultry Production Department, Damanhour, Faculty of Agriculture, Alexandria University, Egypt*

One hundred and eight Khaki-Campbell ducklings 21 d. old were randomly divided into nine treatments. They were fed basal diet or diet supplemented with biogen or dried grounded garlic at levels 0.5, 1.0, 1.5, 2.0 g/Kg (from 3 to 9 weeks of age). A slaughter test was performed to determine carcass quality, endocrine and lymphoid organs and blood parameters.

All biogen supplementations had significantly ($P\leq0.01$) higher body weight (BW), body weight gain (BWG), feed conversion (FC), performance index (PI), absolute carcass weight and relative weight of carcass, giblets, edible parts and liver than those feed garlic. The 0.5 g/kg level from biogen or garlic had superior results than the other levels. As it improved BW, BWG, FC, PI, and increased immunity of the ducks.

Feed intake was higher at the control while it was lower at 2.0 g/kg biogen. The abdominal fat was reduced by (45.9%) with garlic supplementation as compared to biogen supplementation.

Poster PNH1.11

Uptake of aflatoxins by lactic acid bacteria naturally occurring in meat fermentation processes

R. Rullo, D. Balzarano, F. Polimeno, L. Ferrara and G. Maglione, ISPAAM- CNR Via Argine 1085, 80147 Neaples, Italy

Aflatoxins are a group of micotoxins produced by some strains of *Aspergillus flavus* and *parasiticus*. They are worldwide potent carcinogen contaminants, and their inhibition role to the liver protein synthesis has been demonstrated. Many strategies are under investigation to eliminate or inactivate the aflatoxin contaminats from animal feed and human food chain. Aflatoxin may be degraded by physical (heat, ionizing radiation), chemical (oxidizing, hydrolytic agents) or biological methods. Many investigators are taking into account the role that many microorganisms, including bacteria, play in removing these compounds from media. Many *Lactobacillus* strain have been promoted as good probiotics for human and animal. In this study we show how some lactic bacteria strains isolated from fermented meat food, are able to bind aflatoxins and how the uptake was sensitive to the temperature and pH. We also show the similar aflatoxins binding capacity of living and non-living cells in an aqueous solution containing 5µg/ml of aflatoxins. These observation imply that aflatoxins bound initially to the periphery of the cell wall.

Poster PNH1.12

Application of Tannins (Farmatan) in the Feed for Mares and Foals

F. Habe[1], A. Jovan[1], M. Struklec[1], A. Oresnik[1] and I.Rupnik[2], [1]University of Ljubljana, Biotechnical Faculty, Zootechnical Department, Groblje 3, 1230 Domzale, Slovenia, [2]Tanin d.d., Hermanova ulica 1, 8290 Sevnica, Slovenia

The effect of FARMATAN-extracted dry tannins from chestnut wood (Castanea sativa meal) and added to the fodder for brood mares and their foals has been studied on six mares of different race and age and their foals at the Horse Centre Krumperk of Biotechnical Faculty. The experiment and the feeding with experimental mixture (with addition of 0.25 g of tannins to the 1kg of feed) started 14 days before foaling and lasted till the third month after the birth. Behaviour, health status, weight, consumption, changes in oestrus on the mares and foals and on the consistency and changes of faeces and the content of tannins in them were monitored. The addition of tannins did not influence the condition, weight and food consumption and consistency of animal faeces. The group on the additive and tannin pills had shorter periods of diarrhoea caused by oestrus of mares. The analytical results of tannins in food and faeces are presented and discussed.

Poster PNH1.13

A dietary tolerance study on the use of a magnesium bentonite for swine

R. Lizardo[1], J. Brufau[1], S. Sidler[2], G. Gomez[2] and F. Escribano[2], [1]IRTA - Centre Mas Bové, Apartat 415, E-43280 Reus, Spain, [2]TOLSA S.A. -Department of Research and Development, E-28080 Madrid, Spain*

A magnesium bentonite (ATOX(tm)) has been shown as an effective mycotoxin adsorbent. An experiment was conducted to evaluate the tolerance of this bentonite as a feed additive for piglets. Ninety-six weaning piglets were allotted to 4 dietary treatments containing 0, 0.2, 0.5 and 3% ATOX(tm) during 35 days. Feed intake, weight gain and feed conversion were monitored during the trial. Blood samples were collected for hematological determinations and 30 piglets were slaughtered for anatomical and pathological examination at the end of the experiment. No differences (*P>0.05*) were detected in performance between dietary treatments for the all-growing period. Only a tendency towards a lower growth performance was observed for the 3% ATOX(tm) fed group during the first 14days of the experiment (*P*=0.13). Even so, this reduction in growth was lower than expected due to the dietary energy dilution. Hematological parameters fell into the physiological ranges and no differences were observed between experimental treatments (*P*>0.05). No pathological abnormalities were detected in any of the piglets slaughtered.

It was concluded that ATOX(tm) bentonite, as a feed additive for young swine, had no detrimental effects on feed intake, growth performance or physiological and hematological parameters even at 10-fold the recommended dosage of inclusion.

Theatre PNML2.1

Large-scale pig farming systems: opportunities and challenges

S.A. Edwards, School of Agriculture Food & Rural Development, University of Newcastle, Newcastle upon Tyne NE1 7RU, UK

Large-scale pig farming systems are increasing because of the improved economic efficiency which can be achieved in purchasing and marketing, and by the spreading of overhead costs between many animals. There are also advantages for production logistics. The large contemporary batch sizes of breeding sows make it easier to have cost-effective large group gestation housing systems (for example with electronic sow feeding) without repeated mixing of animals. For the farrowing/lactation stage, large numbers offer the possibility of easy operation of all-in all-out systems, justification for dedicated staff allocation during the farrowing period and plentiful scope for effective cross-fostering. For weaned and growing piglets, The ability to operate all-in all-out systems by room, and even by building or site, facilitates health management and phase feeding strategies which can reduce feed cost and excreta pollution potential. Set against these advantages are significant challenges. The presence of many animals in close proximity means that any disease outbreak can have catastrophic consequences. Similarly, the high pig density in one location results in increased risk of adverse environmental impact from odour, noise and manure, with growing EU legislation imposing more stringent restrictions on waste management. However, perhaps the greatest challenge to large-scale production is posed by staffing issues, with recruitment, retention and motivation of high-quality managers and stockpeople being vital for high animal performance and welfare, but increasingly difficult to achieve.

Theatre PNML2.2

Four approaches to additionally reduce costs on large-scale pig finishing farms

J. Zonderland and J. Enting, Applied Research of the Animal Sciences Group of Wageningen UR, Runderweg 6, 8219 PK Lelystad, The Netherlands

Increasing farm size is a well known strategy to reduce cost per kg pork produced. In addition, farmers will look for other ways to reduce production costs. Their main focus will be on feeding, housing and labour. Feeding and housing are the two largest contributors to cost price, which are under direct influence of the farmer. They represent 32% and 12%, respectively, of the cost price under Dutch circumstances. Labour is one of the factors that will be first limiting when increasing farm size. It constitutes approximately 5% of the cost price. The Applied Research division built a new pig facility to investigate four different approaches to additionally reduce costs on large-scale finishing units:

1. Reducing housing cost by building cheap insulated unbedded rooms with natural ventilation and kennels.
2. Minimising labour intensiveness by automatic selection from large groups of pigs ready for slaughter, e.g. using a mobile forelegs weighing system that marks pigs above a set weight.
3. Reducing feeding cost through substitution of part of the concentrate diet by dry or wet by-products, e.g. using a nipple feeding system for wet by-products.
4. Reducing housing cost by increasing the number of pigs per building through the use of two-level pens.

Data-gathering is on pilot-scale, but the results per room will be used to model large-scale finishing facilities. The first findings of this ongoing research will be presented at the congress.

Theatre PNML2.3

Large scale pig farming in Croatia

D. Vincek[1], G. Gorjanc[2], F. Poljak[1], Z. Lukovic[3], S. Malovrh[2] and M. Kovac[2], [1]Croatian Livestock Center, Zagreb, Croatia, [2]University of Ljubljana, Zootechnical Department, Slovenia, [3]Faculty of Agriculture, Department of Animal Production, Croatia

Pig production in Croatia has passed through specific changes after 1990. Between year 1990 and 2003, the number of fattening pigs decreased for 500 000 (20%). At the same time the number of large farms as well as of family farms decreased. The main problem is market irregularity and weak connection between farmers on one hand and slaughterhouse and food industry one the other hand. Data from ten large pig farms, which were included into regular recording scheme, were analyzed. Age at first farrowing was on average 379 days. The number of the piglets born alive was 8.90 for primiparous and 9.80 multiparous sows. Female days per liveborn piglet ranged between 15 and 32 days for gilts, and between 15 and 20 days for sows. On average, lactation length lasted 24 days. Interval between two farrowings ranged between 153 and 163 days. Farrowing to culling interval varied between 75 and 110 days. Meat percentage of slaughter pigs from large farms in year 2003 was 54.67%. In general, productivity of pig industry is low. The farms must improve management in the first place. Possibilities for the management improvements are described in the article.

Theatre PNML2.4

Perspectives in pig production in Slovenia

M. Kovac, S. Malovrh and G. Gorjanc, Darja Cop. University of Ljubljana, Biotechnical Faculty, Animal Science Dept., SI-1230 Domzale, Slovenia*

Slovenian pig production is small producing around 600000 slaughter pigs per year and supplying only 60 to 70% of meat consumed. Before 1990, pig production was well organized: the breeders were running a council on national level deciding on breeding strategies, subsidies, etc. The proportion of hybrid pigs was acceptable (75%). Changes in ownership of large industrial units caused at first some undesirable changes. Uncertainty about ownership reduced investments. Purebreeding has been increased almost to 50%. The main reason for reducing pig productivity has been a rejection of carcass grading. As pigs were paid on live weight the only important criteria were daily gain and feed efficiency. But later, new developments were initiated. Carcass grading has been implemented again in 1996, causing fast increase in lean meat percentage from 52% in 1996 to almost 57% in 2004. Some changes were due to increased usage of Pietrain crosses and genetic changes in other breeds. However, more emphasis was given improving animal husbandry, animal welfare and environmental protection. Furthermore, the paper is discussing changes in pig productivity as well as organisational changes in Slovenian pig production.

Theatre PNML2.5

The changing structure of the United States swine industry

J.A. Sterle, Department of Animal Science, Texas A&M University and Texas Cooperative Extension, 2471 TAMU, College Station, Texas, USA 77843*

The structure of the U.S. Swine Industry has changed drastically in the last two decades. The trend has been away from small, diverse family farms toward larger, vertically integrated operations. While the number of U.S. hog operations has steadily declined, size of operation, retail meat per pig and litter size have increased. The location of these larger operations tends to be in non-traditional hog-producing states. Advantages of larger operations include economies of scale, specialization of labor and market accessibility. However, disadvantages include public perception, environmental sustainability and government regulation, lack of qualified labor force, and biosecurity, health and animal welfare concerns.

One of the biggest challenges to the more traditional, smaller operations is market accessibility. Companies harvesting hogs have also consolidated. Packers may own breeding stock and produce market hogs, or have long-standing contracts with larger farms to guarantee a steady supply of hogs on a daily basis that are often of similar genetics, type and management, leaving smaller producers without shackle space with a packer. Many smaller producers have formed alliances to "act big" and try to capture some of the advantages of larger operations without giving up ownership.

It is apparent that this trend of consolidation and vertical integration will continue. However, total integration, as seen in the poultry industry, will most likely not take place.

Theatre PNML2.6

Relation between system related factors, indicators of energy intake and stress and reproduction performance in commercial herds with group housed non-lactating sows

A.G. Kongsted, Department of Agroecology, Danish Institute of Agricultural Sciences, P.O. Box 50, DK-8830 Tjele*

The number of group housed non-lactating sows is increasing rapidly in Europe as a consequence of changed EU legislation and national extraordinary laws initiated by elevated public concern for animal welfare. However, impaired reproduction has been observed in group housed compared to individually housed non-lactating sows in several on-farm studies. There are indications that social relations causing individual variation in energy intake, stress and fear may be possible causes. Traditional methods for assessing energy intake, stress and fear are not suitable in large-scale on-farm studies or for the individual farmer, who wants to find out whether level of energy intake, stress and fear could be contributing reasons for reproduction problems in group housed sows. A study was conducted to investigate whether indicators of energy intake, stress and fear (e.g. back fat, eating behaviour, skin lesions, aggressions) were suitable to express variation in sows' susceptibility for a good reproduction performance under practical conditions. The study involved 14 commercial herds with herd sizes ranging from 180 to 1000 sows per herd. The results from the study will be presented with emphasis on the relation between the indicators, system related factors (e.g. stocking rate, group size and group composition) and reproduction performance (e.g. litter size and pregnancy rate).

Theatre PNML2.7

Three new feeding systems for group housing of pregnant sows on large scale farms

H.W. van der Mheen and J. Enting, Applied Research, Animal Sciences Groups, Wageningen UR, P.O. Box 2176, 8203 AD Lelystad, the Netherlands*

To improve animal welfare, EU regulations state that all pregnant sows have to be housed in groups as from 2013. Farmers are faced with the complicated task of selecting the ideal housing system for their situation. ESF and feeding stalls are well known feeding systems for sows but also new systems appear on the market. These new systems are less well known. We studied three new feeding systems for group housing of sows; an automatic nipple feeder, a long trough and a feeder that uses time intervals between feed portions to distribute feed between the animals.

Service groups of 42 sows were allocated to one of these feeding systems and were followed during five consecutive cycles. Feeding behaviour, agonistic behaviour, growth, reproduction performance, as well as technical aspects and labour requirements of the systems were recorded.

The three feeding systems differed in many aspects. Variation in individual feed intake was the major concern when using the feeder with intervals. Reproduction results were however not impaired by low feed intake. Group division was the main issue with the trough, while agonistic behaviour related to the nipple feeder.

Reproduction results did not differ between the various feeding systems, and we concluded that farmers own management preferences should determine the choice for a certain feeding system.

 Theatre PNML2.8

Opportunities for increasing sow productivity in Slovenia

D. Cop[1], S. Malovrh[1], Z. Lukovic[2], G. Gorjanc[1] and M. Kovac[1], [1]University of Ljubljana, Biotechnical Faculty, Animal Science Department, Groblje 3, 1230 Domzale, Slovenia, [2]Faculty of Agricultural, Svetosimunska 25, 10000 Zagreb, Croatia*

Regular fertility reports have showed large improvements since 1980, when fertility traits were defined uniquely for all farms and have been used for evaluation of sow efficiency. Reproduction performance was leveled up in the last period. The aim of our study was to analyze reproduction data in order to find possibilities for further improvements. The analysis included complete reproduction data from 1994 to 2003, however we focused on the litter size and empty days. The model for litter size considered age at farrowing as quadratic regression within parity, lactation length, weaning to conception interval, and service boar. The age at farrowing affected litter size, especially in gilts. Results indicated linear relationship between lactation length and subsequent litter size. Among farms, regression coefficients ranged from 0.046±0.019 to 0.127±0.027 liveborn piglets per day, for the interval where more than 80 % sows were weaned. As lactation increased, weaning to the first service and weaning to conception interval decreased. Furthermore, litter size dropped for services on the 6th day and was further decreasing up to the 10th day after weaning. The increased weaning to conception interval caused undesirable changes in sow productivity. Detailed descriptions about possible advances of sow productivity in Slovenia are explained in the article.

Theatre PNML2.9

Carcass traits for prediction of breeding values in pigs

G. Gorjanc, S. Malovrh and M. Kovac, University of Ljubljana, Biotechnical faculty, Zootechnical department, Groblje 3, SI-1230 Domzale, Slovenia*

Covariance components for measurements from slaughter line, dissection of ham, net daily gain, and lean meat daily gain were estimated to study the possibility of including carcass traits in the Slovenian pig breeding program. Data from 2,158 fattened pigs in field environment with known pedigree were analyzed. Fixed effects of sex, genotype and season of slaughter as year-month interaction were fitted. Age at slaughter was also included due to different weight and age at slaughter. The random part of the model included effect of test day, common litter environment, and direct additive genetic effect. All possible two-trait analyses were done, while multiple trait analysis was not feasible due to high correlation between carcass traits. Heritability estimates ranged between 0.22 for net daily gain and 0.46 for lean meat content. Lean meat content (LC) and ham weight (HAM) were proposed for incorporation in selection program. The two traits were chosen on the basis of large phenotypic variability ($\sigma^2_{pLC} = 10.76$ and $\sigma^2_{pHAM} = 1.275$), high heritabilities ($h^2_{LC} = 0.46$ and $h^2_{HAM} = 0.27$), low genetic correlations ($r_G = -0.01$), low costs of taking the measurements and current set of traits in selection program. LC and HAM weight could be recorded with means of sib test in field environment.

Theatre P4.1

The accuracy of Slovenian method for pig carcass classification

M. Candek-Potokar[1], M. Oksama[2] and E.V. Olsen[2], [1]Agricultural Institute of Slovenia, [2]Danish Meat Research Institute

Our objective was to evaluate the accuracy of Slovenian method for meat percentage evaluation of pig carcasses (fat and muscle measurements at the level of *m. gluteus medius* at the carcass split-line using electronic calliper). The study was performed within EU funded project EUPIGCLASS as a part of the study of the accuracy of the on-line methods used in European countries. Therefore, the common experimental plan and ISO 5725 standard were respected. According to later, the accuracy refers to the closeness of agreement between test results under repeatability or reproducibility conditions, and is expressed as standard deviation (sd). The aim was to assess the repeatability sd and reproducibility sd of meat percentage evaluation, and to identify influence of the copy of the equipment and of the operator. For Slovenian method, repeatability sd was 0.4%, indicating we can expect (P=0.95) 0.8% difference in meat % for the same carcass made by the same operator with the same copy of the equipment; reproducibility sd was 0.9%, indicating 1.8% difference in meat % can be expected (P=0.95) for the same carcass due to the factors related to the abattoir. The variation due to the copy of the equipment was minor (sd=0.1%) compared to the operator's effect (sd=0.5%). Slovenian results are situated somewhere in the middle, if related to the results for other methods (an article in preparation). In general, results point out the necessity for regular maintenance of instruments and training of operators in order to maintain the precision at an acceptable level. The recommendations (QAP) have been proposed by EUPIGCLASS project (www.eupigclass.org-.url).

Theatre P4.2

Estimation of the carcass composition of station tested pigs

M. Wiese[1], E. Tholen[1], U. Baulain[2], R. Höreth[3], [1]University of Bonn, Institute of Animal Breeding Science, Endenicher Allee 15, 53115 Bonn, Germany, [2]Federal Agricultural Research Centre (FAL), Institute for Animal Breeding Mariensee, Höltystraße 10, 31535 Neustadt, Germany, [3]Federal Research Centre of Nutrition and Food, Institute of Meat Production and Market Research, EC-Baumann-Str. 20, 95326 Kulmbach, Germany*

Based on a sample of 292 fully dissected carcasses of pure and crossbred pigs the accuracy of the regression formula (Bonner-formula, BF) currently used in station performance testing and the grading systems Fat-O-Meater (FOM) and AutoFOM were verified. While the accuracy of FOM and AutoFOM was sufficient for all lines (RMSE < 2.5%), the BF was reliable only for German Landrace and German Yorkshire pigs. Linear carcass measure-ments, which were recorded at the test station, and 127 AutoFOM basis variables were used to construct new Partial Least Squares (PLS) regression functions.

Using the currently implemented AutoFOM formulas for estimating the proportions of valuable cuts, particularly high (< 5%) relative measurement errors were found for carcass cut loin. Although distinct improvements were achieved by PLS-adapting of these functions using linear carcass measurements or AutoFOM base recordings, relative measurement errors were not below 5%. Because of the strong correlations between the lean meat content of the carcass and the proportions of valuable cuts, specific considerations of ham or loin proportions were only useful in special cases

Theatre P4.3

DNA-based traceability of pork

S. Ostler, R.. Fries and G. Thaller, Technical University of Munich, Chair of Animal Breeding, Alte Akademie 12, 85354 Freising-Weihenstephan, Germany*

The European livestock industry has been affected recently by diseases (BSE) and epidemics (FMD). Therefore, the consumer is asking increasingly for more detailed information concerning quality of pork, which in turn requires a reliable system to trace pork back to the birthplace.
In contrast to cattle, where all animals are identified individually in databases, it is extremely demanding in swine to record each single piglet, for example by ear tags, and economically not feasible to trace pork by direct fingerprints.
In this pilot project, the traceability of pork is based on indirectly assigning the true sow -and thus the farm of origin- out of all possible sows within a defined production system to an individual animal or probe. Preliminary simulations based on a set of ten markers (average PIC=0.67) applied for a production system of 4600 sows showed that the number of markers necessary get out of scale by genotyping sows only. Up to 350 sows came into question. However, by including the boar, the number of possible matings decreased markedly with a correct assignment in 24% of the cases. When additionally using records of effective litters, the true mother was detected with 95%.
Our goal is to balance the expenses for genotyping with the effort to get additional information to yield an economical system that will offer the consumer an accredited, traceable pork.

Theatre P4.4

Feed intake behaviour of different pig breeds during performance test on station

R. Baumung, G. Lercher, A. Willam and J. Sölkner, University of Natural Resources and Applied Life Sciences Vienna (BOKU), Division Livestock Sciences, A-1180 Vienna, Austria*

In Austria pigs are tested for growth and carcass traits centrally at one performance test station in Lower Austria. Pigs are kept in group pens (maximum 13 animals per pen). Each pen is equipped with an electronic feeding station. The identity of the pig, date, feeder entry and exit time and the amount of food consumed are recorded for every visit in the station. Feed intake behaviour of 1,593 pigs of three different breeds (618 Large White, 486 Landrace, 489 Pietrain) was analyzed. Different traits such as visits per hour and per day, time per hour and per day, feed intake per hour and per day, feed consumption rate, time per visit and feed intake per visit were used to describe feed intake behaviour for different observation periods. Observation periods were testing day, testing month and the whole testing period. By using Generalized Linear Models differences in feed intake behaviour between the breeds were assessed. LS-Means were used to develop feed intake patterns for each breed. Another aim of the study was to investigate the effects of MHS-genotype on feeding patterns of Pietrain.

Theatre P4.5

Intraperitoneal electronic identification of piglets labelled (*Suino tipico sardo*)

W.Pinna[1], P. Sedda[1], G. Moniello[1], N. Ferri[2], E. Marchi[2] and I.L.Solinas[3], [1]University of Sassari, Dep. "Biologia Animale", Via Vienna, 2, 07100 Sassari, Italy, [2]IZS dell'Abruzzo e del Molise, Località Campo Boario, 64100 Teramo, Italy, [3]EU - Joint Research Centre, IPSC - Via Enrico Fermi 1, 21020 Ispra, Italy*

The milking piglet's meat is one of the better known and appreciates regional animal production of Sardinia (Italy). Using a steel shot injector 197 injectable transponders (HDX technology, 32.5×3.8 mm) were applied in abdomen cavity of sucking piglets, 2 - 12 days old, 102 males and 95 females of "suino tipico sardo" labelled. The readability, localisation and the health effects of transponders were evaluated in living animals at 1d, 7d, 14d, 28d, 35d and in the slaughterhouse. In 4 piglets (2 males, 2 females) location of transponder in abdomen cavity was verified by radiography at 1d, 7d, 14d, 28d and 35d. The piglets were slaughtered between 28-40 d and average weigh 8,19 kg. During the experiment behaviour and performances of animals were not modified by transponders presence. The employed transponders showed 100% readability during experimental period and were easily recovered in slaughter line. The intraperitoneal injection of the transponders represents a good method for tagging piglets during first day of life and can improve "suino tipico sardo" meat traceability. Farmers and technicians expressed interest for this electronic identification system.

Theatre P4.6

Assessment of the methionine requirement of pigs in the weight range 11 to 23 kg

PB. Lynch[1] and M. Rademacher[2], [1]Teagasc, Moorepark, Fermoy, Ireland, [2]Degussa AG,Hanau, Germany*

The objective of this trial was to determine the optimum ratio of Methionine (MET) to Lysine (LYS) for pigs from 11 to 23kg. A total of 192 pigs were used, with a pair of pigs of the same sex penned together (n = 96) as the experimental unit in a randomised block design. A commercial starter diet was fed for 11days after weaning (26 to 28 days) and the test diets for 24 days. The basal diet contained wheat, barley, full fat soyabeans, peas, corn starch, soya oil, synthetic amino acids, minerals and vitamins. The digestible energy content was 14.9 MJ/kg, crude protein was 166g/kg and total LYS was 12.0g/kg. The basal diet (2.2g/kg MET) was supplemented with DL-MET in increments of 0.4g/kg to give seven diets ranging from 2.2 to 4.6g/kg MET. The ratio MET:LYS varied from 0.183 to 0.383. MET supplementation increased average daily gain (297, 372, 453, 457, 507, 515 and 459g/d; linear - $P<0.01$; quadratic - $P<0.01$) and improved feed:gain ratio (2.55, 2.00, 1.74, 1.75, 1.54, 1.60 and 1.55; linear - $P<0.01$; quadratic - $P<0.01$). Exponential response curves were fitted and the optimum level of MET (95% of the asymptote) was 4.1g/kg for daily gain and 3.7g/kg for feed:gain. It is concluded that the optimum ratio of MET:LYS is 0.31 for feed:gain and 0.34 for average daily gain.

Theatre P4.7

Changes of selected haematological, immunological and biochemical parameters in blood after long term application of Lactobacillus plantarum in piglets

M. Húska[1], L. Hetényi[2], G. Kovác[1], P. Reichel[1], J. Buleca Jr.[1] and T. Zadnik[3], [1]University of Veterinary Medicine, Komenského 73, 041 81 Kosice, Slovakia, [2]Research Institute of Animal growth *Production, Hlohovská 2, 949 92 Nitra, Slovakia, [3]University of Ljubljana, Slovenia*

The experiment was conducted on 24 piglets divided into two groups. Probiotic strain Lactobacillus plantarum LB5 was applied orally in the experimental group during their postnatal period till the 73rd day of experiment. Lactobacillus strain contained 10^{10} bacteria per ml. Blood samples were examined from sinus ophtalmicus. Selected haematological, immunological and biochemical parameters were compared between experimental and control groups of animals. The tested strain did not influence the haematological profile - RBC, WBC values and haemoglobin, haematokrit and MVC concentration. INT test showed higher values in experimental group with significant difference $p<0.05$ on day 42 only. Index of phagocytic activity of neutrophils, phagocytic activity of WBC values were higher in experimental group with signifficant difference $p<0.05$ on day 59 and $p<0.01$ on day 73. Total immunoglobulin concentration showed significant difference $p<0.05$ on days 42 and 49, $p<0.01$ on days 59 and 73. L. plantarum did not influence total proteins, albumin and glucose concentrations, but health status was influenced positively. Positive effect of Lact. plantarum included weight gain stimulation with significant difference ($p<0.05$) on day 59 and ($p<0.01$) on day 73 in experimental group and pH value decrease in the gastro-intestinal tract. Mucosal adherence has been proved by small intestine electron microscopy method.

Theatre P4.8

Interaction of dietary phosphorous and phytase on nutrient digestibility and bone characteristics in growing pigs

M.J. Azain[1] and M.R. Bedford[2], [1]University of Georgia, Athens, GA, [2]Zymetrics, Golden Valley, MN, USA*

A total of 108 pigs (initial wt =9 kg) were assigned to one of 9 dietary treatments (27 pens) in a 3 x 3 arrangement to examine the main effects of dietary phosphorous (P) and phytase. There were 3 levels of available P (0.13, 0.28 and 0.42%). Calcium was maintained at 0.85%. Within each level of P, diets were supplemented with 0, 500 or 12,500 U /kg feed of a novel E. coli phytase expressed in Pichia pastoris. Diets were fed for 28 d. Apparent nutrient digestibility was determined during the last week of the trial. Metatarsal bones were isolated for determination of bone strength and mineral content. Phytase improved growth rate, efficiency and phosphorous digestibility in the 0.13% P diet, but had little effect on the diets with 0.28 or 0.42% P. However, effects of phytase on bone parameters were noted at all levels of dietary P. Increasing diet P and addition of phytase increased bone ash and bone strength. Addition of P or phytase to the diet increased bone content of P and Mg, but decreased Fe and Cu content. There was no effect of diet on bone Ca content. The results suggest effects of phytase in diets with adequate P that may be explained by removal of the anti-nutritional effects of phytin.

Zinc oxide in phytase-low phosphorus diets impairs performance of weanling pigs

R. Lizardo, D. Torrallardona, J. Brufau, IRTA - Centre Mas Bové, Apartat 415, E-43280 Reus, Spain*

The simultaneous utilisation of zinc oxide (ZnO) at a pharmacological dose of 3000ppm and of low phosphorus phytase supplemented diets for piglets was studied. One hundred and twelve piglets were housed in 28 pens during 5 weeks. Four experimental treatments were used, consisting of: a positive control diet with normal P content and with ZnO (T1); a low P diet supplemented with phytase and ZnO (T2); a low P phytase supplemented diet without ZnO (T3); and a low P diet with phytase and ZnO but without the mineral and vitamin premix (T4). There were significant differences between treatments for feed intake (437, 274, 443 and 274 g/d; P<0.001), weight gain (344, 175, 346 and 198 g/d; P<0.001) and feed conversion ratio (1.28, 1.58, 1.28 and 1.40 kg/kg; P<0.001). For all the parameters T1 and T3 were significantly better than T2 and T4. The low P diets resulted in lower serum P concentrations at day 28 and this was further reduced in combination with ZnO (8.0, 3.1, 5.1 and 3.1 mg/100ml; P<0.001). It can be concluded that the use of pharmacological doses of ZnO in low P diets is dangerous for the P homeostasis and that it may affect performance. Although phytase is useful to reduce the dietary P concentration, attention must be paid to possible interactions with other nutrients present in the feed.

Poster P4.10

Fully automatic pig banded chromosomes classification

V.N. Stefanova, E.I. Klykova, M.N. Zenina and V.G. Panteleev, All-Russian Research Institute of Genetics and Farm Animal Breeeding, St-Petersburg, Pushkin (Russia)*

Urgent task of modern pig breeding is revealing and withdrawal boars carrying reciprocal translocations, which are characterized by an essential reduction of fertility. Lack of modern program of pig automatic chromosome classification hampers realization of such researches. The original version of pig chromosome classification was developed using "VideoTesT-Karyo 3.1" computer program. To set up a database of pig chromosome images for the classifier 47 GTG-banded metaphases at different stages of condensation with well spreaded chromosomes from 4 Belgian landrace boars were used. These chromosomes were captured using black and white CCD camera (VT-13CX, 1300X1030 pixels) connected with microscope Axioskop 40, corrected (surface, central axis and centromeres were marked) and identified manually by operator. To check the resulted classifier 16 metaphases selected randomly from another sample of cells were analyzed. The error rate was 3,12±0,26 per metaphase or 8,2% from all chromosomes classified automatically (this value is within acceptable limits for such kind of studies). It should be noted that "VideoTesT- Karyo 3.1" program allows an operator to correct chromosome identification at any stage of metaphase analysis and thus to avoid misclassification in the karyogram resulted. The classifier was successfully checked also on chromosomes of boar with rcp(1p-;11p+) described by us previously. Therefore the developed program can be used in cytogenetic control of pedigree boars and chromosome analysis of longterm pig cell lines used in medical explorations.

Poster P4.11

The growth of young boars and gilts upto two months of age reared in litters standardized to 8 piglets (smaller litters) or 12 piglets (larger litters)

R. Czarnecki, A. Pietruszka, M. Rózycki, M. Kamyczek and B. Delikator, Department of Pig Breeding, Agricultural University of Szczecin, Dr. Judyma 10 str. 71-460 Szczecin, Poland

The aim of the study was a comparison of the growth of young boars and gilts obtained from hyperprolific sows of Polish Line 990, reared in "smaller" (♂n≈300; ♀n≈300) and "larger" litters (♂n≈600; ♀n≈600). The standardization was performed on the first day after birth. In both of groups the sows produced the same total and alive number of piglets per litter, respectively: 13.93; 12.95 and 13.83; 12.96. Weight of boars reared in "smaller" litters at 21; 28 and 63 days averaged, respectively: 5.83; 7.76 and 19.40kg and reared in "larger" litters, respectively: 5.20; 6.91 and 16.76kg. Weight of gilts from "smaller" litters averaged, respectively: 5.59; 7.51 and 18.61kg and reared in "larger" litters 5.19, 6.89 and 16.84kg. In "smaller" litters the young boars and gilts were growing faster. In more favourable rearing conditions ("smaller" litters) the boars grew faster in comparison with gilts. However, the boars could not utilize that the potential to grow faster in worse rearing conditions ("larger" litters).

The study was financed by Committee for Scientific Research (grant No 6 P06E 035 20)

Poster P4.12

Reproductive usefulness of young boars reared in standardized litters of hyperprolific sows of the Polish Synthetic Line 990

R. Czarnecki, M. Kawecka, A. Pietruszka, J. Udala, M. Rózycki, M. Kamyczek and B. Delikator, Department of Pig Breeding, Agricultural University of Szczecin, Dr. Judyma 10 str. 71-460 Szczecin, Poland

The standardization to 8 piglets (76 litters) or 12 piglets in litter (104 litters) was performed on the first day after birth. In the both groups the sows were born the similar total and alive number of piglets in litter, respectively 13.93, 12.95 and 13.83, 12.96. The evaluation of sperm value of the analysed boars was performed at about 7 months of age on a base third ejaculates (86 boars from "smaller" litters and 77 boars from "larger" litters). Weight of boars during weaning at 28 days, reared in "smaller" litters shaped 7,76kg and reared in "larger" litters 6,91kg. Volume of both testes at 180 days (cm^3) shaped, respectively: 262.45 and 248.79; concentration of spermatozoa in cm^3 x 10^6 200.08 and 181.44; total number of spermatozoa x 10^9 21.89 and 19.66; percentage of spermatozoa with major defects 12.73 and 10.50; percentage of spermatozoa with normal acrosome 84.63 and 82.67; ORT test (osmotic resistance test) 67.12 and 66.40. The boars reared in "smaller" litters were havier during weaning, had higher volume of testes and better sperm.

The study was financed by Committee for Scientific Research (grant No 6 P06E 035 20)

Poster P4.13

Effect of own performance of purebred and/or crossbred boars on progeny performance in different environments

I. Bahelka, L. Hetényi, P. Flak and P. Demo, Research Institute of Animal Production, Hlohovská 2, 949 92 Nitra, Slovak Republic

Progeny (160 and/or 182 respectively) from 5 purebred (Slovakian Meaty - SM) and 6 crossbred (SM x Pietrain - PN, Duroc x PN, PN x Yorkshire - Y) terminal boars was tested at control station (CS) Bucany. At the same time progeny (179 and/or 159 resp.) from 5 purebred (Y) and 5 crossbred boars (DU x PN, PN x Hampshire) was evaluated at CS Nitra. Progeny traits were average daily gain (ADG) from 30 to 100 kg live weight, feed conversion ratio (FC), backfat thickness (BF), proportion of prime meaty cuts (PMC), of ham (HAM) and eye muscle area (EMA). Boars were tested in a field conditions and traits were ADG from birth to 100 kg liveweight, BF and lean meat content (LMC).
Significant or highly significant effect of purebred and/or crossbred boars on some progeny traits at CS Bucany was found. At CS Nitra not significant effect of tested boars on progeny performance was found.
Regression coefficients of progeny traits on sire ADG, BF and LMC between purebred and crossbred boars at CS Bucany were very different. Differences between regression coefficients at CS Nitra were smaller, mainly for sire ADG and BF.
The heritability estimates for sire traits were higher in crossbred boars except for ADG at CS Nitra (ADG 0.55/0.98 in purebred/crossbred at Bucany and 0.84/0.79 - Nitra, BF 0.67/0.88 - Bucany and 0.41/0.59 Nitra, LMC 0.66/0.75 - Bucany and 0.42/0.43 - Nitra).

Poster P4.14

The growth and meatiness of young boars and gilts reared in standardized litters of hyperprolific sows depending on RYR1 genotype

A. Pietruszka, R. Czarnecki, M. Rózycki, M. Kamyczek and A. Czechowska. Department of Pig Breeding, Agricultural University of Szczecin, Dr. Judyma 10 str. 71-460 Szczecin, Poland

The study was performed on the young boars and gilts the polish synthetic line 990. The work included only the litters where 12 and more alive piglets were born. The standardization to 12 piglets in litters was performed on the first day after birth. On a base the analyses of polymorphism RYR1 gene, three genotypes were identyfied: $RYR1^{CC}$-50, $RYR1^{CT}$-121 and $RYR1^{TT}$-28 pigs. In all groups was the similar number of young gilts and boars. The animals were weighted at 21, 28, 63 and 180 days of live. The weight of body shaped, respectively: for $RYR1^{CC}$ genotype - 5.88, 7.96, 20.52 and 107.5kg; for $RYR1^{CT}$ genotpye - 5.85, 7.84, 20.52 and 105.7kg and for $RYR1^{TT}$ genotype - 5.39, 7.13, 19.86 and 103.2kg. From 63 to 180 days of life the test on alive pigs was performed. An average daily body weight gain in all groups shaped, respectively: 750,735 and 723g. The percentage meat content in the body was determined on alive pigs. The highest 59.5% meat content in the body reached the pigs with $RYR1^{TT}$ genotype. In the groups $RYR1^{CC}$ and $RYR1^{CT}$ the percentage meat content in the body was 58.6; 59.1.
The study was financed by Committee for Scientific Research (grant No 6 P06E 035 20)

Poster P4.15

Genotype impact on the pig performance realized within commercial herds in the Czech Republic

M. Sprysl, R. Stupka, J. Citek, M. Pour, M. Okrouhla, Czech University of Agriculture, Department of Pig and Poultry Science, Prague, Kamycka, 165 21, Czech Republic

The test was aimed at the verification of the impact of the genotype on partial indicators of production traits of final pig hybrids within commercial breeding in the Czech Republic. The test included 576 final pig hybrids of 8 recommended genotypes ((LWxL)x LW_S, D, PN, (LW_SxBL), (PNxD), (PNxH), Seghers and PIC) and focused on the evaluation of the basic fattening performance and carcass value.

The outcomes have proved that within

- the fattening performance
  - the examined genotypes show highly evidential differences in the growth intensity;
  - they manifest their highest growth potential at the weight/age of approx. 60kg/130 days;
  - they do not show feed intake and feed ratio conversion according to the recommended standards.
- the course of the meat formation
  - they show significant statistical differences between each other with a higher or lower linear decrease of the meat share;
  - genotypes may be divided into those showing a high meatiness in higher slaughter weights exceeding 110 kg [(LWxL)x LW_S, D, (BOxBL), Seghers], and the genotypes that reach the required meat share only up to the weight of 105 kg [(LWxL)xPN, (PNxD), (PNxH), PIC].

Poster P4.16

IgY technology in weaned piglets

P. Danek, R. Becková and M. Rozkot, Research Institute of Animal Production, Prague, Czech Republic

The objective of our experiment was to test the presumption that control of intestinal bacterial flora with yolk antibodies(to major pathogenic agents attacking piglets and a probiotic *Enterococcus faecium* M 74) has a growth stimulating effect. The product was used in four uniform groups of 10 piglets weaned at the age of 28 days., The piglets were fed a commercial diet during the first three weeks after weaning (for 3 experimental groups supplemented with aditivum (dose 0,8 (E1), 1,7 (E2), and 3,1 (E3) g/animal/day), (C) was a control group). Weight loss during the first week was observed in several piglets C (50 %), E1 (30 %) and E2 (20 % piglets), but in none of the group E3. The highest daily weight gain was observed in the group E3 (300 g). Compared with C (217 g), the weigh gain in E2 was higher by 8.7% and in El was lower by 7.5%. Feed consumption per 1 kg of weight gain was in all the experimental groups inversely proportional to the dosage of tested product. The dose 3,1 g eliminated weight loss and assured a considerable increase in weight gains during the first week after weaning and maintained a higher growth rate throughout the experimental period. The results have confirmed the presumption that the application of the IgY technology can enhance the performance of piglets in the post weaning period. (Supported by NAZV Grant QD0100)

Poster P4.17

Drip loss of case-ready meat and its associations with earlier measurements at slaughterhouse and with genetic markers

G. Otto[1], R. Roehe[1], H. Looft[2] and E. Kalm[1], [1]Institute of Animal Breeding and Husbandry, University of Kiel, D-24098 Kiel, Germany, [2]PIC Deutschland GmbH, Ratsteich 31, D-24837 Schleswig, Germany*

Case-ready meat (CRM), a product of increasing importance, from 374 animals was measured for drip loss during each of 7 days after packing at 24h post mortem (CRM_{1-7}) and its associations were analysed with earlier measurements at slaughterhouse using Bag method (BM), EZ-DripLoss method (EZ-DL) and with genetic markers. Mean drip loss of CRM increased substantially from 1.57 to 5.64% during 7 days. At slaughter house, drip loss was 1.8 and 3.11% using BM after 24 and 48h and 4.71% using EZ-DL after 48h measured from samples taken at 24h post mortem. Consecutive measurements using BM showed large correlation with 0.98 and were highly correlated with EZ-DL (0.89). Therefore, the easy to perform and sample standardised method of EZ-DL was recommended. Correlations between drip loss using CRM_{1-7} and those using BM or EZ-DL were large in a range of 0.82 to 0.90, indicating that earlier measurements at slaughterhouse were highly informative. The analysis of 12 DNA-markers revealed three markers to be significant for drip loss with differences between homozygous genotypes of 1.1 to 2.07% determined by EZ-DL method. In conclusion drip loss of case-ready meat can be efficiently improved by reducing drip loss obtained by EZ-DL method at slaughterhouse by using genetic markers.

Poster P4.18

Investigations on the effects of split-weaning

Zs. Benedek, J.P. Polgár, Cs. Rajnai, Z. Lengyel and J. Kovács, Georgikon Faculty of Veszprém University, H-8360, Keszthely, Hungary*

The aim of the present study was to investigate the effects of split-weaning on suckling behaviour and growth of piglets, and on reproduction traits of the examined sows.

9 Large White (LW) sows and their litters (n=103 born piglets) were observed. At day 28 of lactation, piglets in each litter were classified as either "heavy" (7.06 kg mean weight) or "light" (5.00 kg mean weight). Heavy piglets were removed from sows at around day 28 (±2-3days). Light piglets (at least 4 piglets per sow) remained with their mothers for an extra week. The results demonstrate that after removing heavy piglets, most of the light ones suckled more than one teat during the one week period. Their growth increased with the more consumption of milk. During the one week extra lactation period no sows came to heat, but following the final weaning they all returned to estrus within 8 days. The investigations have not showed any negative effects of the split-weaning on the reproductive cycle of the sows yet. Further observations are continuously made on the effects of split weaning.

Poster P4.19

Problems with aggressive and sexual behaviour when rearing entire male pigs

L Rydhmer[1], G Zamaratskaia[2], HK Andersson[3], B Algers[4] and K Lundström[2], Depts of [1]Animal Breeding and Genetics, [2]Food Science, [3]Animal Nutrition and Management, [4]Animal Environment and Health, Swedish University of Agricultural Sciences, Funbo-Lövsta, S 75597 Uppsala, Sweden

Our aim was to evaluate aggressive and sexual behaviour of entire males and their sisters. Aggression level in the pen (ALP) during routine feeding and aggressive interactions between individuals competing for a small amount of feed (IA) were tested on 204 pigs at 130 and 150 days of age. Sexual behaviour (mounting) was recorded from 100 days of age. ALP was lowest in single-sex pens with gilts. ALP was higher at 130 days than at 150 days. A high ALP was related to higher average growth rate. IA increased after the three heaviest pigs in each pen had been slaughtered, although these were the ones that initiated a majority of IA earlier (at 130 days). Males in single-sex pens showed more sexual behaviour than males in mixed pens. Fewer injuries were noted in gilt pens than in mixed and single-sex pens with males. 20% of the males and 10% of the gilts had health problems, almost all related to lameness or injured legs or feet. One male died and three males from single-sex and two males from mixed pens were culled due to lameness or leg fracture. Rearing of entire males might cause welfare problems, due to increased aggression levels and sexual behaviour.

Poster P4.20

Effect of transport and lairage time on various physiological and pork quality traits

Cs. Ábrahám[1,2], M. Mézes[2], M. Weber[2], K. Balogh[2], H. Febel[3], Gy. Huszenyicza[2], M. Vada[4], E. Szücs[2] and K. Ender[1], [1]Research Institute for the Biology of Farm Animals, Wilhelm-Stahl-Allee 2, D18196, Dummerstorf, Germany, [2]Szent István University, Páter K. u. 1., H-5711, Gödöllö, Hungary, [3]Research Institute of Animal Breeding and Nutrition, Gesztenyés út 1., H-2053, Herceghalom, Hungary, [4]Meat Research Institute, Gubacsi út 6/b., H-1097, Budapest, Hungary*

The aim of this research was to study the effect of transport and lairage on pork quality and physiology. A total of 40 halothane-free pigs were transported from pig farm to abattoir. After arrival the animals were divided into two groups: 1 hour lairage and 16 hours one. Blood samples were taken before and after transportation and during exsanguinations. The following physiological measurements were recorded: cortisol, NEFA, lactic acid, glucose along with parameters describing the antioxidant defence system (GSH, GSHPx, ascorbic acid). Meat quality parameters (pH, core temperature, L^*, a^*, b^*, chilling loss) were measured at 45th minute, and 24th hour post mortem.

The transportation causes heavy distress for pigs. All parameters changed more or less significantly during transportation, but no significant difference was found between the two groups either in physiological or meat quality traits.

In conclusion, when using halothane-free animals, the meat quality is largely independent of lairage time. Lairage might have a beneficial effect, but driving to stunning and stunning itself eliminates it.

Poster P4.21

The diameter of muscle fibres of pigs in relation to breed, sex and weight

R. Klimas, Siauliai University, P. Visinskio 25, 76285 Siauliai, Lithuania.

A study was conducted to determine the diameter of muscle fibres (DMF) of purebred Lithuanian White (LW), Finnish Yorkshire (FY), Finnish Landrace (FL) and their crossbred (LW x FY, LW x FL) pigs fattened up to 100 and 130 kg weight. Equal number of gilts and castrates were allotted to each group. Housing and feeding conditions were the same for all pigs. For determination of the DMF, samples were collected from *M.longissimus dorsi* at 9-12 th rib after carcass chilling for 48 hours. The fibres were 280- fold magnified, and the diameter was microscopically measured with eyepiece and objective micrometers. 120 samples (60 for 100 kg and 60 for 130 kg pigs) in total have been analysed.

The highest DMF in the dorsal muscle has been determined for FL pigs and it was higher than that of FY pigs by 20.4 and 7.1%, respectively, for 100 and 130 kg weight ($P<0.001$). The DMF for crossbreds were by 15.5 - 19.5 and 11.5 - 12.3% higher (respectively, for 100 and 130 kg pigs) in comparison with purebred Lithuanian White ($P<0.001$). Besides, this parameter of gilts was higher than that of castrates in most groups. Thus, the DMF is higher for pigs with higher lean meat content in the carcass. The increase of pig weight from 100 to 130 kg has a positive influence on the DMF ($P<0.001$).

Poster P4.22

Assessments of processing suitability of pork meat in function of meat origin and diet

G. Barbieri, A. Pizza, R. Pedrielli, M. Bergamaschi, C. Gianni and M. Franceshini, Stazione Sperimentale per l'Industria delle Conserve Alimentari,43100 Parma, Italy*

Physical and chemical properties of raw meat affect maturation and processing suitability. Samples of shoulders (glicolityc muscles), masseters (oxidative muscles) and thigh trimmings of heavy (Duroc x Large White) pigs, fed with a high linoleic acid diet (HPLA) and with a standard one (HPC), were compared with meats of commercial hybrid light pigs from two different origin (LP1and LP2). All the samples, chilled at 0°C for 5 days and analysed, were partially vacuum-packed and frozen for 13 days, 7 months and 11 months at -16°C. Meat origin and diet factors significantly ($p<0.005$) affected protein solubility and mechanical properties of chilled samples. HPC meats were characterised by greater myofibrillar solubility (8.14±0.13g/100g) and shear force (WB) value (43.53±0.92N) and lower elasticity module MY10 (0.05±0.01Mpa) in comparison to the meats from light pigs and from HPLA. With freezing operation myofibrillar solubility significantly ($p<0.05$) decreased and WB and MY10 values increased for all the samples. Significant differences were observed between the chilled and the first storage time samples, whereas there were few differences between the three different storage time (13 days) for these parameters, since vacuum-packaging probably obviated significant changes ($p>0.05$) in frozen material during storage. Prolonged storage especially increased TBA (0.56±0.09) values in oxidative muscles and MY10 (0.33±0.06) values in shoulders.

Poster P4.23

Characteristics of whole pig carcasses

J. Pulkrábek, J. Pavlík, L. Valis, M. Vítek, Research Institute of Animal Production Praha - Uhríneves, Prátelství 815, 104 01 Czech Republic*

Totally 126 pigs from common pig operations in the Czech Republic were involved in the analysis to determine the predicative ability of whole carcass traits for estimation of general meatiness. The relationships among these traits and lean meat proportion were assessed. First carcass weight was analysed. It was logically negatively correlated with lean meat proportion. However, the relatively low correlation coefficient indicates that current pig hybrids may be able to produce heavy carcasses with a high lean proportion. High correlation coefficients ($r_1 = 0.82 \pm 0.052$; $r_2 = 0.83 \pm 0.050$) were determined when the relationships were evaluated using the estimations of whole carcass lean proportions calculated on the basis of the regression equations RG1 and RG2, respectively. A high predicative ability of the used equations is evidenced by virtually identical estimated average lean proportions. The average lean proportion estimated by RG1 was 54.58 ± 0.331 %. In addition, high correlation coefficients were calculated when some of the measures of fat thickness was involved in the analysed relationship. This may be particularly applied to backfat thickness based on the average of three measurements where the correlation coefficient was $r = -0.77 \pm 0.058$. This fact should be employed in the used methods of pig carcass quality estimation. Only low correlation coefficients were found for the traits characterising carcass length. Thus, differences among pig carcasses from the current types of final hybrids cannot be determined on the basis of these traits.

Poster P4.24

Possibility for estimation of the corporal type at pigs with compactness index

V. Bacila[1], M. Vladu[2], I. Vlad[1], I. Calin [1] and P. Tapaloaga[1], [1]University of Agronomic Sciences and Veterinary Medicine Bucharest, Animal Science Faculty, Technology Department, 59 Marasti boulevard, sect. 1 , 011464, Romania, [2]University of Craiova, Agricultural Science Faculty, Phytotechny - Animal Science Department, 15 Libertatii street, 200583, Romania*

Pig body could be considered a cylinder more or less compresses laterally. On this basis we could estimate the volume of pig body as the produce of the body length and the surface of a vertical section in the pig body. Having an ellipsoidal form, such a surface is committed by multiplying 1/2 of the thorax depth to 1/2 of the thorax width and further to the body length. The data obtained for a number of 138 young boars, six months of age are presented. Every time, the value of the compactness index is over the unit, as it have to be. The former procedure estimates the compactness index as the thoracic circumference/body length ratio and shows some times over unit, or sometimes under unit values, excepting Landrace population where they are always underunit values. The density of a pig body must be an over unit value. There-for the new procedure proposed must be considered a better estimation of the density (or compactness) of a pig body.

Poster P4.26

Perception and preference of individual meat parts of pork by czech consumers

M. Pour and M. Pourová, Czech University of Agriculture Prague, Faculty of Agronomy, Department of Pig and Poultry Science, Czech Republic

The paper is focussed on the perception and preference of pork, its individual meat parts (loin, neck, ham, belly, shoulder and knuckle of pork) and on the purchasing behaviour of the Czech consumers. Primary data was acquired by means of questionnaires and the research carried out at the beginning of 2002 included 151 respondents. All meat parts such as loin, neck, ham, belly, shoulder and knuckle of pork are according to consumers available in the Czech shops and some of them are too expensive (ham, loin). The neck, belly and knuckle of pork were perceived as soft after preparation and at the same time fat. A nice appearance was positively evaluated in the loin, ham and shoulder of pork which are less fat. It is evident that in most cases consumers evaluate better the qualities such as lower content of fat and greater softness after the preparation. Associated with it is also the evaluation of convenience in use and a nice appearance. This proves the appropriateness of breeding of slaughter pigs for a higher share of lean meat.

Poster P4.27

The fatty acid composition of different meat and fat samples of Mangalitsa

*G. Holló[1], K. Horváth[1], *Cs. Ábrahám[2], J. Seregi[1], I. Holló[1], É. Varga-Visi[1], G. Pohn[1] and I. Repa[1], [1]University of Kaposvár, Guba S. 40, 7400 , [2]St. István University, Páter K. 1, Gödöllö 2103, Hungary*

The aim of this study was to establish the fatty acid composition of two muscles, (longissimus, semimmembranosus) back fat and bellies. A group of fattened Mangalitsa (n=10) were divided into two groups (I. 91,64±2,41 kg, II. 114,14±7,70 kg $P<0.001$) according to their slaughter weight. The carcass classification was made by FAT-O-MEATER. The fatty acid composition was determined by a Chrompack CP 9000 gas-chromatograph. The statistical analysis was made by SPSS 10.0.program package. The estimated lean meat content of carcasses was in both groups about 38 %, which is typical for Mangalitsa. Comparing the average back fat thickness (I:4,8 cm vs. II.:7,0 cm) and rib eye area (I.:4,8 cm vs. II.:5,4 cm) of the two groups significant differences can be found. It deserves attention, that the intramuscular fat content of LD in case of other native pig breeds was lower, than that of Mangalitsa (7-9%). The intramuscular fat contains a low polyenoic fatty acid percentage and high oleic acid content compared to commercial breeds. The ratio of stearic acid and the linoleic acid in case of the back fat was favourable: 1,06. Concerning PUFA, there was no significant difference, their ratio both in back fat and bellies were about 15 %, which is a desirable value.

Poster P4.28

Development perspectives of Lithuanian White pigs in direct of lean meat and body conformation

A. Mikelenas[1], Al. Mikelenas[1] and A. Muzikevicius[2], [1]Lithuanian Veterinary Academy, Animal Science Department, Tilzes 18, 47181 Kaunas, Lithuania, [2]Ministry of Agriculture, Republic of Lithuania, Gedmino pr. 19, 2025 Vilnius, Lithuania

The objective of this study was to estimate lean meat of Lithuania White pigs of breeding farm and control test stations at the same time determining the nearest development perspectives of lean meat and body conformation.

Using ANOVA analysis (fixed effect model) was determined that lines of boars and families of sows in the year 2001-2003 influenced increase of results.

Using statistical analysis we ascertained that the biggest influence for lean meat had some line of boars average 55.6% ($P<0.001$).

Good results presented also progenies of our selected families after three years from 52 to 55% lean meat average. Their influence is less than English Yorkshire boars that was used infusing blood and developing lean meat of herd however statistically is significant ($P<0.05$). Unfortunately pig's carcass is shorter executing selection of body conformation, and of lean meat. With regards to control test station results length of carcass was 4.5 cm ($P<0.001$) within these three years.

In three generations it is possible to collect over 60 percent pigs of desired production (54% lean meat and more) in the pedigree pigs herd of 5000 selected animals containing from 30 to 60 percent of desirable traits in annual selection pressure.

Poster P4.29

Effect of diet and age on Δ9-desaturase expression in pigs in relation to intramusclular lipid formation

E.Doran[1], S.K.Moule[2] and J.D.Wood[1], [1]Department of Clinical Veterinary Science, University of Bristol, Langford, Bristol, BS40 5DU, UK, [2]Department of Biochemistry, School of Medical Sciences, University of Bristol, Bristol, BS8 1TD, UK.*

Intramuscular lipid (IL) is an important component of meat quality. The amount of IL increases with the age of animals and can also be modulated by diet. The mechanism of IL formation remains unclear. Activation of expression of lipogenic enzymes might play a significant role in this process. One of the key lipogenic enzymes is Δ9-desaturase (Δ9-d), catalysing the formation of monounsaturated fatty acids. The aim of the present study was to investigate Δ9-d expression in muscles of pigs under conditions known to increase the amount of IL (age and a low lysine diet). Enzyme expression was estimated by Western blotting in the *longissimus* muscle of (i) Meishan x Large White pigs of different age and (ii) Duroc x Large White x Landrace pigs reared on low and high lysine diets supplemented with one of the following oils: palm kernel (PK), soya bean (SB) or palm (P). The Δ9-d level was increased in 174-day old animals when compared with 114 and 144-day old. The low lysine diet supplemented with PK and SB oils also increased the Δ9-d level when compared with the high lysine diet with similar oil supplementation. No effect of the low lysine diet on Δ9-d expression was observed in the presence of P oil. The results suggest that activation of Δ9-d expression might be one of the factors regulating IL formation.

Poster P4.30

Sex-related muscle fibre thickness and ham slaughter value analysis

T. Cervenka, J. Cítek, T. Neuzil, L. Trcová and M. Okrouhlá, Czech University of Agriculture Prague, Faculty of Agronomy, Department of Pig and Poultry Science, Czech Republic

Ham consists of the three basic muscle groups. Two muscles have been chosen for the purpose of our analysis, these are first semimembranosus and second adjuctor. Three tests with hybrid pigs of the firm product were conducted. Those were used for subject to completed VJH experiments. Slaughter pigs with a slaughter weight of 105 kg. 12 (test 1), 14 (test 2) and 24 (test 3) carcass hybrid pigs used in Czech Republic were chosen for our research. Individual muscle samples were measured and evaluated for their thickness. All of measurement results were processed by the SAS program.
Firstly, the average detected thickness of semimembranosus muscle fibres were 39,55 μm (test 1); 57,31 μm (test 2) and 45,14 μm (test 3) in female pigs and male pigs values were 39,16 μm; 52,47 μm and 48,32 μm. Secondly, the average detected thickness of adjuctor muscle fibres were 56,47 μm; 45,86 μm and 59,44 μm and male pigs values were 51,53 μm; 48,21 μm and 49,37 μm. The female pigs demonstrated a bigger muscle fibre thickness compared to male pigs values in two out of three tests. Significant differences were documented between muscle fibre thickness and level of partial index carcass value in carcass pigs.

Poster P4.31

Evolution of the meat/fat accretion and the body composition in growing pigs

N. Warnants, M.J. Van Oeckel and D. De Brabander, Ministry of the Flemish Community, Agricultural Research Centre, Department Animal Nutrition and Husbandry, Scheldeweg 68, 9090 Melle, Belgium

Knowledge of the lean gain is required for the factorial derivation of nutrient requirements. To date there are no compositional data available for the in Belgium very popular Piétrain x hybrid fattening pig. Therefore, the meat/fat accretion and the body composition was studied by the slaughter technique, at 8, 30, 50 and 70 kg live weight. Four barrows and 4 gilts were slaughtered at 8 and 30 kg; 8 barrows and 8 gilts were slaughtered at 50 and 70 kg. Pigs were fed a well-balanced protein rich feed to allow optimal growth. Body, carcass and offal were analysed for protein, fat, ash and phosphorus. From 8 to 30 kg the lean gain amounted to 400 g/d for barrows and gilts, from 30 to 50 kg the lean gain was higher for gilts (506 g/d) than for barrows (491 g/d). Between 50 and 70 kg, gilts increased lean gain with 531 g/d, whereas barrows were already over the top with 450 g/d. The body's fat content was similar for both sexes at 8, 30 and 50 kg with respectively 11, 11 and 15%. At 70 kg the pig's fat content was 23% for barrows and 17% for gilts. This confirms the decrease in lean gain in barrows between 50 and 70 kg.

Poster P4.32

Use of video-analysis in the determination of meatiness of the carcass belly in pigs

J. Citek, M. Sprysl, R. Stupka and M. Pour. Czech University of Agriculture Prague, Department of Pig and Poultry Science, 165 21 Prague 6-Suchdol, Czech Republic

The aim of the work was to verify the possibility of using video-analysis for an objective determination of meatiness of the carcass belly in pigs. The belly part is one of the biggest parts of the carcass. Therefore it is necessary to know the exact composition of this part in terms of the meat: fat ratio.

The work included a detailed dissection of 60 carcass belly in order to identify supporting indicators with a high informative value relating to the share of lean meat in the belly. In practice, the following indicators may be used for the calculation of the share of lean meat in the belly: lean meat area (AM2) and total area of the belly (AT2) in the section between 10th and 11th ribs in mm^2 and the weight of the belly (WEU) in kg: y = 30,41037334 + 0,59364437*(AM2/AT2) - 2,29878959*WEU ($r^2 = 0,784$). Additional indicators may be used for further specification of the estimate: lean meat area (AM3), total area of the belly (AT3) in mm^2 between 7th and 8th ribs and the lean meat area (AM1) in mm^2 behind the last rib: y = 42.63841413 + 0.24603687*(AM2/AT2) - 3.43803239*WEU - 0.00098125*AT3 + 0.00254507*AM3 + 0.00088281*AM1 ($r^2 = 0,857$). The results of the work have proved the possibility of using video-analysis in the estimate of meatiness of the carcass belly.

Poster P4.33

Formation of the lean meat of the belly meat part in relation to the achieved live weight of pigs

R. Stupka, M. Sprysl, J. Citek, M. Pour and M. Okrouhla, Czech University of Agriculture, Department of Pig and Poultry Science, Prague, Kamycka, 165 21, Czech Republic

The tests included 200 final pig hybrids currently produced in the Czech Republic in a balanced sex. Analysis was made of the formation of the lean meat of the belly meat part from the viewpoint of its total percentage in the carcass, the percentage of lean meat and formation of the belly in the monitored part in relation to the live weight.

The results have shown that the increase of the weight does not significantly influence the share of the belly in the carcass, the same trend has been observed also in the percentage of the EU belly in the total belly and in the carcass. The increase of the body weight results in a rapid increase of the total area of the belly which is accompanied by a slower increase of the meat area. Further, the increase of the weight also contributes to the increase of the meat weight but at the same time it decreases the percentage of meat in this part. There occurs a significant decrease of the share of the meat area in the belly approximately up to the live weight of 105 kg, subsequently this decrease slows down. In addition, different deposition of meat and fat in individual sections (1,2,3) has been found out in pigs with a low weight, i.e. a higher share of meat as compared to the animals with a higher weight, i.e. a lower share of meat.

Poster P4.34

Study on the possibilities for direct selection including gene markers for improvement of fattening and carcass traits in pigs

E. Jeliazkov[1], S. Tanchev[1], Ts. Yablanski[1], S. Georgieva[1] and G. Batchvarova[2], [1]Trakia University, Faculty of Agriculture, Department of Animal Genetics, Breeding and Reproduction, Student's town, 6000 Stara Zagora, Bulgaria, [2]"Reproductor" - AD, Yambol, Bulgaria

The aim of the present study was to research the influence of RYR1 and CM genes upon some fattening and carcass traits in pigs. 13 Swedish Large White and New Bulgarian Synthetic Sire Line boars were investigated on the gene markers of HAL-RYR1 and CM. Their offsprings were fattened and tested in the Test Station near to town of Yambol. The following traits were recorded: fattening period (days), age at 100 kg live weight (days), daily gain (kg) and forages per 1 kg gain (kg), slaughtering weight (kg), slaughtering weight (%), backfat tickness CKL_2 (mm), loin eye area (sm2), large and small slaughtering length (cm), weight of neck steak (kg), ham and shoulder weight (kg), loin weight (kg) and salo weight (kg). The fixed ANOVA model were applied.
The genotypes of RYR1 locus have significant effect on slaughtering weight, backfat tickness CKL_2, loin eye area, ham weight, loin weight and salo weight of their offsprins. The different genotypes of CM locus have significant effect on the loin eye area only. There was a tendency, the offsprings of the boars from the New Bulgarian Synthetic Sire Line with heterozygous CM genotype, to have better results of ham and shoulder weight, loin weight and salo weight.
The obtained results are basis for future studies on these kind of relations.

Poster P4.35

Relationships between firmness of dry-cured hams and breeding values for curing weight loss traits

L. Degano, E. Sturaro, M. Noventa, O. Bonetti, L. Gallo and P. Carnier. University of Padova, Department of Animal Science, viale dell'Università, 16, Agripolis, 35020, Legnaro, Padova, Italy

Dry cured ham (DCH) is the most valuable product of the pig industry in Italy. Firmness is an important quality traits affecting slicing. Aim of this study was to investigate the relationship between firmness of DCH (subjective evalution with linear score from 0 to 4) and breeding values (EBV) of pigs for weight loss at the end of curing (WL) and for traits related to the dynamic of weight loss (WLD). Ham weights were recorded at arrival at the ham factory, 15 and 130 day since arrival, and at the end of seasoning (300 days). The dataset included records on left hams from 1602 crossbred heavy pigs (6428 weight records). The dynamic of weight loss was estimated by fitting a nonlinear model (Von Bertallanfy function) and instantaneous rates of weight loss (IS) were determined at different stages of curing (from 10 to 100 d by 10 d steps). Heritability of WL was 25%, whereas heritability of IS ranged from 17 to 30%, with higher values for later curing stages. EBV for WL and IS affected firmness of DCH. High firmness was associated to high EBV for WL and IS.

Poster P4.36

Analysis of firmness of dry cured hams in relation to fresh and dry cured hams traits

M. Noventa, E. Sturaro, O. Bonetti, L. Gallo and P. Carnier, Department of Animal Science, University of Padova, Agripolis, 35020 Legnaro (PD), Italy*

This study aimed to investigate relationships between firmness of dry cured hams and fresh and dry cured ham traits. Measures and scores for firmness of muscles and of subcutaneous fat were obtained for 722 dry cured hams produced by crossbred heavy pigs. Pigs were slaughtered in 11 groups at 163±15 kg of live weight. Herd, management and feeding regime was the same throughout the trial. Firmness was measured by a Hardness Meter MK2. Measures were taken on inner and outer locations of different muscles and of subcutaneous fat. Firmness traits were analysed using a linear model. Firmness was different in different muscles and also in different locations of the same muscle. Variation of muscle firmness was influenced by the slaughter group, sex, carcass weight, some raw ham traits, and weight loss during curing. Firmness of subcutaneous fat exhibited variation similar to that of muscles, and was influenced by the slaughter group and iodine number of subcutaneous fat of raw ham.

Poster P4.37

Correlation between the body weight of young sows on days 21, 28 and 63 of their life and their reproductive value in the first litter

J. Owsianny, R. Czarnecki, B. Matysiak and A. Konik, Departament of Pig Breeding, University of Agriculture, ul. Dr Judyma 10, 71-460 Szczecin, Poland

The primary aim of the study was to determine correlation between the body weight of young sows at their first two months of life and their future reproductive value measured by the number of piglets born and the number and weight of litter on day 21 and weaning day. The study covered 143 young sows of line 990 weighed on days 21, 28 and 63. On day 180, their fattening and meatiness values were determined, which warranted the selection of the best young sows for herd reproduction. Mean body weight obtained by the young sows was, respectively, 5.83 kg, 7.87 kg and 20.32 kg on day 21, 28 and 63. The obtained coefficients of correlation between the body weight of examined sows and their future fertility were low and statistically non-significant (they ranged r=-0.03 between the body weight of sows on day 70 and the number of reared piglets to r=0.12 between the body weight of sow on day 21 of her life and the weight of litter weaned from her on lactation day 28). The obtained results do not warrant the forecast of future reproductive use of sows basing on body weight development in the first two months of their life.
The study was financed by Committee for Scientific Research (grant No 6 P06E 035 20)

Poster P4.38

Computer assisted morphometric analysis of spermatozoa from Large White and Landrace boars

D. Tapaloaga[1], E. Mitranescu[1], P. Rodian Tapaloaga[2], L. Tudor[1] and F. Furnaris[1], [1]University of Agricultural and Veterinary Medicine Bucharest, Faculty of Veterinary Medicine, Splaiul Independentei 105, Romania, [2]University of Agricultural and Veterinary Medicine Bucharest, Animal Sciences Faculty Bucharest, Marasti Avenue No. 59 Romania

Ejaculates from 12 boars from Large White and Landrace breeds were fixed in eosin-nigrosin. The samples were examined using an automated sperm morphometry analysis system for sperm head length and width, tail length and the total length of spermatozoa. Sperm body shape index and sperm head shape index were also calculated.
Sperm head length was lower in Large White boars (9,26 μ ± 0,305) compared to Landrace boars (9,29 μ ± 0,203).
The other data recorded lower values in Landrace boars (sperm head width 5,32 μ ± 0,03 μ vs. 5,35 μ ± 0,025 μ; sperm tail length 41,4 μ ± 1,18 μ vs. 43,88 μ ± 1,37 μ; sperm body shape index 21,34 ± 0,26 vs. 22,305 ± 1,44; sperm head shape index 1,73 ± 0,063 vs. 1,76 ± 0,02).

Poster P4.39

Genetic fertility markers in Norwegian landrace pigs in the Czech Republic

R. Beckova, P. Danek and L.Urbankova, Research Institute of Animal Production, Prague, Czech Republic

The aim of this study is to give information about variability of candidate genes associated with fertility in the population of the mother breed Norwegian Landrace in the Czech Republic. In cultivation rears of the breed and Landrace of the Norwegian provenience (NL) tests for the occurrence frequency of genotypes of markers *RYR, ESR, PRLR, OPN, and MYF4* in the laboratory of the molecular genetics of the MZLU in Brno. A high frequency of N allele in mother breed NL(NN 95.6%) proves a systematic selection for a stress predisposition in the Czech Republic. The genotypes *MYF4 were AA in 97.8.*OPN genotypes in population NL following frequencies: AA =39.1 %; AB =50.0 %; BB =10.9 %.In the population Norwegian Landrace all the sows were homozygote with the genotypes ESR CC. The *PRLP AluI AA* genotype has not been identified. Genotypes *AB* and *BB* occurred in the frequencies 44.4 % and 55.6 %.We present the results of association studies of orientation in NL only in two candidate genes *PRLP* and *OPN* in which the population kept in the Czech Republic is polymorphic. In the homozygote genotype *PRLP BB* there was a higher number of all piglets than in the heterozygote genotype *AB(11.3 vers. 9.6)*. In the marker OPN a higher number of born piglets was in the genotypes AA.
(Supported by the NAZV, QD 0100).

Poster P4.40

Genetic and environmental parameters for growth traits in Norwegian landrace

E. Gjerlaug-Enger, D. Olsen and H. Tajet, Norsvin, P.O.Box 504, NO-2304 Hamar, Norway

The objective of this study was to investigate genetic and environmental correlations of growth for pigs at different ages. Data on age at 100 kg were collected in nucleus herds on 54.000 gilts and boars, and age at 25 kg and days from 25 to 100 kg were measured on 6.000 pigs in boar test and 7.000 gilts and castrates in test stations for meat- and slaughter quality on the boars' half sibs. A Multi-Trait AI-REML animal model was applied using the DMU program (Madsen, P., and J. Jensen. 2003). Phenotypic data from the last four years and pedigree information back to 1993 were included.

All the genetic correlations were positive, but an early growth is not the same trait as the growth after 25 kg (r_g=.35±.08). Selection for age at 100 kg produce a great effect for the growth before 25 kg (r_g=.83±.04) and for the growth from 25 to 100 kg (r_g=.77±.04). There was a negative environmental correlation between early and late growth which supports the theory about compensatory growth. The environmental correlation for herd*year (r_{hy}=-.59±.06) and residual (r_e=-.18±.03) cancel out the positive genetic correlation (r_g=.35±.08) between age at 25 kg and days from 25 to 100 kg, resulting in small phenotypic correlations. Due to this it is important to design multi trait models which include both the genetic and environmental effects when working with growth.

Poster P4.41

Genetic diversity among Vietnamese indigenous pig breeds in comparison with European and Chinese breeds

A.W. Kuss[1], E. Melchinger[1], N.T.D. Thuy[2], N.V. Cuong[2], H. Bartenschlager[1] and H. Geldermann[1], [1]Department of Animal Breeding and Biotechnology, University of Hohenheim, D-70593 Stuttgart, Germany, [2]Institute of Biotechnology (IBT), National Center for Natural Science & Technology, 18-Hoang Quoc Viet Rd, Cau Giay, Hanoi, Vietnam*

Indigenous Vietnamese pig breeds were investigated for their genetic diversity in comparison with breeds from other origins. The experiments were based on material from five Vietnamese autochthonous breeds (Co, Meo, Tap Na, Mong Cai), two exotic breeds kept in Vietnam (Landrace, Yorkshire), three European commercial breeds (German Landrace, Piétrain, Large White), the European Wild Boar and the Chinese breed Meishan. Considering each breed equally, samples and data from about 300 animals were collected and 20 polymorphic microsatellite loci, distributed equally throughout the genome, were selected and genotyped. The Vietnamese indigenous breeds had a higher degree of polymorphism as compared to the other pig breeds and heterozygosity was also markedly higher than in the breeds of European or Chinese origin. This pronounced allelic diversity observed in the Vietnamese autochthonous breeds might be the result of heterogenous environmental influences as well as the distribution in village specific subpopulations. Genetic distances among the breeds reflect their phylogeny and geographical distribution. The results show that Vietnamese indigenous pig breeds constitute a reservoir of considerable genetic diversity for use in livestock bio-conservation.

Poster P4.42

Genetic correlations between exterior traits and stayability in pigs

H. Brandt[1], H. Henne[2], [1]University of Giessen, Institute of Animal Breeding and Genetics, Ludwigstr. 21 B, 35390 Giessen, [2]Züchtungszentrale Postfach 30 40, 21320 Lüneburg*

On a total of 160.000 young sows from two dam lines genetic parameters for 9 exterior traits, liveweight daily gain and backfat were estimated. For 55.000 of these sows fertility traits and the stayability after first and second litter was available to estimate the genetic correlation between the exterior and the fertility traits. For the exterior traits a subjective scoring system with scores from 1 to 5 was used. Low to medium heritabilities between 7 to 25 percent were estimated for these traits. All exterior leg traits showed heritabilities lower then 15 percent. For the stayability the heritabilities ranged from 4 to 8 percent within the lines. Common environmental litter effects for the exterior traits ranged from 2 to 7 percent. Both liveweight daily gain and backfat showed from a breeding point of view a slight negative genetic correlation to stayability. From the 9 exterior traits the side view of the rear leg and the body length showed genetic correlations to the stayability above .25 in both lines. From these results it can be concluded that longer animals will have a reduced stayability, as well as sows with bow hind legs.

Poster P4.43

The anatomical structure of the sow's udder- a different point of view

N. Pospieszny[1] , W. Poznanski[2] , A. Rzasa[3], Z.Zawada[1] , Agricultural University of Wroclaw, [1]Institute of Animal Anatomy, ul. Kozuchowska 1/3, [2]Institue of Pig Breeding, ul.Chelmonskiego 38d, [3]Institute of Immunology and Veterinary Prevency, ul.Norwida 31*

The aim of this work was a fuller study of the anatomical structure of the sow's udder with particular reference to the number of nipple glands, their distribution and size and the number of milk canals in each nipple.

A detailed evaluation of milk glands taken from 8 pbz x wbp sows right after slaughter on the last day of the second lactation was carried out. The obtained material was used to make corrosion preparations by filling milk canals with plastic. In total 83 corrosion preparations were made, unused nipples were disregarded. 2-canal glands were most numerous and 3-canal ones were found in all areas of the udder, i.e. in pectoral, abdominal and inguinal sections.

It is suggested that lactation of particular glands is determined primarily by their position and not by the number of milk canals. On the other hand, the area where SigA can be synthesized is determined by the number of milk canals. It seems that milk from nipples with higher number of canals can be richer in immunoglobulins and, consequently, can be not only a better nutritional agent but also a prophylactic measure in digestive system diseases.

Poster P4.44

Composition of sow's milk: a new point of view

Anna Rzasa[2], Wieslaw Poznanski[1], Jerzy Akincza[1] and Anna Procak[1], Agricultural University of Wroclaw, [1]Institue of Pig Breeding , ul.Chelmonskiego 38d, [2]Department of Immunology and Veterinary Prevency, ul.Norwida 31*

The chemical composition of sow's milk from particular nipples has been estimated many times. Results were differentiated and often contradictory. It testifies to the variability of the parameter, leaving ground for doubts whether everything has been taken into account at its evaluation. The chemical composition of milk depends generally on: nutrition, breed, age of sows but it may also be determined by the number of milk canals in the nipple.

Sow's udders were inventoried with regard to anatomical structure of nipples and then quality of milk on 21-st day of lactation was compared in relation to chosen chemical traits as well as proteinogramme of whey.

Preliminary estimation of chemical composition of milk taken from nipples of different structure shows that 3-canal nipples have milk with slightly higher amount of fat, dry matter and protein as compared to 2-canal ones.

The amount of milk produced does not necessarily go together with its immunological value. Sow's milk differs from milk of other species because of SIga secreted during the whole lactation period. Number of milk canals is consistent with the area where these immunoglobulines can be synthesized, hence future research on anatomical structure of sow's udder and its effect on piglets' rearing results is highly advisable and may contribute a lot to practical pig breeding.

Theatre PL5.1

Herd modeling for improved management: characterizing the herd manager in information space

M.A.M. Commandeur, INRA-LRDE, Quartier Grossetti BP 8, 20250 Corte, Corse, France

Modern livestock herd managers work with a variety of farm objectives in face of their actual situation and future insecurities. There is the daily reality of herd, labour and income that has emerged from the past and with which they have to work. Supported by technology this daily reality is constantly progressing towards further increase of intensity and scale. The future perspective is clouded by a variety of insecure scenarios, of which some are not promising: falling prices, increased risks of disease outbreaks and/or food quality problems, conflicting interests of consumers' demands, etcetera.

Where do farmers take a stand in this turbulent environment? Field surveys have revealed that there is no single answer. Depending on the *rationale* for their current situation and their *ambition* for the future development of their farm, farmers hold a variety of positions. There are various *styles of farming*. Each style of farming represents a specific and logic whole of rationale and ambition, and represents specific objectives and strategies. Making use of examples in pig farming it will be discussed how to analyse styles of farming and which demands each style of farming has with regard to management information. Herd modeling for improved management should therefore relate to the specific objectives and strategies of the various styles of farming.

Theatre PL5.2

Computer based analysis of individual sow herd performance

J. Krieter, Institut of Animal Breeding and Husbandry, Christian-Albrechts-University, 24098 Kiel, Germany*

Increased herd sizes and narrowed income margins are common characteristics of modern swine farming. Therefore good management becomes more and more important for the economic results.
The paper describes a computer based weak point analysis of individual sow herd performance. Three stages were distinguished: (1) Tracing deviations between farm performance and a given standard in order to detect trends in the production process. It is shown that a modified exponentially weighted moving average control charts are an effective tool in detecting small performance shifts. (2) Weighting the deviations by calculating the statistic and economic relevance allows the ranking of different traits independently of scales und units. (3) Finding the causes for the performance shifts. Decision tree algorithm was investigated to gain more insight in the critical points of production. A decision tree starts with the root node representing the traits which mostly influenced the target attribute (critical point), followed by internal nodes and leaf nodes. The generated graphical decision trees are transparent and the outputs are easy to interpret for the farm manager or the consultant. Methods were applied to simulated and real sow herd datasets.

Theatre PL5.3

Herd modeling for improved management: practical applications

H. Jubbega, FARM software,Keizersveld 43b, 5803 AM Venray, The Netherlands

Management of commercial pig farms is rapidly becoming more complicated with the increasing scale of pig production. Automated recording systems for growing pigs, and especially sows, have been widely used for about 15 years now, initially mainly to provide the farm manager with the data to base his day-to-day tactical decisions on. But the information captured there can also be exploited for automated optimization of the production unit, which provides the farm manager with a list of optimized actions for today, based on his pre-defined long-term strategies. This paper deals with the state-of-the-art in this respect of current software, and options for application in commercial production, using the FARM package as an example. FARM has been developed since 1985 with Power Builder using a Sybase SQL database, which allows for a stable multi-user and multi-tasking environment under Windows. Because of the international perspective (available in more than 20 languages) the software has specific country settings; changing the language is a one-key process. FARM provides the technical and financial part as well. Reports and analyses are easily made and user-defined. Future developments beside the individual management program are an internet-based product designed for Tracking & Tracing. This will allow for connections with genetic, feed, veterinary, abattoir and other databases, so that information from those sources can be consolidated locally and analysed to provide management support. Validation of data takes place at the moment of input. Controls and settings can be done by the organisation, but it is the user's responsibility to enter the correct data.

Theatre PL5.4

Decision support model for pig production management

S. Lungu[1], C. Lazar[2], [1]UASV Bucharest, [2]ICDA Fundulea, Romania

The relations between environmental conditions, biological potential of animals and economic data, represent the core of the animal husbandry model. Within farm, the possibility of collecting and using technical information is increasing and for a decision support, this information flow offers the advantage of monitoring of the animals.
The objective of this article is to investigate and develop a method for monitoring the production of sows, of principal events from technological flow of pigs, using time related observations.
A model describing the dynamics of population pigs from different age categories is proposed. It was chosen to focus on following problems: fecundity of sows in correlations with environmental conditions and dynamic evaluation of growth rate in continuous slaughter pig production.
The following aspects were investigate in detail: the number of pregnant sows is generated each ten days on a basis of a polynomial relationship with temperature and available space per animal, the prolificity is simulated through a random function which generate a distribution similar with parameters obtained from observed data, the distribution of the weaners into different classes of initial weight is considered to depend on the prolificity, the growth rate is a function of initial weight and available space. The main conclusion that was drawn: every series of fattening pigs that leave the farm for slaughtering, allows us to estimate growth rate for each delivery to slaughter, in a continuous production efficiency of using space.

Theatre PL5.5

The principal components analysis as a tool for a better management of bovines farming system

J.C.-R. Santos and E.V. Barros, Divisão de Produção Animal (DRAEDM) - Rua Franca, n.º 534, 4800-875 S. Torcato, Portugal*

The principal components analysis method leads us to select a group of discriminating variables and reassembles the analyzed farm based on its parameterization. We studied a group of 35 farms where Minhota is the main local breed. This cow-calf breed is raised in the Northwest of Portugal, within a system where the main characteristics are its reduced number of animals per farm (1-6), permanent lodgement (83% are kept in the cattle-shed) and traditional feeding system (oat+rye-grass+rye: 1,7-16,9 Kg DM/cow.day; maize straw: 2,6-22,0 Kg DM/cow.day; maize silage: 1,1-21,0 Kg DM/cow.day), being the farms constituted by many patches (3-33) with reduced dimension (0,06-0,40 m^2). The farmer have an advanced age (31% are above 65 years), low literacy degree (11% cannot read and write) and low qualification level (66% without technical courses). We choose 10 variables (from an initial group of 50) that presents significant statistical correlations ($P<0,05$) and explains more than 50% of the total variance. Based on these 10 variables, the ACP built the following equations that will allows us to reassemble the 35 farms according to its characteristics.

$Y_1=(0{,}487_{AMED}+0{,}423_{PAR}+0{,}415_{CABO}-0{,}093_{PNUM}-0{,}122_{AMAC}+0{,}168_{NAVE}+0{,}105_{IPAR}+0{,}396_{ISIL}-0{,}0921_{ÁRCA}+0{,}428_{AREJ})$

$Y_2=(0{,}234_{AMED}+0{,}29_{PAR}-0{,}146_{CABO}-0{,}173_{PNUM}+0{,}472_{AMAC}+0{,}531_{NAVE}-0{,}066_{IPAR}-0{,}248_{ISIL}+0{,}454_{ÁRCA}-0{,}181_{AREJ})$

Although it is necessary further studies, we can conclude that ACP can be used as a tool in order to decide the best strategy to improve bovine herd's management.

Modeling the management of production in sheep

S. Cournut[1] and B. Dedieu[2], [1]ENITAC Agricultures et Espaces 63370 Lempdes, [2]INRA-TSE 63122 Saint-Genès-Champanelle, France

The development of lamb quality signs is one of the major evolutions of the French lamb meat market. It requires a controlled distribution of the quality carcasses offer throughout the year based on adapted rearing systems with two and more lambing sessions per year. Producing models able to take account of different rearing strategies is a challenge for the development of decision support tools. It includes considerations on production level, production distribution and replacement. In this paper, we propose a decision model referring to the conceptual framework of "management of production" coming from industry, considering the livestock project, the strategic steering and the operational action program. The rules of strategic steering define and coordinate functional collective entities (batch, batch production cycle) and connect the reproductive herd to the cohorts of products and ewe lambs. The validation of this "management of production model is made with experts and bibliography. The implications on the herd dynamics simulation techniques are discussed: i) various animal entities (from the flock, the batch to the ewe) have to be taken into account as connected objects of the model design ; ii) the simulator must include either the calendar time of the farmers actions or the biological time of the animal responses. Illustration is given with the TUTOVIN discrete events simulator which models the effect of different management rules of the accelerated -3 lambing per 2 years- rearing system on the lamb numbers per fortnight.

Poster PL5.7

The use of herd-test-day solutions of the random regression test-day model in dairy herd management web-tool "Maitoisa"

M. Koivula[1], J.I. Nousiainen[1], J. Nousiainen[2] and E.A. Mäntysaari[1], [1]MTT, Agrifood Research Finland, Animal Production Research, FIN-31600 Finland, [2]Valio Oy, FIN-00039 Valio, Finland*

A random regression test-day model considers test-day records as repeated observations within parity and as different traits across lactations. Moreover, instead of herd-year (hy) classification of management groups, the herd-test-day (htd) classification in the test-day model allows a better interpretation of the short-term environmental variations in the production- and SCC traits between test-day months. This is one of the important advantages of the test-day model compared to the animal model. The herd-test-day solutions of milk yield (milk deviation, kg/day), protein and fat concentration (protein and fat deviation, %) and SCC (SCC deviation) are used in the dairy herd management diagnostic web-tool "Maitoisa" (in English "Milky"). It helps to recognize several management problems e.g. in feeding. Besides deviations from the population mean, modelled deviations of herd-test-day solutions assist users to identify unusual test-days, repeated seasonal variation, and enables the prediction of probable deviations in the subsequent year. The solutions are plotted by test-day months and displayed as graphs within internet-browser. For recognition of patterns and unusual test-days, a time series type model of three random sine curves is fitted to the solutions. In addition to management effects of his own herd, the farmer can request the country or region quartiles to be displayed in the graphs. The Maitoisa service has been offered to farmers and dairy advisors since 2001.

Theatre PG6.1

Statistical analysis of longitudinal data in genetics

L. Varona, Area de Producció Animal, Centro UdL-IRTA, Av. Rovira Roure 191, 25198 Lleida, Spain

In recent years, longitudinal traits are attracting the interest of scientific community in statistical genetics. These traits evolve along time, and consequently, different genes can be involved in its regulation at each point of the production process. Traditionally, these traits have been analysed by splitting the data into several traits (e. g. weight at different ages). However, it is also possible to include, into the model of analysis, a statistical description of the data as a function of time. This approach allows the genetic modification of the shape of the function of the data along time. Two main alternatives have been proposed recently: 1) Bayesian analysis of hierarchical models of productions functions and 2) Random regression models. Both procedures are review and compared in the scope of pig production traits, such as body weight, body composition or food intake. A simulated example was presented to illustrate this comparison. Finally, new possibilities such as the influence of prior information, analysis of categorical data, detection of QTL, or calculation of residual food intake are also discussed.

Theatre PG6.2

Genetic analysis of growth curve parameters for beef cattle using Markov Chain Monte Carlo estimation methods

E. Venot[1], D. Laloë[1], M. Piles[2], A. Blasco[2], G. Renand[1], A. Vinet[1] and F. Jaffrézic[1], [1]Institut National de Recherche Agronomique, Département de Génétique, Domaine de Vilbert, 75352 Jouy-en-Josas Cedex, France, [2]Universidad Politécnica de Valencia, Departamento de Ciencia Animal, Valencia 46071, Spain*

Performances taken into account nowadays in French genetic evaluation of beef cattle consist mainly of young growth characteristics (weight at birth, at 120 and 210 days). However, selection for better young growth rate generally implies a correlated increase of mature animal weight. Modeling the entire animal growth will allow the breeders to include selection criteria on different parts of the growth curve and therefore better manage the animal selection.
Data used in this study came from an experimental Charolais herd, created in 1985-1987 out of 300 pure bred Charolais heifers mated with 60 Charolais bulls. 913 females born in a 16 years period were weighed monthly. Eventually, 65% of these animals have more than 50 weights recorded, associated with their pedigree information and other individual factors.
Growth curves were modeled with the three parameter Brody function, each parameter estimated by means of a mixed model including both animal random effect and systematic effects (such as twinning, dam age at calving and year of birth). Heritabilities and genetic correlations for these parameters were obtained by a Bayesian approach using Gibbs sampling.

Additive genetic and phenotypic aspects of feed intake and feed efficiency using random regression

B. Nielsen[1], P. Berg[2] and L. Damgaard[1,2], [1]Danish Bacon and Meat Council, Axeltorv 3, 1609 Copenhagen V, [2]Danish Institute of Agricultural Sciences, Dept. Animal Breeding and Genetics, PO.Box 50, 8830 Tjele, Denmark

Linear models with fixed and random regression coefficients were used to describe the cumulative feed intake as a function of live weight in the period from 30 kg to 115 kg for performance tested boars (DanBred) from the Danish test station. For each animal data consist of periods with complete daily feed intake measurements and weekly body weight measurements. Random regression coefficients of both additive genetic and permanent animal effects were estimated. Based on the models applied, the change in additive genetic, permanent, and phenotypic variability over the growth period are presented. Correlations between feed efficiency in sub periods and mean feed intake during the total test period is presented.
The estimated fixed regression coefficients in the model allow calculating the mean daily efficiency from the derivative of the cumulative feed intake curves. The result shows how the feed efficiency of growing pigs increases during the growth period. Implications of random regression models for selection for feed efficiency are discussed.

Theatre PG6.4

Covariance functions for modelling weight gain in pigs

Spela Malovrh[1], Eildert Groeneveld[2] and Milena Kovac[1], [1]University of Ljubljana, Biotechnical Faculty, Animal Science Dept., SI-1230 Domzale, Slovenia; [2]Federal Agricultural Research Centre (FAL), Institute for Animal Science Mariensee, D-31535 Neustadt, Germany*

Performance tested boars were analysed for repeated measurements of body weight. The data set consisted of 3,819 boars of four lines with 22,907 records from the Federal Hybrid Crossbreeding Scheme. Boars were housed in groups, fed with automatic feeders and weighed six times during the test. The pedigree consisted of 7,406 animals. The fixed part of the model was composed of class effect of test day (day of weighing), linear regression on age at start within successive weighing, and quadratic regression on time on test within line. Covariance components for random regression coefficients were estimated by the REML method using the VCE5 software package. From a linear to quartic Legendre polynomials for time on test were used as regression for weight gain for additive genetic and different environmental effects. Animal permanent environmental effect for weight gain over time accounted for 35 to 40% of phenotypic variation. The proportion of common litter effect decreased with time (0.24 to 0.15), direct heritability was around 0.40, and maternal heritability had a value of 0.03 at the start and 0.01 later. Correlation between direct and maternal additive genetic effect was positive. It changed from 0.33 at the beginning of trajectory over close to zero to 0.42 at the end. Eigenfunctions and corresponding eigenvalues showed that 90% of the total genetic variability was explained by the constant term in regression, while 10 % was genetic variability in the shape of growth curve.

 Theatre PG6.5

Random regression analysis of cattle growth path

H.R. Mirzaei, A.P. Verbyla, M.P.B. Deland and W.S. Pitchford, University of Adelaide, Roseworthy Campus SA 5371 Australia, SARDI, Struan, Naracoorte SA 5271 Australia

A cubic random regression analysis was conducted to model growth of crossbred cattle from birth to about two years of age. Hereford cows (581) were mated to 97 sires from Angus (AxH), Belgian Blue (BxH), Hereford (HxH), Jersey (JxH), Limousin (LxH), South Devon (SxH) and Wagyu (WxH), resulting in 1144 steers and heifers born over 4 years. The model for ln(wt) included fixed effects of sex, sire breed, age (linear, quadratic and cubic), as well as two-way interactions between the age parameters and sex or breed. Random effects were animal (genetic relationship), dam, residual between animal, linear age by animal, age by dam, management group, management by age as linear, quadratic and cubic. At birth, AxH, JxH and WxH were 7, 18 and 11% lighter and than purebred Hereford calves. Growth rate increased during the pre-weaning period with a change in the rank of breeds during the first 6 months. Relative growth rate was higher in the smaller breeds (JxH and WxH) than the heavy breeds. Genetic (r_G) and maternal permanent environmental (r_C) correlations between birth weights (constant) and linear growth rate was strong and negative (-0.87 ± .04 and -0.96 ± .04, respectively). The model is being used to predict the effect of growth path on carcass quality.

Theatre PG6.6

Genetic parameters for number of piglets born alive using random regression model

Z. Lukovic[1,2], G. Gorjanc[1], S. Malovrh[1] and M. Kovac[1], [1]University of Ljubljana, Biotechnical Faculty, Zootechnical Department, Groblje 3, 1230 Domzale, Slovenia, [2]University of Zagreb, Faculty of Agriculture, Department of Animal Production, Svetosimunska 25, 10000 Zagreb, Croatia*

Genetic parameters for number of piglets born alive (NBA) were estimated using random regression model. Litter records data from three large Slovenian pig farms from the first to the tenth parity were used for analyses. Fixed part of the model included genotype of sow, mating season, parity and weaning to conception interval as class effects. The age at farrowing was adjusted as quadratic regression nested within parity. Previous lactation length was fitted as linear regression. Random regressions on Legendre polynomials of standardized parity were included for direct additive genetic, permanent environmental and common litter environmental effect. Legendre polynomials from the linear power to cubic power were fitted. Heritabilities on all three farms ranged from 0.09 to 0.14. Permanent environmental effect as ratio increased along trajectory from 0.03 to 0.12. Magnitudes of common litter effect were generally small (0.01 to 0.02). The eigenvalues of covariance functions showed that between 10 and 20 % of genetic variability was explained by individual genetic curve of sows. This proportion was mainly covered by linear and quadratic coefficients. The solutions for the random regression coefficients were used to compute breeding value for NBA along trajectory.

Theatre PG6.7

The use of allometric and nonlinear functions for longitudinal data analysis of physical and chemical body composition in pigs

R. Roehe[1], S. Landgraf[1], M. Mohrmann[1], P. W. Knap[2], H. Looft[2] and E. Kalm[1], [1]Institute of Animal Breeding and Husbandry, University of Kiel, D-23098 Kiel, Germany, [2]PIC International Group, Ratsteich 31, D-24837 Schleswig, Germany*

Allometric and sigmoidal functions were used to analyse changes in total body composition (soft tissue, bone, viscera; primal carcass cuts and their components; organs; chemical composition) during growth of pigs, obtained from a three generation full-sib design used to identify QTL. Allometric functions showed high goodness of fit (R^2 = 0.98) for the mass of chemical body protein, lipid, ash and water in relation to empty body weight with allometric growth coefficients b = 0.99, 1.68, 0.99 and 0.86, respectively. However, increase in lipid deposition at higher weight (120-140 kg) deviated from the allometric pattern. If this growth period is of interest, nonlinear functions were more appropriate. Additionally, we analysed protein and lipid accretion with two-stage nonlinear models, fitting Gompertz functions in relation to age for individual pigs. This resulted in estimates for maximum protein accretion rate at 132 and 135 g/d and for mature protein mass at 27.9 and 28.0 kg for females and castrated males, respectively. Allometric functions of empty body weight and nonlinear Gompertz functions of age mostly fitted our data well; their few and biologically interpretable parameters can therefore be used to identify genomic regulation of growth with DNA markers.

Theatre PG6.8

Estimation of Gompertz growth curve parameters by a nonlinear mixed effects model

K. Vuori, I. Strandén, M-L. Sevon-Aimonen and E.A. Mäntysaari, Agrifood Research Finland, Animal Production Research, 31600 Jokioinen*

Simulated growth data with parameters reasonable for pigs were generated through a 2 step process. First, a linear mixed model was used to produce Gompertz growth curve coefficients for each pig. Then, 30 weekly weights were produced using the coefficients. Simulated pedigree had 3 generations without selection: 10 unrelated grandsires and 100 sires, each with 12 tested offspring. All mates were assumed unknown. A method based on Taylor series approximation was applied to estimate breeding values for curve parameters and the corresponding (co)variance components. In every iteration, a working variate (sum of predictions by linearized model and a non-linear model residual) was utilized to solve location and dispersion parameters of the linearized model. Approach required only minor changes to existing software for random regression models. In addition to data from full growing period, a truncated time trajectory data was analyzed to test the approach when animals are slaughtered prematurely. Truncation was at slaughter weight 110kg, which reduced the number of measurements to 11 per pig on average. From 50 replicates with full data the bias of estimates of three genetic variances were 2%, 9% and 3% of the parameter values. Corresponding values for the truncated data were 11%, 9% and 30%. Results of this pilot study indicated that the linearization approach was able to estimate original parameters satisfactorily for full data, but not as well for truncated data.

Theatre PG6.9

Comparing body weight measured as single time point versus modelled as a longitudinal trait for QTL detection in pigs

A.J. Buitenhuis, M.S. Lund, A.H. Petersen, L.B. Madsen, F. Panitz, R. Labouriau and C. Bendixen, Danish Institute of Agricultural Sciences, Department of Animal Breeding and Genetics, P.O. Box 50, 8830 Tjele, Denmark*

Genes involved in body weight measured at different time points may be differentially expressed because each time point represents a different developmental stage in the growth curve of an animal. Often body weight is analysed as a trait at each time point whereas growth is often taken as a difference between two time points. However, in reality this is a continuous process. By modelling body weight as a longitudinal trait the growth curve is taken into account. In addition, using this method may increase the power to detect time-dependent QTL effects, because differential expressed QTL have likely a small average effect. In this study a pig population consisting of 12 half-sib families with a total of 10,000 animals was phenotyped for growth measured at four different time points and genotyped with SNP markers. The aim of study was to compare the QTL results of single and multiple trait analysis regarding growth as separate traits at the four time points versus modelling growth as a longitudinal trait using candidate regions on the porcine genome.

Theatre PG6.10

The relationship between heritability and genotype-by-environment interaction using reaction norm analyses in Merino sheep traits

G. E. Pollott[1] and J. C. Greeff[2]. [1]Imperial College London, Department of Agricultural Sciences, Wye, Ashford, Kent, TN25 5AH, UK., [2]Great Southern Agricultural Research Institute, 10 Dore Street, Katanning, WA 6317, [3]Australia*

The use of reaction norm analyses to investigate genotype by environment (GxE) interactions also results in estimates of the way heritability varies with the 'environmental variable'. It is postulated that traits with a constant heritability across the environmental range will show no evidence of a GxE effect. The analyses of 11 production traits of Merino sheep were used to investigate the relationship between GxE and heritability in different environments. Linear and quadratic polynomial random regressions of sire breeding value on an environmental variable were used for the investigation. The environmental variable was the contemporary group mean for the trait being analysed. A range of patterns for the way heritability varied over the environmental range was found. For example, for greasy fleece weight, clean fleece weight and liveweight heritability was high in 'poor' environments, remained constant over the middle range of environments and then fell again in 'good' environments. Other traits had a constant heritability across the environmental range. Traits with a constant heritability all had a GxE effect of less than 0.01 of the phenotypic variance. The slope variance of the reaction norms gave an indication of the consistency of the GxE effect across sires.

Theatre PG6.11

Estimates of covariance function for lifetime growth of Australian Angus cattle

K. Meyer, Animal Genetics and Breeding Unit, University of New England, Armidale NSW 2351, Australia

Records for weights of Australian Angus cows and their contemporaries from birth to 3000 days of age were analysed fitting a random regression (RR) model on age at recoding, using Bayesian analysis via Gibbs sampling. Data comprised 450,191 records on 108,267 animals, but only 87,470 records represented weights for 18,017 cows which had mature cow weights recorded. Including pedigree information added genetic effects for 51,020 parents. The model of analysis fitted contemporary groups and cubic regressions on age at recording, nested within sex, dam age group, calving status and birth type classes, as fixed effects. Random effects were modelled through regression on age at recording, using Legendre polynomials or quadratic B-splines as base functions. The former modelled animals genetic and permanent environmental effects through cubic and quartic polynomials, respectively, and fitted a quadratic regression for both genetic and environmental maternal effects. The spline analysis fitted 7 knots (at 0, 300, 600, 900, 1500, 2100 and 3001 days) for both animal effecs, and 3 knots (at 0, 300 and 601) days for both maternal effects, assuming no maternal influnec after 600 days of age.

Theatre PG6.12

Use of linear splines to simplify longitudinal analyses

I. Misztal[1], J. Bohmanova[1], S. Tsuruta[1] and H. Iwaisaki[2], [1]University of Georgia, Athens, 30605, USA, [2]Niigata University, Niigata-shi 950-2181, Japan

This study investigated the use of linear splines as alternatives to polynomials in random regression models. With linear splines, parameters for all effects except permanent environment and residual can be the same as in multiple trait models, simplifying validation and computations. Also, artefacts at boundaries are less likely. One comparison involved simulated data in beef cattle involving weights at days 1, 205±50 and 365±50. Models included were multiple trait, random regression with cubic Legendre polynomials, and random regression with linear splines and 3 knots. Variance components in the three models were equivalent at days 1, 205 and 365. The multiple trait model was the least accurate because it did not did not account for variability in days for random effects. Both random regression models gave nearly identical results, but the model with splines was simpler and converged much faster. In another comparison involving field data on beef cattle, variance components for a similar model were estimated by a multiple trait and by a random regression model with linear splines. Large percentage of records for birth and weaning weights were missing. The model with splines gave more realistic estimates of heritabilities and correlations. Random regression models with linear splines are simpler and safer alternatives to models with polynomials.

Poster PG6.13

Curve of weight gains in bulls

J. Pribyl[1], H. Krejcová[2] and J. Pribylová[1], [1]Res.Inst.Anim.Prod., P.O.Box 1, Uhríneves 10401, Czech Rep., [2]MLU Halle-Wittenberg, Adam Kuckhoff Strasse 35, 06108 Halle (Saale), Germany*

Legendre Polynomials (LP) with three to seven parameters were used to evaluate daily gains of 6 484 young bulls of Czech Fleckvieh cattle in performance-test stations by the age of 420 days. The model comprises effects: station-test day STD, fixed LP within the year-test station classes, random LP of the permanent environment of an individual and random genetic LP. Each animal was weighed on average 11 times in one-month intervals. Weight gain was determined in relation to the test day, GAINTO from preceding weighing, GAINAFTER until subsequent weighing and AVERGAIN as the average of gainto and gainafter. Components of variance were determined by the REML method (REMLF90 programme). Residual variance decreased with the increasing size of the polynomial, values of -2LogL and information criterion AIC decreased at the same. The residual variance of GAINAFTER (77 955 - 89 584) was on average slightly higher than that of GAINTO (74 882 - 82 292), GAINAFTER heritability was slightly lower. AVERGAIN had on average lower residual variance (48 214 - 71 838) and approximately twofold heritability (h^2= 0.07 - 0.12) compared to GAINTO and GAINAFTER (h^2 = 0.03 - 0.05). Heritability slightly increased with the size of polynomial, but both the curves of heritability and the curves of particular components of variance fluctuated. Heritability of cumulative gain ranged on average from h^2= 0.41 to 0.68.

Poster PG6.14

Estimates of genetic covariance functions for growth from birth to 550 days of age in Tabapuã cattle

L.G. Albuquerque[1,2], L.T.Dias,[1,2], H. Tonhati[1,2], [1]Faculdade de Ciências Agrárias e Veterinárias - UNESP, Jaboticabal-SP, 14884-980, Brazil, [2]CNPq/FAPESP, Brazil*

A total of 21,762 weights from 4,221 Brazilian Tabapuã cattle, from birth to 550 days of age, recorded every 3 month, were analysed using random regression models. The model of analysis included fixed effects of contemporary groups, age of dam as a quadratic covariable and a cubic regression on orthogonal polynomials of animal age. Genetic direct (A) and maternal (M), and direct (R) and maternal (Q) permanent environmental effects were modelled by random regression on Legendre polynomials of age at recording. Changes in residual variances with age were considered through a variance function. Orders of polynomial fit were from 2 to 5, resulting in up to 51 parameters to be estimated. Based on Bayesian Information Criteria, a model with orders of fit of 4, 3, 5 and 2 for A, M, R, and Q, respectively, and a quintic variance function for residuals fitted best. Variance estimates were similar to those from correspondent univariate analysis. Direct heritability estimates were highest for weights at birth and 550 days of age. Maternal heritability estimates were low, increasing slightly from birth to a peak around 160 days of age. Direct genetic correlation estimates between weights at standard ages (birth, weaning, yearling and final weights) were moderate to high while maternal genetic correlations, in general, were high but those with birth weight.

Theatre H2.1

Developmental Orthopaedic Disease

C.W. McIlwraith and B. Cox, Anthony University Chair of Orthopaedic Research, Colorado State University, USA

The term "developmental orthopaedic disease" (DOD) was coined in 1986 to encompass all orthopaedic problems seen in the growing foal. It is a term that encompasses all general growth disturbances of horses and is therefore non-specific. The spectrum of diseases was earlier designated "metabolic bone disease," which is inappropriate. Rather than referring specifically to bone, the entities of DOD usually involve joint and growth plate conditions. DOD includes osteochondritis dissecans (OCD), subchondral cystic lesions, angular limb deformities, physitis, flexural deformities, cuboidal bone malformation, juvenile arthritis and cervical vertebral malformation. The most important clinical problems are OCD and subchondral cystic lesions, and most of these are associated with osteochondrosis (dyschondroplasia). While all OCD and subchondral cystic lesions used to be considered manifestations of defective endochondral ossification, studies have shown otherwise. However, malformations in the epiphyseal and metaphyseal growth cartilages are the most significant clinical entity. Factors involved in the pathogenesis include genetic predisposition, growth and body size, mechanical stress and trauma, defects and vascularization, nutrition, endocrine factors and an association between sight, vulnerability and traumatic factors. The biochemical and molecular processes that might be involved in the initial lesion of defective endochondral ossification have been studied but not completely clarified. They will be reviewed.

Theatre H2.2

Genetic backgrounds of Osteochondrosis

A. Ricard, Station d'Amélioration Génétique des Animaux, Institut National de la Recherche Agronomique, Auzeville BP 27, 31326 Castanet-Tolosan, France

A review of the genetic basis of osteochondrosis on horses will be presented. Actually, few works give measurable results on the inheritance of this disease. First of all, the frequency of the measurement of disease varies from populations, according to the scale of measure in each country, which make comparisons difficult. Second, heritability is at present estimated on a small number of horses on major studies or on selected sampling. However, it seems that abnormality visible on X-rays is heritable with different values due to the location of the lesion and on breed (sport horses or trotters). For navicular disease, heritability is relatively homogeneous among works about 0.20-0.30 (Winter *et al.*, Willms *et al.*, Van Heelsum *et al.*, Ricard *et al.*). Osteochondrosis in hock seems heritable in Trotters (Phillipson *et al.*, Grondhal *et al.*, Ricard *et al.*) but not in sport horses (Winter *et al.*, Van Heelsum *et al.*). For fetlock homogeneous results were found with heritability lower than for navicular disease, between 0.15-0.20. For carpus, the interesting work of Dolvik and Klemetsdal suggest a higher heritability (0.65) when considering only bilateral lesions rather than all lesions.
The present challenge is to associate to this research a molecular approach, with the help of known map on mice or human.

 Theatre H2.3

Selection strategies for radiographic findings in German Warmblood horses

K.F.Stock, H. Hamann and O.Distl, School of Veterinary Medicine Hannover, Department of Animal Breeding and Genetics, Bünteweg 17p, 30559 Hannover, Germany*

The results of a standardized radiological examination of 5,928 Hanoverian Warmblood horses selected for sale at auction were analyzed for their genetic background. They were further used to predict relative breeding values (RBV) for the most prevalent radiographic findings, i.e., osseous fragments in fetlock (OFF; 20.8%) and hock joints (OFH; 9.1%), deforming arthropathy in hock joints (DAH; 11.7%) and pathologic changes in navicular bones (PCN; 24.7%). A total index radiographic findings (TIR) was derived from these RBV. The TIR was combined with the officially published total indices for dressage (TID) and jumping (TIJ) to develop selection strategies that account for orthopaedic health traits and for performance criteria.
Genetic parameters were estimated with REML under both linear animal and linear sire models. After transformation to the underling liability scale the heritability estimates for radiographic findings were in the range of $h^2 = 0.14$-0.46. They were correlated additive genetically with $r_g = -0.34$ to 0.24. If selection of sires was half-and-half based on performance criteria and radiological state, the percentages of above-average sires were almost the same as in the case of exclusively performance-based selection. However, all total indices significantly increased in the selected sires, and the prevalences of radiographic findings decreased in their offspring. Therefore, the prediction of breeding values for radiographic findings can be recommended to improve the horses' radiological state.

Theatre H2.4

Genetic analysis of the wither growth in Lipizzan horse using random regression

M. Kaps[1], I. Curik[1] and M. Baban[2], [1]University of Zagreb, Faculty of Agriculture, Animal Science Department, Svetosimunska 25, 10000 Zagreb, Croatia, [2]University J.J. Strossmayer, Faculty of Agriculture, Zootechnical Institute, Trg Sv. Trojstva 3, 31000 Osijek, Croatia*

Longitudinal records of wither height (taken at birth, six, 12, 24 and 36 months of age) were analyzed for 879 Lipizzan horses born in Croatia between 1931 and 1999. The corresponding pedigree file consisted of 4733 horses. Covariance functions were estimated by using bivariate animal models with ages defined as separate traits and by using random regression model directly on the data. In all models, sex, age of dam and stud-year-season interaction were considered as fixed effects while individual horses and residual errors were considered as random effects. The random regression model also included a permanent environment effect due to repeated measurements. Polynomial functions of order three were adequate to explain additive genetic functions. Estimated heritabilities from bivariate and random regression models ranged from 0.14 to 0.25 and from 0.12 to 0.23, respectively. Genetic correlations between longitudinal records ranged from 0.76 to 0.99, and corresponding phenotypic correlations ranged from 0.12 to 0.49. The estimated covariance functions from this study demonstrated heterogeneity of covariances of wither height in Lipizzaner horse and proved to be adequate in estimating covariances and correlations between heights at any two age points from birth to 36 months of age.

Theatre H2.5

Effects of body weight gain on skeleton development in growing sport horses

C. Trillaud-Geyl[1], G. Fleurance[1], G. Bigot[3], M. Donabedian[3], G. Perona[2] and W. Martin-Rosset[3], [1]Haras nationaux, France, [2]University of Torino, Italy, [3]INRA Centre research of Clermont/Theix,63122 Saint genes Champanelle, France*

2 groups of 12 foals of French breeds were subjected to a high (H group) or limited (L group)growth from weaning to 3 years of age. In each group, 6 foals were either of heavy (h) or light (l) body weight (BW) at weaning. The foals were adjusted to H or L growth according to INRA 1990 recommendations; and groups of the foals were adjusted to meet the same BW at 3 years according to a linear (L): BW = a + bX or a curvilinear (H): BW = aX^c models. Variations in body size and their components were determined by photometry.

At 3 years of age BW were: 559 kg (Hh); 515 kg (Lh); 492 kg (Hl) and 473 kg (Ll). Height at withers (HW) of Hh and Lh are not statistically different whereas HW of Hl and Ll are statistically higher. HW growth curves are highly correlated to BW curves in all cases. HW gain of Hl is statistically lower at 18 months of age than in Ll group. Similar figures are observed for all the other parameters either in the forelimb or hind limb. The thickness of the fore cannon bone is 9p100 higher in h group than in l group but there is no influence of BW curve.

Skeleton development is proportional to BW curve with a strong interaction of BW met at weaning.

Theatre H2.6

Influence of management and nutrition on growth in young thoroughbred horses: a case study on

N. Miraglia[1], M. Polidori[1] and M. Niccolucci[2], [1]Molise University, Departement of Animals, Vegetables and Environmental Sciences, Via De Sanctis, 86100 Campobasso, Italy, [2]Pian del Lago Stud Farm, Strada del Pian del Lago 6, 53100 Siena, Italy*

In race horses both training and racing start when growth and bone development are not completed: this situation affects considerably osteoarticular pathologies, generally depending on nutrition imbalances and mistakes in feeding planning. Management and feeding practices should be adjusted to support a high body development and to prevent the appearance of Developmental Orthopaedic Disease(DOD). An experimental controlled and balanced nutrition planning was carried out in young thoroughbred horses from weaning till 18 month (sales period) over a 12 years period. At the beginning of this experiment many negative aspects concerning management and nutrition were observed and they involved a high percentage of developmental diseases, scarce growth and poor performances in races. The changes carried out progressively in the stud concerned the use of specialised technicians in horse nutrition, monthly checks of horses for general welfare conditions, live weight, clinical conditions, parasites controls feet conditions, ration formulation in relation to monthly controls and improve in breeding planning. The success of this experiment permitted to balance the ration in relation to growth and to improve considerably the quality of these horses as demonstrated by the evident decrease in DOD and by the increase of sales and breeder's prices during these years.

Theatre H2.7

Effect of exercise and diet on the incidence of DOD

P. Harris[1], W. Staniar[2] and A. Ellis[3], [1]Equine studies group, WALTHAM Centre for Pet Nutrition, UK, [2]Middleburg Agricultural Research and Extension Center, Virginia Tech. USA, [3]Independent Nutritionist, The Netherlands

It has been suggested that that the clinical signs of Developmental Orthopaedic Disease (DOD) occur only after a progression of events that begin with a disturbance in the normal development of the cartilage followed by the superimposition of physical stresses. Due to the multifactorial nature of DOD, no single cause is likely to result in expression. Factors that may contribute include a genetic disposition, biomechanical trauma, mechanical stress through inappropriate exercise, obesity, rapid growth as well as inappropriate or imbalanced nutrition. Different combinations may be involved in different cases. Environmental or managemental factors most likely determine if expression occurs (i.e. provide the final triggering factor[s]). The way that young horses are fed and managed, therefore, may play a major role in the incidence of DOD by either helping increasing or decreasing the risk.

By reviewing the various studies that have investigated the role of the diet (in particular energy; protein, calcium, phosphorus and copper content) and exercise (both type and amount) this paper provides valuable pointers into the potential importance of these various factors in this condition. This helps provide support for the recommendations that can be made today to help reduce the risk of DOD as well as insights into future productive areas of research.

Theatre H3.1

Molecular methods and equine genetic diversity

P. Cunningham, Trinity College Dublin, Ireland

Since RFLPs became available in the 1970's, a number of developments (microsatellites, snips, sequencing) have greatly increased the power of molecular methods to describe and manage genetic variability. This paper reviews results of these applications in horse populations. Mitochondrial sequence data have clarified the nature of the domestication process in horses, and show a pattern different from that in other domesticated species. Recent results point to a very restricted recruitment of males in the domestication process. Microsatellite (and increasingly SNIP) data have clarified relationships between breeds worldwide. Deep pedigree structures have been studied closely in some populations (Lippizaner, Thoroughbred) using molecular methods. Finally, a new method, in which pedigree structures and molecular data are combined to give an estimate of genetic variability in founder populations, is presented.

Theatre H3.2

Genetic distance as a tool in the conservation of rare horse breeds

E. Cothran, University of Kentucky, Department of Veterinary Science, Lexington, KY, 40546-0076 USA

Genetic distance analysis of gene marker data can be a useful tool in the identification of unique breeds or populations that should be considered for preservation. However, there are a variety of types of genetic markers that can be tested and this raises the question of which type of genetic marker type is most useful for the identification of unique horse breeds. I have analyyzedf genetic distance among horse breeds based upon blood group and biochemical genetic data, microsatellite data, and mtDNA sequence data. Each provides useful information about the genetic distance among the horse breeds but each type of data also provides different insights. I will discuss the merits and disadvantages associated with each type of genetic analysis.

Theatre H3.3

Review on the methods of parentage and inbreeding analysis with molecular markers

B. Langlois, INRA-CRJ-SGQA 78352 Jouy-en-Josas France

In horse populations there is a great concern for pedigree. Genetic markers are commonly used for exclusion procedures to assess the right sire and dam of the foal .However pedigree information is limited because the total genetic history of an animal or a population can not be traced from the beginning. In this paper we try to review how genetic markers can help us to overcome these difficulties Formulae in the literature for estimating F from the state of markers consider the two causes that make sorting two genes alike. They are either identical by descent or alike by state. All authors agree that estimators for pairwise relatedness or individual inbreeding coefficients need a lot of independent co-dominant marker loci where alleles are balanced in frequencies in order to reach a minimum accuracy in estimations. In this perspective the development of a kit of SNP satisfying these conditions would be a tool of great interest to address the problems connected with parentage, inbreeding and genetic diversity in horse populations where the good management of pedigree information appears insufficient to do it properly.

Theatre H3.4

History of Lipizzan horse maternal lines based on mtDNA analysis

P. Dovc[1], T. Kavar[1], G. Brem[2], J. Soelkner[3], F. Habe[1], [1]Department of Animal Science, Biotechnical Faculty, University of Ljubljana, Groblje 3, SI-1230 Domzale, Slovenia; [2]Ludwig Boltzmann-Institut für immuno-, zyto- & molekulargenetische Forschung, Veterinärmedizinishe Universität Wien, Veterinärplatz 1, A-1210 Wien, Austria; [3]Department of Livestock Science, University of Agricultural Sciences, Gregor Mendel Strasse 33, A-1180 Wien, Austria*

The history of Lipizzan maternal lines is well known, but until recently it was based mainly on historical- and pedigree data. Additional information could be provided examining the nucleotide sequences of the mtDNA control region. The mtDNA control region of 212 representatives of Lipizzan maternal lines from eight traditional Lipizzan studs was sequenced and revealed 37 distinct haplotypes, clustering into almost all haplotype subgroups present in domestic horses. According to historical data, numerous Lipizzan maternal lines originating from founder mares of different breeds were established during the breed history. Haplotypes, identical to haplotypes of early domesticated horses, were also found in Lipizzans. Comparison of Lipizzan haplotypes with 56 pedigree maternal lines showed disagreement of biological parentage with their pedigree data for at least 11% of Lipizzans. Distribution of haplotype-frequencies was unequal (0.2%-26%), due to pedigree errors and due to sharing of haplotypes among some founder mares. Biased distribution of haplotypes among eight Lipizzan studs, including many private haplotypes and only one haplotype represented in all studs, suggested that the independent status of studs was preserved till today.

Theatre H3.5

Genetic diversity of the Akhal Teke horse breed in Turkmenistan based on microsatellite analysis

A. Szontagh[1], B. Bán[2], I. Bodó[3], E.G. Cothran[4], W. Hecker[1] and Cs. Józsa[2], [1]University of Kaposvár, Department of Cattle Breeding, H-7400 Guba Sándor u. 40., Hungary, [2]National Institute for Agricultural Quality Control, Laboratory of Immunogenetics, H-1024 Budapest, Keleti Károly u. 24., Hungary, [3]University of Debrecen, Department of Animal Breeding and Nutrition, H-4032 Debrecen, Böszörményi út 138., Hungary, [4]University of Kentucky, Department of Veterinary Science, Lexington, KY 40546, USA*

A sample of 48 individuals of the Turkmenian stock of the ancient Akhal Teke horse breed, indigenous in Turkmenistan and regarded to be a genetic resource, were genotyped for 12 DNA microsatellites. To place results in context, DNA samples of the following breeds were also analysed: Standardbred, Thoroughbred, Turkoman horses from Iran and Akhal Teke populations from the USA and Iran. The mean number of alleles per locus (MNA) was 6.1 with the effective number of alleles per locus being 3.821, revealing that the breed has a relatively high allelic diversity. The average genetic diversity measured as expected heterozygosity (H_e) was 0.69. The mean F_{IS} value, used for estimating the inbreeding, came to 0.033 showing a negligible deficit of heterozygotes. Despite of the separation and long history of the Akhal Teke horses, compared to other breeds its genetic diversity appears not to have reduced. The data gained from the analysis of DNA samples of non-Turkmenian Akhal Teke populations included in the study also supports this conclusion.

Theatre H3.6

Genetic characterisation of Pentro young horses by microsatellite markers

D. Iamartino, M. Fidotti, N. Miraglia and F. Pilla, Dip. S.A.V.A., Università degli Studi del Molise, 86100 Campobasso, Italy

The Pentro horse is an autochthonous population from breeding area characterized by climatic and geographic peculiarity and today it counts about 350 heads. Because of its peculiar adaptive ability, the "Pentro Horse" is a genetic resource which needs to be preserved as demonstrated by the recent constitution stud book. The genetic structure of Pentro population was studied by means of microsatellite markers (Bruzzone *et al*, 2003) with the aim of the analysis of its biodiversity and relationship versus other horse breeds. The data introduced in this work regard the genetic and morphological assessment of Pentro young horses assigned to the reproduction demonstrating again the uniqueness and the identity of the Pentro group. A total of twelve microsatellite loci were used to score 23 Pentro young horses and then the obtained data were used to compare the genetic structure of the Pentro horse to other six Italian horse breeds (Maremmano, Murgese, TPR, Halflinger, Trottatore and Bardigiano). The genetic distance among individuals was calculated as the proportion of shared alleles, (Dps=1-Ps) according to Bowcock *et al.* (1994) using the Microsat software (Minch *et al*, 1995). The Neighbour-Joining tree computed on individual genetic distance showed that all Pentro horses clustered together.

Theatre H3.7

Fhenotypic characterization of equine northeastern breed in Pernambuco state, Brazil

M.N. Ribeiro, A.E.V. Travassos, J.C.V. Oliveira, and L.L. Rocha, University Federal Rural of Pernambuco State, Brazil

The present work was carried out to make phenotypic characterization of the northeastern equine breed, through morfometrics quantitative traits (live weight (LW), thoracic perimeter (TP), withers height (WH), body length (BL), height (H), were measured. Data of 173 had been evaluated (64 females and 109 males), chosen in the bigger concentration areas in Pernambuco State, Brazil. The mean and standard deviation of LW, H, WH, BL and TP of 0 to one year of age, had been respectively 131,00 ± 5,03, 1,152 ± 0,03, 1,152 ± 0,03, 1,172 ±0,03 and 1,177±0,05; of one the two years had been: 218,90 ±45,65, 1,276 ± 0,07, 1,271±0,08, 1,284±0,06 and 1,380± 0,13; of two the three years had been: 249,90±42,54, 1,304±0,05, 1,323±0,04, 1,284±0,07 and 1,417±0,04; for animals above of three years they had been: 276,30±54,66, 1,313±0,06, 1,317±0,06, 1,320±0,08 and 1,504±0,11. To males, these values had been in the same order: 134,18±31,70, 1,097±0,08, 1,118±0,07, 1,042±0,16, 1,150±0,08, 0,490±0,05 and 0,230±0,02; 223,80±34,48, 1,273±0,05, 1,278±0,05, 1,275±0,08, 1,426±0,14, 0,550±0,03 and 0,244± 0,02; 278,10±54,41, 1,336±0,05, 1,340±0,05, 1,325±0,08, 1,523±0,11, 0,571±0,03 and 0,268±0,06; 304,87±54,43, 1,366±0,04, 1,362±0,04, 1,377±0,05, 1,573±0,09, 0,574±0,02 and 0,262±0,03. The results suggest change in breed standard due to changes in many traits accurred during several years of natural and morphological selection.

Theatre H3.8

Rare horse breeds in Northern Europe

M.T. Saastamoinen[1] and M.Mäenpää[2], [1]MTT Agrifood Research Finland, Equine Research, Varsanojantie 63, 32100 Ypäjä, Finland, [2]Suomen Hippos, Tulkinkuja 3, 02650 Espoo, Finland

There are 17 rare Nordic and Baltic horse breeds on a list upkept by the Nordic Gene Bank Farm Animals (NGH, Nordisk Genbank Husdyr). The horse breeds are cold bloods, small horses or ponies. The main breeds and the size of the current population are as follows: Döle (3000), Estonian native horse (500), Finnhorse (20000), Fjord (5500), Gotlandruss (800), Icelandic horse (74000), Nordland (2300), North Swedish horse (4500), Norwegian coldblood (1500) and Swedish Ardennes (4500). The importance of the breeds has previously been mainly in farm and forest work or transportation. The number of horses in each breed has decreased dramatically after the second world war because horses are not needed for their traditional use. The breeds have adapted well to the cold climate (winter) and other circumstances in the countries of their origin. Consequently, they are still ideal horses in Nordic conditions used for various purposes. The horses are nowadays used mainly in pleasure riding, trekking and driving. Some breeds are also used for trotting races. Many of the breeds are of same origin and closely related with each others. In some breeds the population size is very small and avoiding inbreeding is difficult. A PC-programme (EVA) developed by DIAS and financed by NGH has demonstrated in Finland for Finnhorses in purpose to plan mating and to minimize inbreeding.

Poster H3.9

The study of the gene pool of aboriginal and thoroughbred horse breeds using DNA-technology

G.E. Sulimova, M.V. Kostyuchenko, I.G. Ulina and I.A. Zakharov, N.I.Vavilov Institute of General Genetics, RAS, Moscow, Russia

To conserve the biological variety of aboriginal breeds, their genetic diversity must be estimated by means of modern biotechnology approaches. Therefore, the goal of our work was to estimate genetic diversity of five aboriginal breeds by RAPD-PCR, variation of STR and D-loop of mtDNA, including comparative analysis of D-loop variation in other equine breeds. Aboriginal breeds (Vyatskaya, Yakutian and Mongolian) were shown to have higher level of heterozygosity compared with thoroughbred breeds (Orlov and Arabian). The range of genetic differentiation of horse breeds was determined by RAPD analysis, STR variation of two loci (VIAS-H39 and MPZ001) and variation of D-loop fragment of mtDNA. According to mtDNA D-loop variation, the time of differentiation of Mongolian breed is 0.36 Mya, thus, suggesting that accumulation of heterogeneity has begun long before the domestication of horses. Gene pool of Mongolian horses can be considered as descendant from one of the main Asian sources of horse breed origination.

Poster H3.10

Genetic characterisation of the Croatian autochthonous horse breed based on polymorphic blood proteins and mtDNA data

A. Ivankovic[1], P. Dovc[2], P. Caput[1], P. Mijic[3] and M. Konjacic[1], [1]University of Zagreb, Faculty of Agriculture, Department of Animal Production, Svetosimunska 25, 10000 Zagreb, Croatia, [2]University of Ljubljana, Biotechnical Faculty, Department of Animal Science, Groblje 3, 1230 Domzale, Slovenia, [3]Faculty of Agriculture, Department for Zootechnics, Trg Svetog Trojstva 3, 31000 Osijek, Croatia.*

Cold-blooded horses breeds in Croatia, differentiated in three breeds (Croatian cold-blooded, Muinsulaner horse and Posavina horse), belong to the group of autochthonous endangered breeds. Genetic polymorphisms at the protein- and DNA-level can be successfully used for population analysis and for establishing relationships within and among breeds (populations). Mitochondrial DNA has been frequently used as a marker system in population and evolutionary studies. In this study, by analysing polymorphic blood proteins and mitochondrial DNA, we investigated the genetic structure of the Cold-blooded horses in Croatia, made an attempt to determine genetic distance between populations and within other cognition horse breeds. Mitochondrial DNA was isolated from blood samples of thirteen unrelated animals. Our results show relatively high genotypic diversity within and between three autochthonous horse breeds in Croatia. Our molecular data nicely support phenotypic differentiation between Posavina horse on one side and Croatian cold-blooded and Muinsulaner horse on the other side. These results are the first indicators for genetic profile of mentioned breeds and will stimulate further research and application of molecular data for selection.

Poster H3.11

Analysis of inbreeding in the genetic resource „Old Kladruber horse" in the period from 1993 to 2003

V. Jakubec[1], J. Volenec[2] and I. Majzlík[1], [1]Czech University of Agriculture, 165 21 Prague, [2]National Stud of Kladruby nad Labem, Czech Republic*

The Old Kladruber horse, which was established by the end of the 18th and at the beginning of the 19th century, is the most important genetic resource in the Czech Republic. The structure of the breed was in 2003: 39 stallions, 350 mares and effective population size (N_e) 140. The breed was closed against immigration in 1992 and since that time circular group mating and mating of non-related animals within the frame of the breeding scheme were applied. In 1993 and 2003 average coefficients of inbreeding ($F\overline{x}$) were calculated from 5 parental generations for stallions and mares in the whole breed, white and black varieties and sire lines within the varieties. From 1993 to 2003 decreased the $F\overline{x}$ (%) for stallions in the: breed from 7,16 to 5,47, white variety from 6,06 to 5,10, black variety from 8,21 to 5,94 and for mares in the: breed from 7,9 to 5,05, white variety from 7,29 to 4,17 and black variety from 8,4 - 5,86. The $F\overline{x}$ (%) fluctuated in the sire lines between 2,01 and -7,9 for stallions and between 2,21 and -5,72 for mares. The majority of sire lines showed a reduction in the $F\overline{x}$.

Poster H3.12

Characterisation of several Connemara Pony populations

D.M. Feely, P.O. Brophy and K.M. Quinn, University College Dublin, Department of Animal Science and Production, Faculty of Agriculture, Belfield, Dublin 4, Ireland.*

The Connemara Pony is one of Irelands few indigenous equine breeds. It is a numerically small breed, with approximately 2,000 breeding females and 250 breeding males in Ireland. The Department of Agriculture and Food in Ireland have classified the breed as endangered. Traditionally, the Connemara was a working pony and enjoyed a prominent role in agricultural life in the West of Ireland. In the 1940's an extensive export market developed for the breed and Connemara Ponies were exported worldwide. There are now 16 countries, outside Ireland, that have formed their own Breeders' Societies.
The results generated from a characterisation of the population in Ireland indicate that past breeding practices have caused a substantial loss in the breeds' genetic diversity. Paternal family sizes were highly unbalanced, with 10% of sires producing 55% and 30% of the breeding females and males respectively. The average inbreeding and relationship coefficient, for foals born between 1998 and 2001 inclusive, were 4.65% and 10.66% respectively. Contributions made by ancestors and founders were largely unbalanced.
As a consequence of this project a study is currently being undertaken to characterise and compare the genetic diversity of Connemara Pony populations outside Ireland. We will present the results of this study, which is aimed at identifying sources of genetic variability that could widen the gene pool of the Irish Connemara Pony population.

Poster H3.13

Mitochondrial DNA haplotypes in Slovenian Lipizzan mare family lines

T. Kavar[1], F. Habe[1] and P. Dovc[1], [1]Department of Animal Science, Biotechnical Faculty, University of Ljubljana, Groblje 3, SI-1230 Domzale, Slovenia*

Sequence analysis of mitochondrial DNA from Slovene Lipizzan horses was performed. In 53 Slovenian Lipizzan horses 17 distinct mtDNA haplotypes were found: ten haplotypes (Capriola, Allegra, Monteaura, Slavina, Batosta, Gratiosa, Wera, Betalka, Dubovina and Gaetana) in classical mare families, haplotype Thais in Slovenian mare family Rebecca and haplotypes C, M, Strana, Trompeta, Boka and Y in mare families of Croatian, Romanian and Hungarian origin. Prevalence of "classical" haplotypes was expected, due to the use of Lipizzans from classical mare family lines for breeding Slovenian Lipizzans after 1945. However, two or three mare family lines might have been re-established with Lipizzans with wrong maternal origin. In addition, more than one haplotype was found within four families, suggesting wrong maternal origin, documented in the pedigrees. The geographic origin of Lipizzan haplotypes remains uncertain. However, haplotypes Allegra and Monteaura could have Spanish origin due to the high frequency of such haplotypes in Iberian (Andalusian and Lusitano) and North African horses (Barbs). Haplotypes Strana and Gratiosa suggest common matrilineal ancestors with northern European ponies (Exmoor, Fjord, Icelandic and Scottish Highland). Kladrubian origin of haplotypes Batosta and Slavina is supported by the presence of both haplotypes in present Kladruber horses; and Arabic origin of haplotype Gaetana is supported by high frequency of this haplotype in Arabian horses.

Theatre H4.1

Effects of stallion and mare behaviour on conception in intensively managed mating in thoroughbreds

M. Kennedy, Writtle College, Department of Equine Studies, Chelmsford, UK

19 matings across 3 stallions were observed at a Thoroughbred stud. Mares were assessed for oestrus through their behaviour on presentation to a stallion and veterinary examination. An upper-lip twitch, bridle and felt boots were applied to all mares. One handler controlled the stallion, another restrained the mare, and a third assisted intromission. Lutenising hormone was administered after mating in an attempt to ensure ovulation.

10 matings were identified as resulting in conception by ultrasound scan at 16 days post-mating. 2 of 5 matings to stallion A resulted in conception, as did 3 of 9 to stallion B and 5 of 5 to stallion C. This suggests an effect of stallion on success of breeding.

Analysis of mare behaviour during mating identified a non-significant trend for increased clamping of the tail over the perineum where mares did not conceive ($P=0.079$, mean frequency = 2.2 vs. 1.0, Mann-Whitney U test) suggesting that they were not in full behavioural oestrus. Analysis of stallion behaviour identified a significantly greater latency to ejaculation after the final mount in matings where mares did conceive ($P<0.05$, mean and st. dev.= 18.3s ± 2.7 vs 16.0s ± 2.8, Mann-Whitney U test). This may be due to longer stimulation of the mare's reproductive tract facilitating successful conception. There was no significant difference across stallions in latency to ejaculate after the final mount.

Theatre H4.2

Estimation of genetic distance between traditional horse breeds in Hungary

S. Mihók[1], Beata Bán[2], Csilla Józsa[3], E. Takács[2] and I. Bodó[1], [1]University of Debrecen Faculty of Agronomy Department of Animal Breeding and Nutrition H-4032 Debrecen Böszörményi út 138, [2]National Institute for Agricultural Quality Control H-1024 Budapest, Keleti Károly-u. 24, [3]Veszprém University Georgikon Agricultural Faculty Department of Animal Breeding and Nutrition H-8360 Keszthely Deák Ferenc-u 16, Hungary*

Modern methods immuno- and molecular genetics are adapted for preservation of traditional domestic animal breeds, horses incl..

Comparison of blood goups, polymorphic systems and DNA structure of Nonius, Gidran and English Thoroughbred was carried out. These breeds are original Hungarian breeds established at the Mezöhegyes military stud two centuries ago Samples of 55 Gidran, 48 Nonius stallions and mares were compared to those of English Thoroughbred and Standardbred horses. The homozygozity of Gidran and Nonius population at the VHL-20, HTG-4, HTG-6, HTG-7 loci were 21 - 25%, 60 - 56%, 33 - 36%, and 49 - 38% respectively. The frequency of alleles at the HTG10, AHT4, AHT5, ASB2, HMS2, HMS3, HMS6, HMS7 loci of Gidrans were 20, 19, 26, 21, 21, 24, 24, 21 % respectively

Statistical analysis of molecular genetic results helps to estimate gene drift of endangered rare populations Based upon these results proposals can be elaborated for breeding and mating technics in order to avoid close inbreeding and genetic loss.

Estimation of genetic distance between breeds, lines and families is important for breeding strategies and long term decisions.

Theatre H4.3

Results of a contingent valuation survey to evaluate the breeding objective for the South German Heavy Horse

C. Edel[1,2], L.Dempfle[1], [1]Technical University of Munich, Department of Animal Science, Alte Akademie 12, 85354 Freising/Weihenstephan, Germany, [2]Bayerische Landesanstalt für Landwirtschaft, Institute for Animal Breeding, Prof.-Dürrwächter-Platz 1, 85586 Poing/Grub, Germany

To define the breeding objective for the South German Heavy Horse, a Contingent Valuation survey was conducted using a stratified random sample drawn out of all registered breeders in Bavaria. The main purpose of the study was to derive weighing factors for all relevant conformation and performance traits. A total of 62 personal interviews each lasting about an hour were carried out and analysed. The results show a remarkably stable preference structure across all strata. Highest weights were expressed for the quality of the legs and hoofing followed by temperament traits like 'pleasantness' and 'coolness'. High weights were also identified for the quality of the gait 'walk' and the trait 'willingness to work'. These results reflect strengths and weaknesses of the breed and indicate the breeders' attempt to preserve and improve typical characteristics of the breed. Moreover, strong preferences for certain colours were found. We also found strong commitments to the local private or cooperative stallion owners. The familiarity with the local stallion owner in combination with the apparent logistic advantages is the crucial criterion in selecting a breeding stallion after it has passed its breeding registration. As a result it seems advisable to support these established structures and recognize their relevance for the breeding process.

Theatre H4.4

Genetic variation and pedigree structure in two endangered Norwegian horse breeds

H.F. Olsen[1], G. Klemetsdal[1], J. Ruane[1,2] and T. Helfjord[3], [1]Department of Animal and Aquacultural Sciences, Agricultural University of Norway, P.O. Box 5003, N-1432 Ås, Norway, [2]Food and Agricultre Organization of the United Nations (FAO), Viale delle Terme di Caracalla, 00100 Rome, Italy, [3]Norwegian Equine Centre, Starum, N-2850 Lena, Norway

The Norwegian Norland and the Døl are both on the World Watch List for domestic animal diversity, with an average of 171 and 117 registered animals per year, in the reference population, born 1990-1998. Based on pedigree information on a total of 31,142 and 1,987 individuals for the Døl and the Nordland, respectively, an average animal of the reference population had pedigree complete for 10.5 and 7.3 generations, respectively. The level of inbreeding was about 12 % in both breeds, but with a rather different evolution over time in the two breeds. Effective population sizes were found to be 158 in the Døl and 62 in the Nordland. The effective numbers of founders were 48 and 14, respectively, extremely imbalanced relative to the actual number of founders; 770 and 49, respectively. Correspondingly, the effective numbers of ancestors were only 12 and 7, respectively, while number of founder genome equivalents were found as 2.7 and 1.8, respectively. All these parameters indicated a loss of genetic diversity in the two breeds, and address the need for careful planning of future breeding .

Theatre H4.5

Genetic diversity of the hair colour in horses

Zs. Tóth[1], I. Bodó[1] J. Sölkner[2], I. Curik[3], [1]University of Debrecen, Faculty of Agronomy, Böszörményi 138, 4032 Debrecen, Hungary [2]University of Natural Resources and Applied Life Sciences, Gregor Mendel 33, 1180 Vienna, Austria [3]University Zagreb, Faculty of Agriculture, Svetosimunska 25 10000 Zagreb,*

Coat colour variation is an important characteristic of all mammalian species. Although traditionally well accepted, the categorical approach to coat colour brings some difficulties. In theory, colour is a wave and is measured physically by spectrophotometers or chromameters on a continuous scale. Traditional Hungarian horse breeds Lipizzan, Gidran, Arabian, Nonius horses are involved in the current study. For quantitative measurement of coat colour we use the Minolta Chromameter applying the CIE L*a*b colour system. Coat colour was measured on four body parts (neck, shoulder, croup, belly) for each horse. Within the group of Shagya, Pure Bred Arabian horses (n=64) from Bábolna Stud, an average L (lightness) level of these four parts $\bar{x}$=63.83 was obtained. The same measurements on Lipizzan horses (n=69) from Szilvásvárad Stud gave a mean of $\bar{x}$=68.40. Quantitative analysis of eu- and pheomelanin was carried out by a spectrophotometric method. The main aim of this study is to estimate variance components (additive variances, environmental variances, heritabilities and genetic correlations) for the coat colour treated as quantitative trait, as well as to determine environmental factors influencing coat colour variation and shades. To determine non-genetic factor that are responsible for the coat colour we use modules and procedures of the SAS statistical package.

Theatre H4.6

Quantitative inheritance of the coat greying process in horse

I. Curik[1], M. Seltenhammer[2], S. Toth[3], G. Niebauer[2] and J. Sölkner[4], [1]University of Zagreb, Faculty of Agriculture, Animal Science Department, Svetosimunska 25, 10000 Zagreb, Croatia, [2]University of Veterinary Medicine (VMU) Vienna, Clinic for Surgery and Ophtalmology, Veterinärplatz 1, A-1210 Vienna, [3]University of Debrecen, Faculty of Agronomy, Böszörményi 138, 4032 Debrecen, Hungary, [4]BOKU - University of Natural Resources and Applied Life Sciences, Gregor-Mendel-Str. 33, A-1180 Vienna, Austria*

The quantitative inheritance of grey level was analyzed on 706 Lipizzan horses born in five state studs (Austria, Croatia, Hungary, Slovakia and Slovenia). A total of 1196 measurements (one to four per horse) of coat grey level, defined as L* parameter of the CIE L*a*b system, were taken by Minolta Chromameter CR210 during a period of four years (2000-2003). After analysis of the greying dynamics (the four parameter Richards growth equation provided the best fit) and variance heterogeneity, three data sets were formed; a) horses younger than seven years (377), b) horses older than six years (352) and c) all horses (706). On all three data sets we estimated (co)variance components by REML animal models with repeated records. The estimated heritabilities were 0.790±0.087 (young horses), 0.579±0.029 (old horses) and 0.494±0.055 (all horse) while the estimated permanent environmental effect was 0.107±0.083 (young horses), 0.000±0.000 (old horses) and 0.345±0.050 (all horses), respectively. The obtained estimates demonstrate that the speed and level of greying involve a very large heritable component. Genetic relationships with melanoma and vitiligo are investigated.

Theatre H4.7

Teaching equine sciences and technics in Italy

F. Martuzzi, A.L. Catalano and S. Filippini, University of Parma, Veterinary Medicine Faculty, Department of Animal Production, via del Taglio 8, 43100 Parma, Italy*

Italy has a horse population of about 300.000 horses. The need of educated people in the sector was noticed since long time. With the reform of the Italian University system many new Animal Science courses were born. Among these the first and only Equine Sciences and Technics three-year course was created in the University of Parma. First year of activation was Academic Year 2002-03. The degree is obtained with 180 credits. Teaching is organized in theoretical and practical hours. The course is intended to prepare students for a career in the equine world, to deepen the knowledge of people involved in that world already, like breeders, instructors, agonists or for those pursuing their passion in horses. 75 students are admitted to the first year. Due to more than 150 requests each year an admission test was necessary. A questionnaire was distributed to the first and second year students to investigate their characteristics (e. g.: provenience area, previous studies, employment, major interest areas, job expectations) and to better finalize the teaching programs. Regarding the different horse categories, the students are mainly interested in saddle horses for show jumping, then in order: thoroughbred for gallop racing and trotters. Among the subjects taught during the course, main interest areas are: genetics, nutrition and training technics. Job expectations are mainly toward equine physiotherapy.

Poster H4.8

On breeding value of the Estonian Native Horse and Tori Horse

H. Peterson, Ph.D. Associated prof. of the Institute of Animal Science, Estonian Agricultural University, Estonia

The objective of the study was the determination of breeding value of the Estonian Native- and Tori Horse, expressed by indexes.

The study is based on the results of experiments carried out in 1995-2003. According to the stallions used in both breeds, the average indexes of 7-8 years for the Estonian Native and Toric, have been calculated.

The most important qualities have been considered: general index of stallions, indexes of type, body, feet, pace, trot, and free jump. Average indexes of different horse breeds were also calculated per each year as well as for the whole period.

Based on 176 tests, the best general indexes among Estonian Native had Aroon 656 E (index 110.96), Aku 684 E (110.94), and the lowest value had Rokkar 713 E (86.68). Among main 19 Toric stallions, and acording to the 165 tests, the best improvers were Premium (index 110.91), Gimalai (109.38), and Casanova (108.72). Impairers appeared to be Reval and Hermelin (the last one only in 1999).

Poster H4.9

Morphometric descricption of the Noriker horse breed in Austria

T. Druml and J. Sölkner, Division Livestock Sciences, University of Natural Resources and Applied Live Sciences, BOKU, Gregor-Mendel-Strasse 33, 1180 Vienna, Austria*

Morphological measurements of Noriker horses from six different departments in Austria, representing 6 breeding areas, were recorded to characterise the base population of Noriker horse breeding in Austria. A total of 31 distance and angular measures were taken from 482 breeding mares and 103 stallions. Significant differences between areas were found for 10 traits. Considering the breeding value classes derived from linear type description, significant differences in 19 traits have been found. Multivariate analyses were applied separately to mares and stallions. Mahalanobis distances based on all 37 measurements were significant for pairwise comparison of all breeding classes and breeding areas. An analysis of the differences of sire lines showed no significance. The results reflect at least partially different breeding goals, and showed that there is no morphologic distinction between coat colour groups as professional breeders of the Noriker horse maintain.

Poster H4.10

The Bohemian-Moravian Belgic horse breed in the Czech Republic

M. Hajkova and Z. Matousova -Malbohanova, Czech University of Agriculture Prague, Faculty of Agronomy, Department of Cattle Breeding and Dairying, 16 21 Prague 6 - Suchdol, Czech Republic

The Bohemian-Moravian Belgic horse is one of three heavy horses breeds in the Czech Republic. This breed belongs to the genetic resource of the Czech Republic.

The analysis of fourteen body measurements was done in 85 mares and 19 stallions, three years old and older. Data were analysed by LSA using GLM procedure. The linear model included fixed effects of the sex, age, and origin (Bohemia, Moravia). Correlations between body measurements were estimated. Significant diferences among traits according to sex and age were found. Traits evaluated according to the origin did not show any significant diferences.

 Poster H4.11

The Noric and the Silesian Noric horse breeds in the Czech Republic

M. Hajkova and Z. Matousova -Malbohanova, Czech University of Agriculture Prague, Faculty of Agronomy, Department of Cattle Breeding and Dairying, 16 21 Prague 6 - Suchdol, Czech Republic

The Noric (N) and Silesian Noric horse (SN) together with the Bohemian-Moravian Belgic horse are the most important heavy horses breeds in the Czech Republic. The Silesian Noric horse belongs to the genetic resource of the Czech Republic.
The analysis of fourteen body measurements was made in mares and stallions, three years old and older (65 SN, 60 N). Data were analysed by LSA using GLM procedure. The linear model included fixed effects of the breed, sex and age. Correlations between body measurements were estimated. Significant diferences among traits according to breed, sex and age were found.

Poster H4.12

The Franches-Montagnes: a living Swiss cultural heritage

K.Wohlfender[1], G. Obexer-Ruff[2], C. Gaillard[2], M.-L. Glowatzki[2], W. Pfister[2] and P.-A.Poncet[3], [1]Federal Office for Agriculture, Switzerland, [2]Institute for Genetics, Nutrition and Housing, University of Berne, Switzerland, [3]National Stud Avenches, Switzerland

The Franches-Montagnes (FM) is the only indigenous horse breed of Switzerland. These horses originating from the Jura mountains are of the light cold blood type. They are robust, early mature, easy to handle and of multipurpose use. In former days mainly used as carriage- and working horses in agriculture, forestry and the army, it is today primarily a hobby horse showing excellent characteristics for riding and driving. Breeding of FM has a longstanding tradition and belongs to the Swiss cultural heritage and is supported by the federal government. The National stud is providing infrastructure and know-how. Stallions are educated and are placed at disposals to breeders during the breeding season. Crossing with foreign stallions were applied to some extent to reach breeding goals faster. Since 1950, primarily Warmblood stallions of various breeds have been used. The proportion of foreign blood reaches approximately 10 %. In 1997, the over 100 year old herdbook has been closed by the FM breeding association. This association put a great emphasis to maintain a large variety of blood lines. Possible negative influences of "foreign" blood on the typical characteristics of the FM and its development towards the Warmblood type gives rise to discussions among breeders. In a study of the University of Berne with molecular markers it could be demonstrated that the FM is a breed of its own that differentiated distinctly from the Warmblood.

Poster H4.13

Biometric observations of the pure bred Arabian population, breeding in the Czech Republic

Jan Navrátil, Czech university of Agriculture Prague, Czech Republic, Department of Cattle and Horse Breeding and Pure Bred Shagya-arab Society CZ

In the Czech Republic arabian horses are breeding in the three forms - as Shagya-arab, Full-blooded Arab and Pure Bred Arab (see previous reports to EAAP). In connection with the successive typology analysis of Shagya-arabs, a similar investigation of Full-blooded Arabs (32) was conducted. Values of the body measurements were obtained and campared with the results form Janow Podlaski (Poland, JP) and Tersk (USSR, T)to find out the statistical significant differences. The mares breeding in the Czech Republic are taller, but with lighter bones in comparison with the mares form Janow Podlaski. They have lighter bone and longer body size then the mares form Tersk. The sires form JP and T have bigger size of the thorax depth (statistical significant difference), but here is possible the subjective mistake of the mensurements.
At the same time the hipometric indexes were deternined as used in CZ and the results were compared with the data from JP and T.

Poster H4.14

Linkage of body measurements in Thoroughbred mares

I. Majzlík, V. Jakubec, and M. Hájková, Department of Genetics and Animal Breeding Faculty of Agronomy, Czech University of Agriculture, 165 21 Prague, Czech republic

Conformation of Thoroughbred stud Napajedla in CR (96 mares) was studied using 14 body measurements. All morphometric measures were analysed by least squares analysis using GLM procedure, correlations between traits were estimated for assessment of linkage between body measurements. The mean values of basic measurements are (in cm) : Height at withers =160,3, Body length=162,2, Width of chest=46,5, Depth of chest=74,5, Circumference of chest=194,5, Circumference of cannon bone=19,9.The results show also that mares are nearly of squared (height at withers : body length) format.
Linkage of the body measurements expressed by correlation showed significance for most relations with height of withers (stick), body length and depth of chest. The measures of pelvis showed also very good and significant linkage to most of body measurements studied. On the contrary height of withers (tape) showed no valuable relations to another measurements.

Poster H4.15

Onset and duration of oestrous period in Lipizzan fillies

N. Cebulj-Kadunc, V. Cestnik and M. Kosec, University of Ljubljana, Veterinary Faculty, Gerbiceva 60, 1000 Ljubljana*

The mare is a seasonal breeder, showing a series of spontaneous oestrus cycles at regular intervals during spring, summer and autumn months (from April to November), first commencing at 12 to 18 months of age. Well-fed stabled mares sometimes show regular oestrus cycles as late as in January. The aim of our work was to determine the onset and duration of ovarian activity in Lipizzan fillies ageing one to three years. To this purpose, we measured progesterone concentrations in Lipizzan fillies, ageing one (Group 1), two (Group 2) and three (Group 3) years (n = 15, 14 and 17, respectively) with a commercial EIA kit. Progesterone concentrations were measured twice a month (with a 10-day interval) for 13 months, starting with January. Progesterone fluctuations, indicating onset of cyclic ovarian activity, were first noticed in February for Group 3 (23% of animals), in March for Group 2 (7,14% of animals) and in May for Group 1 (13,13% of animals). In Group 1, cyclic ovarian activity was present from May to September in 33% of animals. From June to October all fillies of Groups 2 and 3 revealed oestrus cycles which were still present in January (in 35,71% and 11,76% of animals, respectively). Results of our study indicate the possibility of persistent cyclic activity in some 2 or 3 years old Lipizzan fillies without seasonal anoestrus during the winter months.

Poster H4.16

Variations of fatty acids in Ragusana ass's milk during lactation

B. Chiofalo, V. Azzara, L. Venticinque, D. Piccolo, L. Chiofalo, Sect. Animal Production, Dept. MO.BI.FI.P.A., University of Messina, Polo Annunziata, 98168 Messina, Italy

The study was carried out on ass's milk of the Ragusana race, which is still quite well represented in Sicily, with the aim of deepening the knowledge and of exploiting the role and the importance of this milk characterised by nutritional and extra-nutritional properties, reconsidering the zootechnic role of the donkey. Acidic composition of individual milk samples of 14 asses were investigated during the lactation (from 30th to 180th day after parturition). No significant differences were observed for saturated and monounsaturated fatty acids, while polyunsaturated fatty acid percentages showed a significant variability ($P = 0.026$) between 5th (15.93 ± 1.21) and 6th (20.27 ± 1.45) month of lactation; the n6 and n3 content were higher than the levels observed in milk of other animal species. The essential fatty acids (linolenic and linoleic acids) were the most represented components of the PUFA class showing higher values than human milk and that of other animal species. The quality indices (Atherogenic and Thrombogenic) were lower than the milk of different animal species. These results confirm the bio-dietetic value of ass's milk, very similar to human milk, and suggest the potential economic value of the asinine breeding, designing new valorisation perspectives for this species with positive effects for the protection of its natural habitat, mostly represented by "marginal areas", as well as for animal biodiversity.

Poster H4.17

Mare milk EFA composition and effects in human nutrition

J. Goracci, M.C. Curadi and M. Orlandi, University of Pisa, Veterinary Medicine Faculty, Animal Production Department, Italy*

Mare's milk seem to contain alpha-linolenic (ALA) and linoleic (LA) acids amounts respectively precursors of ω-3 and ω-6, higher than in cow's milk. Most infant formulae contain only the precursor essential fatty acids (EFA), alpha-linolenic acid and linoleic acid, from which formula-fed infants must respectively synthesize their own docosahexaenoic acid (DHA) and arachidonic acid (AA). The purpose of our studies concerns the evaluation of mare's milk quality, since some our current researches seem indicate interesting EFA amounts and variability connected with breed and with administered feedstuff.

Mare milk samples from 22 Haflinger multiparous mares were collected during the first 105 days of lactation, in order to evaluate the fatty acids content's variation, saturated/unsaturated ratio and, particularly, linoleic/alpha-linolenic ratio. Collection time was 30,60,90,105 days. Linoleic acid amount was 10.89%, as average, on total fatty acids, in 30 days samples, with a significant ($p<0.05$) decrease until the end of the observation period (8.54%). Alpha-linolenic significantly increased ($p<0.05$) until 105 days milk samples from 5.56% and 6.66% on total FA. EPA and DHA were present only as traces. Linoleic/alpha-linolenic ratio was about 2:1 in 30 days samples and pointed out a progressive decrease, with an increasing alpha-linolenic acid amounts during lactation. Saturated/unsaturated ratio was favorable to the first ones as a ratio between 1.2:1 and 1.3:1.

Poster H4.18

Influence of sex on pH values from horse meat in Southern Italy

F. Nicastro, S. Dimatteo, L. Zezza, A. Pagone and R. Gallo, Department of Animal Production, University of Bari, Via G. Amendola, 165/A, 70126 Bari, Italy

Fourteen colts, seven male and seven female, were fed on straw, bran, oat grains, ground cereals (corn and oat) and milk serum *ad libitum*.

After slaughtering (age 24-26 months), pH values have been measured from shoulder and ham muscles and from *Longissimus dorsi*. From a sample of this muscle, stored at 4°C, pH values have been measured after 24, 48 and 168 hours of storage.

Some differences have been noted between sex, probably owing to different glycogen content and/or different metabolic activity. At slaughtering, females showed significantly lower pH values ($P<0.05$) from *Ld* and ham. Subsequently pH value from males *Ld* dropped faster, significantly ($P<0.05$) after 48 and 168 hours of storage. This trend may be due to a different *post mortem* glicolitic fall rate.

In conclusion, pH values trend, during storage phase, has been influenced by sex effect. In fact female meat, with an initial pH lower than the male one, showed, over the storage period, a lesser drop to such a point that pH in the last measurement (168 hours) was higher than the male one.

Poster H4.19

Evaluation of gene expression in endurance horses by Reverse Transcription Polimerase Chain Reaction (RT-PCR)

K.. Cappelli, A. Verini-Supplizi and M. Silvestrelli, Sport Horse Research Centre, University of Perugia, Via San Costanzo 4, 06126 Perugia, Italy*

The knowledge of molecular mechanisms of stress response is necessary in the athlete horses management in order to plan definite and high-grade training to obtain better performance preserving the horse from the so called ''overtraining syndrome". It is well known that excessive muscular exercise could lead to a number of responses influencing hypothalamic-pituitary-adrenocortical axis and sympathetic system functions which may be associated to modification of the mRNA levels for a number of genes. In our previous researches cDNA-AFLP technique was applied to two Arab horses under different stressing conditions. In the present study we focalised our attention on four cDNA-AFLP modulated genes, never sequenced in horse. They were full-lenght cloned using RACE-PCR and included in GenBank database (Equus caballus interleukin 8; Equus caballus retinoblastoma binding protein 6 horse mRNA; Equus caballus eukaryotic initiation factor eIF4G; Equus caballus heat shock protein Hsp90). Moreover, we analysed the expression of these genes in other two horses during an endurance race (140 Km) to confirm the observed variations. Blood samples were collected just before the ride and 0, 24 and 48 hours after the competition. Peripheral blood mononuclear cells (PBMCs) were isolated and RT-PCR was carried out with specific primers of cDNA-AFLP fragment on total RNA. The evidenced individual differences to the same relative or absolute exercise intensity, suggest that interpreting the average effect of exercise on a group of animals may results on difficult interpretation if the sample would not be representative.

Poster H4.20

Effects of road transport stress on thyroid function of stallions submitted to journey there and back after the breeding season

P. Medica, E. Fazio, V. Aronica and A. Ferlazzo, Department of Morphology, Biochemistry, Physiology and Animal Productions - Unit of Physiology - Faculty of Veterinary Medicine, Polo Universitario Annunziata, 98168 Messina, Italy

Aim of this study was to compare the effects of transport stress on thyroid function of stallions (n=27) submitted to journeys there (March-April) and back after the breeding season (June-July) of different lengths (50-250 km). Thyroid response was evaluated before and after transport on the basis of length of journey (50-100; 100-150; 150-200; 200-250 km) and of the differentiated effects of transport phases. Significative increases of T_3, T_4, fT_3 and fT_4 levels *after journey there* ($p<0.001$) and of T_3, T_4 and fT_4 levels after *journey back* ($p<0.001$) were obtained. Significative lower post-transport T_3 levels ($p<0.05$) in journey back than in the journey there were detected. Significatly higher fT_3 and fT_4 levels before ($p<0.001$) and after ($p<0.005$) the journey back in comparison to data of journey there were obtained. These results confirm that transport stress induces a significant involvement of thyroid function in horses and suggests that stallions are more stressed during a journey back than a journey there, independently of length. It is concluded that the degree of transport stress is greater in stallions after the breeding season (journey back) than before (journey there). Investigations on relationships between thyroid (T_3 and fT_3) and sexual hormones of horses during transport stress will be interesting.

Poster H4.21

Effect of orientation during confinement and transport on β-endorphin, ACTH and cortisol levels of stallions

F. Cusumano., E. Fazio, P. Medica and A. Ferlazzo, Department of Morphology, Biochemistry, Physiology and Animal Productions - Unit of Physiology - Faculty of Veterinary Medicine, Polo Universitario Annunziata, 98168 Messina, Italy

Aim of the research was to distinguish the effect of orientation on β-endorphin, ACTH and cortisol levels of horses studied (Group A: N=6 facing forward; Group B: N=6 facing backward) during confinement in a stationary vehicle for different times (1-3 hours) and then, two week later submitted to road transport of different lengths (<50 and >150 km). Significant higher levels of β-endorphin during transport than during confinement were recorded in both groups. Higher β-endorphin levels were detected after confinement ($p<0.02$) than basal. Significant lower basal ACTH levels in transported than confined were obtained only in Group B ($p<0.05$). Cortisol levels showed significant increases in post-transport than in post-confinement both in group A ($p<0.01$) and B ($p<0.02$) and also than pre-transport in both groups ($p<0.02$). No significative differences of β-endorphin, ACTH and cortisol levels related to orientation in both experimental conditions were found. A significative effect of confinement in stallions facing backward only on β-endorphin levels ($F= 5.61$; $p<0.05$) was shown. A significative effect of transport on cortisol levels was obtained in stallions both facing forward ($F=17.94$; $p<0.01$) and backward ($F=16.16$; $p<0.01$). It is concluded that orientation does not affect hormonal levels of horses both in confinement and transport conditions, although they undergo a major stress condition during transport than confinement. β-endorphin levels seem to be conditioned to a greater extent by confinement than by transport.

Poster H4.22

An investigation into patterns associated with dressage scores at selected novice events

T.C. Whitaker and J. Hill, Faculty of Applied Science and Technology, Writtle College, Chelmsford, Essex CM1 3RR, UK

The dressage scores at four British Eventing Novice Events held in September 2003 were analysed. Results from 468 horses formed the population studied. All horses were scored using standard scoring methods by one dressage judge, however judges varied between individual events. The data for all four competitions, Upton House, Purston Manor, Berkshire College and Witton Castle were normal in distribution (skewness of -0.171, 0.161, 0.326 and 0.434 respectively). The variation in mean scores between events was less than two penalty points. Upton House recorded a mean of 38.91 ±5.56 (n=73), Purston Manor 37.81±4.94 (n=190), Berkshire College 37.86 ±7.91 (n=64) and Witton Castle 37.26 ±6.38 (n=141). The ranges of scores at each event were examined and demonstrated considerable variation. The highest range of 38.1 points was at Witton Castle, whilst the lowest of 23.3 points was at Berkshire College. The results from this study suggest evidence of commonality in scoring patterns by judges. It is however apparent that there is greater divergence in scoring patterns and the reasons for this require further investigation. Relevant techniques may need to be employed if competition records are to be used to formulate decisions over the potential genetic worth of a horse to the population.

Poster H4.23

Statistical analysis of the competition data from equine eventing in the UK in 2000

T.C.Whitaker and J.Hill, Faculty of Applied Science and Technology, Writtle College, Chelmsford, Essex CM1 3RR, UK

During the year 2000, 9,387 horses were registered to compete at affiliated eventing with the national governing body British Eventing. Competition data is stratified reflecting the ability of the horse, level of competition and position achieved. Horse are graded in relation to lifetime winnings at novice level (0-20 points won), intermediate level (21-60 points) and advanced level (>60 points). The range of points won within the studied population was 0-1461 with a mean of 24.04 (±75.41), however the median was 0 points. The distribution of data displayed skewness of 8.35 and kurtosis of 100.7. Attainment of points by horses showed 50.8% (n=4770) of the population had failed to gain any points, 25.8% (n=2423) were graded at novice level, 13.2% (n=1235) were intermediate level and 10.2% (n=959) were advanced level. The results from the study show a high degree of non-normality in the data distribution. Currently a large number of horses are graded at the same non achievement level of 0 points. Additionally it is apparent that at the highest grade of competition horse are performing substantially better than the 60 point threshold with the leading horse performing 24 times better than this level. These characteristics of the population make further investigations of measured performance difficult. The structure and manner in which performance levels are recorded needs to be carefully reviewed.

Poster H4.24

The effect of gender on the performance of event horses at selected pre-novice events in the UK

T.C.Whitaker and J.Hill, Faculty of Applied Science and Technology, Writtle College, Chelmsford, Essex CM1 3RR, UK

Performance defined by penalty points attained in competition for competing horses (n=879) was collated from the six pre-novice events run between May-July 2001. The population studied represented 44% of all pre-novice event horses performing in the UK during this period. Geldings contributed to 70.8% (n=622) of the competing population, mares 24.2% (n=213) and stallions 0.5% (n=4) a further 4.6% (n=40) provided no information on the gender of the horse. Low variations between horse gender and the various phase and final penalty scores were observed. Stallions however score 12.2 points ($P<0.01$) lower (better) than mares and 11.4 points ($P<0.001$) lower than geldings. The greatest difference in scoring within phases were seen in the dressage phase with stallions scoring 5 points ($P<0.001$) than mares. In the show jumping phase the greatest difference was observed between stallions and geldings, 4.3 penalties ($P<0.001$). Stallions only scored 1 penalty lower than geldings and 1.7 penalties lower than mares for cross country jumping. Stallions scored 4 penalties ($P<0.001$) lower than both mares and geldings for cross country time penalties. The number of stallions competing was extremely small however they performed better in competition than mares and geldings. The common perception that mares are generally more unsuitable for event competition and perform worse once in competition was not evident from this study.

Theatre HG5.1

INTERBULL guides through the labyrinth of national genetic evaluations

W.F. Fikse, Interbull Centre, P.O. Box 7023, 750 07, Uppsala, Sweden*

Ten years ago the first international genetic evaluation of dairy bulls were released by INTERBULL as a tool for comparison of genetic merit of dairy bulls across countries. The Nordic countries were the first to participate with data on Holstein and Ayrshire bulls. Today 25 countries subscribe to the service that is offered for six breeds (Ayrshire, Brown Swiss, Guernsey, Holstein, Jersey, and Simmental) and four trait groups (production, conformation, udder health, and longevity). Evaluations are computed four times per year, and considers almost 125 000 bulls. The international genetic evaluation is based on a joint statistical analysis of national genetic evaluation results of participating countries. International genetic evaluation results are expressed on the scale of each country, and reflect locally prevailing conditions. Performance of imported bulls has been shown to be predicted accurately by INTERBULL evaluations. Participating countries can thus choose among domestic as well as foreign bulls to select superior animals to breed the next generation, enhancing the possibilities for genetic progress. Activities of INTERBULL have also had a large impact on the development of national genetic evaluations systems. Exchange of information is a major benefit for member countries. In addition, the requirements for participation in INTERBULL evaluations have had a positive influence on the quality of national genetic evaluations results.

Theatre HG5.2

INTERSTALLION - on the way to an international genetic evaluation of sport horses

E.W. Bruns[1], A. Ricard[2] and E.P.C. Koenen[3], [1]Institute of Animal Breeding and Genetics, Albrecht-Thaer-Weg 3, 37075 Göttingen, Germany, [2]INRA, Castanet-Tolosan, France, [3]NRS, P.O. Box 454, 6800 AL Arnhem, The Netherlands*

Many horse breeding organisations aim at improving the genetic ability of sport horses based on various types of performance tests for young horses and sport competition data relating mainly to dressage, jumping and eventing. Over the past ten years there has been a dramatic increase in the use of artificial insemination within and across breeding organisations, countries and even continents. Therefore, information on testing and genetic evaluation of sport horses has to become more transparent which was the basis for founding INTERSTALLION as a working group of the WBFSH, EAAP and ICAR. INTERSTALLION has published details on a number of breeding populations and programmes in a standardised format on www.wbfsh.org. Reviews on breeding objectives, testing and evaluation methods have been prepared as pre-studies of improved genetic evaluations across countries. The on-going studies investigate the similarities in genetic proofs of stallions used in various European countries and their genetic connectedness. For comparing the genetic proofs across breeding organisations genetic correlations between similar traits recorded in different countries are to be estimated and the availability and usage of data on internationally performing horses are to be tested.

Theatre HG5.3

International genetic evaluations of the Icelandic horse

T. Arnason[1] and A. Sigurdsson[2], [1]International Horse Breeding Consultant AB, Knubbo, S-74494 Morgongava, Sweden, [2]Farmers Association of Iceland, Baendaholl v/Hagatorg, IS-107 Reykjavik, Iceland*

The International Federation of Icelandic Horse Associations (FEIF) has harmonized a common breeding goal and a standard form of evaluating the conformation and riding abilty traits of Icelandic breeding horses on a scale of linear scores. An important step towards global genetic evaluation of the Icelandic horse has been initiated. A global pedigree database containing 190,000 horses with unique identification numbers, whereof 20,000 horses have records from FEIF breeding tests were breeding horses have been evaluated by authorized judges. The material has been used for estimation of genetic parameters and for studying genetic connectedness across country. The genetic correlations between traits evaluated in different time periods and countries have been studied. The genetic model was validated by regressions of solutions from the complete data on solutions from randomly subsampled data (method R). An animal model MT-BLUP using records from Iceland, Denmark, Sweden, Norway and Finland has been implemented for genetic evaluations of all horses included in the data base. In total 30 traits were included. The information on pedigree, color, owner, test-results and EBVs is stored in a SQL database and can be accessed on the Web by Java servlets for subscribers on URL: http://www.worldfengur.com

---

Theatre HG5.4

International comparisons of breeding values: Use of French EBVs in the Canadian Swine Improvement Program as an example

L. Maignel[1], P.K. Mathur[1], I. Delaunay[2] and B.P. Sullivan[1], [1]Canadian Centre for Swine Improvement, Central Experimental Farm, Bldg54 Maple Drive, Ottawa, Ontario K1A0C6, Canada, [2]InstitutTechnique du Porc, BP35104 La Motte au Vicomte, 35651 Le Rheu Cedex, France*

International evaluations and comparisons have been widely used in cattle breeding schemes for many years. In swine breeding programs, the use of foreign genes is less common, and few examples of international comparisons exist. Imports of animals and semen, and more recently, embryo transfers, allowed the use of French Large White and Landrace genes in Canada. A major issue was that the imported boars or semen didn't have Canadian breeding values before having progeny tested in Canada, which could take a long time, especially for sow productivity traits. To address this problem, an agreement was made between the French Pig Technical Institute and the Canadian Centre for Swine Improvement, to exchange EBVs for animals with evaluation results in both countries. Several methods, among those used for dairy cattle, were investigated in order to develop a suitable method to convert French EBVs into Canadian equivalents useful in the Canadian genetic evaluation system. The analysis included 136 boars with progeny in France and in Canada. Conversion formulas were computed regarding EBVs for number of piglets born per litter, age and backfat thickness at 100 kg.

There is evidence for a growing interest in this kind of study, in order to improve the reliability and accuracy of swine genetic evaluations, in the context of growing international exchange of genetic material.

Theatre HG5.5

Estimation of the total merit index for foreign bulls

H. Jorjani[*,1] and W.F. Fikse[1], [1]Interbull Centre, Department of Animal Breeding & Genetics, Swedish University of Agricultural Sciences, P.O.Box 7023, S-75007 Uppsala, Sweden*

Selection of dairy bulls in many countries is based on the total merit index (TMI), which is a combination of a potentially large number of traits. Some of the traits included in any country's TMI may not have any counterparts in other countries, and even for those traits that have counterparts in other countries an international genetic evaluations may not be available. Here we present the results of some simple methods for estimation of TMI for foreign bulls. These methods included single and multiple regression of foreign evaluation(s) on domestic evaluation (with and without adjustments for the reliabilities) and an alternative that conceptually lies between selection index (SI) and multi-trait across country evaluation (MACE), referred to as Simple-MACE. The data used in this study included more than 12000 Holstein bulls from Sweden (with information on 74 traits, including TMI and its part indices) and more than 32000 Holstein bulls from the USA (with information on 29 traits). A cross-validatory method was used to compare different methods by assessing the correlation between converted breeding and actual breeding values and by the mean squared error of prediction. Preliminary results indicate that the correlation of breeding values between the Swedish TMI with the two US traits, net merit dollars and cheese merit dollars, were rather high (0.73-0.74).

Poster HG5.6

Multiple-trait across country evaluations of weakly linked bull populations

T. Mark[1*], *P. Madsen*[2], *J. Jensen*[2] *and W.F. Fikse*[1], [1]*Interbull Centre, Uppsala, Sweden.* [2]*Danish Institute of Agricultural Sciences, Foulum, Denmark*

International genetic evaluations of dairy bulls are performed using the Multiple-trait Across Country (MACE) model, which considers traits measured in different countries as different, but correlated traits. The estimation of the necessary (co)variance components is difficult due to the majority of bulls having daughters in only one country. The problem of poor connectedness is much more severe for Ayrshires compared to Holsteins. Bayesian methodology provides an elegant framework to incorporate prior information while taking into account uncertainty of location and dispersion parameters. The aim of this study was to compare estimated genetic parameters and predictive ability of Bayesian MACE (B-MACE) and MACE assuming a weighted average of prior and estimated genetic correlations (W-MACE). National genetic evaluation results for fore udder attachment from nine Ayrshire populations were considered for this purpose. Genetic correlations from Holstein populations from the same countries were used as priors. Posterior means of genetic correlations obtained from B-MACE were on average 0.2 units higher than those obtained by REML, but the differences in estimated genetic correlations were mainly for country-pairs with no or few direct genetic links. The overall predictive ability slightly favoured B-MACE compared with W-MACE. Sire rankings differed depending on method of analysis, but mainly for bulls with low reliabilities. The degree of prior belief did not have much impact on sire rankings and predictive ability.

Poster HG5.7

Estimation of across country genetic parameters based on daughter yield deviations or deregressed proofs with an approximate REML algorithm

Z. Liu, F. Reinhardt and R. Reents, VIT, Heideweg 1, 27283 Verden, Germany

Current international bull comparison is based on deregressed national estimated breeding values (EBV) and effective daughter contribution (EDC). The deregressed EBV or equivalently daughter yield deviations (DYD) are trait values adjusted for all other but additive genetic effects in national evaluation. EDC associated with DYD is scalar for single trait model. However, for multiple trait models it becomes a matrix, which must be correctly considered in parameter estimation based on DYD or deregressed proofs. The formula for estimating genetic (co)variances via the Expectation Maximization (EM) REML expressed in Mendelian sampling term is extended to multiple trait models. This study shows how genetic (co)variances can be reasonably well estimated without inverting the coefficient matrix of mixed model equations, which is usually infeasible for large scale parameter estimation. The multiple trait EDC method, developed for approximating reliabilities of EBV under multiple trait models, is incorporated in the approximate EM-REML method. The new method is capable of using EDC matrices or scalars as weights of the data. National parameters are fixed during the estimation of across country parameters. Due to low usage of computing resources, the new method can be applied to large data sets, which eliminates the need of bull- or country subsettings. Simulation study and application to real data are needed to verify the accuracy and efficiency of the new parameter estimation method.

Poster HG5.8

The effect of breed of the sire in the Polish Halfbred Horse (SP)

D. Lewczuk, Institute of Genetics and Animal Breeding Polish Academy of Science Jastrzebiec, 05-552 Wolka Kossowska

Nowadays the Polish Halfbred Horse (SP) is the most popular breed in Poland. This breed was created for horses intended to be sport ones and which did not fulfil pedigree requirements for pure Polish halfbred horses (Wielkopolski and Malopolski breeds). The aim of the study were to investigate: the effect of the sire's breed on the conformation of SP horses recorded in the 1-st Stud Book and the influence of breed of stallions that will be used for SP on performance test's results in Poland. Height of the withers, circumference of the cannon and the circumference of the chest as well as the total scores for conformation were analysed (498 horses). The influence of breed on the performance test was studied on results from the years 2001-2003 (355 stallions in total). The traits were adjusted for the group of the age for the conformation results and for the year, term and place of the test for the performance results. The effect of the sire's breed was statistically significant for all investigated traits. Horses which sires were Hannoverian, KWPN or Holstein differ in basic measurement from Polish sires, which progeny was smaller. In most cases the scores for conformation were higher for foreign breeds than for the Polish sires. Analysis of the performance test results showed that imported foreign stallions coming from KWPN, Hannoverian, Holstein received better results then Polish Halfbred Horses. However SP had higher performance results then Wielkopolski and Malopolski breeds.

Theatre H6.1

Horse breeding in Slovenia

F. Habe[1], J. Rus[2], N.Gorisek[2] and M. Kosec[2], [1]University of Ljubljana, Biotechnical Faculy, Zootechnical Department, Groblje 3, 1230 Domzale, Slovenia, [2]University of Ljubljana, Veterinary Faculty, Gerbiceva 60, Ljubljana, Slovenia

Slovenia has quite a long horse driving tradition and a widely known breeding of Lipizzan horses (until 1580) and trotters (until 1884) while breeders' organizations of other horse breeds were reorganized after the World War in 1962. There used to be 60,000 horses but now there are only about 20,000 with varied breed structures from cold-blood to sport warm-blood breeds. Eleven breeders' associations are registered now: Association of Slovenian Cold-blood and Haflinger breed is the oldest association, following by Associations for breeding of Lipizzan, Posavje horse, Slovenian warm-blood horse, Associations for breeding of Arab horses, Thoroughbred horses, Island horses, Quarter horses and Trotter Union of Slovenia. Stud books and selection for all the named breeds are provided by National Horse Breeding Service by Veterinary Faculty and by financial support of Slovenian Ministry of Agriculture. Technical support at testing are offered by Stud Farm Lipica that keeps stud books for Lipizzan breed, and Horse Breeding Centre Krumperk and Centre Prestranek. Equestrian sport and achievements are monitored by FN and by Trotter Union of Slovenia for trotters only. Horseracing has become very popular after the transition. 54 Riding Clubs and 14 Trotter Clubs and 53 Trekking clubs emerged, which means a contribution to the quality and further development of horse breeding in Slovenia.

Theatre H6.2

Organisation of health care in equine reproduction in Slovenia

M. Kosec[1], M. Mesaric[1], J. Rus[1], F. Habe[2], [1]University of Ljubljana, Veterinary Faculty, Gebiceva 60, Ljubljana, Slovenia, [2]University of Ljubljana, Zootechnical Department, Groblje 3, 1230 Domzale, Slovenia

Most of breeding mares in Slovenia are covered naturally by licensed breeding stallions. Mating is provided at breeding stations. On average there are one or two breeding stallions at each station. A relatively low number of quality mares are inseminated with frozen or fresh semen. In the mating season 2003, in Slovenia 337 breeding stallions were licensed. They covered 5868 mares, which means 17.4 mares per stallion.

In Slovenia breeding stallions should match the breeding objectives for a certain breed and veterinary requirements according to the Direction EU 92/65/EEC for stallions in the insemination centres. Before a stallion is approved for reproduction his semen must also be controlled. Two to three ejaculates are taken and analysed. Every year before the mating season diagnostic laboratory tests for the Equine infectious anaemia, Equine viral arthritis and Contagious equine metritis are performed.

Only stallions that respond to breeding objectives, are performance tested according to the breeding programme (depends on breed), are healthy and do not carry infectious diseases, have good libido and satisfactory semen quality can be approved as breeding stalions and are used for reproduction purprose.

Theatre H6.3

Information system for horses

D. Cop, J. Urankar and M. Kovac, University of Ljubljana, Biotechnical Faculty, Animal Science Department, Groblje 3, 1230 Domzale, Slovenia*

The aim of our study was to develop information system (IS) for horses in Slovenia. The computer was rarely used for data storage, therefore the paper documents represented the existing IS. Development of the new IS based on system analysis that assures normalized database structure. Model file, where the database structure is described, enables the connection between the real data flow in horse breeding and computer programs. IS is divided into modules (independent units) that enable optional selection according to the necessity of the users. Modules cover identification and registration, female reproduction, young, growth (genes analyses, exterior traits, working ability), and module base that links all modules together and managing the database. Optional selection of modules gives the opportunity to cover all horse breed specifics and offer enough space to introduce new modules if needed. Due to these aspects the IS satisfies horse societies herds, as well the international organizations. Transition from the old to the new IS is possible by inserting data using screen forms that at the same time enable accessibility to the stored data in the database. IS allows data exchange among participants in electronic form, too. The current the IS mostly comprehends identification and pedigree data and enables review over population. Zootechnical certificates can be issued. In the future the other modules will be filled with the data collected in current population.

Theatre H6.4

Organisation of equestrian sport in Slovenia

A. Cerkovnik[1], G. Pintar[1] and F. Habe[2], [1]Equestrian Federation of Slovenia, Hala Tivoli, 1000 Ljubljana, Slovenia, [2]University of Ljubljana, Biotechnical Faculty, Zootechnical Department, Groblje 3, 1230 Domzale, Slovenia

The organised equestrian sport in Slovenia began in 1875 when Horseracing Association was founded in Ljutomer and started to organise trotter races that has long and interesting tradition as well as history in Slovenia. The disciplines of equestrian sport (the dressage and show jumping) have been developed in the last 20 years especially after transition (1991) while the gallop and other riding sports are at the beginning of expansion. The Equestrian Federation of Slovenia (FN) that was founded in 1949 includes all equestrian sport except trotters and has 54 riding clubs and Trekking Association with 53 trekking clubs. FN besides competitions and licensing of horses and riders and evidencing the results provides training and education for professionals by Slovenian Equestrian Academy and represents Slovenian equestrian sport abroad. The Trotter Union of Slovenia cares for 14 trotter clubs and the trotter calendar. It used to lead the trotter calendar for the whole Yugoslavia. Besides the Trotter Union of Slovenia there is a Trotter Centre. Trotters obtain good results at international competitions. The dressage is a domain of FN and the Stud Farm Lipica that trains horses and organises international performances. The horse show-jumping is the main discipline of riding clubs. Horse driving and horseracing as well as bets are at the beginning of their development. The paper presents development, organisation and current state of equestrian sport in Slovenia.

Theatre H6.5

Lipizzan breeding in Slovenia

J. Rus[1], M. Kosec[1], P. Dovc[2], T. Kavar[2] and F. Habe[2], [1]University of Ljubljana, Veterinary Faculty, Gerbiceva 60, Ljubljana, Slovenia, [2]University of Ljubljana, Biotechnical Faculty, Zootechnical Department, Groblje 3, 1230 Domzale, Slovenia

The Lipizzan horse derives from Lipica in Slovenia. Breeding of this ancient cultural breed started in 1580 when Archduke Karl founded the stud. Through the centuries Lipica provided horses to court and Spanish Riding School in Vienna that made horses famous. Today Lipizzan horses are bred in 8 studs in Europe and in about 30 countries throughout the world. Breeding organisations are organised in Lipizzan International Federation - LIF established in year 1986 in Lipica. In the Stud Farm Lipica there are more than 300 horses and nearly the same number are reared by Slovenian Association of Lipizzan Breeders. Lipica is state stud that keeps stud book and cares for breeding and rearing activities in Slovenia. Stud Farm Lipica has 83 brood mares and maintains 16 mare families and 26 stallions of all six classical bloodlines (C, N, F, M, P, S). Stallions are tested within classical dressage, young mares are tested in carriage and under the saddle. Lipizzaners also take part in sport dressage. Breeding value and results as other horse related activities of Lipica, international centre for these breed, has also been confirmed by COPERNICUS research project on Lipizzan horse that is partly presented in the article.

Theatre H6.6

Trotter breeding in Slovenia

M. Mesaric[1], F. Habe[2], J. Rus[1], J. Slavic[3], V. Habjan[4], L. Blatnik[2], A. Radi[2] and M. Kosec[1], [1]University of Ljubljana, Veterinary Faculty, Gerbiceva 60, 1000 Ljubljana, Slovenia, [2]University of Ljubljana, Biotechnical Faculty, Zootechnical Department Groblje 3, 1230 Domzale, Slovenia, [3]Trotter Union of Slovenia, 9240 Ljutomer, Slovenia, [4]Trotter Centrale, 4207 Cerklje, Slovenia

Trotter breeding in Slovenia is based on Ljutomer trotter beginning in 1884 while first races were performed in 1875. Domestic mares of Gidran type mated by Norfolk, Orlov and later on by American blood gave quality trotter material that via Ljutomer and stud farm Turnisce influenced the quality of trotters in former Yugoslavia. The Trotter Union of Slovenia published Trotter Calendar of Yugoslavia and cared for evidence and register of sport results, which is its duty also in Slovenia. Besides Trotter Union in Slovenia exists also Trotter Centrale. Both include 14 trotter clubs and 11 registered race trucks with about 25 race-days and around 300 licensed racehorses/year. At the moment 510 breeding mares and 38 stallions -all in private hands- are registered. Stallions are mainly of American blood and have 329 (6.6) admits a year. Selection and pedigree service are provided by National Horse Breeding Service by Veterinary Faculty.
Our trotters with best achieved times 1:13.0 race also abroad in Austria, Italy and Germany.

Theatre H6.7

Slovenian warm-blood horse

N. Gorisek[1], F. Habe[2], H. Travner[2], J. Rus[1], and M. Kosec[1], [1]University of Ljubljana, Veterinary faculty, Gerbiceva 60, 1000 Ljubljana, Slovenia, [2]University of Ljubljana, Biotechnical Faculty, Zootechnical Department, Groblje 3, 1230 Domzale, Slovenia

The beginning of modern multipurpose warm-blood riding horse breeding in Slovenia started in 1978 when the Horse Breeding Centre Krumperk of Biotechical Faculty was founded to be a stud farm for Hanoverian breed and to provide genetic material for Slovenian warm blood horses and for former Yugoslavia on the basis of its own breeding, stallions and admits. The Association of Slovenian Warm Blood Horse Breeders was established in 1993 and has 225 members with about 600 horses. The pedigree and selection work and evidence with production control are provided by National Horse Breeding Service by Veterinary Faculty. There are 406 registered mares with about 18 licensed stallions of different origin. Around 170 mares are covered and around 80 foals are registered every year. The association has started to care for testing of mares at the Centre Prestranek. The breed has been forming according to foreign breeding experiences and imported stallions from the EU. The registration of sport achievements is provided by Equestrian Federation of Slovenia (FN). The first good results of young sport horses (dressage, show jumping) of this breed have already been noticed.

Theatre H6.8

Breeding of Slovenian Cold Blood Horse

F. Habe[1], N. Gorisek[2], J. Rus[2], N. Bratina[1], J. Likar[1] and M. Kosec[2], [1]University of Ljubljana, Biotechnical Faculty, Zootechnical Department, Groblje 3, 1230 Domzale, Slovenia, [2]University of Ljubljana, Veterinary Faculty, Gerbiceva 60, 1000 Ljubljana, Slovenia

In Slovenia the cold blood horses prevails and represent about 80 % of all horses.The breeding of the following breeds is organised: Slovenian Cold Blood Horse, Posavje Horse and Haflinger Horse. Slovenian Cold Blood Horse prevails and stud books of it have been kept since 1962. The breed is a result of crossing of domestic cold blood mares with Noric breed stallions from Austria. It is used as a working horse in agriculture and for meat production. The breeding objective is a horse of middle height with about 650 kg body mass. Breeding activities are provided by Association of Breeders of Slovenian Cold Blood and Haflinger Horses and National Horse Breeding Service by Veterinary Faculty.Today about 1187 mares are registered in A and B pedigrees.The 160 stallions -of which 60 Noric breed- have about 4230 admits a year. Most of stallions, about 91 % are a state property. The state breeding service has Centre Briga for rearing of own stallions. Every year exhibitions, control and branding of mares and foals as well as licensing and health control of stallions are provided. The population of it is increasing in number and quality.

Theatre H6.9

Posavje horse

J. Rus[1], N. Gorisek[1], B. Norcic[2] and F.Habe[2], [1]University of Ljubljana, Veterinary Faculty, Gerbiceva 60, Ljubljana, Slovenia, [2]University of Ljubljana, Biotechnical Faculty, Zootechnical Department, Groblje 3, 1230 Domzale, Slovenia

Slovenia intended to preserve autochthonous breeds of local horses hence Posavje horse was restarted to be bred in the nineties. Posavje horse is a cold-blooded horse originated in the North-West part of Croatia. The breeding began with phenotypic selection of animals in the type of ex-Posavje horse in the southern part of Dolenjsko, especially round Krsko and its surrounding areas but some stallions and mares were imported from the neighbouring Croatia. The aim of breeding is to obtain a horse suitable for riding and team carrying, of smaller size - stallions have h.w. 148 cm and mares 143 cm, of firm constitution and merry temperament. Mostly they are brown and bay, rarely black, chesnut or grey. In 1998 the Association of Posavje Horse Breeders was established. It associates 130 members. The Association cares for breeding and rearing activities together with the National Horse Breeding Service by Veterinary Faculty. There are 331 brood mares and 37 licensed stallions registered in the stud book that have covered 315 mares a year. Annually about 53 foals and 45 mares are registered. The breeding has not been consolidated yet. Breeding of Posavje horse is supported by state subsidies.

Theatre H6.10

Haflinger horse breeding in Slovenia

F. Habe[1], J. T.Pucihar[1], J. Rus[2], N.Gorisek[2] and M.Kosec[2], [1]University of Ljubljana, Biotechnical Faculty, Zootechnical Department, Groblje 3, 1230 Domzale, Slovenia, [2]University of Ljubljana, Veterinary Faculty, Gerbiceva 60, 1000 Ljubljana, Slovenia

In Slovenia the breeding of Haflinger horse started in the mountainous regions in the fifties and again in the end of seventies when 181 mares and 29 stallions were imported with governmental support for military and agricultural needs and tourism. The Horse Breeding Centre Krumperk by Biotechnical Faculty was established as a Haflinger stud farm and a promotion centre for Haflinger horses in ex-Yugoslavia. Nowadays there are about 1000 animals in Slovenia. The Slovenian Association of Cold-Blood and Haflinger Breeders -with 230 members-, and the National Horse Breeding Service by Veterinary Faculty organize breeding (stud books and selection). Today 460 mares and 34 stallions are registered (lines: A, B, M, N, S, St and W/V). Young stallions are reared at breeding centre Briga and their working abilities are tested at HBC Krumperk. Most stallions are the state property (97%) and are placed on stallion stations throughout Slovenia. Slovenia is a member of World Association of Haflinger Breeders since 1994. Haflinger horses are regularly taken to exhibitions, tourist and recreation performances and development of tourism will in future certainly expand the breeding of Haflinger horses in Slovenia.

LIVESTOCK FARMING SYSTEMS [L]

Theatre L1.1

Economic consequences of market globalisation on livestock farming systems in Western and Eastern Europe

W. Dunne, Rural Economy Research Centre, Teagasc, 19 Sandymount Avenue, Dublin 4, Ireland

For decades the primary aim of the CAP was food security and the main policy instrument was high prices. This led to intensification, surplus production and excessive resource use with high public cost for EU society. Over the last decade, the CAP has been reshaped and its objectives broadened in response to the:
– accumulation of structural surpluses; – high budgetary costs of these food surpluses; – international trade difficulties arising from the disposal of surplus products; – increasing public apprehension about the effects of intensive production methods and resource use by farmers; – increased affluence of EU society and its impact on the changing societal values placed on resource consumption used in food production and food security; – incorporation of new Member States with diverse economic, social, geographic, and ethnic conditions and additional production capacity.
The reforms have reduced farm product prices, curtailed production, re-balanced the EU market, and provided financial incentives, via direct payments, for farming systems which conserve local resources, and encourage a move towards local food products for both domestic and global markets. Within 50 years the emphasis has shifted from food security to bio-security and food product bio-integrity. Pre-requisites for the full implementation of this policy shift are the definitions and development of compliance criteria for farming systems and the formulation of marketing strategies for EU produced food.

Theatre L1.2

Structural changes in Livestock farming systems in Western Europe countries

M. Gauly[1], J.M.T. Azevedo[2], A. Gibon [3], J. Hermansen[4], A. Pflimlin[5], J. Côrte-Real Santos [6] and K. Peters[7], [1]Institute of Animal Breeding and Genetics, University of Göttingen, D-37075 Göttingen, Germany, [2]Animal Science Department, Universidade de Trás-os-Montes e Alto Douro, Portugal, [3]INRA-SAD-BP 01-78850 Thiverval Grignon, France, [4]Danish Institute of Agricultural Sciences, DK-8830 Tjele, Denmark, [5]Institut de l'Elevage, 75595 Paris, France, [6]Rua Franca 534, 4800-875 São Torcato, Portugal, [7]Institute for Animal Sciences, Humboldt University of Berlin, Germany*

The livestock production systems in Western Europe comprise a mix of intensive large-scale operations, labour-intensive smallholder and crop-livestock operations and various extensive systems. The conditions for expanding livestock production are restrictive in the densely populated regions of Western Europe. However, system changes in sense of farm size, production system, productions organization and technologies over the last two decades were effected by changing economic conditions and agricultural environmental policy. The growth of consumer demands for livestock products and production (animal welfare, hygiene, environment protection) had for example an major impact on the structure of some systems (e.g. beef and poultry sector).
Over the next years trade liberalisation mandated by the World Trade Organisation (WTO) in grains and livestock products will increase the pressure on European producers and therefore impact significantly on the structure in livestock systems. WTO agreements to free trade (globalisation) will open trade borders with consequences for example on product price and production cost structures, product quality and animal welfare standards.
The paper analyses the impact of these likely developments on the structure, technologies and performance of the beef, pig meat, poultry, sheep, goat and dairy systems in Western Europe.

 Theatre L1.3

Trends in the transformation of livestock systems in Central and Eastern European countries

S. Mihina[1] D. Dinev[2], J.Huba[1], V. Juskiene[3], A.Jemeljanovs[4], M. Klopcic[5], V. Matlova[6], E. Orgmets[7] P. Slosarz[8] and F. Szabo[9], N. Todorov[2] , [1]Research Institute for Animal Production, Nitra, Slovak Republic, Trakia University, Stara Zagora, Bulgaria, [3] Institute of Animal Science of LVA, Radviliskio, Lithuania [4] Research centre "Sigra", Sigulda Latvia, [5]University of Ljubljana, Slovenia, [6]Research Institute for Animal Production, Prague, Czech Republic, [7]Institute of Animal Science, Tartu, Estonia, [8]Agricutural University of Poznan, Poland, [9]Georgikon Faculty of Agriculture, Keszthely, Hungary

A marked decrease in the consumption of milk and meat, coupled with continuing imports of these products, caused a decrease in livestock production in Central and Eastern European Countries over the last fifteen years. Numbers of animals and numbers of workers on animal farms decreased rapidly. The opening of product markets demanded higher quality and yields. It also directly influenced changes that have occurred in livestock systems. On the one hand, high-intensity systems on large-scale farms were developed in some countries (Czech Republic, Hungary and Slovakia); and on the other, "animal households" with very few animals and low-input livestock systems were introduced in all evaluated Central and Eastern European Countries. Problems and questions arising from these changes are discussed in the paper.

Theatre L1.4

Transformation trends in livestock systems in Bulgaria

D. Dinev and N. Todorov, Trakia University, Faculty of Agriculture, BG-6000, Stara Zagora, Bulgaria

The aim of this study is to determine the current trends in changes of farming systems in Bulgaria. The study is based mainly on the field survey and the official statistical data. Following wrong political decision, during 1993-1995 the large cooperative and state farms were liquidated and animals spread among former owners of land. As a consequence many small farms appeared and the number of animals decreased in half. Nowadays 83 % of all animals are owned by family farms or private firms. The share of the new production cooperatives is below 2% of all animals. The small farms (households) with less than 10 cows owned 90.6% of the cows in 1995 and 76.7% in 2003. Sheep farms with less than 50 sheep own respectively 89.3 and 77.1%. Pig farms with less than 50 pigs own respectively 44.2 and 53.1% of all pigs in the country. The process of enlargement of farms has already started. However, concentration, specialization, innovation and improvement of efficiency are slower, compared to expectances. The main reasons are the bad economic conditions for animal production, unfavorable input-output price ratio, big variation and low farm gate prices, lack of operative land market and institutions to regulate the transference of ownership, lack of land credit and mortgage, difficult access to bank loans, high bank interest, farmers' uncertainty about the future, etc. The small subsidiary and subsistence farms will continue to exist until the economic situation in the country improves and the robust social security system reduces the need for using farming as a safety net.

Theatre L1.5

Strategic problems of Lithuanian cattle breeding sector

V. Dalinkevicius[1], D. Uchockiene[1], P. Doubravsky[1] and G. Kascenas[2], [1]Lithuanian Cattle Breeders Association, [2]"Litgenas", Kalvarijos 128, 46005 Kaunas-18, Lithuania*

The livestock sector is traditionally the most developed area in Lithuania agriculture. The breeding of cattle in Lithuania goes through a process of reforms at the moment, as well as other agricultural areas. The restructuring of animal breeding sector is still ongoing, having problems in all levels because of lack of suitable strategy.

To define and analyze strategic problems of Lithuanian cattle breeding sector as well as to provide the possible alternatives of their solving, a survey was carried out with 139 respondents from state institutions, private and scientific fields involved in cattle breeding.

The questionnaire included 40 questions about cattle breeding services, state support level, cattle breeding strategy in Lithuania, privatization of breeding sector and legal base. After the calculation of indexes and statistical processing the main problems in Lithuanian cattle breeding were identified. They are following:

- State support. The main problems are insufficient funding of selection program preparation and implementation and sub optimal distribution of state funding.
- Privatization. Respondents negatively estimate the fluency of privatization process and its influence on development of dairy farming.
- Use of additional potential. The possibilities of livestock marketing are limited in Lithuania yet.
- Co-operation. Unfortunately, the associations accomplish only representative functions and the governmental institutions discharge the main part of breeding work.

Theatre L1.6

Current situation and prospects of the Hungarian small scale beef cattle farms

Zs. Wagenhoffer, F. Szabó and Gy. Buzás, Georgikon Faculty of Veszprém University, H-8260, Keszthely, Deák F. u. 16, Hungary*

From 01/05/2004, Hungary will become member of the EU and the CAP will be one of the major key elements to the development of the beef cattle sector. In the last years the only significant direct payment given to the Hungarian farmers was the suckler-cow premium (80€). The eligible maximum number for suckler-cow premium is 117.000 heads while there are about 30.000 registered beef cows in Hungary. Nevertheless around 60 to 70.000 cows (mainly dual purpose Simmental) are kept by small scale farms and individuals that could be switched to suckler-cow production. The aim of our study was to investigate the current situation and future prospects of small scale cattle farms in three given areas (143 farmers interviewed in 57 villages). According to our results the age of farmers is high (55,6), the average herd size is small (7 cows). Cows are mainly (67%) kept indoors even if grassland is available in most of the cases. Average farm size is 29,2 ha of which 48% is grassland, but 35% of the farmers has less than 10 ha. Proportion of land lease is 46%. Income of the farmers originates mainly from cattle production. Milk is sold directly (34%) or to processing units. Most of the bull calves are marketed at a young age (200-300 kg). Prospects of the small scale beef cattle farms depends mainly on herd size (critical herd size is 1-10 cows), age of the farmer (limited number of young farmers) and production capacity (grassland, infrastructure). Ecological potential of the 3 areas is favorable for grazing based beef cattle farming however the different ownership of the fragmented parcels is a crucial element for profitable production. As a local initiative, farmer organizations could be established for grazing the fragmented small sized parcels.

Theatre L1.7

The effect of subsidisation, and market price on the profitability of beef calf production in Hungary

F. Szabó, Gy. Buzás and Zs. Wagenhoffer. University of Veszprém, Georgikon Faculty of Agriculture 8360 Keszthely Deák F. u. 16. Hungary*

Beef cattle stock has decreased in Hungary during the past decade because of the bad profitability situation. In May 2004 Hungary will join the EU and Hungarian beef cattle farmers will receive 55% of EU subsidies (25% from EU and 30% from the national envelope). Based on the subsidy level and the expected weaned calf prices economic analyses was made to evaluate profitability of calf production by suckler cows. Two systems, extensive and semi-intensive of suckling calf production was compared without and with subsidisation in case of different calf price (1.5, 1.8, 2.1, 2.3 €/kg). Extensive system included one calving season (spring), longer grazing period, utilisation of by-products of corn and grain fields after harvesting, by grazing. Only few amount of hay was utilised. The semi-intensive system included two calving seasons, spring and autumn, shorter grazing period on grass, more hay, and some concentrate was fed. Results show better profitability in extensive system, however calf production was a little bit lower than in semi-intensive system. Without subsidy when calf price is law the profit of the extensive method was approximately zero, and negative in semi extensive system. With subsidy the per cow profit in extensive system (1.8 €/kg) was similar to that in semi-intensive system (2.1 €/kg). According to our calculation calf production of approximately 50 cows with susidy gives a reasonable profit that can make a family's living.

Theatre L1.8

With globalisation will low-input production systems prevail in Malawi? The case of rural chicken

T.N. Gondwe* and C.B.A. Wollny, *Institute of Animal Breeding and Genetics, Georg-August Universität Göttingen, Kellnerweg 6, 37077 Göttingen, Germany*

Local chicken are owned by almost all households in rural areas of Malawi. Commercial poultry production supplies meat and eggs to less than 14 % of the human population in urban areas. 134 rural households were monitored to provide information on adequate interventions to improve local chicken production.

The majority of farmers (78 %) supplemented maize bran (ME, 2896 kcal/kg; CP, 10.4 %; CF, 5.0 %) to scavenging. Women (73%) were dominating management of chickens. Off-take was determined by mortality (29%), household consumption (28 %), social (15 %) and income (9 %) needs, predation and theft (14%) and breed stock exchange (4%). Human dwellings were used for roosting (85 %). Newcastle disease vaccination was applied, however, coryza (5 %), chronic respiratory disease (27 %), coccidiosis (8 %) and internal and external parasites (41 %) resulted in losses in 84 % of the flocks. Due to deforestation and loss of indigenous knowledge farmers lacked traditional medicines or treatments. Uncontrolled mating was common practice. Estimated average effective population sizes were low (Ne = 14.9, SD 8.1).

Concluding local free-range low input system chicken production will prevail in the future. A multidisciplinary approach is required to develop and apply optimized technologies. Adopted multi-tier breeding of exotic chickens for crossbreeding have failed. Single tier breeding system allowing full farmer participation is recommended.

Theatre L1.9

Predicted global change effects on livestock performance in Asian countries

R.Tahmasbi, Department of Animal Science, University Of New England, Armidale, NSW, 2350 Australia

The rapid development in recent years of Asia's livestock industry has been matched by the huge increase in the importation of livestock feed. Feed costs are not only a burden on the national budget of nearly every Asian country, where they are often 60% or more of total production costs. Throughout the region, cost and availability of feed are probably the most important constraint to increased livestock production.

Since feed costs are the major part of production costs in Asia, an increase in the use of indigenous feed resources is an important way of helping farmers reduce their costs.

The livestock production systems in Asian developing countries comprise a mix of capital intensive large-scale commercial operations and labour-intensive smallholder peri-urban and crop-livestock operations. Over the next few years, growth in demand for livestock products and trade liberalization mandated by the World Trade Organization (WTO) will impact significantly on these systems.

For developing countries to become significant exporters of livestock products, they would need to achieve productivity gains in livestock production that were considerably higher than those in productivity scenario. This might be enough to generate exports to other developing countries, but for them to develop markets in other industrial countries, they would also need to put in place much higher food safety and quality standards to satisfy the very high consumer standards in industrial country markets.

Poster L1.10

Organisation and economic environment of the sheep meat sector in the North of Tunisia

S. Snoussi, A. Ben Younès, E. Soltani and M. Hammami, Ecole Supérieure d'Agriculture, 7030 Mateur, Tunisia

This study aims to analyze the situation of the sheep meat production and processing system in the North of Tunisia.

It has conducted at a regional scale, in Zaghouan area, where this speculation represents the second agricultural activity after the cereal culture.

It will present the principal indicators characterizing the most important livestock farming systems, the importance of the technical support, and particularly the conditions of the distribution chain. It will present too the evolution of the meat sheep production sector in this area and identify, the constraints known in this sector.

The increase of the stocks of animals doesn't allow satisfying the demand of consumers, especially in some periods.

In spite of one regulation controlled by the professional grouping during the feast period of the "Aïd", the market is not achieved mastery over as well. The continual increase of the prices doesn't favor a veritable integration of the livestock in the farming system, particularly in the big farms. A low performances level is obtained in this households group that possesses the majority of herd.

More structuring and organization are need to improving this sector of sheep meat production.

Poster L1.11

Constraints and prospects of milk production and collecting systems in the new context of the market in Algeria

M.T. Benyoucef, Department of Animal Sciences, National Agronomic Institute, Avenue Pasteur, 16200 El-Harrach, Algiers, Algeria

In Algeria, the milk collecting constitutes a strong constraint for integrating farm milk in the industrial processing. The paper describes the situation of milk production and collecting systems by studying the deliveries to dairy enterprises. It is focused also on the subsidies imputed to the milk and processing systems through the national fund for regulating the agricultural and rural development.
The study reveals that the type of collecting (individual delivery, type of milk containers, transfer duration) has an impact on the connection between milk producers and dairy enterprises (milk price, time of laboratory analysis, agreement duration) and on the milk farm flow.
The emergence of private milk enterprises is favored by the new context of the market. Competition for better milk collecting and processing became effective between different enterprises. It gives another emergence field to research works and professional laboratories for contributing in the diagnosis of the milk sector in Algeria.

Theatre LNCS2.1

Managing grassland for production, environment and landscape. challenges at the farm and the territory level

A. Gibon, INRA Toulouse, UMR 1201 Dynafor, BP 27, 31327 Castanet-Tolosan, France

Grassland “multifunctionality” is increasingly acknowledged in agri-environmental and agricultural and rural development policy. Grasslands and their management are nowadays considered not only in reference to their capacity to contribute to meeting the livestock feeding requirements but also in their various functions with respect to environment and landscape preservation.
The societal expectations support the development of research about grassland biodiversity and other environmental functions and their multipurpose management. They call also for an increased consideration to. This paper addresses the range of questions evocated in reference to grassland multifunctionality and reviews the research orientations currently developed in reference to the spatial dimensions of livestock farming systems, both at the farm and the landscape level.
The necessity of a concomitant account for socio-economic and technical questions for meeting the challenges linked to grassland multifunctionality leads to the development of new frameworks and tools for addressing the spatio-temporal organisation of grassland use. Approaches to livestock farming currently under progress give an insight into new expertise fields requested from practitioners in animal production research and development in order to support the adaptation of livestock farming to societal demands. These are rooted in cooperation with a wide range of disciplines and progress is searched into the use of GIS technologies.

 Theatre LNCS2.2

Management of grassland in intensive livestock farming

T. Kristensen, K. Søegaard and I.S. Kristensen, Danish Institute of Agricultural Sciences, Department of Agroecology, Research Centre Foulum, P.O. Box 50, DK-8830 Tjele, Denmark

Grassland has played a major role in livestock farming through many years. By the intensifying of the livestock production with specialising and larger farm units the role of grassland has changed in many directions, and the importance of the grassland now varied from marginal to major in different farming systems. Aspects of production at crop and herd level, environment effect in terms of nutrient load, emission and methane and to some extent the influence on landscape will be addressed in the paper. The role of grassland is highly dependent on the production system, like intensive grazing versus zero grazing or conventional versus organic farming. Within the systems there are major effects of the management, like the length of grassland in rotation, fertilization level, grassland species and the grazing system used. Also at lot of more specific management elements like grazing intensity, type and level of supplement feeding, and even the daily distribution of nutrients to the animals has major impact on the net productivity. New technology like automatic milking and development in productivity and housing systems has significant impact on the choice between grazing and conservation of grass as silage. Many kinds of animals can utilize both intensive and extensive grassland, which illustrates the great range of different possibilities, and thereby the many ways of grassland utilization that can be used for combining production, environment and landscape issues.

Theatre LNCS2.3

South European grazing-lands: production, environment and landscape management aspects.

I. Hadjigeorgiou[1], K. Osoro[2] and J.P. Fragoso de Almeida[3], [1]Agricultural University of Athens, Department of Animal Nutrition, 75 Iera Odos, Athens, Greece, [2]SERIDA, 33300-Villaviciosa Asturias, Spain, [3]Escola Superior Agrária de Castelo Branco, 6001-909 Castelo Branco, Portugal

Grazing-lands and their management in livestock systems are a matter of special importance in search for sustainability. Sociological and ecological objectives should be considered jointly to animal production development. In addition to the general issues of biodiversity and habitat preservation, the challenges for their management vary according to the regional conditions. In Mediterranean environments, where the past changes in livestock farming led to a general decrease in their use, the questions under study are to find the ways to meet the threats for landscape amenity, biodiversity and the sustainability of local animal feeding resource. Grazing lands and their management is also an important target of EU agri-environmental policy. The multifunctional use of this land, which nowadays is searched for, reinforce the need for animal scientists to consider the use and management of grazing-lands in reference not only to the techno-economical efficiency of animal feeding systems but also in reference to the long term (e.g. biodiversity change) and larger spatial levels (landscape, watershed). An overview of the current challenges attached to grazing-lands and their management in livestock farming systems in Mediterranean environments, and an understanding of the ways to meet jointly production objectives and the realisation of sociological and ecological functions is presented.

An integrated approach to study the role of grazing farming systems in the conservation of rangelands in a protected natural park (Sierra de Guara, Spain)

A. Bernués, J.L. Riedel, M.A. Asensio, M. Blanco, A. Sanz, R. Revilla and I. Casasús, Centro de Investigación y Tecnología Agroalimentaria. Gobierno de Aragón, Apdo. 727, 50080 Zaragoza, Spain*

'*Sierra y Cañones de Guara*' Natural Park (80739 ha, Huesca, Spain) is a protected Mediterranean mountain area dominated by shrub and forest pastures. Traditional agriculture, mainly extensive grazing systems, has decreased in the last decades; concurrently, invasion of shrub vegetation, higher risk of forest fires and landscape changes can be observed.
A study started in 2000 was carried out with two broad objectives: at the farm level, to analyze the farming systems and to evaluate alternative management strategies; at the regional level, to give useful information to conservation authorities for better decision making.
The three parts of the study are described. First, a survey covering all farms that utilized the Park was carried out and Multivariate Analyses were used to characterize farming systems in terms of grazing management, technical, social and economic factors. Second, six representative areas were selected to evaluate, depending on livestock use, grass and shrub vegetation (species, biomass, green/dead ratio, chemical composition) and vegetation dynamics. When comparing non-grazed *vs.* grazed areas, significant differences were found in biomass accumulation and green/death ratio. Third, vegetation and livestock data were analyzed with a Geographic Information System that allowed to detect three priority rangeland areas for intervention.

Theatre LNCS2.5

Sustainable recultivation on karst and hilly land

M. Pogacnik, D. Kompan, M. Kotar and T. Vidrih, University of Ljubljana, Slovenia*

Sheep and goat production can be a valuable part of a sustainable farm. Integration of livestock into the farm system can increase economic and environmental benefits and diversity, thereby making important contributions to the farm's sustainability. Sheep and goats may fit well into the biological and economic niches in a farm operation that otherwise go untapped. They can be incorporated into existing grazing operations with donkey or horse. Goats can also be used for control of weeds and brush to help utilize a pasture's diversity, as long as they are not allowed to overgraze. Livestock produce milk, meat, and fiber using grass and other plants that cannot be digested by humans. In a balanced agro-ecosystem, both animals and plants thrive. Livestock eat the excess plant materials, and animal wastes provide nutrients for the plants. Grazed pastures need less fertilizer than those that are hayed. A primary goal of controlled grazing is to use livestock as a tool to manage forage growth. Animals use up very few of the nutrients from the plants they eat; most minerals are returned in animal wastes and can be considered part of a natural cycling of nutrients. If wastes are evenly distributed throughout the grazing area and biological agents such as earthworms, dung beetles, and soil bacteria are active, the system should be relatively stable.

 Theatre LNCS2.6

Effect of sheep grazing and phosphorus fertilizing on botanical composition and sward productivity of karst pastures

M.Vidrih and F. Batic, Agronomy department, Biotechnical faculty, Jamnikarjeva 101, 1000 Ljubljana, Slovenia

Between 1999 and 2001 sward changes in herbage mass, botanical and floristic composition were determined on the slopes of mountain Vremscica. The sward response on three treatments (1-control, 2-three years fertilized with 30 kg P_2O_5 ha^{-1}, 3-three years fertilized with 90 kg P_2O_5 ha^{-1}, 4-only first year fertilized with 270 kg P_2O_5 ha^{-1}) was measured as available herbage mass at the end of three growth periods (spring, summer, autumn). To compare the effect of different levels of controlled sheep grazing management (high, medium, low, none) on species composition, eight experimental plots (P1-P8) were assigned and in each vegetation aspect a combined assesment of frequency and cover, and the sociability of all species present was taken. The highest herbage mass was obtained at the end of spring growth (3,21 t DM ha^{-1}) and summer growth (2,04 t DM ha^{-1}) in treatment 3. The autumn growth gave the highest herbage mass (1,4 t DM ha^{-1}) in treatment 4 ($p<0,05$). Phosphorus fertilization had the highest effect on legumes increase (treatment 1-10%, treatment 3-33 %) in summer growth period. In 72 phytosociologic surveys 349 different plant species were identified. The highest number of species (121) was found on site P1, which was cleared from hazel shrubs and other bushes 7 years ago. On site P8 (no grazing) the species richness was the most stable.

Theatre LNCS2.7

Free-ranging beef-cattle performances and management implications in the Natural Park of Gorbeia (Basque Country)

N. Mandaluniz, A. Legarra and L.M. Oregui, NEIKER, Apdo. 46. E-01080, Vitoria-Gasteiz, Spain

Traditional mountain system in the Basque Country has changed in last years resulting in an increase in the number of mountain grazing cows. The study of the effects of this change is important to understand both, animal and vegetal responses. Cows grazing patterns and live weight evolution during mountain grazing period (summer and autumn) were studied.
According to the data, cows maintained live weight during mountain grazing season, although this maintenance was a result of live weight gain during summer (140±370 g/day) and loose during autumn (-56±480 g/day). These differences were probably due to pasture seasonal changes and cows modified their grazing strategy increasing lower quality grasses and shrub ingestions in autumn.
Moreover, there was a high heterogeneity in live weight evolution depending on animal physiological state. Cows with higher energetic needs (last gestation and suckling) loose live weight in both periods, summer and autumn, while lower needs ones maintain or gain weight. Results indicate that both, spring and autumn calving cows had similar profits during mountain grazing but spring calving could be more suitable system because grazing period had a higher contribution on the annual energetic needs (56% vs. 40%).
This variability opens multiple management possibilities, for both animal production and environmental maintenance, which will be discussed in the paper.

Plant preferences and grazing behavioural in an old native and a modern dairy cattle breed on mountain pastures

N.H. Sæther[1], O. Vangen[1], H. Sickel[2] and A. Norderhaug[3], [1]Department of Animal Science, Agricultural University of Norway, P.O. Box 5003, N-1432 Ås-NLH, Norway, [2]The Royal Norwegian Society for Development, P.O. Box 115, N-2026 Skjetten, Norway, [3]The Norwegian Crop Research Institute, Kvithamar, N-7500 Stjørdal, Norway*

The objective in this experiment was to study plant preferences and grazing behavioural on mountain pastures of an old native breed of Norwegian dairy cattle, the Black Sided Trønder and Nordland Cattle (STN), in contrast with the dominant dairy cattle breed in Norway, Norwegian Dairy Cattle (NRF).

Three private herds with approximately ten cows in each and with both STN and NRF were studied during three summers. The plant preferences were recorded by analysing samples of manure for remnants of 24 different plants or plant groups. The results show some genotype by environment interaction between the two breeds in two of the grazing areas, this might be of interest when planning management of old semi-natural mountain pastures. The behavioural traits when grazing were recorded every fifteen-minute during the daily grazing time and were either "lying down resting", "lying down sleeping", "standing up", "walking", "grooming" or "playing". Significant differences between the two breeds were observed for "walking" and "playing", the old native breed was both "walking" and "playing" more than the modern dairy breed.

The results will be discussed according to resource allocation theory and the two breeds' selection history.

Theatre LNCS2.9

Grazing as a tool to create sward structure suitable to breeding birds

M. Tichit[1], D. Durant[2] and E. Kernéïs[2], [1]INRA SAD 16 rue Claude Bernard 75231 Paris, France, [2] INRA SAD Domaine Saint Laurent de la Prée, 17450 Fouras, France*

French wet grasslands support important populations of Lapwings and other waders. Grazing management is a key issue since these birds are very sensitive to sward structure (grass height, heterogeneity). During three years, repeated measurements of sward height were conducted in a coastal marsh in order to assess the impact of different grazing regimes on sward during spring. Sward structure was characterised by six variables related to height classes and an index of heterogeneity. Grazing regimes were described by stocking rates per period and nitrogen index. Co-inertia analysis was performed. Total inertia was 0.57 and Monte Carlo simulation gave a probability $p=0.000$ to have a random co-structure between grazing regimes and sward structure. Three groups of fields were characterised. Fields showing a sward structure suitable for lapwings were winter grazed, low fertilised, then heavily grazed in early spring (i.e. strong defoliation limited re-growth until mid spring). Two other groups of fields had a higher height and it was shown that relation between heterogeneity and height was curvilinear. Taking into account wader habitat requirements calls attention to the postponed effects of winter grazing regimes. Thus, heterogeneity emerges as a new characteristic to steer. Its level depends on selective grazing by cattle and influences the sward growth: thresholds thus are to be investigated.

 Poster LNCS2.10

The role of livestock grazing in karst grassland improvement

A. Vidrih and M. Vidrih, Agronomy department, Biotechnical faculty, Jamnikarjeva 101, 1000 Ljubljana, Slovenia

The aim of the work is to develop sustainable karst grassland utilisation. Results of previous research indicate, that livestock grazing can have positive impact on biodiversity of the region and well-kept appearance of the landscape. Limestone and dolomite rocks form karstic character of land, which is expressed through floristic pattern. Grazing livestock greatly affect the composition of pasture plant communities. With proper grazing management animals always cause a pasture to be a more complex mixture of plants than it otherwise would be. This is because animals graze selectively and in patches, and the effects vary in time and space. They do collect the nutrients from wider area and return them more concentrated to the soil. This has great influence on site fertility and vitality of the ecosystem. All these have effect on sward development, which will contain a wide variety of plants adapted enough to survive their different local conditions. The key indicator of a land's stability and productivity is a succession. It is the process of change and development in entire communities; soil, micro-organisms, animals and plant life. To understand the nature of the forces that influence individual plants within a Karst pasture sward, we need to consider what happens in the sward from the plant's point of view. For to maintain karst grassland and prevents it from bush encroachment the grazing of livestock it is of vital importance.

Poster LNCS2.11

Effects of fertiliser n level and grazing management on steer performance

M.G. Keane, Teagasc, Grange Research Centre, Dunsany, Co. Meath, Ireland

Agri-environment schemes such as the Rural Environment Protection Scheme (REPS) in Ireland require reduced fertiliser N usage for beef production. Grazing management can affect animal performance and may interact with N level. Two fertiliser N levels of 227 (Standard) and 57.5 (REPS) kg/ha and three rotational grazing managements were compared in a 2 x 3 factorial experiment using 84 yearling steers (followers) and 90 calves (leaders). The grazing managements were leader/follower (LF), LF but with the calves grazing half a rotation ahead of the yearlings (SR), and LF but with extra herbage for the yearlings (EH). The duration of the grazing season was 208 days for the yearlings after which they were finished for slaughter at about two years of age. Mean liveweight gains during grazing for the Standard and REPS N levels were 996 and 951 (s.e 18.3) g/day, respectively. Corresponding finishing and overall annual gains were 1156 and 1237 (s.e 27.5) and 911 and 915 (s.e 17.9) g/day. There was no difference between the LF and SR, but EH had a higher gain at pasture which was compensated for during finishing. It is concluded that fertiliser N level had no effect on animal performance once herbage supply was adequate and that good animal performance can be achieved with low fertiliser N usage across a range of grazing management systems.

 Poster LNCS2.12

Research concerning the quality of the temporary meadows cultivated on the sandy soils from the south of Oltenia

M. Vladu[1], V. Bacila[2] and I. Calin[2], [1]University of Craiova, Agricultural Science Faculty, Animal Science Department, 19 Libertatii street, 200421, Romania; [2]University of Agronomic Sciences and Veterinary Medicine Bucharest, Animal Science Faculty, Technology Department, 59 Marasti boulevard, sect. 1, 011464, Romania

The research concerning the fodder crops cultivation on the sandy soils aimed to create a new ecological environment, the development of the animal husbandry sector. The possibility of establishment the temporary meadows on these soil types can contribute to their amelioration and to the animal husbandry development.
Among the perennial graminaceous and the leguminous species cultivated in similar conditions, Dactylis glomerata and Medicago sativa showed the best results concerning the yield quantity and quality on the sandy soils.
In order to appreciate the fodder quality, six different alfalfa varieties were studied, fertilised with different nitrogen quantities (N50 and N100), in irrigated conditions. For this purpose, there were determined the leaves percentage and the brut protein content from the green-mass at the second cut from the first, second and third cropping year.
The results showing that through increasing the nitrogen rate from N50 to N100 in the first cropping year, the percentage of leafs from the green-mass and fodder quality increased too. In the next two years, the leaf percentage decreased. The average value of the brut protein content in the studied period, varied between 21,3% and 21,6%.

Poster LNCS2.13

Multi-season production of high quality forage sorghum for ruminants

A. Carmi, N. Umiel, A. Hagiladi and J. Miron, The Volcani Center, ARO, P.O.B. 6, Bet-Dagan 50250, Israel*

Sorghum is strongly increased by hot weather. Consequently it has become a summer crop. This study demonstrates sorghum growing continuously throughout the year, and getting three forage yields. This system became possible due to sorghum survival under moderate Mediterranean winter. Sorghum was sowed on mid August at the end of the summer, irrigated with 250 mm water, got additional 400 mm of winter rainfall, and survived without further irrigation until the next summer. Under these conditions two harvests became possible: in mid November and in May of the next spring. Additional irrigation of 200-250 mm in the dry period, enabled to get a third harvest in next August. This new multi-season concept has significant advantages: 1) Field use is continuous without a fallow period, and forage is available during all over the year. 2) Continuous cropping of sorghum is more profitable, compared to the alternative double-cropping system of sorghum and winter wheat, due to reduction of expanses for tillage and herbicides. 3) Full occupation of ensiling silos is attained over the year. The in vitro dry matter and NDF digestibility values of sorghum plants from the three harvests were similar and high. This multi-system concept is also relevant to other regions of the world, which are characterized by moderate winter and sufficient rainfall.

 Poster LNCS2.14

Concentration of magnesium in soil-plant-animal systems under different grazing intensities

L. Czegledi, J. Prokisch, B. Kovacs, I. Komlosi, M. Arnyasi and B. Beri, University of Debrecen, Centre of Agricultural Sciences, Faculty of Agronomy, 138. Boszormenyi Street, H-4032 Debrecen, Hungary*

Investigation was carried out on a grassland of Hortobagy National Park, Hungary. Hungarian Grey Cattle were kept in free range grazing and grazing pressure was differed on specific part of pasture as overgrazing (watering place), moderate grazing and low grazing intensities.
Soil samples from investigated grassland at depths of 0-20, 20-40 and 40-60 cm, plant (Lolium perenne L.), dung, urine and cattle hair samples were analysed for total magnesium content except soil where plant-available Mg was also measured by ICP-OES.
Soil pH influences plant-available magnesium content of soil. Both total and plant-available magnesium content of soil were significantly the highest in case of overgrazing. The less the proportion of mentioned magnesium forms (total/plant-available Mg), at each sampled soil depth, the higher the grazing intensity.
Different amounts of cattle faeces and urine on grassland soil had no effect on magnesium concentration of ryegrass in late spring.
Magnesium content of hair of each investigated cows were in normal range, consequently magnesium supplement is not necessary except risk of grass tetany.
Further investigations are planned to calculate the amount of moving magnesium in this closed soil-plant-cattle system.

Poster LNCS2.15

Influence of selenium enriched fertilizers on the selenium content in grass and the selenium status in a suckling herd offered selenium enriched grass, silage and barley

I. Dufrasne[1], J.F. Cabaraux[1], M. Coenen[2], H. Scholz[2], L. Istasse[1] and J.L. Hornick[1], [1]Veterinary Faculty, Liege University, Belgium, [2] School of Veterinary Medicine, Hannover, Germany

Selenium is a trace element characterized by antioxidant properties of interest for animals and humans. Permanent pastures were divided in two plots. One plot was fertilised with mineral fertilisers containing selenium and the other one with the same fertiliser without selenium. A total of 15 g selenium was applied per hectare. The plots were grazed by suckling Belgian Blue double-muscled cows and their calves. The animals were offered during the previous winter grass silage and barley grown with or without selenium fertilizer. There were no effects of selenium on the chemical composition of grass with protein and crude fiber contents of 172 and 342g/kg DM. The fertilization with selenium increased the selenium content in grass (0.29 *vs* 0.06mg/kg DM) by almost 5 times, the concentration being much higher at the beginning of the season. Blood selenium measured as glutathion peroxydase activity was significantly higher in the cows and in the calves on the selenium plots (65.9 *vs* 29.5 µg/l, $P<0.01$ and 81.1 *vs* 33.7 µg/l, $P<0.001$). According to the present data considered as first results, it could be concluded that the deficient selenium status of the animals grazed on the control pasture was improved by use of selenium enriched fertilizers.

 Poster LNCS2.16

Zebu cheese quality related to human health

M. Galina, Facultad de Estudios Superiores de Cuautitlan UNAM, México

Zebu cheese quality improves when comes from grazing animals a higher content of mono-terpenes, sesqui-terpenes, unsaturated fatty acids, CLA and the anti-oxidants, than cheese from indoor cattle. Particularly in the dry tropics browsing legum trees. A number of metabolites have been reported from various species most of them good for human health, including amines and alkaloids, cyanogenic glycosides, cyclitols, fatty acids and seed oils, fluoroacetate, gums, non protein amino acids, terpenes (including essential oils, diterpenes, phytosterol and triterpene) hydrolizable tannins, flavonoids and condensed tannins. During 2003 in the spring and summer we monitored for quality related to fatty acids content and therapeutic pharmacological components in cheese made from grazing Zebu cattle and made from indoor animals. Monoterpenes and sesquiterpenes on pasture Zebu cheese were 360 and 520 ng/l respectively. Indoors cattle had 65 and 210 ng/l. Total fat and cholesterol diminished from grazing cattle from 22g/100 g cheese and 245 mg/100 g fat, in garzing ruminants compared to 37g/100 g and 387 mg/100 fat in the indoors cattle made cheese. Garzing accounted to higher tocopherol in cheese from 127 mg/100 mg DM compared to 77mg/100 g DM in indoors cattle cheese. Cheese quality relegated to human health was improved in artisan cheese made from grazing cattle browsing tropical trees. It is concluded that milk from grazing was better than that produced indoors regarding to the feeding system in Zebu cattle.

Poster LNCS2.17

Which structures of shrub cover are ungrazed by flock?

E. Lécrivain, National Institute for Agronomical Research, Ecodéveloppement Unit, Avignon, France

In French Mediterranean area, the increase of overgrown pastoral lands cannot be stopped only with grazing. The purpose is to identify which shrub structures may be explored and where it needs to do mechanical interventions to maintain the accessibility of forage resources. The study was carried out in a paddock of 16 hectares grazed by 400 sheep regularly run on rangelands. Plant community units were identified in terms of percentage of herbaceous (hc), shrub (sc) and tree cover (tc) and height, and mention of the structure of the dominant plant species. The flock's behaviours were followed from daybreak to nightfall every 10 minutes during 5 days. The data were spatialised using a G.I.S. The analyse focused on the structures of areas preferentially grazed.

In overgrown conditions major attractive structural features are herbaceous cover (hc>60%) chosen firstly in open areas (sc<60%). But there are also shrubby areas (sc>80%) of which the structure allows the animals to pass through and permits the development of herbaceous. In opposite, major avoided structural features are areas with a few sward cover (hc=<20%) either opened either with high tree-covered (tc>=90%). Only dense shrubby areas (sc>80%) containing tangles of 20% of thorny plants reaching between 1 and 3 m, or young brambles reaching 30-40 cm, are deserted.

To help sheep exploration and control the shrub invasion, these combined thresholds of the shrub cover structure may be used to plan where to locate mechanical interventions.

 Poster LNCS2.18

Changes in beaver dam streams in north-eastern Poland

P. Janiszewski, W. Szczepanski and L. Janiszewska, University of Warmia and Mazury, Oczapowskiego 5, 10-718 Olsztyn, Poland

Two beaver habitats, occupied by two colonies, were studied. They were located on agricultural land where considerable damage caused by beavers' activity was reported. The experiment consisted in dam removal. The experiment was started on July and completed on August 2002. One of the parameters analyzed in the present experiment was the height of dams. The dam height at research station II is characterized by higher variation, compared with station I. At research station II on July 15 the dam height was 100 cm, and on July 30 - 80 cm. Another parameter analyzed was the depth of a stream and its fluctuations caused by dam removal. At research station I the stream depth in front of the dam varied from 10 to 30 cm; its average value over the experimental period was 17.18 cm. The stream depth behind the dam at this station was 30 to 50 cm. At research station II the stream depth in front of the dam ranged from 10 to 30 cm; its average value over the experimental period was 19.54 cm. The stream depth behind the dam at this station was 30 to 60 cm, on average 45.36 cm. The results obtained allow to formulate the following conclusions: Cyclical dam removal did not result in beaver migration from research stations. Dams were built mostly from new materials. This caused even greater damage, due to cutting down trees.

Poster LNCS2.19

Improvement of degraded calcareous soils using pastures for grazing ewes

O. C. Moreira[1] , J. Ramalho Ribeiro[1], J. Santos[3] and Mª A. Castelo Branco[2], [1]INIAP-Estação Zootécnica Nacional, 2005-048 Vale de Santarém, Portugal, [2]INIAP-Estação Agronómica Nacional, Quinta do Marquês, 2784-505 Oeiras, Portugal, [3]DRAB-Centro Experimental do Loreto, 3000-177 Coimbra, Portugal

The soils located in the Middle Western coast of Portugal are dominantly calcareous, with limiting nutritional factors. The objective of this study was to implant pastures to improve soil quality and to evaluate its nutritive value for grazing ewes (Serra da Estrela breed) from a PDO cheese production region.

In two different locations, five1 ha-experimental swards were implanted in 2001. Different commercial mixtures of legumes and grasses were used. The grazing periods lasted from March to June-July, depending on the pasture availability; afterwards the animals were fed alfalfa hay. During the grazing period, pasture samples were collected for chemical characterisation. Sixty days after parturition, milk was sampled fortnightly, and analysed for fat, protein, casein and nitrogen, Ca and P fractions. The collected data of two years of observations was compared using ANOVA.

Although with a decrease with the evolution of the season, spring pastures were obtained in enough quantity and nutritive value to support the stocking rate (10 ewes/ha) for five months. Milk composition was adequate to produce the typical cheese of that region. The recycle of nutrients from animal origin to the pastures, together with the organic matter of plant origin, would contribute to improve soil characteristics.

 Theatre LMP3.1

Societal expectations of livestock farming in relation to environmental effects

J.A. Milne, Macaulay Institute, Craigiebuckler, Aberdeen, AB15 8QH, UK

The potential environmental effects of livestock farming are mainly associated with intensification of production systems. The major impacts are caused by systems of poultry, pig and dairy cow production where housing can lead to air and water pollution associated with nitrogen and phosphorus from manures. Silage for ruminant feeding also poses a threat to water quality. European society regulates the potential for these types of pollution through a number of European Union directives and national legislation. In grazing systems, nitrogen pollution, associated with the use of nitrogen fertiliser and grass/clover swards, is also legislated against. Perhaps because of this regulation, surveys of the public have found that human food quality and animal welfare are far more important issues than effects on air and water quality when considering livestock systems.
High stocking rates of cattle, sheep and goats, grazing semi-natural or natural pastures can change the structure and composition of vegetation with potential impacts on biodiversity. Through Common Agricultural Policy instruments, maximum stocking rates can be set in order to reduce impacts on biodiversity in Europe. Ruminants contribute to emissions of greenhouse gases through the production of methane. Such stocking rate limits are the only mechanism for regulating ruminant numbers. Surveys of the public have suggested that they are willing to pay for the mitigation of these environmental effects but that they also value the cultural component of grazed livestock systems.

Theatre LMP3.2

Environmental assessment tools for the evaluation and improvement of European livestock production systems

Niels Halberg, Hayo van der Werf and Imke J.M. de Boer, Dept. Agroecology, Danish Institute of Agricultural Sciences, Denmark*

Environmental indicators have been developed to evaluate the environmental impacts of farming systems in Europe on national or EU level for policy and on farm level for the improvement of farm management. The indicator tools vary in their focus on environmental issues and in their basic choice of indicator types. Studies show some interest among farmers and a potential for improvement of farming practices comparing farms but the validity of different indicators should be compared and threshold values for benchmarking between farms should be developed. This paper will give an evaluation and guidelines for efficient indicator use. Environmental impacts such as Green-house-gas-emissions are global why a comprehensive environmental assessment of products should include the whole production chain (e.g. nitrous oxide emissions from tropical soybean cultivation are relevant to the assessment of European livestock production). Environmental gains from improvement in farming systems may be off-set by the amounts and patterns of consumption. Therefore, a product oriented assessment that include the whole production chain and quantify the environmental impacts per kg product delivered is recommended internationally. Life Cycle Assessment is a product oriented assessment tool used to evaluate the environmental efficiency of food production systems. The paper will compare the potential use of LCA and indicator tools for comparing and improving agricultural systems.

 Theatre LMP3.3

Agro-environmental sustainability of small ruminants production in Lebanon

G. Srour[*1,2], M. Marie[1] and S. Abi Saab[2], [1]ENSAIA-INPL-Nancy, Sciences Animales, B.P. 172, 54505 Vandœuvre lès Nancy, France, [2]Université Libanaise, Faculté d'Agronomie, B.P. 5368/13, Horch Tabet, Beyrouth*

Small ruminants production in Lebanon still plays an important role in fighting against human desertification of low economic potential regions and in enhancing the value of pastures, which constitute 29% of the Lebanese agricultural lands. This is done with breeds perfectly adapted to hard life conditions (resistance to diseases and extreme temperatures, ability to consume ligneous plants, ...). The productivity of these systems is facing serious constraints: aridity of the warm season, climatic variations, rising cost of the feeds. On the other hand, this production may be harmful for the vegetation and the environment.

A survey has been conducted in 129 farms, representing 10% of the small ruminants flocks, and originating from the different Lebanese production systems and regions. Sheep were totally of Awassi breed, and 95% of goats were of Baladi breed. From this study, 48 questions led to 19 agro-environmental indicators (animals and plants diversity, space organisation, effluents management, animal welfare, energetic dependency of farms, ...).

If the agriculture practices score is high, space organisation and plants biodiversity are less noted, in particular because Lebanese small ruminants production is mostly associated with monoculture. Ways for improvement of the sustainability of these production systems have to consider crop rotation, management of organic matter, management of forage crops, and protection of water resources.

Theatre LMP3.4

Nitrogen surplus and farm characteristics

A.H. Nielsen[*] *and I.S. Kristensen, Danish Institute of Agricultural Sciences, Department of Agroecology, P.O. Box 50, DK-8830 Tjele*

The objective of this study was to investigate how variation in the farm N surplus could be explained by farm characteristics. Although there is a correla¬tion between stocking rate and farm level N surplus on livestock farms, there is also a con¬sider¬able variation in surplus per hectare for a given stocking rate. The farmer can change the N surplus through a range of factors related to the efficiency and intensity of the production. A part of the surplus may be explained by basic factors deeply embedded in the type of farming (e.g. pig or dairy farming), or correlated with soil quality or climatic conditions. N surplus was measured in one to six years between 1997 and 2002 at 28 conventional dairy farms, 15 organic dairy farms and 21 farms with indoor pig production (225 observations). The average N surplus (kg N ha^{-1} ± std. deviation) was 183±48, 107±21 and 131±57 for the three farms types respectively. Principal component analysis and traditional statistical methods were used to deduce and test relevant statistical models. A model with farm type and stocking rate explained over 80% of the variation in N surplus. However results also showed importance of factors related to feeding practise and choice of crops. Results are discussed in relation to the rather radical changes in environmental regulation of livestock farms as experienced by Danish farmers over the last two decades.

 Theatre LMP3.5

Grazing intensity as a tool to assess positive side effects of livestock farming systems on wading birds

M. Tichit[1], T. Potter[2], O. Renault[2], [1]INRA SAD, [2]Population and Community Ecology, both at INA PG 16 rue Claude Bernard, 75231 Paris, France*

French grasslands support important populations of waders. Grazing intensity is a key issue since these birds are very sensitive to grass height and heterogeneity. On 1594 plots, grazing intensity was characterised as a combination of stocking rate and the proportion of days grazed on each day on which birds were observed. This made it possible to standardized stocking rates such that the various cumulative effects of time were minimised. We modelled the relationship between stocking rate and the proportion of days grazed on plots occupied by birds and those that did not. Birds select plots with a mean spring stocking rate that is significantly lower than the global mean and bird species richness is negatively correlated with intensity of spring grazing. Birds show a degree of specialisation towards grazing intensity: redshank and curlews being two extremes of low and high intensity. Variance in the start date of grazing is only significantly lower than global variance for plots occupied by black tailed godwit and redshank and when only grazed plots are included in the analysis. Those results indicate that a diversity of grazing regimes is necessary to preserve habitat quality for birds. From such analysis useful indicators can be derived to assess positive side -effects of livestock farming systems.

Theatre LMP3.6

Quels indicateurs pour analyser la contribution des systèmes d'élevage ovin à la gestion des milieux pastoraux en montagne méditerranéenne française?

J. Lasseur, INRA-SAD, domaine saint Paul, 84914 Avignon cedex 9, France

Depuis une dizaine d'année, dans le cadre de mesures agri-environnementales, les éleveurs ovin en zone méditerranéenne française sont incités à développer des systèmes d'élevage permettant de lutter contre la fermeture des espaces pastoraux et de contribuer au maintien de la biodiversité de ces milieux. Ces nouvelles préoccupations mettent en question la capacité de gérer en terme dynamique les interactions sur le temps long entre pratiques d'élevage et milieux pâturés. L'analyse des modalités d'évolution des pratiques mises en oeuvre par les éleveurs dans ce contexte devient un objet d'étude central.

En considérant les pratiques dans leurs dimensions socio-techniques, nous avons réalisé des enquêtes auprès d'éleveurs ovins de trois localités dans les Alpes du Sud pour analyser ces processus de transformation des pratiques d'élevage dans un contexte d'autre part marqué par de forts accroissements des dimensions des exploitations. Nous montrerons comment ces changements sont liés à la capacité des éleveurs à reconstruire des systèmes techniques cohérents mais aussi à articuler ces actes techniques aux positions sociales reconnues dans des collectifs professionnels locaux, producteurs de normes sociales définissant les critères d'excellence du métier.

Intégrer cette articulation entre systèmes techniques et conceptions du métier est un enjeu de la production d'indicateurs permettant d'évaluer la capacité des systèmes d'élevage à participer à des opérations de gestion de l'environnement.

 Theatre LMP3.7

The sustainability of Algerian dairy farms and their impact on the environment

N. Bekhouche[1,2], F. Ghozlane[2], H. Yakhlef[2] and M. Marie[1], [1]ENSAIA-INRA, Sciences animales, B.P. 172, 54505 Vandoeuvre, France, [2]INA, Productions animales, El Harrach, Algiesr 16200 Algeria*

A study has been made on fifty dairy farms located in the plain of Mitidja in Algeria to investigate and evaluate the sustainability of the dairy farms and their impact on the environment. According to the first results obtained from environmental indicators (diversity of the productions, organization of space and husbandries), it could be noted that the near total of the surveyed farms (98 %) have a rather high level of diversity of production, which means that complementarities and processes of natural regulation were allowed by complex agro ecosystems. Concerning space organization, the majority of farms present null to negative notations, which is explained by the absence of ecological regulation zone, the absence of an environmental program and a very high stocking density caused by lack of land, which implies a bad management of fodder surfaces. With regard to husbandries, 75 % of our sample of study present notations close to the average, which is explained by a limited use of phytosanitary products and a limited use of water and irrigation systems.

In fact, these indicators inform us of the way in which the nature capital of the farming system (water, soil, biodiversity, air and light) is managed by the production system both on the short and medium term.

Poster LMP3.9

Nitrogen excretion estimates for beef cattle using farm data from intensive Italian herds

L. Gallo[1], S. Schiavon[1], G. Bittante[1], B. Contiero[1], G. Dalle Rive[2] and L. Tondello[2], [1]Department of Animal Science, University of Padova, 35020 Legnaro (PD), Italy, [2]AZOVE, 35045 Ospedaletto Euganeo (PD), Italy*

Nitrogen pollution received increasing attention in the recent years, and existence of limits for spreading manures based on N load per hectare of cultivated land stressed the importance of providing standard figures on N excretion (NE) for different types of livestock and production systems. This study aimed to provide estimates of NE of beef cattle herded in intensive fattening centres typical of northern Italy and to analyse some sources of variation of NE. NE has been calculated as N_{feed}-$N_{animal\ products}$. N_{feed} was evaluated from data involving 524 production cycles, 40 herds and nearly 37000 fattening bulls of Charolaise, Limousine, French crossbreds and Polish Friesian. For each production cycle average initial (346±64 kg) and final (622±70 kg) live weight (LW), length (217±44 d), dietary protein content (14.70±0.85% DM), and feed intake have been recorded. $N_{animal\ products}$ has been calculated assuming an average N content of live weight gain of 2.5% and NE has been expressed as net NE by correcting for a 20% of N lost in atmosphere. Net NE per head/cycle, per place/year and per year/100 kg LW herded averaged 29.5±6.3, 46.4±3.9 and 10.40±1.28 kg, respectively. Variation of NE due to herd, breed and diet effects are discussed.

 Poster LMP3.10

Researches concerning the quality of surface waters in a southern area of Romania

E. Mitranescu[1], L. Tudor[1], D. Tapaloaga[1], P.R. Tapaloaga[2] and F. Furnaris[1], [1] University of Agricultural and Veterinary Medicine Bucharest, The Faculty of Veterinary Medicine, Depatment of Animals' Productions and Public Health, Splaiul Independentei Street, No. 105, Romania, [2]University of Agricultural and Veterinary Medicine Bucharest,Animal Sciences Faculty, Department of Animal Reproduction,Marasti Avenue No. 59, Romania*

The quality of surface waters in Ialomita area was established following the determinations carried out by sampling in eight checkpoints in Ialomita and Prahova rivers. There were established the following: pH, fix residue, dissolved oxygen, CBO_5, CCO-Mn, chlorides, ammonium, total iron, sulphates and also the quality category in every checkpoint.
After the analyse of the obtained results it was found that the water along the inner rivers of Ialomita area is kept at the majority of parameters in the first category of quality in all eight checkpoints excepting:

- for the fix residue, Ialomita river is in the second quality and Prahova river is in the third one;
- concerning the chlorides concentrations, Ialomita and Prahova rivers are in the third category;
- for ammonium and CBO_5, the quality of water in Ialomita river is in the third category and Prahova river in the second one.

The increased quantities of some qualitative parameters of water are due to the industrial activity in this area (chemical industry, building material industry, agriculture, transport etc).
* We mention that Ialomita and Prahova river flow into the **Danube**.

Poster LMP3.11

Life cycle assessment of two production systems of beef

A. Chassot[1], A. Philipp[2] and G. Gaillard[3], [1]Agroscope Liebefeld-Posieux (ALP), Tioleyre 4, 1725 Posieux, Switzerland, [2]Agricultural college, 9230 Flawil, Switzerland, [3]Agroscope FAL Reckenholz, 8046 Zürich, Switzerland*

The environmental impact of two contrasting production systems of beef was analysed by means of life cycle assessment (LCA). An extensive fattening system of steers based on grass (EXT) was compared to an intensive fattening system of bulls (INT). EXT included two grazing periods, one of them on an unfertilised mountain meadow. The two systems differed maximally in the ecotoxic effects on aquatic and terrestrial ecosystems. EXT was more favourable than INT. This was mainly due to the lower use of fertiliser in EXT and to the high impact of the heavy metals contained in fertilisers. To a lesser extent there was also an advantage of EXT on the total eutrophication due to a lower potential leaching of nitrates. This was explained by the higher proportion of grass in the feeding ration of EXT compared to INT. For all other environmental impacts however, there were no marked difference between the two production systems. The feeding ration was the main source of difference between the systems. However, a calculation of two variants for EXT, with or without alternative use of the buildings during the grazing period, showed that the buildings can play a major role on the environmental impact of a system.

 Poster LMP3.12

Development of small mixed farms system in a newly reclaimed area in Egypt

S.M AlSheikh and A.M. Ahmed, Desert Research Center, Mataria, Cairo 11753, Egypt

A linear programming (LP) model was applied to describe crops and livestock enterprises on small mixed farms in a newly reclaimed area in Egypt. The model considered land, labor, livestock, crops pattern and available cash resources as factors affecting agricultural production. Technical coefficients of the model were estimated from a survey data collected from three locations during the period from October 1996 to September 1997. The three locations reflected the different types of producers. Linear programming technique was used to determine the optimum combination of crops and livestock production. One LP model with two scenarios was tested. The first one (LP1) utilized labor, land and available cash resource. While, the second scenario (LP2) was an attempt to meet farmer's needs of basic food along with satisfying animal's requirements under the constraint of availability of LE 10000 as cash resources. Results suggested that, farmers should cultivate 3.25, 2.32 and 2.22 feddan of berseem in winter, and 2.32, 4.12 and 2.43 feddan maize in summer along with 1.61, 1.14 and 1.37 animal units in the three studies locations, respectively, in order to maximize gross margin per farm per year. Combing activities in these prescribed quantities is expected to improved gross margin by 55%, 26% and 42% as compared to real situation in the three studies locatios, respectively.

Poster LMP3.13

Temporal variability of suitable habitats for waders: does grazing management help?

O. Renault[1], T. Potter[1] and M. Tichit[2], [1] Population and Community Ecology [2] INRA SAD, both at INA PG 16 rue Claude Bernard, 75231 Paris, France*

Due to the growing interest in sustaining biodiversity in agricultural landscape, it is now relevant to assess if livestock management can contribute to habitat conservation. In a French coastal marsh of around 5000ha, the impact of livestock systems on wader habitats was studied. Grasslands are mainly grazed by cattle and mowed for hay and landscape is highly heterogeneous despite nearly constant topography. The suitability of the marsh as habitat for five wader species was assessed using Ecological Niche Factor Analysis for each date on which birds were observed in 1996. 25 ecogeographical variables relating to land use and physical characteristics of the marsh were tested. An analysis of covariance was then performed on marginality and specialisation scores for each species to each ecogeographical variable. It was observed that habitat suitability is significantly determined by grazing of permanent pasture. However, no significant relationships between the time or intensity of grazing and habitat suitability were observed. From these results we conjecture that at the landscape level, grazing intensity and timing are insignificant in determining the suitability of habitat for the wetland avian community. However, we are currently testing the hypothesis that heterogeneity at the landscape-level in the intensity and timing of grazing is likely to be of significance in determining avian habitat suitability.

Poster L4.1

Productive performance of Carmagnola Grey rabbits from birth to weaning

C. Lazzaroni[1] and F.M.G. Luzi[2], [1]Department of Animal Science, University of Turin, Via Leonardo da Vinci 44, 10095 Grugliasco, Italy, [2]Istituto di Zootecnica, Facoltà di Medicina Veterinaria, Via G. Celoria 10, 20133 Milano, Italy*

To evaluate the improvement of productive performance of Carmagnola Grey rabbits, an endangered breed indigenous to Northern Italy (Piemonte region) under selection since 1982, a research has been carried out from 2001 to 2003 on 673 litter. According to previous works, effect of parity (from 1 to 6 and more) and birth seasons (spring, summer, autumn, winter) were studied on the number of total and alive born and on the mortality rate at birth, while the effect of parity, weaning seasons and age at weaning (between 29 and 49 days of age) were studied on the number of weaned, the mortality rate at weaning and the litter and average individual weight at weaning. The results showed a good number of born alive (8.0), low mortality rate at birth (3.36 %) and at weaning (14.99%), and a good weight at weaning, both as litter (7069 g) and as individual value (1020 g). There was a seasonal effect on the most interesting productive parameters. These results, with the performance already achieved, allowed us to continue in improving the Carmagnola Grey rabbit, a rabbit suitable for meat production and to be used as bucks in hybrid production, which performance are comparable to the commercial lines.

Poster L4.2

Preliminary results on the use of electronic identification transponders for the traceability and management of rabbits

F. Chiesa[1], O. Ribò[2], M. Zecchini[1] and F.M.G.Luzi[1], [1]Istituto di Zootecnica, Facoltà di Medicina Veterinaria, Via G. Celoria 10, 20133 Milano, Italy, [2]Datamars SA, Via ai Prati, 6930 Bedano - Lugano, Switzerland*

A total of 60 hybrids (40 fattening rabbits of 55 days old and 20 nullipare does) were injected with 23mm passive FDX-B transponders (Datamars, Switzerland). Two body locations for transponder's injection were studied: laterally to the neck (n= 30) and armpit (n=30). Readings of the transponders were performed using portable ISO transceivers, before injection, after injection, the day after, after one week, after one month, before the slaughtering process in the case of the fattening rabbits, and every month during the productive life of the nullipare does. A 100% readability of transponders was obtained in the nullipare does 2 months after injection. Readability of transponders in the fatteners until slaughterhouse was 100%. One animal died because of the armpit injection. Recovery of transponder during the slaughtering process was easy with 100% of transponders recovered. At slaughtering, 79.5% of transponder were recovered near the injection site; 12.8% of transponder migrated; 5.1% were recovered into the scapular fat and 2.6% into the scapular muscle. No harmful effects of the transponder injection on the animals were observed. Preliminary results showed that transponders can be used as a method for traceability and monitoring of live rabbits, allowing the improvement of the management of a farm.

Poster L4.3

Effect of transgenesis on the reproductive capabilities of male rabbits

P. Chrenek[1], D. Vasicek[1], A.V. Makarevich[1], R. Jurcik[1], K. Suvegova[1], M. Bauer[1], J. Rafay[1], J. Bulla[1], L. Hetenyi[1], J. Erickson[2] and R.K. Paleyanda[2], [1]RI.A.P., Nitra, Slovak Republic, [2]A.I.B., Inc., Woodstock, U.S.A*

We have generated several lines of rabbits transgenic for human Factor VIII, using the mouse whey acidic protein promoter. To increase transgenesis efficiency we used conventional microinjection of DNA into a single pronucleus (SM), as well as double microinjection (DM) of male and female pronucleii. Our results suggested that double microinjection may reduce both time and cost in producing transgenic rabbits.
However, the effect of double pronuclear microinjection on the reproductive capabilities of rabbits is not known. We observed that the volume of ejaculate and the concentration of sperm did not differ much between control non-transgenic and transgenic males. The number of motile sperm did not vary significantly, with transgenic males having about 85-95% motile sperm. Our preliminary data show that the ability of sperm from transgenic males to fecund eggs ranged between 90-100%; while the number of transgenic embryos developed ranged from 45% for founder DM males, to 54% for SM males of the F_1 generation. About 39% of live offspring derived from F_0 generation DM males were transgenic, while 50% of the offspring from SM F_1 males were transgenic. Studies in progress with F_1 and F_2 generation animals will determine whether this is a founder effect or connected to double microinjection.

Poster L4.4

Sterilität als eine der Ursachen für die Verschlechterung der Reproduktionsleistungparameter von Chinchillas in Betrieben des Nordpolens

B. Seremak, M. Sulik and B. Lasota, Landwirtschaftliche Universität Szczecin, Polen, 71-460 Szczecin, ul. Dra Judyma 6, Poland

Die Untersuchungen umfassten 4 Chinchillabetriebe im Nordpolen in den Jahren 2000-2002. Es wurden folgende Kennziffer der Reproduktionsleistung ausgewertet: Fruchtbarkeit, Zahl der geborenen Nachkommen, Anteil von sterilen weiblichen Tieren. Die Untersuchungen zeigten, dass diese Kennziffer bei einem durchschnittlichen Niveau lagen und sie wichen von den in der Fachliteratur zu findenden Angaben nicht ab. In den Betrieben wurden 1,18-1,36 Würfe je Jahr registriert. Diese Ergebnisse sind in Anbetracht der viel größeren potentiellen Möglichkeiten nicht zufriedenstellend. Dieses Ergebnis wurde durch einen großen Anteil an sterilen Weibchen, der im Bereich 16,8-24,6% lag, beeinflusst. Histopathologische Untersuchungen der Geschlechtsorganen von sterilen Weibchen und dieser mit verminderten Fruchtbarkeit zeigten, dass ein großer Anteil an krankhaften Veränderungen am Uterus, seltener an Ovarien, bei diesen Tieren festgestellt wurde, was sie für immer aus weiterer Reproduktionsnutzung eliminierte. Demzufolge wird vorgeschlagen, die Tiere mit einer über 1 Jahr anhaltenden Sterilität aus der Herde zu beseitigen.

Poster L4.5

Evaluation of partial substitution of soybean meal by some non-conventional plant protein sources on Nile Tilapia fingerling

Samia M. Hashish[1], O. El-Husseiny[2], F.Hafez[3] and A. Elwaly[1], [1]N.R.C., Cairo, Dokki; [2]Fac.of Agric. CairoUni; [3]Agric. Ministry

This work aimed to evaluate the 50% replacement of dietary soybean meal protein by some Non-conventional plant protein sources (guar meal, sweet lupin meal, nigella meal, leuceana leaf meal and sesbania leaf meal) on growth response and performanceof Nile tilapia fingerlings. The formulated diets were isocaloric (3000Kcal ME/kg diet) and isonitrogenous (35 %crude protein). Six concrete ponds (24 m^2/pond) were used, O. niloticus fingerlings were distributed into the ponds to represent 6 nutritional treatments.
Deits were fed to triplicate random groups of 16 fingerlings for 98 days. Every 2 weeks fish were weighed, their growth performance and feed utilisation parametersdetermined.
Digestibility coefficients of the different diets and groo body composition of tilapia fish were analysed.
Tilapia fingerlings fed diet contained 50% nigella meal protein instead of soybean meal protein recorded highest values of live body weight,average daily gain, conditional factor, specific growth rate and feed conversion ratio per fish, while the lowest values of the same parameters were recorded for fish received 50% sesbania leaf meal protein instead of 50% soybean meal protein.
The feed utilisatiln parameters (PER,PPV% and ER%) of the group fed nigella meal were significantly surpassed all other treatments. It was concluded that the diets contained nigella meal gave the best fish performance results, while sesbania leaf meal gave the worst results.

Poster L4.6

The estimation of contamination in alimentary products with aquatic origin by non-choleric Vibrio species

L. Tudor[1], E. Mitranescu[1], D. Tapaloaga[1] and F. Furnaris[1], [1]University of Agricultural and Veterinary Medicine Bucharest, The Faculty of Veterinary Medicinet, Department of Animals' Productions and Pulich Health, Splaiul Independentei Street, No. 105, Romania*

This study was developed in some city of Romanian country: Bucharest, Constanta, Tulcea, Medgidia, Cernavoda.
The investigations have searched the contamination level of alimentary products with aquatic origin, from some offers center of (free trade zones, restaurants, depot units, markets with alimentary profile, etc.). The substantiations had harvested either from fresh products or from conserved products by low temperatures: refrigerated or frizzed.
The study was performed for a seven years period (1997, 1998, 1999, 2000, 2001, 2002, 2003); there was harvested in total 1427 samples from fresh products with aquatic origin, 574 samples from refrigerated products with aquatic origin and 927 samples from fresh frizzed products with aquatic origin.
Following the bacteriologic exams, from the 2928 samples harvested have been segregate 127 bacterial stems of 4 non-choleric Vibrio species: 92 stems segregated from samples from fresh products, 28 stems segregated from samples from refrigerated products and 7 stems segregated from samples of frizzed products.

Poster L4.101

The effect of floor heating and feed protein level on the incidence of footpad dermatitis in turkey poults

C. Berg[1,2] and B. Algers[2], [1]Swedish Animal Welfare Agency, POB 80, SE-532 21 Skara, Sweden, [2]Department of Animal Environment and Health, SLU, POB 234, SE-532 23 Skara, Sweden*

In temperate climate where fattening turkeys are reared in climate-controlled houses efforts are made to keep the litter dry to minimize the risk of footpad dermatitis (FPD). Nevertheless, FPD is a considerable bird welfare and product quality problem for many producers. This study aimed at evaluating the effect of using floor heating (FH) to decrease the prevalence of FPD by improving litter quality, and also the effect of lower protein levels on the prevalence of FPD, via the correlation with nitrogen levels in the droppings. The study showed a significant effect of using FH on the prevalence of FPD, still present at 8 weeks of age (FH+ 21.5 ± 3.7%, FH- 45.0 ± 7.1%, $p<0.05$), and also a significant difference in the prevalence of FPD with different crude protein levels (a 25%, b 24% and c 23.5%) in the feed during the first 3 weeks of life. At one week of age, the prevalence of severe FPD was 14.0%, 7.5% and 6.5% in the three feed groups respectively ($p<0.05$ for a versus b and c). The difference between the groups later decreased. However the low and medium protein groups in the first replicate suffered an increased early mortality due to cannibalism. This aspect will have to be further investigated.

Poster L4.102

Effect of zinc-lysine supplementation on the performance and zinc concentration of some tissues in broilers

A.Ö. Yıldız and O. Yazgan, Department of Animal Science, Faculty of Agriculture, University of Selçuk 42031 Konya, Turkey*

This study was carried out to determine the effects of diets containing different levels of ZnL on performance, carcass characteristics and liver and plasma Zn concentration of broilers. In this study, 160 1-d of age Avian Farmt broiler chicks were fed four levels of ZnL for 6 weeks. The chicks were randomly divided into four experiment groups of forty birds each, and each treatment was replicated four times with ten birds *per* replicate. The dietary treatments consisted of the supplementation of the basal diets (starter and grower) with 0 (control), 20, 40 and 60 mg/kg Zn supplied from ZnL. Feed and water were given as *ad libitum*. Data obtained from the trial were analyzed by a one way analysis of variance for the level of supplemental Zn in the diet and means were compared using Duncan's multiple range test. Supplemental Zn from ZnL effected BWG and FI at 0-6 weeks of age ($P<0.05$). The supplemental Zn did not effect FCR, liver and plasma Zn concentration, and neck, wing, thigh and back weights. But ZnL levels were significantly effect carcass and breast weights ($P<0.05$). In conclusion, supplementing the basal diets used in this study with from 0 to 60 mg/kg Zn in the form of ZnL had minimal effect on performance of broilers.

Poster L4.103

The effect of yellow lupin meal and extracted rapeseed meal on the carcass quality and meat quality characteristics of meat type ducks of the A44 strain at the age of 7 weeks

B. Witak, J. Górski and B. Biesiada-Drzazga, University of Podlasie, Poultry Breeding Department, 14, Prus Street, Siedlce, 08-110, Poland

The aim of the study was to define the effect of feed mixtures containing, instead of extracted rapeseed meal, different levels of yellow lupin meal and extracted rapeseed meal on the carcass traits and meat quality characteristics of ducks of A44 strain. At the age of 7 weeks of rearing health of the experimental ducks, in comparison with ducks of the control group was very good, despite increasing in weight and content of thyroid gland, liver pancreas in ducks fed with mixtures containing yellow lupin and extracted rapeseed meal.

Meat ducks of the experimental groups, in relation to the control, were characterized by lower acidity pH_{24} of thigh and drumstick muscles in the 7th week of life. However, the values of the meat quality characteristics were not beyond the norms. Meat colour did not change under the influence of feed mixtures of the experimental groups. The results suggested that in feed mixtures KB-1 and KB-2 soybean meal can be replaced with extracted rapeseed meal '00' at the level of 5% and yellow lupin meal at the level of 2.5/7.5; 5/10 or 10/15, in the first and second period of rearing, respectively.

Poster L4.105

Effects of the application of DL- methionine and enzymatic preparations in diets for broilers' chickens

T. Banaszkiewicz, University of Podlasie, Department of Animal Nutrition, Prusa Street 14, Siedlce, 08-110, Poland

The aim of the study was to define the effect of application DL- methionine and enzymatic preparations to wheat - soyabean and wheat - rapeseed diets on indices production of broiler chickens. Chickens Hybro were fed of loose isoenergetic and isonitrogen diets ad libitum, with free access to water , in agreement with standard methodology. The group I and II were fed one of control diets, wheat - soybean (group I) and wheat - rapeseed (group II), without animal fodders. Diets for remaining groups were supplemented suitably DL- methionine (III and IV), DL-methionine and xylanase preparation (V and VI) or complex of carboxydases (VII i VIII) or both enzymatic preparations together (IX and X) The addition of DL- methionine had a profitable effect on productive results in the groups fed wheat- soybean and wheat - rapeseed alike, but more profitable in relation to the wheat- soyabean diet. Supplements of enzymatic preparation caused a further increase in profitability, but the most effective was the xylanaza preparation (applied separately).

Poster L4.106

Influence of glutamine and prebiotic on egg shell quality, small intestine structure and bone mineralization

B. Kostanjevec[1], T. Frankic[2], R. Flere[2], A. Pogacnik[3], M. Vrecel[3] and J. Salobir[2], [1]Perutnina Pluj d.d., Potrceva 10, 2250 Pluj, Slovenia, [2]University of Ljubljana, Biotechnical Faculty, Zootechnical Department, Groblje 3, 1230 Domzale, Slovenia, [3]University of Ljubljana, Veterinary Faculty, Gerbiceva 60, 1000 Ljubljana, Slovenia

The aim of the study was to determine whether the supplementation of broiler breeders feed with glutamine or prebiotic at late laying stage has a favorable effect on the digestive tract morphology that would lead to improved productional parameters, better egg shell quality and improved bone mineralization. 90 broiler breeder hens (ROSS 308) in the 37th week of lay, weighing 3,8 kg were penned individually and after a 7-days adaptation period divided in to five groups: a control group (without supplementation), three groups with different glutamine supplementation (0,65%, 1,30% and 1,95 %) and a group supplemented with prebiotic (0,15 %). Experimental period lasted for 28 days. The glutamine and prebiotic supplementation had no effect on the feed consumption, egg production, egg weight, egg shell weight and egg shell quality. Glutamine supplementation significantly increased jejunal villous width and villous area, while a significantly decreased intestinal crypt depth was observed in both glutamine and prebiotic supplemented groups. A positive effect of glutamine on bone mineralization was observed only in animals with higher rate of lay.

Poster L4.107

Growth of broilers fed with bioprepared Lactosil

M. Parvu[1], C. Dinu[1], A. Soare[2] and A. Marmandiu[3], [1]University Spiru Haret, Veterinary Medicine Faculty, Bucharest, Jandarmeriei 2, sector 2, [2]Institute of Biology and Animal Nutrition, Balotesti, sos. Bucuresti-Ploiesti km.18, [3]USAMV, Veterinary Medicine Faculty, Bucharest, Splaiul Independentei 210, Roumanie*

The study examined the growth of broilers giuven feed supplemented with the biopreparation Lactosil. The 600 one-day-old chickens were divided in four lots: one lot control and three experimental lots. The birds were housed on the floor. Experimental duration was 42 days. The chickens were given diets that were isocaloric and isoproteic diets. At experimental lots Lactosil was included at 5, 10, 15% in the ration. The lot with 10% Lactosil obtained the best results: compared with the control, daily gain increased by 11,5%, feed conversion decreased by 5,5%. The rate of mortality was 16,2% in the control and 3,4% in the lot with 10% Lactosil. NTG (Escherichia coli) decreased by 14%, number of lactic bacteria increased by 16,4%. The biopreparation Lactosil included at 10% in the broiler diet fortified the microbial functional barrier and increased immune activity.

Poster L4.108

Effect of different growth promoters on the cecal microflora and performance of broiler chickens

S.A. Denev, Department of Microbiology, Agricultural Faculty, Trakian University, P. O. Box 208, 6000 Stara Zagora, Bulgaria

The objective of this study was an evaluation of the effects of different growth promoters on the cecal microflora and performance of broiler chickens. The experiment was conducted using four feeding groups: Group I: Basal diet without supplement (Untreated control); Group II: Basal diet supplemented with probiotic Lacto-Sacc® (*Alltech Inc.,USA*) (1kg/t); Groups III and IV: Basal diet supplemented with antimicrobial growth promoters Pharmastim (0,3 kg/t) and Avilamycin (5,0 mg/kg), respectively.
Lacto-Sacc® significantly increased the number of lactobacilli and enterococci in the cecum of the broiler chickens and depressed the number of coliforms by the low pH of cecal content. There was no significant difference in the number of lactobacilli, enterococci, and coliforms present in the cecum of the control, Pharmastim and Avilamycin-fed groups. Lacto-Sacc® enhanced growth of beneficial microorganisms, regulated the microbial environment in the cecum and significantly increased the body weight of broiler chickens, compared to the untreated control and the groups fed Pharmastim and Avilamycin. The supplements Lacto-Sacc®, Pharmastim and Avilamycin improved feed efficiency by 8.8; 1.1 and 1.9% (compared to the control), respectively.The results suggested that feed additives based on microorganisms provide an alternative to antimicrobial substances in animal nutrition. The probiotics and especially Lacto-Sacc® (*Alltech. Inc.*) can reduce or replace antibiotics used for growth promotion in broiler chickens.

Poster L4.109

The estimation of growth and body confirmation of W31 geese till 10 week of life

B. Biesiada-Drzazga, B. Witak and J. Górski, University of Podlasie, Poultry Breeding Department, 14, Prus Street, Siedlce, 08-110, Poland

The investigation was carried out in 2003. 60 Koludzkie White geese of W31 strain (30 males and 30 females) were used as an experimental material. The birds were reared as broiler geese in the intensive system until the age of 10 weeks. During the experiment the geese were kept in a closed poultry house. They were fed ad libitum only with dry mash. During the rearing period the birds were weighed every week, and some zoometrical measurements were conducted at the age of 3, 8 and 10 weeks. The obtained results were used to calculate massiveness, compactness and high-leg indexes.
At the end of the rearing period the geese average body weight amounted to 5259 g. Males had significantly larger body weight than females (5468 and 5015 g). Growth rate of males and females body weight in the suceessive weeks: 0-1, 1-2, 2-3, 3-4, 4-5, 5-6, 6-7, 7-8, 8-10 amounted to 117.9, 52.2, 64.7, 40.8, 26.2, 17.7, 13.5, 9.5 and 5.5%,respectively.
At the 10 week of life males, in comparison with females, were characterized by slightly larger body length (56.8 and 55.5 cm), trunk length (35.2 and 34.4 cm) and longer shank (10.5 and 10.1 cm), similar sternum length (19.3 and 19.5 cm) and significantly larger chest circumference (48.2 and 45.3 cm). Average values of massiveness, compactness and high-leg indexes amounted to 15.1, 134.5 and 18.3%, respectively and their values were larger in males than in females.

Poster L4.110

Effect of inorganic and organic selenium in the diet of turkey-cocks on biological ability of the sperm during liquid storage

S. Dimitrov, V. Atanasov and S. Denev, Agricultural Faculty, Trakian University, Stara Zagora 6000, Bulgaria*

The objective of this study is evaluation of the effect of different selenium sources of the turkey-cock's diet on biological ability of the sperm during liquid storage.The control (C) and experimental (E) groups were supplied standard diet with 0.3 ppm/kg inorganic Se as Sodium selenite and 0.3 ppm organic Se as Sel-Plex™ (*Alltech Inc.*) respectively. After 30 weeks of feeding, the semen samples (n=6) were collected twice a week by abdominal massage. Pooled semen samples were diluted 1+1 (v/v) with *TUR-2* diluent and storaged in a water bath (10-15° C) for 6 h. The motility, viability and lipid composition in seminal plasma were examined before and after storage of the semen. The fertilizing ability of storaged sperm was assessed by artificial insemination of 30 turkeys per group with dose consists 200 $x10^6$ spermatozoa. The values of examinated semen after dilution (0h) were in similar levels, and the differences between C and E groups were nonsignificant. After 6h of storage, the motility of spermatozoa decreased significantly in the C group-8.70 % ($P<0.05$) and nonsignificantly in the E group-3.95 %. The proportion of the live spermatozoa after storage was lower in the C group (68.86 %) in comparison with the E group (69.04 %). The positive effect of organic Se was observed also on the lipid composition of storaged semen ($P<0.05$). The fertilizing ability of storaged semen was 90.46 % in the E group against 88.00 % in the C group. In conclusion, organic Se supplementation in the turkey-cock diet is an effective manner to increase the sperm motility and membrane integrity during *in vitro* liquid storage and artificial insemination in comparison with sodium selenite.

Poster L4.111

Cadmium accumulation in the organs and tissues of laying hens

A.D. Matiosov, Tomsk State Agrarian Institute, Russia

In Sverdlovskaya region (Russia) the hen cross "Radonite" was derived from the cross "Lotann x Brown" (1996). It is distinguished by high egg laying (315 eggs), egg weight (21.0 kg) and fud conversion (2.1 kg/1 kg of egg eveight). Cd content in the organs and tissues of laying hues aged 180 days was examined on the poultry farm Tuganskoy of Tomsk region in Siberia. It was identified that in ecologically safe areas the Cd level was higher in liver and kidneys (0.42 - 0.39 mg/kg) and lower in muscles and heart (0.28 - 0.24 mg/kg). With the environment highly polluted by Cd its concentration in muscles and heart was 0.9 - 0.8 mg/kg and kidneys and liver - 25.0 - 21.0 mg/kg. Thus, when the environment anthropogenically polluted by Cd, the toxicant is accumulated less in muscles and heart and more in kidneys and liver reaching concentrations hundreds of times.

Poster L4.112

Evaluation of commercial laying hens performance in floor pens and conventional cages under south valley conditions in Egypt

T. El-Sheik[1] and A.-B. Ali[2], [1]South Valley University, Sohag, Egypt, [2]Assiut National Company for Poultry and Eggs, Egypt

This study was investigated to evaluate the performance of commercial Hy-line laying hens in floor litter compared to conventional cages during summer and winter seasons. Two flocks of Hy-line were evaluated under cage and floor management systems from the age at sexual maturity up to the end of laying season. Hens were housed in cages batteries by 5 hens per cage or in litter floor pens. Pullets were randomly distributed into two replication of 2 treatments groups. Floor hens showed significantly higher egg quality, viability and least cost production compared caged hens. While feed conversion, egg weight and hen day production were higher on the cages compared floor pens. Dirty eggs were higher in floor system, while broken and creaked eggs were higher in cages. Feed consumption, egg number, egg mass, feed conversion and mortality rates were 34.979 kg/hen, 272 egg /hen, 16.944 kg egg/hen, 2.064 kg feed/kg egg and 15.7% during the period of 20 to 72 weeks of age for hens housed in cages compared to 34.888 kg/hen, 256 egg /hen, 15.842 kg egg/hen, 2.202 kg feed/kg egg and 10.5%, respectively for hens housed in floor pens. Shell and yolk percentage of floor pens were significantly higher than cages one.

Poster L4.113

The effect of layer age, storage and Strain of hen on egg quality during summer season under Sohag conditions

T.M. El-Sheikh, Faculty of Agriculture, South Valley University, Sohag, Egypt

A total of 600 eggs from 32 and 60 weeks old of Hy-line-White, Hy-Line -Brown and Fayoumi hens were used after lay and after periods of storage of 0, 3, 6, 9 and 12 d at room temperature. Longer periods of storage resulted in lower albumen weight and albumen height and higher albumen pH within the two ages. Eggs from Hy-line-Brown hens had more albumen and shell than those from Hy-line -White and Fayomi hens. The lowering albumen weight and height was not pronounced in Fayoumi eggs compared to white and brown Hy-line. Within each line and storage period, the egg weight was more closely associated with albumen weight than with yolk or shell weight. The albumen height of eggs from Hy-line- White hens was lower than those of HY-line- Brown and Fayomi hens at all storage times, but the albumen pH was higher in either Hy-Line white and brown compared to Fayoumi. The loss in weight from shell, albumen and yolk was increased as age increases; therefore the changes in egg composition are solely the consequence of weight loss from the albumen.

Effect of different hatcher temperatures during plateau and pipping stages of incubation on hatching results in commercial broiler eggs

I. Yildirim, A. Aygun and V. Sariyel, University of Selcuk, Faculty of Agriculture, Department of Animal Science, 42031, Campus, Konya, Turkey*

Incubation one of the most important stages of development for the broiler and the effect of hatching cabinet temperature for last six days of incubation upon hatching results and Ca-P levels in tibia ash was studied in the experiment. Eggs from a 32 week old Ross x Ross 308 breeder flock were incubated under standard terms for the first 15 days (37.6°C and at 58% relative humidity). Viable embryos were then divided and subjected to either control (37.2°C) or high temperatures (H treatment group - 38.3°C) in separate hatchers for the next 6 days. These temperatures are in the range that can be found under commercial conditions. No significant differences were noted in total hatchability and some selected organ weights (relative weights of heart, liver, yolk sac and lung) between groups at pull time. Additionally, the differences for Ca and P levels in tibia ash between groups were not evident. The temperature applied in H treatment group (38.3°C) might be included as a normal temperature and tolerated by the embryo for the last stages of incubation in commercial broiler eggs.

AUTHORS INDEX

C

D

E

F

Page

Page

H

I

L

M

P

Q

R

S